PSA 1988

VOLUME TWO

PSA 1988

PROCEEDINGS OF THE 1988
BIENNIAL MEETING
OF THE
PHILOSOPHY OF SCIENCE
ASSOCIATION

volume two

Symposia and Invited Papers

edited by

ARTHUR FINE

&

JARRETT LEPLIN

1989
Philosophy of Science Association
East Lansing, Michigan

Library of Congress Catalog Card Number 72-624169

Cloth Edition: ISBN 0-917586-28-X

ISSN: 0270-8647

Manufactured in the United States of America
by Edwards Brothers, Ann Arbor, Michigan

CONTENTS

CONTENTS (Continued)

CONTENTS (Continued)

CONTENTS (Continued)

Preface

This is the second of two volumes comprising the proceedings of the 1988 Biennial Meeting of the Philosophy of Science Association held in Evanston, Illinois. The first volume, consisting of the program and the contributed papers, was published in advance of the meeting in October, 1988. The second volume consists of the presidential address and those invited papers and symposia that were made available to us by their authors. Northwestern University provides generous support for the preparation of these volumes

The Program Committee, with Jarrett Leplin as its chair, arranged for the symposia and invited papers. Micky Forbes, the Assistant Editor of these Proceedings, supervised the editing and processing of the papers to produce uniform, camera ready copy. The PSA Business Office saw the copy through to publication.

Thanks are due to the Program Committee and the authors. We are grateful to Northwestern University for financial support and to its Academic Computing Center for the use of their facilities. Special thanks are due Wendy Ward whose expertise and imagination in using these facilities shows up on every page of the volume.

Jarrett Leplin
Department of Philosophy
University of N. Carolina
Greensboro

Arthur Fine and
Micky Forbes
Department of Philosophy
Northwestern University

Synopsis

The following brief summaries, arranged here alphabetically by author, provide an introduction to each of the papers in this volume.

1. *On the Possibility that the Present Quantum State of the Universe is the Vacuum.* **David Z. Albert.** It is inquired how much an observer can ascertain of the quantum state of a system of which he and his measuring apparatus form a part; how much, for example, observers like ourselves can ascertain of the quantum state of the Universe. It turns out that no practicable experiment (and: perhaps, no experiment whatever) can establish that that state is not the vacuum. Some of the implications of this curious result are discussed.

2. *Backward Induction without Common Knowledge.* **Cristina Bicchieri.** A large class of games is that of non-cooperative, extensive form games of perfect information. When the length of these games is finite, the method used to reach a solution is that of a backward induction. Working from the terminal nodes, dominated strategies are successively deleted and what remains is a unique equilibrium. Game theorists have generally assumed that the informational requirement needed to solve these games is that the players have common knowledge of rationality. This assumption, however, has given rise to several problems and paradoxes. Most notably, it has been shown that the common knowledge assumption makes the theory of the game inconsistent at some information set. The present paper shows that a) no common knowledge of rationality need be assumed for the backward induction solution to hold. Rather, it is sufficient that the players have a number of levels of knowledge proportional to the length of the game, and b) it is also necessary that the number of levels of knowledge is finite and proportional to the length of the game. For a higher number of levels of knowledge, inconsistencies arise.

3. *From Micro to Macro: A Solution to the Measurement Problem of Quantum Mechanics.* **Jeffrey Bub.** Philosophical debate on the measurement problem of quantum mechanics has, for the most part, been confined to the non-relativistic version of the theory. Quantizing quantum field theory, or making quantum mechanics relativistic, yields a conceptual framework capable of dealing with the creation and annihilation of an indefinite number of particles in interaction with fields, i.e. quantum systems with an infinite number of degrees of freedom. I show that a solution to the standard measurement problem is available if we exploit the properties of the infinite quantum models available in this broader conceptual framework.

4. *Sets and Point-Sets: Five Grades of Set-Theoretic Involvement in Geometry.* **John P. Burgess.** The consequences for the theory of sets of points of the assumption of sets of sets of points, sets of sets of sets of points, and so on, are surveyed, as more generally are the differences among the geometric theories of points, of finite point-sets, of point-sets, of point-set-sets, and of sets of all ranks.

5. *Albert Einstein Meets David Lewis.* **Jeremy Butterfield.** I reject Norton and Earman's hole argument that spacetime substantivalism is incompatible with determinism. I reconcile these both technically and philosophically. There is a technical definition of determinism that is not violated by pairs of models of the kind used in the hole argument. And technicalities aside, the basic idea of determinism is not violated if we claim that at most one of the two models represents a possible world. This claim can be justified either by metrical essentialism (advocated by Maudlin), or by denying transworld identity for points: I prefer the latter.

6. *A Case Study in Realism: Why Econometrics is Committed to Capacities.* **Nancy Cartwright.** It is common, following Quine, to look to what theories say to determine the ontological commitments of a scientific discipline. But methods and practices are equally telling. This paper considers early doctrines in econometrics. It argues that what is directly confirmed in tests of the theory will not support the applications to which the theory is to be put unless we can assume a kind of stability and atomism characteristic of capacities. The leap from confirmation to application will only be valid in a world where capacities are at work.

7. *Some Public Policy Problems with the Science of Carcinogen Risk Assessment.* **Carl F. Cranor.** Government agencies and private risk assessors use (quasi) scientific risk assessment procedures to try to estimate or predict risk to human health or the environment that might result from exposure to toxic substances in order to take steps to prevent such risks from arising or to eliminate the risks if they already exist. In this paper I discuss several ways in which the "science" of carcinogen risk assessment differs from ordinary scientific enterprises. I also consider several ways in which normative policy considerations infect this regulatory science. Scientists, philosophers of science, moral philosophers and policy makers should address these issues forthrightly in order to serve better the aims of science and regulation.

8. *Ceteris Paribus Conditions as Prior Knowledge: A View from Economics.* **Neil de Marchi and Jinbang Kim.** We interpret *ceteris paribus* conditions as the conditions necessary to conducting an experiment. "*Ceteris paribus*" is thus not a hold-all for whatever we do not know, but a listing of the various decisions taken in moving from a theoretical hypothesis to a testable proposition. The decisions range from modeling in a certain way to selecting a particular functional form or estimation technique. They embody best knowledge/best practice. Debate about the meaning and importance of any test result must center on these decisions; hence they should be laid bare. We give a detailed example from recent macroeconomics, Lucas' test of the natural rate hypothesis.

9. *Interpreting Science.* **Arthur Fine.** Using episodes in the history of the interpretation of the psi-function, this paper addresses the question of how the understanding of science sought by philosophy of science relates to the understanding sought by science itself. This leads to a conception of the discipline of philosophy of science as an historical entity. The paper concludes by drawing out the implications of that conception for our role in the humanities, and our relationship to the sciences and to ongoing scientific work.

10. *Lorentz Invariant State Reduction, and Localization.* **Gordon N. Fleming.** In this paper I will present conceptions of state reduction and particle and/or system localization which render these subjects fully compatible with the general requirements of a relativistic, i.e. Lorentz invariant, quantum theory. The approach consists of a systematic generalization of the concepts of initial data assignment at definite times, initiation and completion of measurements at definite times, and dynamical evolution as time dependence, to the concepts of initial data assignment on arbitrary space-like hyperplanes, initiation and completion of measurements on arbitrary space-like hyperplanes, and dynamical evolution as space-like hyperplane dependence, respectively. I also briefly discuss the superluminal propagation which emerges from the localization study and the manner in which causal anomalies are nevertheless avoided.

11. *Multiple Constraints, Simultaneous Solutions.* **Peter Galison.** In the 1960s, the history and philosophy of science made common cause in the search for universal patterns of theory change: philosophers provided models, historians offered examples. But the two enterprises pulled apart during the 1970s. Now there is a new arena of joint concern. Historians and philosophers are searching for the conditions under which standards of theoretical and experimental demonstration are established.

I argue against the picture of these standards as independent of (or reducible to) the context of their introduction. Instead, I suggest that we think of scientific developments as simultaneous solutions to multiple constraints; the constraints issue from domains as diverse as the technical, the aesthetic and the political.

12. *Descartes and Method in 1637*. **Daniel Garber.** This paper attempts to characterize the method that Descartes put forward in the *Discours de la methode* of 1637 and the earlier *Regulae ad Directionem Ingenii*. It is argued that because if important changes in Descartes' scientific and epistemological programs, Descartes abandons the method of his earlier years at just the moment that he makes it public in the *Discours*.

13. *Non-Bayesian Confirmation Theory, and the Principle of Explanatory Surplus*. **Donald A. Gillies.** This paper suggests a new principle for confirmation theory which is called the principle of explanatory surplus. This principle is shown to be non-Bayesian in character, and to lead to a treatment of simplicity in science. Two cases of the principle of explanatory surplus are considered. The first (number of parameters) is illustrated by curve-fitting examples, while the second (number of theoretical assumptions) is illustrated by the examples of Newton's Laws and Adler's Theory of the Inferiority Complex.

14. *Geometry, Time and Force in the Diagrams of Descartes, Galileo, Torricelli and Newton*. **Emily R. Grosholz.** Cartesian method both organizes and impoverishes the domains to which Descartes applies it. It adjusts geometry so that it can be better integrated with algebra, and yet deflects a full-scale investigation of curves. It provides a comprehensive conceptual framework for physics, and yet interferes with the exploitation of its dynamical and temporal aspects. Most significantly, it bars a fuller unification of mathematics and physics, despite Descartes' claims to quantify nature. The work of his contemporaries Galileo and Torricelli, and of his successor Newton, illustrates conceptual possibilities Descartes left aside, due to his attachment to method.

15. *Philosophers of Experiment*. **Ian Hacking.** This paper surveys a decade of philosophical discussion of laboratory science, and concludes with a bibliography. Among its topics are: (1) The historical emergence of distinct styles of experimental reasoning and practice; the relation of this to constructionalist theses. (2) The extension of Duhem's thesis to instruments and apparatus; not only are theory and observation malleable resources, but also the *matériel* with which one works. (3) The demarcation of science not by method or content, but by product; the creation of phenomena. (4) The disunity of science; science conceived of as a motley of autonomous activities, not as hierarchy. (5) The need to rethink the idea of experiment as intervention; to replace the model of experimenter as master of nature by that of investigator as collaborator.

16. *Decisions, Games and Equilibrium Solutions*. **William Harper.** This paper includes a survey of decision theories directed toward exploring the adequacy of alternative approaches for application to game theoretic reasoning, a review of the classic results of von Neumann and Morgenstern and Nash about equilibrium solutions, an account of a recent challenge to the idea that solutions should be equilibria, and, finally, an explicit reconstruction and defense (using the resources of causal decision theory) of the classic indirect argument for equilibrium solutions.

17. *Science, Certainty, and Descartes*. **Gary Hatfield.** During the 1630s Descartes recognized that he could not expect all legitimate claims in natural science to meet the standard of absolute certainty. The realization resulted from a change in his physics, which itself arose not through methodological reflections, but through developments in his substantive metaphysical doctrines. Descartes discovered the metaphysical foundations of his physics in 1629-30; as a consequence, the style of explanation employed in his physical writings changed. His early methodological conceptions, as preserved in the *Rules* and sketched in Part Two of the *Discourse*, pertained primarily to his early work in

optics. By the early 1630s, Descartes was concerned with new methodological problems pertaining to the postulation of micro-mechanisms. Recognition of the need to employ a method of hypothesis led him to lower the standard of certainty required of particular explanations in his mature physics.

18. *Ceteris Paribus Clauses and Causality in Economics.* **Daniel M. Hausman.** In this paper I distinguish the kind of *ceteris paribus* qualifications that often attach to derivative generalizations from those which typically attach to fundamental laws and argue that the latter are typically more tractable. I provide a sketch of a semantics for qualified generalizations and an account of how they may be justified. In addition I argue that legitimate uses of *ceteris paribus* qualifications must satisfy specific causal conditions.

19. *The Many Worlds Interpretation of Set Theory.* **Geoffrey Hellman.** Standard presentations of axioms for set theory as truths *simpliciter* about actual—objects the sets —confront a number of puzzles associated with platonism and foundationalism. In his classic (1930), Zermelo suggested an alternative "many worlds" view. Independently, Putnam (1967) proposed something similar, explicitly incorporating modality. A modal-structural synthesis of these ideas is sketched in which obstacles to their formalization are overcome. Extendability principles are formulated and used to motivate many small large cardinals. The use of second-order logic as a coherent and clear framework for set theory is supported.

20. *From Logical Formalism to Control Structure: The Evolution of Methodological Understanding.* **C.A. Hooker.** The thesis of this paper is that scientific method is to be thought of as a complex many-leveled regulatory hierarchy of principles, interacting with theory also viewed as a complex many-leveled hierarchy. This conception of method is illustrated in particular through one episode in the contemporary development of plasma physics, and related to others. It provides for method-theory interaction and for the development of method itself as science develops.

21. *Theoretical Models, Biological Complexity and the Semantic View of Theories.* **Barbara L. Horan.** In this paper I discuss how, given the complexity of biological systems, reliance on theoretical models in the development and testing of biological theories leads to an uncomfortable form of anti-realism. I locate the source of this discomfort in the uniqueness and hence diversity of biological phenomena, in contrast with the simplicity and uniformity of the subject matter of physics. I have argued elsewhere that the use of theoretical models creates an unresolvable tension between the explanatory strength and predictive power of hypotheses, and I review this argument again here. My discussion parallels that of Nancy Cartwright, who claims that the use of *ceteris paribus* laws in physics creates an antagonism between truth and explanation that requires theoretical models to figure centrally in scientific explanation, thereby precluding realism. I argue instead that in biology it is the use of theoretical models that creates this conflict, and conclude that adequate biological explanation cannot rely on the modeling approach alone. Finally, I claim that if we accept the semantic view of theories, which makes theoretical models an integral part of our conception of theories, we must accept anti-realism as well.

22. *Accommodation, Prediction and Bayesian Confirmation Theory.* **Colin Howson.** This paper examines the famous doctrine that independent prediction garners more support than accommodation. The standard arguments for the doctrine are found to be invalid, and a more realistic position is put forward, that whether evidence supports or not a hypothesis depends on the prior probability of the hypothesis, and is independent of whether it was proposed before or after the evidence. This position is implicit in the subjective Bayesian theory of confirmation, and the paper ends with a brief account of this theory, and answer to the principal objections to it.

23. *Scientific Realism in Real Science.* **Roger Jones.** Pre-analytically, we are all scientific realists. But both philosophers and scientists become uncomfortable when forced into

analysis. In the case of scientists, this discomfort often arises from quite practical difficulties in setting out a carefully described set of objects and their properties which adequately account at least for the phenomena with which they and those in their research specialty are concerned. I offer a set of representative examples of these difficulties for contemporary physicists. These examples challenge the traditional realist vision of mature scientific activity as struggling toward a clear and ontologically well-defined world picture.

24. *Formal Learning Theory and the Philosophy of Science.* **Kevin T. Kelly.** Formal learning theory is an approach to the study of inductive inference that has been developed by computer scientists. In this paper, I discuss the relevance of formal learning theory to such standard topics in the philosophy of science as underdetermination, realism, scientific progress, methodology, bounded rationality, the problem of induction, the logic of discovery, the theory of knowledge, the philosophy of artificial intelligence, and the philosophy of psychology.

25. *Confirmation, Complexity and Social Laws.* **Harold Kincaid.** I defend the prospect of good science in the social sciences by looking at the obstacles to social laws. I criticize traditional approaches, which rule for or against social laws on primarily conceptual grounds, and argue that only a close analysis of actual empirical research can decide the issue. To that end, I focus on problems caused by the *ceteris paribus* nature of social generalizations, outline a variety of ways those problems might be handled, and then examine in detail the work of Paige on agrarian revolutions. Paige's work, I argue, handles its problems roughly as well as does some of the best work in evolutionary biology. The upshot is that some social laws can be relatively well confirmed.

26. *The Semantic Approach and its Application to Evolutionary Theory.* **Elisabeth A. Lloyd.** In this talk I do three things. First, I review what I take to be fruitful applications of the semantic view of theory structure to evolutionary theory. Second, I list and correct three common misunderstandings about the semantic view. Third, I evaluate the weaknesses and strengths of Horan's paper in this symposium. Specifically, I argue that the criticisms leveled against the semantic view by Horan are inappropriate because they incorporate some basic misconceptions about the semantic view itself.

27. *The Essence of Space-Time.* **Tim Maudlin.** I argue that Norton & Earman's hole argument, despite its historical association with General Relativity, turns upon very general features of any linguistic system that can represent substances by names. After exploring various means by which mathematical objects can be interpreted as representing physical possibilities, I suggest that a form of essentialism can solve the hole dilemma without abandoning either determinism or substantivalism. Finally, I identify the basic tenets of such an essentialism in Newton's writings and consider how they can be updated to apply to the case provided by General Relativity.

28. *Toward a More Objective Understanding of the Evidence of Carcinogenic Risk.* **Deborah G. Mayo.** I argue that although the judgments required to reach statistical risk assessments may reflect policy values, it does not follow that the task of evaluating *whether a given risk assessment is warranted by the evidence* need also be imbued with policy values. What has led many to conclude otherwise, I claim, stems from misuses of the statistical testing methods involved. I set out rules for interpreting what specific test results do and do not say about the extent of a given risk. By providing a more objective understanding of the evidence, such rules help in adjudicating conflicting risk assessments. To illustrate, I consider the risk assessment conflict at the EPA concerning the carcinogenicity of formaldehyde.

29. *Modest Realism.* **William Newton-Smith.** Realism as an explanatory theory of science (faded realism) is not convincing. However, neither "internal realism" nor instrumentalism are plausible. Assuming common sense realism a non-explanatory form of sci-

entific realism (modest realism) can be defended. Modest realism has affinities with Fine's NOA. To NOA it adds a descriptive thesis about scientific progress towards truth or verisimilitude. In addition it adds a concern with purely philosophical issues which arise in reflections on the nature of science. However, there is little to say about truth over and above what is conveyed by Tarski/Davidson semantics.

30. *Truth or Consequences? Generative Versus Consequential Justification in Science.* **Thomas Nickles.** Pure consequentialists hold that all theoretical justification derives from testing the consequences of hypotheses, while generativists maintain that reasoning (some feature of) the hypothesis from we already know is an important form of justification. The strongest form of justification (they claim) is an idealized discovery argument. In the guise of H-D methodology, consequentialism is widely supposed to have defeated generativism during the 19th century. I argue that novel prediction fails to overcome the logical weakness of consequentialism or to render generative methodology superfluous. Specifically, Bayesian consequentialism is not an alternative to generativism but reduces to an instance of it.

31. *The Hole Argument.* **John Norton.** I give an informal outline of the hole argument which shows that spacetime substantivalism leads to an undesirable indeterminism in a broad class of spacetime theories. This form of the argument depends on the selection of differentiable manifolds within a spacetime theory as representing spacetime. I consider the conditions under which the argument can be extended to address versions of spacetime substantivalism which select these differentiable manifolds plus some further structure to represent spacetime. Finally, I respond to the criticisms of Tim Maudlin and Jeremy Butterfield.

32. *Finite Axiomatizability and Scientific Discovery.* **Daniel N. Osherson and Scott Weinstein.** This paper provides a mathematical model of scientific discovery. It is shown in the context of this model that any discovery problem that can be solved by a computable scientist can be solved by a computable scientist all of whose conjectures are finitely axiomatizable theories.

33. *Common Knowledge and Games with Perfect Information.* **Philip J. Reny.** The usual justification for Nash equilibrium behavior involves (at least implicitly) the assumption that it is common knowledge among the players both that the Nash equilibrium in question will be played by all and that all players are expected utility maximizers. We show that in a large class of extensive form games, the assumption that rationality is common knowledge cannot be maintained throughout the game. It is shown that these can have serious consequences on traditional extensive form solution concepts (such as Selten's (1965) notion of subgame perfect Nash equilibria).

34. *Modern Physics and the Philosophy of Science.* **Dudley Shapere.** This paper examines some sources of the concepts of existence, explanation, and force (together with some related ideas) in ancient thought, and shows how those ideas have been altered in fundamental ways in modem physics. Some lessons for the philosophy of science, in particular implications for its methodology, are considered.

35. *Risk Assessment and Uncertainty.* **Kristin Shrader-Frechette.** The "prevailing opinion" among decision theorists, according to John Harsanyi, is to use the Bayesian rule, even in situations of uncertainty. I want to argue that the prevailing opinion is wrong, at least in the case of societal risks under uncertainty. Admittedly Bayesian rules are better in many cases of individual risk or certainty. (Both Bayesian and maximin strategies are sometimes needed.) Although I shall not take the time to defend all these points in detail, I shall argue (1) that there are compelling reasons for rejecting Harsanyi's defense of the Bayesian strategy under uncertainty; (2) that it is more rational, in specific types of situations, to prefer the maximin strategy; and (3) that calibrating expert opinions is superior to using the equiprobability assumption or subjective probabilities.

36. *Ultimate Explanations: Comments on Tipler.* **Lawrence Sklar.** Tipler has previously argued that the nature of the universe is a matter of contingency rather than necessity. Now he argues that the existence of the universe can also be demonstrated to be a matter of necessity. I argue that both arguments are fatally flawed, and that neither supports the conclusion it is intended to establish.

37. *Learning Simple Things: A Connectionist Learning Problem from Various Perspectives.* **Edward P. Stabler, Jr.** The performance of a connectionist learning system on a simple problem has been described by Hinton and is briefly reviewed here: a finite set is learned from a finite collection of finite sets, and the system generalizes correctly from partial information by finding simple "features" of the environment. For comparison, a very similar problem is formulated in the Gold paradigm of discrete learning functions. To get generalization similar to the connectionist system, a non-conservative learning strategy is required. We define a simple, non-conservative strategy that generalizes like the connectionist system, finding simple "features" of the environment. By placing an arbitrary finite bound on the number and complexity of the features to be found, learning can be guaranteed relative to a probabilistic criterion of success. However, this approach to induction has essentially the same problems as many others that have failed.

38. *String Theory, Quantum Gravity and Locality.* **Cyrus C. Taylor.** Department of Physics. String theory is, at present, the only quantum mechanical theory of gravity interacting with matter which is not known to be inconsistent. While the theory is still rather far from making testable predictions, it does offer the prospect of realistic model-building in the not-too-distant future. Perhaps more importantly, it can serve as a theoretical testing ground for developing our ideas about what concepts like locality and general covariance mean in a quantum mechanical framework.

39. *Explanation in the Semantic Conception of Theory Structure.* **Paul Thompson.** During the last ten years John Beatty, Elisabeth Lloyd and I have argued that the semantic conception of theories is, in the context of biological theorizing, a richer conception of theory structure than the syntactic ("received view") conception. Specifically, I have argued semantic conception of theory structure better represents the structure of evolutionary theory and the relationship of this theory to phenomena. One aspect of the semantic conception that is in need of greater attention is the nature of explanation on this conception. In this paper, I argue that the semantic conception provides a richer mo accurate account of scientific explanation, in particular, of evolutionary explanations. In essence, I argue that explanation involves an appeal to numerous theories in addition to the putative explanatory theory. Employment of these theories is not formally possible in a syntactic conception but is in a semantic conception.

40. *The Anthropic Principle: A Primer for Philosophers.* **Frank J. Tipler.** An outline of the three basic versions of the Anthropic Principle—the Weak, the Strong, and the Final Anthropic Principles—is given from a philosophical point of view.

41. *A Philosopher Looks at String Theory.* **Robert Weingard.** In this paper I first describe some simple, but interesting string theory. Then I discuss string field theory and suggest that even though we do not have a complete mathematical formulation, we can get an idea of some of its ontological implications. Next, the significance of supersymmetry and superspace in string theory is briefly considered. Lastly, I consider the question of whether there is, in fact, (good) reason to think string theory may (or will) emerge to replace quantum field theory.

Part I

PRESIDENTIAL ADDRESS

Interpreting Science[1]

Arthur Fine

Northwestern University

1. Interpreting the Psi-Function

In the quantum theory the state of an evolving system is represented by a function that depends on time and perhaps on some other parameters. It is generally called the "psi-function", and it evolves according to an equation called the Schrödinger equation. Indeed Schrödinger introduced the psi-function into physics. In the beginning (i.e., in 1926) Schrödinger thought that the psi-function referred to a fuzzy bit of reality. That is, he pictured an electron, whose state was given by a certain psi-function, as a pulsating bit of electricity, something like an electrical cloud or a patch of electrical fog. In short order, however, Schrödinger saw difficulties with this interpretation, and after trying out more sophisticated refinements he abandoned the whole project. What he abandoned was the program of trying to interpret the psi-function as directly representing some spatio-temporal features of the object whose state it characterized. The central difficulty for this program of direct representation was brought out by Einstein a few years later in one of his distinctively simple and clever examples. Writing to Schrödinger on August 8, 1935 Einstein called Schrödinger's attention to the following case to show why one cannot give a direct interpretation to the psi-function.

> The system is a substance in chemically unstable equilibrium, perhaps a charge of gunpowder that, by means of intrinsic forces, can spontaneously combust, and where the average life-span of the whole set-up is a year. In principle this can quite easily be represented quantum-mechanically. In the beginning the psi-function characterizes a reasonably well defined macroscopic state. But, according to your equation, after the course of year this is no longer the case at all. Rather, the psi-function then describes a sort of blend of not-yet and of already-exploded systems. Through no art of interpretation can this psi-function be turned into an adequate description of a real state of affairs; [for] in reality there is just no intermediary between exploded and not-exploded. (Fine1986a, p. 78)

Although phrased in the language of interpreting the psi-function, it is actually something slightly different that is at issue.

In classical mechanics one thinks of the state of a system as an assignment of exact values of position and linear momentum. In quantum mechanics the state is represented by the psi-function. So, can we (analogously) think of that function as assigning values to all the relevant quantum mechanical quantities? Schrödinger's original idea was to think of the value of a quantity (e.g., of position or energy) as inexact—something fuzzy or blurred (*verwashen*)—and then to pursue the analogy with classical mechanics, so that the psi-function would assign "fuzzy values" to the quantities of the theory. A micro-object then, on this scheme of interpretation, would be a bundle of fuzzy bits (fuzzy positions, momenta, energy, spin, etc.). There is certainly nothing incoherent in such a scheme for, as Schrödinger noted, "There is a difference between a blurred or out-of-focus picture, and a [focused] photograph of clouds and patches of fog" (Schrödinger 1935, p. 812). What Einstein's example shows, however, is that this interpretation of the quantities as fuzzy-valued cannot be restricted to the microcosm. When applied to the macrocosm, however, it becomes patently absurd precisely because, in reality, there *is* (for example) no fuzzy intermediate between exploded and not-exploded.

Schrödinger, of course, agreed. Indeed by 1935 the physics community had already settled on quite a different way of understanding the psi-function and the quantum mechanical quantities. That way, which is ours today, also pursues the analogy with classical mechanics and treats the psi-function as an assignment to the quantities. But (in general) what it assigns to each quantity is not an amount (or value) of that quantity, not even a fuzzy one, but rather the psi-function assigns a probability distribution over exact values (or amounts). Because what is assigned is not fuzzy, the probability distribution is not to be thought of as indicating an existing variable density of the quantity in question, for to do so would just be to introduce Schrödinger's idea all over again. Instead we use the assigned probability distribution to tell us how likely we are to find the various exact values of the quantity when we measure it. In the case of Einstein's exploding gun powder, after a year has gone by the psi-function for the system assigns a certain probability to the explosion. This does not tell us "the degree to which the powder is exploded," nor even the likelihood of an explosion. But we can use the assigned probability as the odds for an explosion to occur if we (or others) bother to look for one.

In a perfectly straightforward sense we can say that the quantum theory treats the psi-function instrumentally. It assigns probabilities that we can use as reliable guides to certain actions. Moreover, students of the theory learn how to prepare systems to have certain psi-functions, and Schrödinger's equation tells one how those functions evolve under reasonably general conditions Thus we have entrance rules, rules of play, and exit rules for dealing with psi-functions. To be sure, not every problem is solved (I have, for instance, conveniently glossed over the measurement problem), nor are the rules, perhaps, as general as one would like. But, insofar as meaning is determined by use, and insofar as understanding involves grasping the meaning we certainly understand the psi-function.

2. Understanding Science

When asked what they are up to philosophers of science are fond of saying that they are trying to understand science. I have often said this myself although, I confess, most often with a certain niggling worry. My worry was that presumably scientists, and perhaps pre-eminently scientists, were also concerned with understanding science. So my question is what do *we* do that *they* don't? Part of the answer, and I think a substantial part, is connected with the idea of interpretation. For typically philosophers are concerned with questions of significance and with "making sense of" this or that (sometimes, science as a whole). For instance in sketching a defense of his own particular brand of anti-realism Bas van Fraassen writes

> [T]here is also a positive argument for constructive empiricism—it makes better sense of science, and of scientific activity, than realism does and does so without inflationary metaphysics. (van Fraassen 1980, p. 73)

But as my little story about the psi-function shows, scientists themselves are concerned with significance, interpretation and making sense of their theories and practices. This is not an extra-curricular concern, either, but rather it is an integral part of ordinary scientific practice. Variants of "What does this mean?" are heard as much in everyday scientific life as they are in everyday philosophical life. Are, then, philosophical concerns with interpretation different from the concerns of science?

The quantum theory is a wonderful arena in which to study this question for there the philosophical game of interpretation has been playing for over half a century, and a good deal of it has been focused on the interpretation of the psi-function. Let me single out for consideration just one line of philosophical investigation that connects easily with the story I have already told. This line simply pursues the original Schrödinger program of direct interpretation. Under a sequence of different names—"spreads" (Fine 1971), "inexact values" (Teller 1979), "disjunctive facts" (Stairs 1983)—several philosophers have revived the idea that quantities do not have point-values, but rather have as values some more extended sort of thing, and that these things are what the psi-function assigns. The purpose and motivation of these different revivals differs. In my case (in part) I was trying to oppose the idea that the concept of probability used in the quantum theory was somehow deviant and non-classical. I put my "spreads" to work in order to construct a general (and classical) probabilistic framework, one whose articulations applied as well to classical statistical mechanics as it did to quantum mechanics. Stairs and Teller had different problems. Stairs was trying to deal with a problem internal to the program of realist quantum logic; namely, with the concept of a property in that program, and with the realist demand that "having a property" not depend on acts of measurement. Teller, on the other hand, was concerned with the mathematical fact that continuous quantities (like position or momentum) always have zero probability associated with every point-value. How, then, can we talk about the values of such quantities, even on measurement? His way out of this puzzle was to introduce "inexact values," which generally have non-zero probability. I think it fair to say that all three of us were discontent with the instrumentalist path followed by the physicists and were trying to see whether one could address our particular problems by treating the quantities of the theory (and hence the psi-function) in a more realist way. The particular problems themselves were set against recognizably philosophical backgrounds. In my case the background concern was over inter-theory relations and whether the transition from classical to quantum physics amounted to a "revolution," even with regard to framework concepts like that of probability. The background to Stairs' commitment to quantum logic appears similar. With Teller the background issue seems to be the topic of idealizations and approximations in the application of science (a difficult and untidy subject, I might say).

This description of these interpretations of the psi-function (to put it loosely) suggests that what distinguishes the philosophical brand of interpretation from the scientific is a matter of philosophical attitudes. The solution displayed here involves constructing an interpretation that relieves the philosophical tension, or at least to try. Thus the understanding of science that philosophers seek would seem to be the sort of understanding that derives from an interpretation that assimilates a new idea to a favored framework, in this case to a specifically philosophical one. While this resolution of the issue may contain some element of the truth of the matter, I believe that (in general) it is a rather small part of the whole truth. For if we recall the story of Schrödinger and Einstein then I think we must acknowledge that their approaches were set against backgrounds no less philosophical than were those of the three contemporary philosophers I have discussed. Moreover, in a part of the story I have not told, both Schrödinger and Einstein go on to offer their own interpretations of the psi-function, each in aid of philosophical realism (and in Schrödinger's case a kind of holism as well). In so far as understanding does amount to assimilation to a favored framework, the understanding of new science that scientists strive for is bound to be similar in kind to the understanding we seek as philosophers. Since, moreover, the frameworks that confer understanding are precisely the ones that will have been the subject of philosopher's attention and reflection, there is hardly much room to escape from the conclusion that the understanding so achieved by science will have a philosophical flavor. Of course, one might suggest that this flavor may only be appreciated by those who already have developed a special taste for it.

This is the suggestion that primarily philosophers will reflect on and appreciate the philosophical character or aspects of science. Perhaps it is a kind of division of labor. Perhaps. Certainly there is a genuine difference in professional perspectives between philosophers and scientists, a difference that shows up in the degree of precision, rigor and fluency with which scientists handle our vocabulary (and we theirs!). Every philosopher who reads Bohr or Heisenberg, for example, has to work out what conception of "definition" is at work when these scientists tell us that position and momentum cannot be defined simultaneously. But I am not sure whether this division of professional training cuts all the way down to a real division in understanding. After all no-one can read the Einstein-Schrödinger correspondence, from which my gunpowder example was extracted, without recognizing that it constitutes a philosophical conversation of the most sensitive and profound character. Moreover if one set out to compile a short list of the most important literature on the interpretation of the quantum theory older works by Einstein, Schrödinger, Heisenberg, Bohr and Born would be musts, as would recent papers by Kochen and Specker, and by Bell. These are all scientists. Thus the suggestion that philosophers are after a distinctively philosophical understanding of science, something different in kind from the understanding sought and gained by scientists themselves, does not seem to me to be born out in practice. Professional affiliation just does not seem to divide up the labor of philosophical reflection along professional lines.

Here is a more promising division. Philosophers think about science. Scientists too think about science. But scientists also *do* science, unlike philosophers. Surely there is something *to* this, although if publishing in the professional journals is a criterion for doing science then some of us in the philosophy-of-science business will have to be counted as doers as well as thinkers. Moreover the doing/thinking-about distinction itself is not so clear, because scientists who think about science may well be doing science (and conversely for philosophers who do science). Perhaps the division looks more compelling if we take science as empirical science, emphasizing the hands-on, data-gathering, laboratory-tinkering, experimental aspects. With very few exceptions I think this divides philosophers from scientists pretty well. Even more, this division corresponds to the seemingly natural stopping point in many philosophical papers, the point where the discussion of a topic is terminated because, we are told, to go any further would require empirical investigation. However, although this division is neat and applicable, I don't think it applies to the problem at hand. Recall that we are looking for what philosophers *do,* by way of understanding science, that scientists do *not* do. The distinguishing feature here singled out is that philosophers (*qua* philosophers) do *not* do the empirical work of science. Thus if this suggestion were correct then what we would fail to gain in stopping short of empirical scientific work would be precisely that understanding of science that comes (so to speak) from getting our hands dirty. Now I have every reason to think that there is an understanding that comes from practice, and I also think that in philosophical discussions of science we usually fail to get it. But this is a failing. That is, among the understanding philosophy of science requires is this very understanding-that-comes-from-practice, much of which we do not get. This may afford us a perspective from which to criticize our own enterprise or, if you like, a growing point for it. I, for one, would like to see philosophers of science also doing some empirical science. But this attitude toward practical understanding simply acknowledges its professional relevance, and the fact that, in empirical work, delicate questions of meaning and interpretation pervade nearly every move. Thus the focus on empirical science may well point to an understanding of science that philosophers often miss. It does not, however, single out something that we, as philosophers, are aiming for that is different from what scientists themselves get by way of understanding.

3. The Situation of Philosophy

Let me summarize the discussion so far. If we consider what scientists do as part of their daily trade then we find that they are engaged in serious reflection on the meaning and interpretation of their theories, norms and practices. That reflection, moreover, frequently involves attempts to relate aspects of science to recognizable epistemological and metaphysical concerns (or traditions). Thus various plausible ways of separating philosophy of science from science, in terms of what we do in trying to understand science that scientist do not do, fail. I particular neither (a) a basic concern with questions of meaning and interpretation, nor (b) a central concern with relating science to recognizably philosophical issues, nor (c) an emphasis on reflection and "second-order" thinking does the job of separation well enough. Moreover the fact that scientists, unlike philosophers, do the empirical work of science, only points to a kind of understanding that philosophy of science can not be content to pass over.

To the extent to which my argument is correct, however, it must certainly, somehow miss the mark. For certainly we all do feel a difference between our own approach to understanding science and the way of the scientists themselves. Perhaps the mistake is to think of that difference primarily in terms of *doings*, or how things are done (e.g., "philosophically" as opposed to "scientifically"). Perhaps the difference resides more in what one might call the respective *situations* of philosopher and scientist. We are, after all, situated in philosophy: we are generally trained in philosophy, teach in Departments of Philosophy, and write for an audience of philosophers. The problems we pursue, the techniques and vocabularies we employ, and the styles we exhibit reflect the history of our discipline, and the sociology of norms and practices that are its current embodiment. We can illustrate what difference *situation* makes by returning to the story about the psi-function.

The initial concern about interpreting the psi-function was Schrödinger's concern to understand the tool that he himself had invented. In those early days no one really understood how to use the psi-function, where its use was likely to be reliable, or whether such reliability as there was derived from something perhaps even more fundamental. The relevant physics community engaged in lively explorations and discussions of all these things, eventually settling on the instrumentalist interpretation with regard to use and reliability, and giving up the search for any deeper (especially realist)foundation. Some of the most important figures involved in the early investigation of quantum phenomena (including Einstein and Schrödinger) could not reconcile themselves to this instrumentalist, community judgment, and the statistical nature of the theory that emerged. Thus in 1935, when the dust was beginning to settle, Einstein and Schrödinger began going over the rationale for that community judgment, trying to see whether there were different paths that the theory might plausibly have taken — or might even yet take. This was not primarily a rear guard action, I think, but rather an attempt to make some conceptual room for a new and differently directed vanguard. (In character the examination by Einstein and Schrödinger was no different from the look that "superstring" theorists today take toward the last decade of work in so-called "grand unified theories.") Neither Einstein nor Schrödinger succeeded in finding room for a progressive theoretical refinement. Schrödinger basically stopped doing physics, and Einstein turned to steady (and frustrating) work on a unified field theory, hoping for a breakthrough there that would lead the way around again to a better account of quantum phenomena.

The original interpretive deliberations over the quantum theory took place in an intellectual atmosphere heady with the rise of logical positivism. There is little doubt that the neo-positivist emphasis on observation, and the anti-metaphysical orientation of that movement, actually helped to open up the possibility for physics to embrace the loose pragmatism of the quantum theory, and the instrumentalist reading of the psi-function. Conversely, reflections on the path taken by physics quickly became integrated into the neo-positivist literature, affecting its concern with questions of interpretation and meaningfulness. The legacy of this in the succeeding generation of philosophers was to reserve a place in the standard philosophical canon for problems over the interpretation of the quantum theory.

While there also continued to be some literature in physics on the broad questions of interpretation, these issues no longer occupied a central part of live physics in the generation following that of the founders of quantum theory. Thus the philosophical discussions fed mostly on themselves, rather than on pressing questions for physics. In particular philosophers did not share the context of Einstein's or Schrödinger's involvement with realism; namely, the context of a search for a new and better physics. Rather philosophical developments fostered an interest in realist interpretations of the quantum theory as an end in itself. Indeed, not only were the broad issues of interpretation largely severed from on-going physics, at the other end, the issue of realism was also cut off from on-going examination of that issue in epistemology and metaphysics. Thus it was a largely unexamined realism with which philosophers of science were engaged. This is the situation in which, nearly fifty years after its refutation, it was possible for philosophers to re-introduce the original Schrödinger program for how to understand the psi-function, and the quantities of the quantum theory.

My point here is not that the concept of "understanding" is indexical, or context dependent. Very much to the contrary, I believe that understanding is understanding is understanding. (Just as NOA contends that truth is not relative, and neither correspondence nor acceptance; just simply truth. See Fine 1986a and 1986b.) Precisely on that account, however, if we want to see how "philosophical understanding" differs from "scientific understanding" then we have to attend to the different communities of inquirers and the different contexts of inquiry. What makes *all* the difference is how the project of understanding is differently set. In the case of the psi-function this should be especially clear. In 1926 Schrödinger was really looking for an interpretation; shortly thereafter the community found one. It follows that fifty years later one could only be looking for a *re-interpretation*. But why do this? In 1935 one could have answered that question in Einstein's terms. It was a last look for a better direction for physics to take. After fifty years of continuing progress and success, however, that answer is no longer available. Hence history forces a change in the project of understanding the psi-function, from interpretation to re-interpretation, and it also forces a change in motivation, one that largely severs the project from its original engagement with the construction of physics.

I believe that the story of the psi-function is a fairly typical story of how philosophers go about understanding science. What I hope it illustrates is how the character of that understanding is really the same in kind as what scientists themselves are after. The differences in understanding, I suggest, derive from the different contexts of inquiry, including historical periods, and extend to the motivations that promote the inquiry. This way of approaching things, to look for unity in the operative concept and diversity among its contexts of deployment, seems to me relevant to assessing other philosophical tasks. That is, instead of inquiring about the "nature" of a philosophical enterprise, and its distinguishing marks, I think it is often more sensible to begin by supposing a unity (or at least a continuity) between what philosophers do and what others do under the same banner. This, then, directs the inquiry to features of history and context that may contribute the felt differences. I would even suggest, more radically, that the same program applies to philosophy as a discipline. That is, that what distinguishes philosophy is not a set of distinguishing characteristics but rather only a history of practice, and the contexts of that practice.

4. Some Morals for Philosophy of Science

If one adopts an historical (and contextual) point of view, then questions about the connections between philosophy and the other disciplines will only have historical answers. One can point out, for example, how philosophical concerns with the interpretation of quantum theory started with live concerns in physics but eventually turned in on themselves, becoming largely divorced from such concerns. One could even continue the story and tell how John Bell's work in 1964 began a period in which some connections were re-established between philosophy of physics and physics itself, which would show how contingent the relationship is between past practice and future developments. I do not believe that one can project out underlying traits of the discipline that actually constrain the practice over time. To be sure, one can certainly reorganize the history of practice so as to make certain features stand out, and then one can promote those features as norms and urge their projection into the future. This is what most reflections on the nature of the discipline seem to me about. This does not differ from the way, in our private lives, we all periodically take stock (usually in periods of personal stress or crisis), re-evaluate ourselves, and determine how to go on. Certainly, there is nothing wrong with this unless we also fool ourselves into thinking that we have now, finally, got to the real and essential *us*. Similarly with philosophy. If we are going to take stock of our discipline, and determine how to go on then let us not fool ourselves about the character of the reflective exercise. What the historical perspective suggests is that anything at all is a possible way to go on, provided it can grow out of the present, and hence connect with the past. These privisos are not very strong. For example, they would allow both for a turning to do empirical scientific work while still wearing one's philosophical hat, or for a returning to an interest in the metaphysical foundations of science in a way divorced almost completely from scientific practice.

It may well be that certain external constraints are more important than those we generally promote in a rational reconstruction of our practice. In particular I believe that the location of our subject among the humanities, and our function as teachers of the humanities, may impose some very strong constraints indeed (at least in the short run for, of course, that location too could change). Our position as teachers of humanities to undergraduates forces us to connect our specialities to broader and less technical concerns (and "force" is correct, since it often seems to go against our will). As a practical matter, we are only permitted to teach philosophy (never mind philosophy of science) if we connect it to what the academy, and its funding bodies, hold as valuable from a humanistic point of view. Of course, we know that in this regard we often get away with murder. But, especially where undergraduates are concerned, I believe that we feel the constraints in a way that influences the curriculum. In turn, our teaching has to affect our thinking when we move from the classroom back to the study. The result is this. The institutional setting of philosophy makes it difficult for our subject to move outside the boundaries of traditional humanistic studies in any really dramatic way. (Of course, those boundaries themselves shift over time.)

This fact enters into how philosophy of science approaches its subject. By training, and by their place in the workforce of the academy, philosophers of science are connected to a humanistic tradition that sees itself as unscientific, sometimes rather glorying in that image. By interest, of course, philosophers of science are concerned with science. One possible resolution of the bind that this creates is to turn one's interest in an historical direction, either by showing a concern for understanding the science of the past (e.g., "interpreting" the psi-function half a century later), or by showing an active concern with the history of science or with the history of the philosophy of science itself. Since concern with understanding the past, one way or another, is on the traditional humanistic agenda, this historical orientation provides a convenient release from the bind. Clearly it is not only this historical approach that shows an institutional effect. For it is always an option for a philosopher of science to turn in toward the general problems of philosophy proper, promoting these general problems to the top, and then doing philosophy of science in a top-down way. (There is a whole literature on conceptual change in science, for example, that has this character.) The harder option is to engage with active science on a more or less equal basis. This is difficult because of training. We all feel that *they* can do it much better than we can. But it is also difficult because of institutional barriers. One would expect that in the less highly developed sciences there would be more opportunity to cross those barriers, and also less to learn. Indeed, first in biology and now in the cognitive sciences we have begun to see a real working relationship develop between philosophy of the subject and the subject itself.

I am probably overemphasizing the case in tracing so much of the practice of philosophy of science to its setting in the academy. Nevertheless that may be a good antidote to the usual essentialist focus on the "nature" of the discipline. In any event, there is a great deal to be learned by viewing our discipline from an historical and institutional perspective. One of the things it may help us see is how the approach to our subject, and the style of our work, is entangled with questions of validation and with community norms (and sanctions). Kuhn should have taught us long ago that what counts as good work (even possible work) is a function of history and community. In just this regard philosophers of science have a unique opportunity. For, in effect, we play to two audiences — the humanists and the scientists. This affords us the opportunity to amalgamate the standards of the two communities: that is, to do work judged good by both. I have my own rather vague formula for this amalgam: it is to make sure that our philosophy of science engages on-going science.

This formula is certainly too vague to function as a standard. Yet it may be a workable rule of thumb, promoting some problems and issues over others precisely because they more clearly connect with on-going science. One implication I should want it to carry is that we see ourselves as engaged cooperatively with scientists. (Cooperative does not mean uncritical!) That, in turn, implies that their orientation toward constructive work (i.e., toward making science) should be an orientation that in some measure we share. This idea, I think, may be a very important one for philosophy of science. For it suggests that begging off thinking just at the point where science is to be done would no longer be an easy way out. The labor would no longer divide so conveniently. Moreover, if philosophy of science began to share the scientists' cooperative orientation toward doing constructive work then I think there would have to be a corresponding shift in the overall ethos of the profession. For we now employ a private ownership model of our work, and the correlative model of criticism—namely, each against all. But if we came to feel that criticism should be in aid of constructive output in the context of a cooperative enterprise, then we might even come to see the proposals and ideas of others as, perhaps, not-yet-done rather than as half-baked.

I mentioned some trends in the profession, especially among those concerned with the developing sciences, that already actively engage philosophy with on-going science. My rule of thumb would encourage these trends and it would value that sort of work over the opposite. In support of the rule I can only offer a few considerations (and no arguments). The first is a growing sense of sterility, for example, at playing the game of interpretation when the context does not really connect with live scientific concerns. Since I have written rather a lot about this recently in connection with the realism/anti-realism debates, I won't press the points again here. (See, for example, Fine 1986a and 1986b) Let me just suggest that the sense of sterility may come from the fact that when we cut ourselves off from active science we severely reduce the constraints on the game of interpretation. This too radical underdetermination may make us feel that anything clever will do. Over time, that feeling is liable to turn into a sense of sterility. The second consideration in favor of grading it higher the more a work connects with live science is simply the challenge of doing that higher grade of work. It seems to me that with regard to growth as an individual, or as a profession, a more challenging environment is better than a less challenging one. Finally, I would remind you of the potential in science itself for addressing virtually all the sorts of interpretive questions and issues that philosophy traditionally pursues. Thus, connecting with active science probably entails not much loss and, potentially, large gains for philosophy, and also for science.

I have argued that the historical background and the contemporary setting of our discipline allows considerable latitude for how we are to get on with our job. To be sure, our relationship with the traditions of the humanities, and our role as teachers of the humanities, imposes a practical obligation to connect our interest in science with humanistic concerns more generally. I would emphasize that this is an obligation that derives from practice, and not from the so-called "nature" of the humanities. In effect, we have to show our work to our humanistic colleagues and trust that they will find something of value in it. To engage their interest in our work, however, surely means that we must pay attention to theirs. I would count this interchange a good thing. It is the necessary background to our distinctive engagement with science, and it provides the institutional framework that frees us to connect with scientific life as it is actually lived, without fear of thereby losing our own identity. Moreover our engagement with our humanistic colleagues on behalf of our interest in science also seems to me a healthy thing for the development of the humanities. The humanities too could turn challenge to growth.

My perspective in this paper has been historical and situational. My argument has been that such a perspective does not undermine the values of our subject but, rather to the contrary, it provides a realistic setting for assessing how those values can grow and develop, and change. All we have left to do is to determine which directions of change are best. My personal preferences, I think, are clear enough.

Notes

[1] An earlier version of this paper was read at a conference on the autonomy of philosophy of science held at the University of Virginia. I want to thank the participants there, and especially Paul Humphreys, for useful comments.

References

Fine, A. (1971), "Probability in Quantum Mechanics and in Other Statistical Theories," in M. Bunge (ed.) *Problems in the Foundations of Physics*. New York: Springer-Verlag, pp. 79-92.

_ _ _ _ . (1986a), *"The Shaky Game: Einstein, Realism and the Quantum Theory*. Chicago: University of Chicago Press.

_ _ _ _ . (1986b), "Unnatural Attitudes: Realist and Instrumentalist Attachments to Science", *Mind* 95: 149-79.

Schrödinger, E. (1935), "Die gegenwärtige Situation in der Quantenmechanik", *Naturwisssenschaften* 23: 807-12.

Stairs, A. (1983), "Quantum Logic, Realism and Value Definiteness", *Philosophy of Science*: 50: 578-602.

Teller, P. (1979), "Quantum Mechanics and the Nature of Continuous Physical Quantities", *Journal of Philosophy* 76: 345-61.

van Fraassen, B. (1980), *The Scientific Image*. Oxford: Clarendon Press.

Part II

HERBERT FEIGL REMEMBERED

Herbert Feigl: 1902-1988

C. Wade Savage

University of Minnesota

Herbert Feigl was born December 14, 1902 in Reichenberg, Austria, and he died in Minneapolis, Minnesota, on June 1,1988. His wife, Maria, followed him in death on February 13,1989. He is survived by a son, Eric O. Feigl, who is professor of physiology at the University of Washington.

Feigl's parents were Jewish but not religious; indeed, his father was an outspoken atheist. His father was a weaver and later an influential textile manufacturer, who would have preferred the son to become an industrial chemist. His mother was a devotee of the arts and an amateur pianist, who would play Beethoven sonatas over and over at the young son's request. His high school interests were philosophy, science fiction, physics, and chemistry (which was his major). Moritz Schlick became his favorite philosopher after he read the latter's *Allgemeine Erkenntnisslehre*. His chief intellectual hero was Einstein. The description of his first contact with relativity will strike a familiar chord in many philosophers of science.

I had happened, at the age of sixteen or seventeen, upon an article...on the special theory of relativity... My first reaction was the suspicion that the theory must be in error. Promptly I set about to refute it. But in trying to do that I learned a great deal of physics and mathematics and, of course, found out after a few months of diligent work just who was wrong! My attitude then changed completely and Einstein became my number one intellectual hero. (Feigl 1980, p. 2.)

Anti-semitism in German universities led him in 1922 to the University of Vienna, where in 1927 he earned a doctorate in philosophy, with a dissertation entitled *Zufall and Gesetz*. (The full title in English: "Chance and Law: An Epistemological Analysis of the Roles of Probability and Induction in the Natural Sciences".) Some of the ideas in the dissertation appeared as a paper in *Erkenntnis* (Feigl 1930), which has been translated into English (Feigl 1980, ch. 6). In 1929 his first book, *Theorie und Erfahrung in der Physik*, was published. Chapter III of the book has also been translated into English (Feigl 1980, ch. 7).

While a student at Vienna Feigl became a member of the "Vienna Circle", an informal group of philosophers, mathematicians, and scientists organized by his teacher Moritz Schlick. Among its members were Otto Neurath, Hans Hahn, Victor Kraft, Philipp Frank, Rudolf Carnap, Karl Menger, Kurt Gödel, Friedrich Waismann, and Gustav Bergmann. Among its frequent visitors were Philip Frank, C.G. Hempel, Alfred Tarksi, A.J. Ayer,

PSA 1988, Volume 2, pp. 15-22

and A.E. Blumberg. Several of the members (including Feigl) were in frequent contact with Ludwig Wittgentstein and K. Popper. Under the leadership of Neurath the group would later undertake but only partly complete a monumental Encyclopedia of Unified Science (Neurath, et. al. 1938-69), whose monographs have become classics in the philosophy of science. While the Vienna Circle was in formation so was the Berlin Society for Scientific Philosophy, led by Hans Reichenbach and Richard von Mises. These two groups and their associates would become the most influential and controversial collection of scientists and philosophers since the encyclopedists of the French enlightenment, and would come to be called "the logical positivists".

The Vienna group's principal texts were Mach's writings, Russell and Whitehead's *Principia Mathematica*, and Wittgenstein's *Tractatus Logico-Philosophicus*. Its general philosophical position was Humean empiricism in modern logical dress: statements were held to be cognitively meaningful if and only if they are (i) analytic, that is, logically true or false, or (ii) synthetic a posteriori, that is, testable in principle by observation. Synthetic a priori statements—factual statements that transcend experience—were held to be meaningless. The positivist meaning criterion generated enormous controversy; for it was widely employed to argue that ethical, theological, metaphysical, and many (if not most) philosophical statements are cognitively meaningless, and that true statements of pure mathematics are tautologies (albeit non-trivial ones).

Feigl came to the United States late in 1930 on an International Rockefeller Fellowship, and spent nine months at Harvard University, a group of whose faculty (Bridgeman, Stevens, Perry, Quine, etc.) he regarded as the American equivalent of the Vienna Circle. In 1931 he and A.E. Blumberg published a paper in the *Journal of Philosophy* entitled: "Logical Positivism, A New Movement in European Philosophy", apparently christening the movement and introducing it to American philosophers. Like many others, he came to prefer the name "logical empiricism", to distinguish his scientific realist position from the phenomenalism of some of the other positivists. He returned to Vienna in 1931 to marry Maria Kasper, a philosophy student, and bring her with him to the University of Iowa, at Iowa City, where he had been appointed lecturer in philosophy. He became assistant professor at Iowa in 1933, and associate professor in 1938. Although other positivists would follow him in their flight from the Third Reich, Feigl was the first to settle in the United States. (Schlick had visited Stanford University for a term in 1930.) He thus became the new philosophy's first representative—"missionary" in Feyerabend's (1966) phrase—to the United States, a role somewhat comparable to that in Britain of A.J. Ayer, author of the widely read, *Language, Truth, and Logic* (1936).

Feigl was awarded a second Rockefeller Fellowship to work at Harvard and Columbia Universities in 1940, and in that same year he accepted a position as professor of philosophy at the University of Minnesota. His three decades at Minnesota were highly productive, both in publications and in activities that advanced the philosophical position he represented; and they were important factors in the formation of philosophy of science as an academic specialty. In 1949 he and his colleague Wilfrid Sellars edited *Readings in Philosophical Analysis*, which became a standard text of analytic philosophy and logical empiricism, and after enduring success was published in a new edition twenty three years later. In that same year he and Sellars, with colleagues May Brodbeck, John Hospers, and Paul Meehl as co-editors, founded the journal, *Philosophical Studies*. In 1953 he and Brodbeck edited *Readings in the Philosophy of Science*, which immediately became and for many years remained the standard anthology in the field.

In 1953 Feigl obtained a grant from the Hill Foundation (later supplemented by grants from the Carnegie Corporation and the National Science Foundation) to establish the Minnesota Center for Philosophy of Science. It was the first institution of its kind in the country, and perhaps first in the world. Philosophers of science throughout the world were brought to the Center to participate in workshops and collaborative research. Many of the fruits of this research were published by the University of Minnesota Press in the series, *Minnesota Studies in the Philosophy of Science,* whose first three volumes (1956, 1958, 1962) contained seminal work by many of the world's leading philosophers of science, and furnished the Center with a global reputation. The second of these volumes contained Feigl's essay, "The 'Mental' and the 'Physical'", which served as a stimulus to Center research on the mind-body problem and philosophy of psychology generally, and in 1967 was expanded into a book under the same title. He believed that history of science was important to philosophy of science, and he brought historian Roger Stuewer to the University of Minnesota in 1967. In 1972 Stuewer founded the University's program in history of science and technology. On his retirement in 1971 Feigl was succeeded as director of the Center by his associate and former student, Grover Maxwell. Maxwell served as director until his death in 1981. The Center has expanded, and continues to produce and subsidize research in the philosophy of science and to publish *Minnesota Studies in the Philosophy of Science.*

Feigl was a frequent organizer of formal and informal conferences, and was a member of several professional associations. In 1959 he and Maxwell organized a prototypical session on philosophy of science for the American Association for the Advancement of Science (of which he was then vice-president), and the proceedings were published two years later as *Current Issues in the Philosophy of Science.* He served on the governing board of the Philosophy of Science Association, and was a founding member (1934) of the editorial board of *Philosophy of Science,* which later became the official journal of the Association. He was president of the Western Division of the American Philosophical Association during 1961-62.

Feigl was among the University's most successful teachers. He was unpretentious and supportive of students, and he promoted the search for truth instead of intellectual competition. His ability to explain abstruse matters in plain terms for students, colleagues, and laypersons was legendary. So was his facility with philosophical slogans and phrases, for example "The power of positivistic thinking". So also was his ability to improvise at the piano and to play the works of his favorite composers by ear. He was appointed Regents Professor of the University of Minnesota in 1967.

After his retirement in 1971, Feigl encouraged colleagues to visit him at his home in Minneapolis for philosophical conversation and his favorite drink, cranapple juice, until a longstanding disease of the nerves in his legs made the visits impossible. Nonetheless he continued to work and publish. In 1972 a new edition of his and Sellars' 1949 *Readings* was published. In 1974 he published two papers summarizing and defending his philosophical position (1974a, 1974b). The year 1980 saw the appearance of *Inquiries and Provocations*, an extremely useful collection of his papers edited with his collaboration by Robert S. Cohen. He provided the editor with a typically overmodest self-appraisal: "I'm more of a catalyst than producer of new and original ideas...with just a few exceptions *perhaps*" (Feigl 1980, p. ix). His eyesight began to fail; but he continued to read in specially prepared large type, and be read to. His hearing also failed, but he continued to listen to music with electronic aids, and to instruct those entrusted with his care in the virtues of the classical composers, especially the Austrian. A few months before his death from cancer he attended with great difficulty his last concert—a performance of the Second Symphony of his beloved Mahler by the Minnesota Orchestra. He had said in an autobiographical essay in 1974: "I consider truly great music *the* supreme achievement of the human spirit...I am inclined to think that music expresses (even more than poetry) what is inexpressible in cognitive and especially in scientific language" (Feigl 1980, p. 5). Such was the faith of one of our most ardent and judicious humanists, a thinker who, under the terms of the logical empiricism he helped fashion and propagate, believed passionately in both art and science.

Feigl began his philosophical career as a logical positivist, soon converted to logical empiricism (realist positivism), and remained faithful to the latter position to the end of his life, with a few adjustments he seems to have regarded as minor. I will discuss his philosophical contributions under four headings.

Induction and Probability. Feigl's doctoral dissertation dealt with the roles of probability and induction in the natural sciences, and the topic was a major concern for the rest of his career. He was among the first—if not the first—to argue that probability does not afford a logical solution to Hume's problem of the unjustifiability of induction, and to propose instead a pragmatic solution to the problem—a "vindication" of induction rather than a "validation". His argument was roughly as follows. The probability of an event is empirically meaningful only if it is taken to be the relative frequency of the event in some reference class. However, no directly observable, finite frequency is exactly the probability, and so we identify it with the limit of an infinite sequence of such frequencies, if one exists. Whether the infinite sequence has a limit, and if so what its limit is, can only be inferred from an observed, finite segment of the sequence; and this inference is of course an induction. Thus, the determination of probabilities presupposes induction, and consequently cannot replace induction or establish its reliability.

In spite of this difficulty, Feigl felt that induction could be vindicated by the consideration that it is an essential ingredient of scientific practice, that science as we understand it would be impossible without induction. He believed that in some of his attempts at vindication he had partly anticipated the now familiar Reichenbach-Salmon pragmatic justification of induction (Feigl 1980, pp. 27, 101). Either the sequence of relative frequencies of the event converges or not. If not, no inference from the observed to the limiting frequency (probability) of the event is correct, and science is impossible. If so, then any inductive rule of inference that takes the observed frequency to be in some way representative of the limiting frequency will provide estimates that converge to the probability in the long run, which is the best science can offer and perhaps good enough for practical purposes. (See Salmon 1967.)

Mind-Body Identity. Feigl's second life-long concern was the mind-body problem. Perhaps his best known work is his development and defense of the thesis that mental events are contingently identical with physical (neurophysiological) events. He had inherited a generally monistic view from Schlick (1918) and reinforced it with the writings of Russell. His version of the view was first presented in detail in his essay entitled "The 'Mental' and the 'Physical'" (Feigl 1958). Its central feature was that the identity of mental and physical events is a theoretical identity that must be justified like any scientific theory on the basis of how well it explains the facts. He compared it with the identification of heat and molecular motion in the kinetic theory of gases, which is justified on the ground that it provides the simplest, best-unified theory that entails the observed data. This thesis was initially quite controversial. For the dominant mind-body views at the time were variants of behaviorism: either the operationist, according to which mental events are completely definable in terms of behavior; or the instrumentalist, on which mental events are taken to be theoretical fictions useful in predicting behavior. However, the *Zeitgeist* was moving against behaviorism, and toward theoretical realism, and Feigl was soon joined in his defense of the identity theory by other philosophers—most notably, the Australian physicalist J.J.C. Smart (1959). Today the mind-body identity theory is the prevailing view among psychologists and philosophers of psychology, though it remains controversial among philosophers generally, at least as regards its contingency.

Realism. As noted above, Feigl called his general philosophical position "logical empiricism" to distinguish it from instrumentalist and operationist versions of positivism that were commonly held by positivists in the early phases of the movement (Carnap of the *Aufbau* and Neurath, for example). These other versions maintain that theoretical terms are either completely definable by observation terms, or else refer to calculational devices and/or useful fictions. Feigl's position, like that of Schlick, Reichenbach, and Popper, was scientific realist: he held that the theoretical terms of scientific theories—

terms such as "force", "radiation", "electron", "gene", "ego", etc.—are almost never completely definable by observation terms and yet do refer in approximately true theories to existing, "real" entities. Sometimes he called it "critical realism" to distinguish it from the "naive realism" that holds some physical properties of physical objects can be directly known ("known by acquaintance"). Sometimes he called it "structural realism" to indicate that only the "structural" features of physical entities can be known in the strict sense ("known by description"). He saw no conflict between empiricism and scientific realism.

The liberalized meaning criterion allows for hypothetico-deductive tests (i.e., confirmations and disconfirmations), for inductive interpolations and extrapolations, for analogical conceptions and inferences [for example, inferences to other minds]. Hence I think that a critically realist position is justifiable. (Feigl 1974b, p. 13.)

Empiricism. Feigl's summing up, "No Pot of Message" (Feigl 1974b—which has been frequently quoted above—is a restatement of vintage Humean empiricism. As regards the two types of meaningful statements, he says:

> I remain unconvinced by the clever arguments of Quine that are intended to show that there is no sharp line of distinction between the purely formal truths (e.g., of arithmetic) and the factual truths (e.g., of physics). I retain the distinction between the purely formal and the factual type of meaning. But I do not object if the formal type of meaning is regarded as the 'null case', or extreme lower limit of factual significance. The tautologies, i.e., the logical truths, or statements whose form— once *definienda* have been replaced by their *definientia*—boils down to logical truth do not require observations for their validation.... (Feigl 1980, p. 12)

As regards the criterion of factual meaningfulness, he says:

> The much debated and often revised testability criterion of factual meaningfulness seems to me useful and, even, indispensable. Unless some of the concepts appearing in our statements are connected, no matter how indirectly, with some data of immediate experience, those statements would at best have formal significance but they would be devoid of factual meaning. I think the enormous amount of debate and quibbling that concerned the meaning criterion has been largely a waste of time and energy. (Feigl 1980, pp. 12-13)

He then refers—as he often does in his later writings—to the "liberalized [meaning] criterion". He concedes that "much of traditional metaphysics and even of theology is perfectly meaningful" on this criterion, but he insists that there are no good reasons to believe that absolute space, time, and substance, or a personal god, exists. It appears that the positivist anathemas are no longer held to be meaningless, but merely (probably) false.

Feigl's reference to "the liberalized meaning criterion" may suggest that he had in mind a precise criterion of the kind that Hempel finally despaired of finding (Hempel 1950). But no such criterion is presented in any of his writings, and in a 1969 paper he seems to have wondered whether one is achievable.

> ...some of the early formulations [of the positivist meaning criterion] were too drastic in that they eliminated difficult questions along with nonsensical ones. This was remedied by later more circumspect and more liberal formulations. Discussion and dispute concerning the very feasibility of an adequate formulation continues. (Feigl 1980, p. 78).

I conjecture that although he may have continued to hope that an acceptable, precise meaning criterion would be formulated, he believed empiricism did not require it, for the following reason. Scientists in obviously successful sciences such as physics and chemistry evaluate theories by deriving observable consequences from the theory, verifying or falsifying the consequences by observation, and using the resulting data as evidence for or against the theory. Whatever the specific features of the method employed, no theory is factually meaningful unless it can be evaluated by a method of this general type. Alternatively, if the proponent of a theory—such as evolution or creationism—is unable to identify observation statements that are evidence for the theory and others that are evidence against it, then that theory has no factual meaning for the proponent. This I suggest is "the liberalized meaning criterion" he deems indispensable. It is the necessary condition for empirical meaning that a working scientist might employ, not the necessary and sufficient condition of a logician.

In "Empiricism at Bay" (1974a) he discusses the modifications in empiricism that may be required. He notes that the *discovery*, or origin and development, of scientific theories is an important topic, but points out that empiricism is a doctrine concerning the *justification* of scientific theories and is unaffected by discoveries about the origin and development of theories. He admits that the distinction between analytic and synthetic statements is difficult to apply to actual scientific theories, but he maintains that it was intended to apply to reconstructions of theories in an ideal language that are intended to throw light on the structure and testing of actual theories, though not on their origin and development. He concedes that most observation statements scientists actually employ as data are theory-laden to some degree, and that the correction of data by theory is more pervasive than some empiricists realized, but he continues to hold that observational data are the basis of all factual knowledge. He reminds us that there are many well-established empirical generalizations in the physical sciences that cannot reasonably be doubted, and that any acceptable theory must explain them. His final conclusion is that "empiricism, though in need of renovation, will remain a fruitful and adequate philosophy of science" (Feigl 1980, p. 285).

Feigl wrote these words at a time when he perceived the fortunes of empiricism to be at a low ebb. The frontispiece of *Inquiries and Provocations* is a 1973 photograph of him and Hempel labelled "The Last Two Empiricists". I think he was able to remain steadfast because he believed that empiricism is the only adequate philosophy for experimental science. Though he became a philosopher instead of a chemist, he never lost the perspective, and the scientific commonsense, of a practicing scientist. He was, in the paradigmatic sense, a philosopher *of science*.

References

Ayer, A.J. (1936), *Language , Truth, and Logic*, London: Oxford University Press. Second revised edition, Gollancz, London, 1946.

Blumberg, A.E. and H. Feigl (1931), "Logical Positivism, A New Movement in European Philosophy", *Journal of Philosophy*, 28: 281-296.

Carnap, R. (1928), *Der Logische Aufbau der Welt*, Weltkreis-Verlag, Berlin-Schlactensee. Translated by R.A. George, *The Logical Structure of the World*, University of California Press, Berkeley and Los Angeles, 1969.

Feigl, H. (1927), *Zufall und Gesetz*, Unpublished doctoral dissertation, University of Vienna.

_ _ _ _ . (1929), *Theorie und Erfahrung in der Physik*, G. Braun, Karlsruhe. English translation of chapter III in Feigl 1980, pp. 116-144.

_ _ _ _ . (1930), *Wahrscheinslichkeit und Erfahrung"*, *Erkenntnis*, 1: 249-259. Translated as "Probability and Experience" in Feigl 1980, pp. 107-115.

_ _ _ _ (1958), "The 'Mental' and the 'Physical'". In Feigl and Scriven, pp. 370-497.

_ _ _ _ (1967), *"The 'Mental' and the 'Physical' : The Essay and a Postscript"*, Minneapolis: University of Minnesota Press.

_ _ _ _ (1969), "The Wiener Kreis in America". In D. Fleming, and B. Bailyn (eds.), *The Intellectual Migration 1930-1960*, Cambridge: Harvard University Press, pp. 630-673. Reprinted in Feigl 1980, pp. 57-94.

_ _ _ _ (1974a), "Empiricism at Bay?". In R. S. Cohen and M. W. Wartofsky (eds.), *Methodological and Historical Essays in the Natural and Social Sciences*. Boston Studies in the Philosophy of Science, Volume XIV, Dordrecht: Reidel, pp. 1-20. Reprinted in Feigl 1980, pp. 269-285.

_ _ _ _ (1974b), "No Pot of Message". In P. Bertocci (ed.), *Mid-Twentieth Century Philosophy: Personal Statements*. New York: Humanities Press, pp. 120-139. Reprinted in Feigl (1980), pp. 1-20.

_ _ _ _ (1980), *Inquiries and Provocations: Selected Writings 1929-1974*. Edited by Robert S. Cohen. Vienna Circle Collection, Volume XIV, Dordrecht: Reidel.

_ _ _ _ and W. Sellars (eds.) (1949). *Readings in Philosophical Analysis*. New York: Appleton-Century-Crofts .

_ _ _ _ and M. Brodbeck (eds.) (1953). *Readings in the Philosophy of Science*. New York: Appleton-Century-Crofts.

_ _ _ _ and M. Scriven (eds.) (1956). *The Foundations of Science and the Concepts of Psychology and Psychoanalysis*. Minnesota Studies in the Philosophy of Science, Volume I. Minneapolis: University of Minnesota Press.

_ _ _ _ , M. Scriven, and G. Maxwell (eds.), *Concepts, Theories, and the Mind-body Problem*, Minnesota Studies in the Philosophy of Science, Volume II. University of Minnesota Press, Minneapolis, 1958.

_ _ _ _ and G. Maxwell (eds.) (1961). *Current Issues in the Philosophy of Science*. New York: Holt, Rinehart, and Winston.

_ _ _ _ and G. Maxwell (eds.) (1962). *Scientific Explanation, Space, and Time*. Minnesota Studies in the Philosophy of Science, Volume III. Minneapolis: University of Minnesota Press.

_ _ _ _ , W. Sellars, and K. Lehrer (eds.) (1972). *New Readings in Philosophical Analysis*. New York: Appleton-Century-Crofts.

Feyerabend, P.K. and G. Maxwell (eds.) (1966). *Mind, Matter, and Method: Essays in Philosophy of Science in Honor of Herbert Feigl*. Minneapolis: University of Minnesota Press.

Hempel, C.G. (1950), "Problems and Changes in the Empiricist Criterion of Meaning", *Revue Internationale de Philosophie*, 11: 41-63.

Neurath, O., R. Carnap, and C. Morris (eds.) (1938-69). *International Encyclopedia of Unified Science*, Volumes 1 and 2. Chicago: University of Chicago Press.

Popper, K. (1935), *Logik der Forschung,* Vienna: J. Springer. Translated by the author. *The Logic of Scientific Discovery*, New York: Basic Books, 1959.

Reichenbach, H. (1938), *Experience and Prediction.* Chicago: University of Chicago Press.

Russell, B. (1948), *Human Knowledge: Its Scope and Limits.* Simon and Schuster.

Salmon, W. (1967), *The Foundations of Scientific Inference.* Pitsburgh: University of Pittsburgh Press.

Schlick, M. (1918), *Allgemeine Erkenntnislehre.* Berlin: J. Springer. Second edition, 1925. Translated by A.E. Blumberg, with an introduction by Blumberg and Feigl. *General Theory of Knowledge.* New York: J. Springer, 1974.

Smart, J.J.C. (1959), "Sensations and Brain Processes". *Philosophical Review*, 68: 141-156.

Herbert Feigl

Bruce Aune

University of Massachusetts

Herbert Feigl was one of the most admirable people I have known. I was fortunate to have him a a friend and as a teacher (he directed my PhD dissertation). He was a very special friend and a very special teacher. To round out what is being said here, I want to share my impressions of him as a teacher and as a person.

I don't think I ever took a course from Herbert; he taught me informally as a thesis-director and as a helpful older friend. When he guided me through the writing of my thesis, I was almost unaware fo the teaching he was doing. It is only in recent years that I have come to appreciate how important it was for me.

In comparison with other teachers, Herbert was extremely permissive—the opposite of coercive. For him, nothing was off-limits. As it happened, he disagreed with much of what I was saying, but he didn't trample me: he let me go my own way. He guided me, all right, but his guidance was indirect, and it brought out the best in me. "Such and such book (or article) has a bearing on this argument of yours; you ought to read it," he'd say. And then, after I'd read it, he'd ask "What did you think? Was the writer correct, or not? Tell me." In seeking my opinion as he did, he was rewarding me for my efforts, treating me as a serious philosopher and making me, in the process, more serious and more of a philosopher. This treatment continued after I had finished my degree and was out in the world. When I would see him on visits to Minneapolis (my home town), he would ask me what I was working on—and really be interested in hearing what I had to say. He did the same with other visitors, and his practice in this regard partially accounts, I believe, for his success as director of his Center. Instead of attacking a visitor's paper, trying to refute it (as philosophers commonly do), his aim was to make sense of what the speaker was saying to see if any light was being shed on serious problems. His approach was encouraging, not daunting; and people were stimulated by it. The Center was an exciting, gratifying place to visit.

In my experience Herbert was a thoroughly positive person. He was always kind, never carping, captious, or unpleasant; always considerate, never superior or malicious; always encouraging, never destructive; always extremely modest (admitting to Tuesday-Thursday doubts about even his pet theories); and never hostile to those who did not agree with him about the most serious things in the world, science and philosophy. I never heard him refer to anyone as a dunderhead or fool.

PSA 1988, Volume 2, pp. 23-24

The image of him that I am left with is that of a real philosopher who, like a ruler in Plato's Republic, wants understanding rather than honor or power, and who is, in personal relations, kind and good and encouraging. Herbert was a serious, civilized, cultivated human being who will always be a role-model for me and, I am sure, for many others.

Part III

NATURAL PHILOSOPHY

The Anthropic Principle: A Primer for Philosophers[1]

Frank J. Tipler

Tulane University

1. Introduction: The Weak Anthropic Principle

Let's begin with a

Definition: **The Anthropic Principle** is the drawing of scientific inferences from a consideration of Man's Place in Nature.

There are various versions of the Anthropic Principle. The most conservative version of the Anthropic Principle is nothing but a systematic working out of the fact that the astrophysical (and other scientific) data we have is self-selected due to the fact that *Homo sapiens* is a particular type of intelligent being. This conservative version of the Anthropic Principle is called the

> **Weak Anthropic Principle** (WAP): The observed values of all physical quantities are not equally probable, but rather take on values restricted by the fact that these quantities are measured by a carbon-based intelligent life-form which spontaneously evolved on an earthlike planet around a G2 type star.

Again, the Weak Anthropic Principle is just a warning to take into account a grandiose type of selection bias when interpreting data. But selection bias is familiar to scientists. For an astrophysics example, suppose we want to know the fraction of all galaxies that lie in particular ranges of brightness. It is not sufficient to list all the galaxies seen according to their brightness, for the simple reason that many galaxies are too faint to be seen, or are not big enough to be distinguished from stars. Thus our observations are biased toward the very bright galaxies. We must therefore correct for this selection bias. All instruments are subject to some sort of selection bias. The Weak Anthropic Principle merely says that we — *Homo sapiens* — are also a type of measuring instrument, and it is necessary to take into account our special properties (given above) when interpreting data. WAP does *not* claim that our form of life (based on carbon, etc) is the only possible form of life. One can easily imagine non-carbon based forms of life, and indeed, as we shall see, a very speculative form of the Anthropic Principle assumes that non-carbon forms of life can exist. Such forms of life, if they do exist, are obviously not subject to the same selection biases that *Homo sapiens* is. But equally obviously, *Homo sapiens* is subject to the selection biases of *Homo sapiens*. Thus the Weak Anthropic Principle must be accepted, for it is just an application of standard scientific logic.

PSA 1988, Volume 2, pp. 27-48
Copyright © 1989 by the Philosophy of Science Association

The Dicke-Wheeler Relation

But the Weak Anthropic Principle is not trivial, for it leads to unexpected relationships between observed quantities that appear to be unrelated! The best example is the Dicke-Wheeler relation, which gives the measured age of the universe in terms of elementary particle time scales and fundamental interaction constants.

Now there is no *a priori* reason for believing in the existence of any definite unique relation between the age of the universe t_U and the various fundamental constants. In fact, if the Universe is open or flat (as the data suggests) t_U will eventually sweep out all values between zero and infinity, so *all* possible relations will hold at some time. However, the Weak Anthropic Principle says we (*Homo sapiens*) must measure t_U to have a value consistent with our carbon-based species having evolved on an earthlike planet. This implies that there must have been sufficient time for the elements heavier than helium to form in stars, and be scattered by supernovae; thus t_U must be greater than lifetime t_S of a typical main sequence star.

Let us make an estimate of main sequence stellar lifetime. The luminosity of a massive star is typically the Eddington Luminosity $L_E = 4\pi G M m_p c / \sigma_T$, where M is the stellar mass, m_p is the proton mass, and σ_T is the Thomson cross-section. Thus if $\eta \leq 10^{-2}$ is the fraction of the star's mass than can be released through nuclear burning, t_S is roughly

$$t_S \approx \eta M c^2 / L_E = (\eta M c^2 \sigma_T)/(4\pi G M m_p c) = \eta(\alpha_e^2/\alpha_G)(m_p/m_e)^2 t_p = [\eta(\alpha_e^2(m_p/m_e)^2]\alpha_G^{-1} t_p$$

where $t_p = \hbar/m_p c^2 \approx 10^{-23}$ seconds is a typical strong interaction timescale, $\alpha_e \equiv e^2/\hbar c \approx 1/137$ is the fine structure constant (electromagnetic force strength), $\alpha_G \equiv G m_p^2/\hbar c \approx 5 \times 10^{-39}$ is the gravitational fine structure constant, and the electron mass is m_e. The collection of constants in brackets is of order unity, so

$$t_S \approx \alpha_G^{-1} t_p \approx 10^{10} \text{ years.}$$

Further, t_U can't be too many orders of magnitude larger than this, because otherwise the gas in the interstellar medium will be used up and G2 stars can't form. (This can be made more precise. The amount of gas available for the formation of new stars is, in the standard theories of star formation in the galaxy, decreasing exponentially. The exponential time constant for the exhaustion of the interstellar gas — a few billion years — implies that essentially no G2's will be formed after 10^{12} years, On the other hand, no G2's *can* be formed before 10^8 years. Probability theory tells us (e.g., Alder and Roessler 1964, pp. 33-34) that for processes characterized by a constant ratio between consecutive numbers — exponential decay is such a case, since (Feller 1968, p. 458) it is approximated by a geometric progression— the most likely time is the geometric mean of 10^8 years and 10^{12} years, namely 10^{10} years.) Thus to within a few orders of magnitude, the expected age of the Universe that our particular form of life would measure is 10^{10} years, which should also be the Hubble time H_0^{-1}, which is roughly the age of the universe So we have

$$H_0^{-1} \approx \alpha_G^{-1} t_p \approx 10^{10} \text{ years.} \tag{1}$$

Thus the Weak Anthropic Principle has allowed us to express Hubble's constant in terms of the fundamental constants, and even to compute it to a few orders of magnitude. Again, it is important to note that (1) need not hold, and indeed won't hold much earlier and later in Universal history, since Hubble's "constant" changes with time. [(1) was derived by Dicke in 1961, and Wheeler later pointed out that it implied the Universe must be at least 10^{10} lightyears across (that is, $H_0^{-1}c$) in order to have a *single* intelligent species like us. This is why I've termed (1) the Dicke-Wheeler relation.]

Dicke *could* have used (1) to rule out the steady state theory, for in this theory, there is no necessary relationship between Hubble's constant and the main sequence lifetime, and the above derivation assumed the Big Bang. (Missed prediction!) This shows that the Weak Anthropic Principle can be used to rule out theories.

The Carter Inequality on the Age of the Earth

The Weak Anthropic Principle can also make predictions. Carter (see Barrow and Tipler, 1986 Section 8.7) has pointed out the additional temporal coincidence:

$$t_e \approx t_{ms} \tag{2}$$

where t_e is the length of time to evolve *Homo sapiens* (4.6 billion years) and t_{ms} is the length of time the Sun will remain on the main sequence (10^{10} years). There is no reason to expect this approximate equality; indeed quite the reverse. We have seen that t_{ms} is determined by physical constants, while we would think t_e to be biologically based. Thus we would expect *a priori*

$$t_e \ll t_{ms} \qquad \text{or} \qquad t_e \gg t_{ms} \tag{3}$$

rather than (2), which is actually observed.

Carter has shown we can explain (2) from $t_{av} \gg t_{ms}$ and the Weak Anthropic Principle, where t_{av} is the *average* time needed to evolve an intelligent species on an *immortal* earth-like planet. Now the Earth is not immortal as a life-containing planet, for biological evolution will cease when the Sun leaves the main sequence. If intelligent life doesn't evolve by then, it will never evolve. Thus there is a least upper bound t_{lub} to the time evolution can proceed on an earthlike planet. But the longer life exists, the more likely it is for intelligence to evolve. If $t_{av} \gg$ tms, this means intelligence is more likely to appear near t_{lub} than near the beginning of an earthlike planet, so most likely te $\approx t_{lub} \leq t_{ms}$. (Note that we need not have the equality $t_{lub} = t_{ms}$; something other that the death of the Sun may destroy life on Earth.)

This rough argument can be made quantitative. Suppose there are n crucial steps in the evolution of intelligence, each so improbable it is unlikely to occur until long after t_{lub}. Suppose also that the n steps are statistically independent and that the likelihood that the ith step occurs at any given time instant is independent of how long evolution has been going on. The time independence of the ith step is justified as follows. Suppose that the ith step is some property that is coded by a gene or a collection of genes. Each generation the genome changes, by both reassortment and mutation. If the rate of genetic change is roughly constant, and in addition if which genes/codons appear is completely random, then the likelihood that a given collection of codons or genes will arise at any given instant will be time independent. Of course, this time independence will be only an approximation, for a given mutation may be much more likely to give a viable organism if other mutations have occurred first. But I think it is a reasonable approximation. The same argument can be used to justify the statistical independence of the n steps. Now it can be shown (Feller 1968, pp. 328-329, 458-460 of Volume I and p. 8 of Volume II) that the time independence of the ith step implies that the probability it will occur by time t is given by the exponential distribution. Thus the probability that intelligence will evolve on an immortal earthlike planet by time t is given by the product of n exponential probability distributions

$$p(t) = \prod_{i=1}^{n} (1 - \exp[-t/\alpha_i]) \approx \beta t^n$$

where α_i is the expectation time for the occurrence of the ith improbable step. (We have $1 - \exp[-t/\alpha_i] \approx t/\alpha_i$, since $t \ll \alpha_i$ for all $t \leq t_{lub}$.) From p(t), we can compute the conditional probability that intelligence evolves at time t, given that it *must* evolve on or before t_{lub}. This conditional probability is

$$p(\text{intelligence arises by } t \mid \text{it definitely occurs before } t_{lub}) = \gamma t^n$$

where the normalization $\gamma = t_{lub}^{-n}$ is chosen to make the conditional probability equal to one when $t = t_{lub}$. The expectation $<t>$ for the time to evolve intelligence given that it does evolve in $(0, t_{lub})$

$$<t> = \int_0^{t_{lub}} t \, dp = \int_0^{t_{lub}} t(t_{lub}^{-n}) nt^{n-1} dt = t_{lub} n/(n+1)$$

We would expect $<t> \approx t_e$, so we finally obtain

$$t_{lub} - t_e \leq t_e/n$$

which is known as Carter's Inequality.

This inequality is testable in principle; we need only know t_{lub} and n. Unfortunately, modern evolutionary biology is unable to give us an estimate for n. But qualitatively, Carter's argument is based on intelligent life being improbable, so SETI (The Search for Extraterrestrial Life) is actually a test of Carter's argument. If the evolution of one-celled organisms happens to be one of the improbable steps (we don't know whether it is or not), then it follows that we are unlikely to find any sort of life elsewhere in the Solar System.

It is extremely important to note that the Dicke-Wheeler relation and the Carter inequality require the *actual* existence of an ensemble out of which we self-select our observations. For the Dicke-Wheeler relation, we are by our existence selecting one t_U out of (0, t very large). For the Carter inequality, we are by our very existence selecting one earthlike planet out of an actual infinity (in the open or flat case), or out of a near infinity (closed case). (In fact, it is another prediction of the Carter argument that the Universe is huge. If it were small, then the ensemble would not actually exist.)

Domain Universes and the Cosmological Constant

Many modern cosmological models assume that the "total" universe (= everything that exists) is divided up into what we might call "domain universes" of varying sizes in which the various physical constants and initial conditions differ. For example, Linde's Eternal Chaotic Inflation model (Linde 1989) assumes that the total universe is composed of an infinity of domain universes, many of which begin as a Planck-sized (10^{-33} cm) quantum fluctuation inside another domain universe, and inflate to Hubble size (10^{10} lyr). These "child universes" can have different physical constants, and will appear to have different initial conditions. However such domain universes arise, in the ensemble of all of them — i.e., in the total universe — all possibilities are realized. But only in very special subsets is our form of life possible. In this way, WAP self-selection avoids the problem of initial conditions. In fact, any cosmological theory which allows different initial conditions (or constants) must *generally* invoke WAP somewhere. The best example of this is the WAP solution to the Cosmological Constant Problem.

The Cosmological Constant Problem is this (Weinberg 1989). Due to quantum fluctuations, the vacuum has energy, and by Lorentz invariance, the stress-energy tensor of vacuum must have the form $<T_{\mu\nu}> = <\rho>g_{\mu\nu}$, where $<\rho>$ is a constant. This gives an effective cosmological constant $\lambda_{eff} = \lambda + 8\pi G<\rho>$, where λ is the classical cosmological constant; equivalently, the total vacuum energy density is

$$\rho_V = <\rho> + \lambda/8\pi G = \lambda_{eff}/8\pi G \tag{4}$$

This vacuum energy density generates a universal gravitational force field: if $\lambda_{eff} > 0$ the force is repulsive and if $\lambda_{eff} < 0$ the force is attractive. Astronomical observations show that $|\rho_V| < 10^{-29}$ gm/cm^3 $\approx 10^{-47}$ GeV^4. On the basis of dimensional arguments, we would expect the energy density of vacuum to be the Planck density $\rho_{Pk} = 5 \times 10^{93}$ gm/cm^3, in which case the two terms in (4) must cancel to 122 decimal places. There are many mechanisms which can give a vacuum energy density, and most require similar cancellations. For instance, if the spontaneous symmetry breaking in electroweak theory is real, then we must have enormous λ ($\geq 10^8$ GeV^4) before the electroweak phase transition in order to have $\lambda_{eff} \approx 0$ today.

The WAP solution to the Cosmological Constant Problem is this. Let us assume the existence of an ensemble of domain universes with λ different in different domain universes. Averaged over all domain universes, the expectation value of λ is huge (the Planck density, say), but only when λ is very small — as in *our* domain universe — can intelligent life develop. We can obtain a lower bound on ρ_V as follows (Barrow and Tipler 1986, section 6.9). This lower bound will be negative. If $\rho_V < 0$, Universe expands and recollapses in time $T \leq 2\pi[8\pi G|\rho_V|/3]^{-1/2}$, independent of whether it is closed, flat, or open (Tipler 1976). As discussed above, we need $T \geq 10^9$ years for our type of intelligence to evolve; i.e., $T \geq 0.1 H_0^{-1}$; where $H_0^{-1} = [8\pi G\rho_{M_0}/3]^{-1/2}$ is the value of the inverse Hubble constant in an approximately flat universe and ρ_{M_0} is the non-vacuum energy density now. This gives a lower bound of

$$10^3\rho_{M_0} > |\rho_V|$$

The upper (positive) bound on ρ_V has been obtained by Weinberg (1987), (1989). If $\rho_V > 0$, then ρ_V generates a repulsive gravitational force which can prevent gravitational condensations (formation of stars and planets). But as long as the non-vacuum energy density is less than ρ_V, repulsion has little effect. We know condensations started in our universe at $z_c \geq 4$, where z is the redshift, when the density was larger than ρ_{M_0} by the factor $(1 + z_c)^3$, so if $\rho_V < 100\rho_{M_0}$, the evolution of our type of life is definitely possible, since we in fact evolved in our (domain?) universe. On the other hand, if $\rho_V > 10^4\rho_{M_0}$, then galaxies are disrupted very early (several star generations are not allowed), and our type life is impossible. Summarizing, WAP self-selection requires:

$$- 10^3\rho_{M_0} < \rho_V < 10^4\rho_{M_0} \tag{5}$$

Now the expectation value $<\rho_V>$ computed from the ensemble of domain universes in face of above WAP constraints depends on the distribution of ρ_V's. For instance, if it were a Gaussian distribution with peak and standard deviation equal to *positive* Planck density, then WAP puts us in the far tail of the distribution, and the distribution is very flat over WAP range. Thus any value in range (5) equally likely. On the other hand if Nature gives us an exponential distribution of ρ_V's, with zero probability for negative ρ_V's, then we would then expect an ρ_V nearer the upper bound, as happened in Carter's explanation of the age of the Earth.

The crucial point was made by Weinberg (1987, 1989): if ρ_V is actually measured to be non-zero, the *only* plausible explanation for such a state of affairs is WAP self-selection! The reason is that quantum field theory seems most unlikely to give 122 decimal cancellation and leave a residual. It is a corollary that if we actually measure ρ_V to be non-zero, this measurement would be experimental evidence for other domain universes where ρ_V is different. The current opinion in the physics community (an opinion which I share) is that in fact some quantum field mechanism will be found which gives a cosmological constant of exactly zero. The most promising proposal to date is the Hawking-Coleman "baby universe" mechanism (see Weinberg (1989) for a discussion). But if this fails, I would suggest that particle physicists begin to calculate distributions for ρ_V.

2. The Strong Anthropic Principle

In contrast to the self-selection aspects of Man's Place in Nature, consider the possibility that in some way, intelligent life is essential to the Universe. This idea is called **The Strong Anthropic Principle** (SAP). Note that there is no ensemble in SAP! In fact, the existence (or lack of) an ensemble is the basic difference between WAP and SAP. Let me warn the reader that any version of SAP is *VERY* speculative! To emphasize this, Carter (1989) has suggested that we call SAP The Strong Anthropic *Proposal* to distinguish it from WAP, which is a genuine principle of physics that we *must* accept. A *Proposal*, on the other hand, we need not accept; it is merely put forward for consideration. Why should we consider SAP? Let me remind the reader of two important facts:

(1) there is no evidence for intelligence in the Universe today; this strongly suggests it is unimportant to the Universe.

(2) due to WAP selection, we are viewing the Universe very early in its history.

Thus (2) could counter the suggestion of (1) if the true significance of life is made manifest only in the far future. But if life dies out (ultimately wiped out by the Heat Death, say), it is hard to see any essential role for life to play. This leads to

The Final Anthropic Principle (FAP): the Universe is sufficiently benign so that once intelligence first evolves, the laws of physics *permit* its continued existence forever.

I think scientists should take the FAP seriously because we have to have *some* theory for the future of the physical universe — since it unquestionably exists — and the FAP is based on the most beautiful physical postulate: that total death is not inevitable. *All* other theories of the future necessarily postulate the ultimate extinction of everything we could possibly care about. I once visited a nazi death camp; there I was reinforced in my conviction that there is nothing uglier than extermination. We physicists know that a beautiful postulate is more likely to be correct than an ugly one. Why not adopt the Postulate of Eternal Life — FAP, that the extinction of everything we could possibly care about is not inevitable — at least as a working hypothesis? (See Linde (1989), Abramowicz and Ellis (1989), and Barrow (1989) for more discussion of FAP.)

The Omega Point Theory

In order to investigate whether life can continue to exist forever, I shall need to define "life" in physics language. I claim that a "living being" is any entity which codes "information" (in the sense this word is used by physicists) with the information coded being preserved by natural selection. (I justify this definition in section 8.2 of Barrow and Tipler 1986). Thus "life" is a form of information processing, and the human mind — and the human soul — is a very complex computer program. Specifically, a "person" is defined to be a computer program which can pass the Turing Test (See Hofstadter and Dennett 1981, 69-95 (Chapter 5) for a detailed discussion of this test).

There is actually an astonishing similarity between the mind-as-computer-program idea and the medieval Christian idea of the "soul". Both are fundamentally "immaterial": a program is a sequence of integers, and an integer — 2, say — exists "abstractly" as the class of all couples. The symbol "2" written here is a *representation* of the number 2, and not the number 2 itself. In fact, Aquinas (following Aristotle) defined the *soul* to be "the form of activity of the body" . In Aristotelian language, the *formal* cause of an action is the abstract cause, as opposed to the material and efficient causes. For a computer, the program is the formal cause, while the material cause is the properties of the matter out of which the computer is made, and the efficient cause is the opening and closing of electric circuits. For Aquinas, a human soul needed a body to think and feel, just as a computer program needs a physical computer to run.

Aquinas thought the soul had two faculties: the agent intellect (*intellectus agens*) and the receptive intellect (*intellectus possibilis*), the former being the ability to acquire concepts, and the latter being the ability to retain and use the acquired concepts. Similar distinctions are made in computer theory: general rules concerning the processing of information coded in the central processor are analogous to the agent intellect; the programs coded in RAM or on a tape are analogues of the receptive intellect. (In a Turing machine, the analogues are the general rules of symbol manipulation coded in the device which prints or erases symbols on the tape vs. the tape instructions, respectively.) Furthermore, the word "information" comes from the Aristotle-Aquinas' notion of "form": we are "informed" if new forms are added to the receptive intellect. Even semantically, the information theory of the soul is the same as the Aristotle-Aquinas theory.[2]

The "mind as computer program" idea is absolutely central to this paper; indeed, it forms the basis of a revolution now going on in mathematics, physics, and philosophy. The best defense of the idea can be found in Hofstadter and Dennett's *The Mind's I* (1981), particularly pages 69-95; 109-115; 149-201; and 373-382.

In the language of information processing it becomes possible to say precisely what it means for life to continue forever.

Definition: I shall say that life can continue forever if:

(1) information processing can continue indefinitely along at least one world line γ all the way to the future "c-boundary" of the universe; that is, until the end of time.

(2) the amount of information processed between now and this future c-boundary is infinite in the region of spacetime with which the world line γ can communicate.

(3) the amount of information stored at any given time τ within this region can go to infinity as t approaches its future limit (this future limit of τ is finite in a closed universe, but infinite in an open one, if τ is measured in what physicists call "proper time").

The above is a rough outline of the more technical definition given in Section 10.7 of Barrow and Tipler 1986. (See also (Tipler 1986) and (Tipler 1988)) But let me ignore details here. What is important are the physical (and ethical!) reasons for imposing each of the above three conditions. The reason for condition 1 is obvious: it simply states that there must be at least one history in which life (=information processing) never ends. (See below for more on what "c-boundary" means. For now, think of it as meaning "the end of time.").

Condition 2 tells us two things: First, that information processed is "counted" only if it is possible, at least in principle, to communicate the results of the computation to the history γ. This is important in cosmology, because event horizons abound. In the closed Friedmann universe, which is the standard (but over-simplified) model of our actual universe (if it is in fact closed), every comoving observer loses the ability to send light signals to every other comoving observer, no matter how close. Life obviously would be impossible if one side of one's brain became *forever* unable to communicate with the other side. Life is organization, and organization can only be maintained by constant communication among the different parts of the organization. The second thing condition 2 tells us is that the amount of information processed between now and the end of time is potentially infinite. I claim that it is meaningful to say that life exists forever only if the number of thoughts generated between now and the end of time is actually infinite. But we know that each "thought" corresponds to a minimum of one bit being processed. In effect, this part of condition 2 is a claim that time duration is most properly measured by the thinking rate, rather than by proper time as measured by atomic clocks. The length of time it takes an intelligent being to process one bit of information — to think one thought — is a direct measure of "subjective" time, and hence is the most important measure of time from the perspective of life. A person who has thought 10 times as much, or experienced 10 times as much (there is no basic physical difference between these options) as the average person has in a fundamental sense lived 10 times as long as the average person, even if the rapid thinking person's chronological age is shorter than the average.

The distinction between proper and subjective time crucial to condition 2 is strikingly similar to a distinction between two forms of duration in Thomist philosophy. Recall that Aquinas distinguished three types of duration. The first was *tempus*, which is time measured by change in relations (positions, for example) between physical bodies on Earth. *Tempus* is analogous to proper time; change in both human minds and atomic clocks is proportional to proper time, and for Aquinas also, *tempus* controlled change in corporeal minds. But in Thomist philosophy, duration for *incorporeal* sentient beings — angels — is controlled not by matter, but rather is measured by change in the mental states of these beings themselves. This second type of duration, called *aevum* by Aquinas, is clearly analogous to what I have termed "subjective time". *Tempus* becomes *aevum* as sentience escapes the bonds of matter. Analogously, condition 2 requires that thinking rates are controlled less and less by proper time as τ approaches its future limit. *Tempus* gradually becomes *aevum* in the future. (The third type of Thomist duration is *aeternitas*, which can be thought of as "experiencing" all past, present, and future *tempus* and *aevum* events in the universe all at once. But more of *aeternitas* later.)

Condition 3 is imposed because although condition 2 is necessary for life to exist forever, it is not sufficient. If a computer with a finite amount of information storage — such a computer is called a *finite state machine* — were to operate forever, it would start to repeat itself over and over. The psychological cosmos would be that of Nietzsche's Eternal Return. Every thought and every sequence of thoughts, every action and every sequence of actions, would be repeated not once but an infinite number of times. It is generally agreed (by everyone but Nietzsche) that such a universe would be morally repugnant or meaningless. Only if condition 3 holds in addition to condition 2 can a psychological Eternal Return be avoided. Also, it seems reasonable to say that "subjectively," a finite state machine exists for only a finite time even though it may exist for an infinite amount of proper time and process an infinite amount of data. A being (or a sequence of generations) that can be truly said to exist forever ought to be physically able, at least in principle, to have new experiences and to think new thoughts.

Let us now consider whether the laws of physics will permit life/information processing to continue forever. John Von Neumann and others have shown that information processing (more precisely, the irreversible storage of information) is constrained by the first and second laws of thermodynamics. Thus the storage of a bit of information requires the expenditure of a definite minimum amount of available energy, this amount being inversely proportional to the temperature (See Section 10.6 of Barrow and Tipler 1986 for the exact formula). This means it is possible to process and store an infinite amount of information between now and the Final State of the universe only if the time integral of P/T is infinite, where P is the power used in the computation, and T is the temperature. Thus the laws of thermodynamics will permit an infinite amount of information processing in the future, provided there is sufficient available energy at all future times.

What is "sufficient" depends on the temperature. In the open and flat ever-expanding universes, the temperature drops to zero in the limit of infinite time, so less and less energy per bit processed is required with the passage of time. In fact, in the flat universe, only a *finite* total amount of energy suffices to process an infinite number of bits! This finite energy just has to be used sparingly over infinite future time. On the other hand, closed universes end in a final singularity of infinite density, and the temperature diverges to infinity as this final singularity is approached. This means that an ever-increasing amount of energy is required per bit near the final singularity. However, most closed universes undergo "shear" when they recollapse, which means they contract at different rates in different directions (in fact, they spend most of their time *expanding* in one direction while contracting in the other two!). This shearing gives rise to a radiation temperature difference in different directions, and this temperature difference can be shown to provide sufficient free energy for an infinite amount of information processing between now and the final singularity, even though there is only a *finite* amount of proper time between now and the end of time in a closed universe. Thus although a closed universe exists for only a finite proper time, it nevertheless could exist for an infinite subjective time, which is the measure of time that is significant for living beings.

But although the laws of thermodynamics permit conditions 1 through 3 to be satisfied, this does not mean that the other laws of physics will. Its turns out that although the energy is available in open and flat universes, the information processing must be carried out over larger and larger proper volumes. This fact ultimately makes impossible any communication between opposite sides of the "living" region, because the redshift implies that arbitrarily large amounts of energy must be used to signal (This difficulty was first pointed out by Freeman Dyson.). This gives the

First Testable Prediction of the Omega Point Theory: the universe must be closed.

However, as I stated earlier, there is a communication problem in most closed universes — event horizons typically appear, thereby preventing communication. But there is a rare class of closed universes which don't have event horizons, which means by definition that every world line can always send light signals to every other world line. Now Roger Penrose has found a way to define precisely what is meant by the "boundary" of spacetime, where time ends. In his definition of the "c-boundary", world lines are said to end in the same "point" on this boundary if they can remain in causal contact unto the end of time. If they eventually fall out of causal contact then they are said to terminate in different c-boundary points. Thus the c-boundary of these rare closed universes without event horizons consists of a single point. For reasons given in Section 10.6 of Barrow and Tipler 1986 (see also Section 3.7), it turns out that information processing can continue only in closed universes which end in a single c-boundary point, and only if the information processing is ultimately carried out throughout the entire closed universe. Thus we have the

Second Testable (?) Prediction of the Omega Point Theory: the future c-boundary of the universe consists of a single point; call it the *Omega Point*. (Hence the name of the theory.)

It is possible to obtain other predictions. For example, a more detailed analysis of how energy must be used to store information leads to the

Third Testable Prediction of the Omega Point Theory: the density of particle states must diverge to infinity as the energy goes to infinity, but nevertheless this density of states must diverge no faster than the square of the energy .

The Omega Point has an interesting property. Mathematically, the c-boundary is a completion of spacetime: it is not actually in spacetime, but rather just "outside" it. If one looks more closely at the c-boundary definition, one sees that a c-boundary consisting of a single point is formally equivalent to the entire collection of spacetime points. In effect, all the different instants of universal history are collapsed into the Omega Point; "duration" for the Omega Point can be regarded as equivalent to the collection of all experiences of all life that did, does, and will exist in the whole of universal history, together with all non-living instants. This "duration" is very close to the idea of *aeternitas* of Thomist philosophy. We could say that *aeternitas* is equivalent to the union of all *aevum* and *tempus*.

This identification of the Omega Point with the whole of the past, present, and future universal history is more than a mere mathematical artifact. *The identification really does mean that the Omega Point "experiences" the whole of universal history "all at once!"* For consider what it means for us to "experience" an event. It means we think and emote about an event we see, hear, feel, etc. Consider for simplicity just the "seeing" mode of sensing. We see another contemporary person by means of the light rays that left her a fraction of a second ago. But we cannot "see" a person that lived a few centuries before, because the light rays from said person have long ago left the solar system. Conversely, we cannot "see" the Andromeda Galaxy as it now is, but rather we "see" it as it was 2 million years ago. So we experience as "simultaneous" the events on the boundary of our past light cone (for the seeing mode; it is more complicated for all other modes of sensing, for we experience as simultaneous events which reach us at the same instant along certain timelike curves from inside our past light cone).

But all timelike and lightlike curves converge upon the Omega Point. In particular, all the light rays from all the people who died a thousand years go, from all the people now living, and from all the people who will be living a thousand years from now, will intersect there. The light rays from those people who died a thousand years ago are not lost forever; rather these rays will be intercepted by the Omega Point. Or to put it another way, these rays will be intercepted and intercepted again, by the living beings who have engulfed the physical universe near the Omega Point. All the information which can be extracted from these rays will be extracted at the instant of the Omega Point. The beings existing at that last instant of time can experience the whole of time simultaneously just as we experience simultaneously the Andromeda Galaxy and a person in the room with us. (I should warn the reader that I have ignored the problem of opacity and the problem of loss of coherence of the light. Until these are taken into account, I cannot say exactly how much information can in fact be extracted from the past. But at the most basic ontological level, *all* the information from the past [= all of universal history] remains in the physical universe and is available for analysis.)

The preceding analysis was entirely classical. However, since the existence of life for infinite subjective time requires the universe to actually achieve the final singularity — the Omega Point — the Omega Point theory is necessarily a quantum cosmological theory: almost all of subjective experience occurs after the universe has recollapsed past the Planck size! I shall briefly outline how one quantizes the Omega Point Theory.

In classical general relativity, a spacetime is generated from its initial data in the following manner. One is given a *3-dimensional* manifold S, and on S the non-gravitational fields F (and their appropriate derivatives F'), and two tensor fields h and K, with (F, F', h,K) satisfying certain equations called constraint equations. The constraint equations say nothing about the time evolution; rather they are to be regarded as consistency conditions amongst the fields (F, F',h,K) which must be satisfied at every instant of time. The physical interpretation of h is that of a spatial metric of the manifold S, and so S and (F, F',h,K) together represent the entire spatial universe at an instant of universal time. S and (F, F',h,K) are called the initial data. We now try to find a *4-dimensional* manifold M with metric g and spacetime non-gravitational fields F such that (1) M contains S as a submanifold; (2) g restricted to S is the metric h, and (3) K is the "extrinsic curvature" of S in M (roughly speaking, K says how rapidly h is changing in "time"). The manifold M and the fields (g,F) are then the whole of physical reality, including the underlying background spacetime (that is, (M,g)), the gravitational field (represented by the spacetime metric g), and all the non-gravitational fields (given by F). There will be infinitely many such M's and g's, but one can cut down the number by requiring that g satisfies the Einstein field equations everywhere on M, and that the Einstein field equations reduce to the constraint equations on S.

But even requiring the Einstein equations to hold everywhere leaves infinitely many spacetimes (M,g) which are generated from the *same* initial data at the spacetime instant

S. To see this, suppose we have found a spacetime (M,g) which in fact has S and its initial data as the spatial universe at some instant t_0 of universal time. Pick another universal time t_1 to the future of t_0 and cut away all of the spacetime in (M,g) to the future of t_1 (including the spatial instant corresponding to t_1.) This gives a new spacetime (M',g) which coincides with (M,g) to the past of t_1, but which has absolutely nothing — no space, no time, no matter — to the future of t_1. Clearly, both (M,g) and (M',g) are spacetimes which are both generated from S and its initial data. Furthermore, the Einstein equations are satisfied everywhere on both spacetimes. There are infinitely many ways we can cut away (M,g) in this way, so there is an infinity of (M',g)'s we can construct. True, the universe (M',g) ends abruptly at t_1, for no good reason. But what of that? The point is, the field equations themselves cannot tell us that the physical universe should continue past the time t_1. Rather, in classical general relativity one must impose as a separate assumption, over and above the assumption of the field equations and the initial data, that the physical universe must continue in time until the field equations themselves tell us that time has come to an end (at a spacetime singularity, say).

It is possible to prove (Hawking and Ellis 1973, Chapter 7) that there is amongst all the mathematically possible (M',g) —we might call these "possible worlds" — a *unique* "maximal" spacetime (M,g) which is generated by the initial data on S. "Maximal" means that the spacetime (M,g) contains any other (M',g) generated by the initial data on S as a proper subset. In other words, (M,g) is the spacetime we get by continuing the time evolution until the field equations themselves won't allow us to go further. This maximal (M,g) is the natural candidate for the spacetime that is actualized, but it is important to keep in mind that this is a physical assumption: all of the (M',g) are possible worlds, and any one of these possible worlds could have been the one that really exists.

Once we have the maximal (M,g) generated from a given S and its initial data, there is an infinity of other choices of 3-dimensional manifolds in M which we could picture as generating (M,g). For example, we could regard the spatial universe and the fields it contains now as "S with its initial data", or we could regard the universe a thousand years ago as "S with its initial data". Both would give the same (M,g) since the Einstein equations are deterministic. Everything that has happened and will happen is contained implicitly in the initial data on S. There is nothing new under the sun in a deterministic theory like general relativity. One could even wonder why time exists at all since from an information standpoint it is quite superfluous. (I'll suggest an answer to this question in the next Section.) None of the infinity of initial data manifolds in (M,g) can be uniquely regarded as generating the whole of spacetime (M,g). Each contains the same information, and each will generate the same (M,g), including all the other initial data manifolds.

Even in deterministic theories, relationships between physical entities are different at different times. For example, two particles moving under Newtonian gravity are now 2 meters apart (say), and a minute later 4 meters apart. This is true even though given the initial position and velocities when they were 2 meters apart, it is determined that they will be 4 meters apart a minute later. The question is, will the totality of relationships at one time become the same (or nearly the same) at some later time? If this happens, then we have the horror of the Eternal Return. As is well-known, it is possible to prove that the Eternal Return will occur in a Newtonian universe provided said universe is finite in space and finite in the range of velocities the particles are allowed to have. It is possible to prove that in classical general relativity (Tipler 1979; 1980), the Eternal Return *cannot* occur. That is, the physical relationships existing now between the fields will never be repeated, nor will the relationships ever return to approximately what they now are. What happens is that the Einstein field equations will not permit the gravitational equivalent of the "range of velocities" to be finite: the range simply must eventually become infinite. Thus history, understood as an unrepeatable temporal sequence of relationships between physical entities, is real.

Since I am interested here in discussing quantum cosmology, I am virtually forced into adopting the Many Worlds Interpretation, because only in this interpretation is it meaningful to talk about a quantum universe and its ontology. The Copenhagen Interpretation assumes that a process called "wave function reduction" eliminated quantum effects on cosmological scales an exceedingly short time after the Big Bang, so the universe today is not quantum except on very small scales. The problem with this assumption is that the wave function reduction process is almost entirely mysterious — we have no rules for deciding what material entity can reduce wave functions — so it is impossible to give a sharp analysis of contingency when this process is operating. The Many Worlds Interpretation does not suffer from this drawback: there is no reduction of the wave function, physical reality is completely described by the wave function of the universe, there is an equation (the Wheeler-DeWitt equation) for this wave function, and the universe is just as quantum now as it was in the beginning. Of course, the Many Worlds Interpretation may be wrong; most physicists think it is (most physicists think it's nonsense). But the overwhelming majority of people working on quantum cosmology subscribe to some version of the Many Worlds Interpretation, simply because the mathematics forces one to accept it. The mathematics may be a delusion, with no reference in physical reality. Or the situation may be similar to that of early 17th century physics: astronomers believed the Earth went around the Sun, because the mathematics of the Copernican system forced them to. But few other scholars or ordinary people believed the Earth moved. Their own senses told them it did not. I shall adopt the Many Worlds Interpretation in what follows. For a more detailed defense of this interpretation see Section 7.2 of Barrow and Tipler 1986.

In quantum cosmology the universe is represented by a wave function $\Psi(h,F,S)$, where as in classical general relativity, h and F are respectively the spatial metric and the non-gravitational fields given on a 3-dimensional manifold S. The initial data in quantum cosmology are not (h,F) given on S as was the case in classical general relativity, but rather $\Psi(h,F,S)$ and its first derivatives. From this initial data, the Wheeler-DeWitt equation determines $\Psi(h,F,S)$ for all values of h and F. In other words, the wave function, not the metric or the non-gravitational field, is the basic physical field in quantum cosmology. It is the initial wave function (and appropriate derivatives) that must be given, but once given, it is determined everywhere. What we think of as the most basic fields in classical general relativity, namely h and F, play the role of coordinates in quantum cosmology. But this does not mean h and F are unreal. They are as real as they are in classical theory. But it does mean more than one h and F exist on S at the same time! To appreciate this, recall that the classical metric h(x) is a function of the spatial coordinates on the manifold S. This metric has (non-zero) values at all points on S; that is, for the entire range of the coordinates as they vary over S, which is to say, as we go from one point to another in the universe. Each value of h(x) is equally real, and all of the values of h at all of the points of $_S$ exist simultaneously. Similarly, the points in the domain of the wave function $\Psi(h,F,S)$ are the various possible values of h and F, each set (h,F) corresponding to a complete universe at a given instant of time. The central claim of the Many Worlds Interpretation is that each of these universes actually exists, just as the different h(x) exist at the various points of S: Quantum reality is made up of an infinite number of universes (worlds). Of course, we are not aware of these worlds — we are only aware of one — but the laws of quantum mechanics explain this: we must generally be as unaware of these parallel worlds as we are of our motion with the Earth around the Sun. (In extreme conditions, for instance near singularities, it is possible for the worlds to effect each other in a more obvious way than they do now.)

To fix the classical initial data, we pick a *function* h(x) out of an infinite number of possible metric functions which could have been on S. All of these possible worlds comprise a function space. To fix the quantum initial data, we pick a *wave function* $\Psi(h,F,S)$ out of an infinite number of possible wave functions which could have been on the classical function space (h,F). Remember, however, that all values of the function space (h,F) *really are* on S simultaneously. In quantum cosmology, the collection of all possible wave functions forms the set of the possible worlds; what is contingent is which single unique universal wave function is actualized. But the possible worlds of classical cosmology — the space of all physically possible (h,F) on S — are no longer contingent. All of them are actualized.

In classical deterministic general relativity, we had a philosophical problem with time: since everything that did or will happened was coded in the initial fields on S, time evolution appeared superfluous. What was the point of having time? The problem is solved in quantum cosmology: *there is no time!* The universal wave function Ψ(h,F,S) is all there is, and there is no reference to a 4-dimensional manifold M or a 4-dimensional metric g in the wave function. At the most basic ontological level, time does not exist. Everything is on the 3-dimensional manifold S. How can this be? Of course we see time! Or do we? What we see is relationships between objects — configurations of physical fields —in space. In the discussion of the Eternal Return, I argued that time and history could be truly real only if the spatial relationships between the various fields never returned to a previous state. In quantum cosmology, there is no spacetime in which the spatial relationships between fields can change. Rather, all we have is paths (trajectories) in the collection (h,F) of all possible relationships between the physical fields on S. But this is enough, because each such path defines a history, a complete spacetime.

To understand this, imagine that we are at a point P in (h,F), and have selected a particular path g in (h,F) starting at P. Each point, remember, corresponds to an entire universe (spatially). As we go along g, the relationships between the physical fields vary smoothly from their values at P. *This variation would appear as temporal variation* from inside the path γ, because each point on γ is a complete spatial universe, and thus the sequence of points constitute a sequence of spatial universes. But this is exactly the same as the classical 4-dimensional manifold M with its spacetime metric γ and spacetime fields F, which in the above classical analysis we obtained as an extension of S and its fields! Each path in (h,F) thus is an entire classical universal history, an entire spacetime.

All paths in (h,F) really exist, which necessarily means that all — and I mean all — histories which are consistent with the "stuff" of the universe being (h,F) really exist. In particular, even histories which are grossly inconsistent with the laws of physics really occur! Closed paths in (h,F) obviously exist, so there are histories in which the Eternal Return is true. There are also real histories leading to our presently observed state of the universe (the point P in (h,F)) in which real historical characters — for instance Julius Caesar — never existed. What happens in such a history is that the physical fields rearrange themselves over time (more accurately, over the path corresponding to this strange history) to create false memories, including not only human memories but also the "memories" in a huge number of written records and in massive monuments. Just as there is an infinity of actual pasts which have led to the present state, so there is an infinity of really existing futures which evolve from the present state. So every consistent future is not only possible but it really happens. But not all futures are equally likely to be seen. That is, there is one path in (h,F) leading from a given point P which is overwhelmingly more likely to follow from P than all the others. This path is called *the classical path*. Along this path, the laws of physics hold, and memories are reliable. A classical path in (h,F) very closely resembles a classical spacetime (M,g) obeying the Einstein equations.

So far I have not said what the wave function Ψ itself does. But it must do something physically detectable, something not coded in the fields (h,F) alone. If it did not exert some physical effect, we could just omit it from physics; it would have no real existence. But I claimed above that Ψ was a *real* field, something as real as the fields (h,F).

What Ψ does is determine the set of all classical paths, and also the "probabilities" which are associated with each point and each path in (h,F). A wave function is a complex function, and all complex functions are actually two functions, a "magnitude" and a "phase". The classical paths are by definition those which are perpendicular to the surfaces of constant phase. The square of the magnitude at a point P in (h,F) is the "probability" of that point. The physicists Heisenberg and Mott showed mathematically that if "probability" has its usual meaning, then given the fact that we are (approximately) at P, the conditional probability of going to a nearby point Q is maximum if Q lies along the classical path through P. The relative probability is very close to 1 on the classical path, and it drops rapidly to 0 as one moves away from the classical path connecting P and Q. (See Section 7.2 of Barrow and Tipler 1986 for details about how this works.)

What must be shown is that the square of the magnitude is in fact a "probability" in the usual sense. This is done as follows. We obviously can't get hold of the wave function of the entire universe, but we can prepare in the laboratory a number N of electrons with the same spin wave function. Suppose we measure the vertical component of the electron spin. It turns out that this component can have only two values, spin up and spin down. If the wave function is not in what is called "an eigenstate" of spin up or spin down — in general the electron wave function would not be in an eigenstate, so let's suppose it's not — then each time we measure the vertical component of an electron in our ensemble of N electrons, we will get a different answer. Some of the electrons will be found to have spin up, and the others will have spin down. We can't predict before the measurement what the vertical component of that particular electron will be. But it can be shown that if we compute the relative frequency with which we get spin up, then this number approaches the square of the magnitude of the wave function evaluated at "spin up" as the number N of electrons in the ensemble approaches infinity. And experimentally, this is what we see.

All the physics is contained in the wave function. In fact, the laws of physics themselves are completely superfluous. They are coded in the wave function. The classical laws of physics are just those regularities which are seen to hold along a classical path by observers in that classical path. Along other paths, there would be other regularities, different laws of physics. And these other paths exist and hence these other laws of physics really hold; it is just extremely unlikely we will happen to see them operating. The Wheeler-DeWitt equation for the wave function is itself quite superfluous. It is merely a crutch to help us to find the actual wave function of the universe. If we knew the boundary conditions which the actual universal wave function satisfied, then we could derive the Wheeler-DeWitt equation, which is just a particular equation (among many) which the wave function happens to satisfy. Thus in quantum cosmology, there is no real contingency in the laws of physics. Any law of physics holds in some path, and the law of physics governing the universal wave function can be derived from that wave function. All the contingency in quantum cosmology is in the wave function, or rather, in the boundary conditions which pick out the wave function which actually exists. The well-known Hartle-Hawking boundary condition, which says that "the universal wave function is that wave function for which the Feynman sum over all the paths (classical and otherwise) leading to a given point P is over paths that have no boundaries (more precisely, the 4-dimensional manifold corresponding to a given path is a compact manifold whose only boundary is P)" is one such boundary condition. I should like to propose the

Teilhard Boundary Condition for the universal wave function:

The wave function of the universe is that wave function for which all classical paths terminate in a (future) Omega Point, with life coming into existence along at least one classical path and continuing into the future forever all the way into the Omega Point.

The Teilhard Boundary Condition is enormously restrictive. For example, since classical paths are undefined at zeros of the wave function, we immediately have

Fourth Testable Prediction of the Omega Point Theory: the universal wave function must have no zeros in the spacetime domain.

It turns out (as one might expect) that the Hartle-Hawking boundary condition does not satisfy the Teilhard boundary condition. I have a rough argument that one can construct simple quantized Friedmann cosmological models in which all classical paths terminate in an Omega Point, but I don't know yet what the existence of life requires of a wave function. So at present I can only conjecture, not prove, that a wave function satisfying the Teilhard Boundary Condition in its full generality exists mathematically. (This is not unusual; there is also no general existence proof yet for the Hartle-Hawking boundary condition.)

3. The Universe Necessarily Exists: A Strong Anthropic Ontological Argument

Ever since Kant showed that "existence is not a predicate", most philosophers have felt ontological arguments to be invalid; that is, they have believed it impossible to prove the existence of anything by means of logic alone . I want to claim this is incorrect; I think you can prove that the universe necessarily exists. The proof will be based on an analysis of what the word "existence" means.

Let us begin with some computer metaphysics. Much of computer science is devoted to making *simulations* of phenomena in the physical world. In a simulation, a mathematical model of the physical object under study is coded in a program. The model includes as many attributes of the real physical object as possible (limited of course by the knowledge of these attributes, and also by the capacity of the computer). The running of the program evolves the model in time. If the initial model is accurate, if enough key features of the real object are captured by the model, the time evolution of the model will mimic with fair accuracy the time development of the real object, and so one can predict the most important key aspects which the real object will have in the future.

Suppose we try to simulate a city full of people. Such simulations are being attempted now, but at a ludicrously inaccurate level. But suppose we imagine more and more of the attributes of the city being included in the simulation. In particular, more and more properties of each individual person are included. In principle, we can imagine a simulation being so good that every single *atom* in each person and each object in the city and the properties of each atom having an analogue in the simulation. Let us imagine, in the limit, a simulation that is absolutely perfect: each and every property of the real city, and each and every real property of each real person in the real city is represented precisely in the simulation. Furthermore, let us imagine that when the program is run on some gigantic computer, the temporal evolution of the simulated persons and their city precisely mimics for all time the real temporal evolution of the real people and the real city.

The key question is this: do the simulated people exist? As far as the simulated people can tell, they do. By assumption, any action which the real people can and do carry out to determine if they exist — reflecting on the fact that they think, interacting with the environment - the simulated people also can do, and in fact do do. There is simply no way for the simulated people to tell that they are "really" inside the computer, that they are merely simulated, and not real. They can't get at the real substance, the physical computer, from where they are, inside the program. One can imagine the ultimate simulation, a perfect simulation of the entire physical universe, containing in particular all people which the real universe contains, and which mimics perfectly the actual time evolution of the actual universe. Again, there is no way for the people inside this simulated universe to tell that they are merely simulated, that they are only a sequence of numbers being tossed around inside a computer, and are in fact not real.

How do we know we ourselves are not merely a simulation inside a gigantic computer? Obviously, we can't know. But I think it is clear we ourselves really exist. Therefore, *if* it is in fact possible for the physical universe to be in precise one to one correspondence with a simulation, I think we should invoke the Identity of Indiscernibles and identity the universe and all of its perfect simulations. (For more discussion of whether a simulation must be regarded as real if it copies the real universe sufficiently closely, see (Hofstadter and Dennett 1981), particularly pages 73-78, 94-99, and 287-320.)

But is it possible for the universe to be in precise one-to-one correspondence with some simulation? I think that it is, if we generalize what we mean by simulation. In computer science, a simulation is a program, which is fundamentally a map from the set of integers into itself. That is, the instructions in the program tell the computer how to go from the present state, represented by a sequence of integers, to the subsequent state, also represented by a sequence of integers. But remember, we don't really need the physical computer; the initial sequence of integers and the general rule (instructions or map) for replacing the present sequence by the next is all that is required. But the general rule can itself be represented as a sequence of integers. So, if time were to exist globally, and if the most basic things in the physical universe and the time steps between one instant and the next were discrete, then the whole of spacetime would definitely be in one to one correspondence with some program. But time may not exist globally (it doesn't if standard quantum cosmology is true), and it may be that the substances of the universe are continuous fields and not discrete objects (in *all* current physical theories, the basic substances are continuous fields). So, if the actual universe is described by something resembling current theories, it cannot be in one to one correspondence with a standard computer program, which is based on integer mappings. There is currently no model of a "continuous" computer. Turing even argued that such a thing is meaningless! (There are definitions of "computable continuous functions," but none of the definitions are really satisfactory.)

Let's be more broad minded about what is to count as a simulation. Consider the collection of all mathematical concepts . Let us say that a perfect simulation exists if the physical universe can be put into one to one correspondence with some mutually consistent subcollections of all mathematical concepts. In this sense of "simulation" the universe can certainly be simulated, because "simulation" then amounts to saying that the universe can be exhaustively "described" in a logically consistent way. Note that "described" does not require that we or any other finite (or infinite) intelligent being can actually find the description. It may be that the actual universe expands into an infinite hierarchy of levels whenever one tries to describe it exhaustively. In such a case, it would be impossible to find a Theory of Everything. Nevertheless, it would still be true that a "simulation" in the more general sense existed if each level were in one to one correspondence with some mathematical object, and if all levels were mutually consistent ("consistency" meaning that in the case of disagreement between levels, there is a rule — itself a mathematical object — for deciding which level is correct). The crucial point of this generalization is to establish that the actual physical universe is something in the collection of all mathematical objects. This follows because the universe has a perfect simulation, and we agree to identify the universe with its perfect simulation. Thus at the most basic ontological level, the physical universe is a concept.

But of course not all concepts exist physically . But *some do. Which ones? The answer is provided by our earlier analysis of programs. The simulations which are sufficiently complex to contain observers — thinking, feeling beings — as subsimulations exist physically.* And further, they exist physically by definition: for this is exactly what we mean by existence, namely, that thinking and feeling beings think and feel themselves to exist. Remember, the simulated thinking and feeling of simulated beings are real. Thus the actual physical universe — the one in which we are now experiencing our own simulated thoughts and simulated feelings, exists necessarily, by definition of what is meant by existence. Physical existence is just a particular relationship between concepts. Existence is a predicate, but a predicate of certain very, very complex simulations. It is certainly not a predicate of simple concepts, for instance "100 thalers."

With equal necessity, many different universes will exist physically. In particular, a universe in which we do something slightly different from what we actually do in this one will exist (provided of course that this action does not logically contradict the structure of the rest of the universe). But this is nothing new; it is already present in the ontology of the Many-Worlds Interpretation. Exactly how many universes really exist physically depends on your definition of "thinking and feeling being". If you adopt a narrow definition — such a being must have at least our human complexity — then the range of possible universes appears quite narrow: *The Anthropic Cosmological Principle* (Barrow and Tipler 1986) is devoted to a discussion of how finely tuned our universe must be if it is to contain beings like ourselves. (Although the above discussion of existence comes entirely from physics, namely the physics of computer simulation and the MWI, the conclusions are essentially the same as those of the philosophers Plantinga (1974) and Lewis (1986, p. 73) on the necessary existence of all possible worlds. Plantinga and Lewis are motivated by an analysis of the meaning of modal logic.)

What happens if a universal simulation stops tomorrow? Does the universe collapse into non-existence? Certainly such terminating simulations exist mathematically. But if there is no intrinsic reason visible from inside the simulation for the simulation to stop, it can be embedded inside a larger simulation which does not stop. Since it is the observations of the beings inside the simulation that determines what exists physically, and since nothing happens from their view point at the termination point when the terminating simulation is embedded in the non-stopping simulation, the universe must be said to continue in existence. It is the maximal extension which has existence, for by the Identity of Indiscernibles we must (physically) identify terminating programs with their embedding in the maximal program. (One could use a similar argument for asserting the physical existence of the maximal evolution from given initial data in the classical general relativity evolution problem.) Furthermore, if it is logically possible for life to continue to exist forever in some universe, this universe will exist necessarily for all future time.

4. Philosophical Implications of the Existence (?) of the Omega Point

Suppose the Omega Point really exists. Then even on the most materialistic level, the future existence of the Omega Point would assure our civilization of ever growing total wealth, continually increasing knowledge, and quite literal eternal progress. This perpetual meliorism is built into the definition of "life existing forever" given above. Of course, it is a consequence of physics that although our civilization may continue forever, our species *Homo sapiens* must inevitably become extinct, just as every individual human being must inevitably also die. For as the Omega Point is approached, the temperature will approach infinity everywhere in the universe, and it is impossible for our type of life to survive in this environment. (The non-existence of the Omega Point would not help us. If the universe were open and expanded forever, then the temperature would go to zero as the universe expanded. There is not enough energy in the frigid future of such a universe for *Homo sapiens* to survive. Also, protons probably decay, and we are made up of atoms, which require protons.)

But the death of *Homo sapiens* is an evil (beyond the death of the human individuals) only for a racist value system. What is humanly important is the fact we think and feel, not the particular bodily form which clothes the human personality. If the Omega Point exists, the advance of civilization will continue without limit into the Omega Point. Our species is an intermediate step in the infinitely long temporal Chain of Being (Lovejoy 1936) that comprises the whole of life in spacetime. An essential step, but still only a step. In fact, it is a logically necessary consequence of eternal progress that our species become extinct! For we are finite beings, we have definite limits. Our brains can code only so much information, we can understand only rather simple arguments. If the ascent of Life into the Omega Point is to occur, one day the most advanced minds must be non-*Homo sapiens* . The heirs of our civilization must be another species, and their heirs yet another, *ad infinitum* into the Omega Point. We must die — as individuals, as a species — in order that our civilization might live. But the contributions to civilization which we make as individuals will survive our individual deaths. Judging from the rapid advance of computers at present, I would guess that the next stage of intelligent life would be quite literally information processing machines. At the present rate, computers will reach the human level in information processing and integration ability probably within a century, certainly within a thousand years.

Many find the assurance of the immortality of life as a whole cold comfort for their death as individuals. But recall my discussion of Thomist *aeternitas*. I pointed out that all the information contained in the whole of human history, including every detail of every human life, will be available for analysis by the collectivity of life in the far future. In principle at least (again ignoring the difficulty of extracting the relevant information from the overall background noise), it is possible for life in the far future to construct, using this information, an exceedingly accurate simulation of these past lives: in fact, this simulation is just what a sufficiently close scrutiny of our present lives by the beings of the future would amount to. And I emphasized above that a sufficiently perfect simulation of a living being would *be* alive! Whether the beings of the future would choose to use their power to do this simulation, I cannot say. But it seems the physical capability to carry out the scrutiny would be there. Furthermore, the drive for total knowledge — which life in the future must seek if it is to survive at all, and which will be achieved only at the Omega Point — would seem to require that such an analysis of the past, and hence such a simulation, would be carried out.

I should emphasize that this simulation of people that have lived in the past need not be limited to just repeating the past. Once a simulation of a person and his/her world has been formed in a computer of sufficient capacity, the simulation can be allowed to develop further — to think and feel things that the long dead original person being simulated never felt and thought. It is not even necessary for *any* of the past to be repeated. The beings of the future could simply begin the simulation with the brain memory of the dead person as it was at the instant of death (or 10 years before, or 20 minutes before, or ...) implanted in the simulated body of the dead person, the body being as it was at age 20 (or age 70, or ...). This body and memory collection could be set in any simulated background environment the future beings wished: a simulated world indistinguishable from the long-extinct society and physical universe of the revived dead person, or even a world that never existed, but one as close as logically possible to the ideal *fantasy* world of the resurrected dead person. Furthermore, all possible combinations of resurrected dead can be placed in the same simulation and allowed to interact. For example, the reader could be placed in a simulation with *all* of his/her ancestors and descendents, each at whatever age (physical and mental, separately) the future beings please. The intelligent beings of the future could interact — speak, say — with their simulated creatures, who could learn about them, about the world outside the simulation, and about other simulations.

The simulated body could be one that has been vastly improved over the one we currently have: the laws of the simulated world could be modified to prevent a second physical death. We could call the simulated, improved and undying body a "spiritual body", for it will be of the same "stuff" as the human mind now is: a "thought inside a mind" (in Aristotelian language, "a form inside a form."; in computer language, a virtual machine inside a machine). The spiritual body is thus just the present body (with improvements!) at a higher level of implementation.[3] Only as a spiritual body, only as a computer simulation, is resurrection possible without a second death: our current bodies, implemented in matter, could not possibly survive the extreme heat near the final singularity.

Although computer simulation resurrection overcomes the physical barriers to eternal life of individual human beings, there remains a logical problem, namely, the finiteness of the human memory. The human brain can store only about 10^{15} bits (Barrow and Tipler 1986, p. 136) [this corresponds to roughly a thousand subjective years of life], and once this memory space is exhausted, we can grow no more. Thus it is not clear that the undying resurrected life is appropriately regarded as "eternal". There are several options. For example, the beings of the future could guide us to a "perfection" of our finite natures. Whatever "perfection" means! Depending on the definition, there could be many "perfections". With sufficient computer power, it should be possible to calculate what a human action would result in without the simulation actually experiencing the action, so the future beings would be able to advise us on possible perfections without us having to go through the trial and error procedure characteristic of this life. If more than one simulation of the same individual is made, then *all* of these options could be realized simultaneously. Once an individual is "perfected", the memory of this perfect individual could be recorded permanently — preserved all the way into the Omega Point. The errors and evil committed by the imperfect individual could be erased (or also permanently recorded). The perfected individual personality would be truly eternal: she would exist for all future time. Furthermore, when the perfected personality reached the Omega Point, it would become eternal in the sense of being beyond time.

In his *On the Immortality of the Soul*, David Hume raised the following objection to the idea of a general resurrection of the dead: "How to dispose of the infinite number of posthumous existences ought also to embarrass the religious theory." (Hume 1755; reprinted in Flew 1964, p. 187). Hume summarized the argument in a later interview with the famous biographer James Boswell: ".. [Hume] added that it was a most unreasonable fancy that he should exist forever. That immortality, if it were at all, must be general; that a great proportion of the human race has hardly any intellectual qualities; that a great proportion dies in infancy before being possessed of reason; yet all these must be immortal; that a Porter who gets drunk by ten o'clock with gin must be immortal; that the trash of every age must be preserved, and that new Universes must be created to contain such infinite numbers." (Hume 1776 [1977], p. 77).

The ever-growing numbers of people whom Hume regarded as trash nevertheless could be preserved forever in our single finite (classical) universe if computer capacity is created fast enough. By looking more carefully at the calculations summarized above, one sees that they also show it is physically possible to save *forever* a certain constant percentage of the information processed at a given universal time. Thus, the computer capacity will be there to preserve even drunken porters, (and perfected drunken porters) provided only that the beings of the future wait long enough before resurrecting them. Even though the computer capacity required to perfectly simulate is exponentially related to the complexity of entity simulated, it is physically possible to resurrect an actual infinity of individuals between now and the Omega Point, even assuming the complexity of the average individual diverges as the Omega Point is approached, and guide then *all* into perfection.[4] Total perfection of all would be achieved at the instant of the Omega Point.

But this preservation capacity has an even more important implication: *it means that the resurrection is possible even if sufficient information to resurrect cannot be extracted from the past light cone.* Since the universal computer capacity increases without bound as the Omega Point is approached, it follows that if only a bare bones description of our current world is stored permanently, then there will inevitably come a time when there will be sufficient computer capacity to simulate our present-day world by simple brute force: by creating a simulation of *all* logically possible variants of our world. For example, the human genome can code about 10 to the 10^6 power possible humans, and the brain of each could have 2 to the 10^{15} power possible memories. With the computer power that will eventually become available, the beings of the future could simply simulate them all. Just the knowledge of the human genome would be enough for this. And even if the record of human genome is not retained until the computer capacity is sufficient, it would still be possible to resurrect all possible humans, just from the knowledge it was coded in DNA. Merely simulate all possible life forms that could be coded by DNA (for technical reasons, the number is finite) and all logically possible humans will necessarily be included.. Such a brute force method is not very elegant;[5] I discuss it only to demonstrate that resurrection is unquestionably physically possible. And if there is no other way, it almost certainly will be done by brute force in the drive toward total knowledge. In our own drive to understand how life got started on our planet, we are in effect trying to simulate — resurrect — all possible kinds of the simplest life forms which could spontaneously form on the the primitive earth.

Eternal worldly progress and the hope of individual survival beyond the grave, usually pictured as polar opposites, turn out to be the same. An interesting feature of the Omega Point Theory is that it provides a plausible physical mechanism for a universal resurrection[6] (for more discussion of the physics of the resurrection see (Tipler 1989)). As Wolfson (1965) has pointed out, resurrection has been inconsistent with the accepted physics of the day for the past two thousand years. That has now changed.

Notes

[1]I am grateful to Frank Birtel, Michael Heller, Wolfhart Pannenberg, and John Wheeler for their comments on an earlier version of this paper. This work was supported in part by the National Science Foundation under grant number PHY-86-03130.

[2]Unfortunately, Aristotle ruined his own idea of the soul by soiling it with Platonic dualism. This mistake led to Aquinas' contradictory notion of "substantial form". Both ideas suggest that the personality survives death naturally. See Flew (1964, pp. 16-21) and 1987, pp. 71-87). Modern physics tells us however that personality does not survive death. When you're dead, you're dead until the resurrection, if it indeed occurs.

[3]See (Hofstadter and Dennett 1981, pp. 379-381) for a very brief discussion of the extremely important computer concept of "levels of implementation."

[4]This depends in a crucial way on the fact that there will be an actual infinity ($\aleph_0$) of information processed between any finite time and the Omega Point. It is an example of what Bertrand Russell (1931, p. 358) has termed the Tristram Shandy paradox. Tristram Shandy took two years to write the history of the first two days of his life, and complained that at that rate, material would accumulate faster than he could write it down. Russell showed that, if Tristram Shandy lived forever, nevertheless no part of his biography would have remained unwritten. In the case of the Omega Point, which literally does live forever, all beings that have ever lived and will live from now to the end of time can be resurrected and remembered, even though the time needed to do the resurrecting will increase exponentially, a much worse case than Tristram Shandy faced. It is important that at any given time on a classical trajectory, there is only a finite number of possible beings which could exist. If this were not true, then the number of beings that would have to be resurrected between now and the Final State might be the power set of $\aleph_0$, which is higher order of infinity than $\aleph_0$, and thus resurrecting all possible beings via the brute force method might be impossible because only $\aleph_0$ bits can be recorded between now and the Final State.

[5]One could also worry about the morality of such brute force resurrection: not only are the dead being resurrected, but also people who never lived! However, the central claim of the Many-Worlds physics and the Many-Worlds metaphysics discussed above is that all people and all histories who could exist in fact do. They just don't exist on our classical trajectory, and so we have no record of them. So the resurrected dead would presumedly not care which classical trajectory they are resurrected in — their own trajectory or another one — so long as they *are* resurrected.

[6]The version of eternal life discussed here is not attractive to everyone. What is happening is that an exact replica of ourselves is being simulated in the computer minds of the far

future. The philosopher Antony Flew, for example, considers it ridiculous to call this "resurrection", and he puts forward the 'Replica Objection': "No replica however perfect, whether produced by God or man, whether in our Universe or another, could ever be — in that primary, forensic sense — the same person as its original." (Flew 1987, p. 12). "To punish or to reward a replica, reconstituted on Judgement Day, for the sins or the virtues of the old Antony Flew dead and cremated, perhaps long years before, is as inept and as unfair as it would be to reward or to punish one identical twin for what was in fact done by the other." (Flew 1987, p. 9). Flew is wrong about our legal system. It does in fact equate identical computer programs. If I duplicated a word processing program and used it without paying a royalty to the programmer, I would be taken to court. A claim that "the program I used is not the original, it is merely a replica" would not be accepted as a defense. I could also be sued for using without permission an organism whose genome has been patented. Identical twins are *not* identical persons. The programs which are their minds differ enormously: the memories coded in their neurons differ from each other in at least as many ways as they differ from the memories of other human beings. They are correctly regarded as different persons. But two beings who are identical both in their genes *and* in their mind programs *are* the same person, and it is appropriate to regard them as equally responsible legally. I am surprised that an empiricist philosopher like Flew would make the claim that entities which cannot be empirically distinguished, even in principle, are nevertheless to be regarded to be utterly different. Any scientist would think that two physically indistinguishable systems are to be regarded as the same, both physically and legally.

References

Abramowicz, M. and Ellis, G.F.R. (1989), "The Elusive Anthropic Principle", *Nature* 337: 411- 412.

Alder, H.L. and Roessler, E.B. (1964), *Introduction to Probability and Statistics*. San Francisco: Freeman.

Barrow, J. D. (1989), "Anthropic Principle", *Nature* 339: 196.

_ _ _ _ _ _ _. and Tipler, F.J. (1986), The Anthropic Cosmological Principle. Oxford: Oxford University Press.

Carter, B. (1989) in *The Anthropic Principle* , U. Curi (ed.). Cambridge: Cambridge University Press.

Feller, W. (1968), *An Introduction to Probability Theory and Its Applications*, Volumes I & II. New York: John Wiley & Sons.

Flew, A. (ed.) (1964), *Body, Mind, and Death*. New York: Macmillan.

_ _ _ _ _. (1984), *God, Freedom and Immortality: A Critical Analysis*. Buffalo: Prometheus.

_ _ _ _ _. (1987), *The Logic of Mortality*. Oxford: Blackwell.

Hawking, S. W. and Ellis, G.F.R. (1973), *The Large Scale Structure of Space-Time*. Cambridge: Cambridge University Press.

Hume, D. (1977), *Dialogues Concerning Natural Religion*. Norman Kemp Smith (ed.). Indianapolis: Bobbs-Merrill.

Hofstadter, D. R. and Dennett , D.C. (1981), *The Mind's I*. New York: Basic Books.

Lewis, D. (1986), *On the Plurality of Worlds*. Oxford: Blackwell.

Linde, A.D. (1989), "Particle Physics and Cosmology", In *Proceedings of the XXIV International Conference on High Energy Physics*, R. Kotthaus and J. Kühn (ed.). Heidelberg: Springer-Verlag.

Lovejoy, A.O. (1936), *The Great Chain of Being*. Cambridge: Harvard University Press.

Plantinga, A. (1974), *The Nature of Necessity*. Oxford: Clarendon Press.

Russell, B. (1931), *Principles of Mathematics*. New York: Norton.

Tipler, F.J. (1976), "Singularities in Universes with Negative Cosmological Constant", *The Astrophysical Journal* 209: 12-15.

_ _ _ _ _ . (1979), "General Relativity, Thermodynamics, and the Poincaré Cycle." *Nature* 280: 203-205.

_ _ _ _ _ . (1980), "General Relativity and the Eternal Return." In *Essays in General Relativity*, F.J. Tipler (ed.). New York: Academic Press, pp. 21-37.

_ _ _ _ _ . (1986), "Cosmological Limits on Computation." *International Journal of Theoretical Physics* 25: 617-661.

_ _ _ _ _ . (1988), "The Omega Point Theory." In *Physics, Philosophy, and Theology: A Common Quest for Understanding*, by R.J. Russell, W. Stoeger, and G. Coyne, pp. 313- 331. Notre Dame: University of Notre Dame Press.

_ _ _ _ _ . (1989), "The Omega Point as Eschaton: Answers to Pannenberg's Questions for Scientists", to appear in the July issue of Zygon .

Weinberg, S. (1987), "Anthropic Bounds on the Cosmological Constant", *Physical Review Letters* 59: 2607-2610.

_ _ _ _ _ _ _. (1989), "The Cosmological Constant Problem", *Reviews of Modern Physics* 61: 1-23.

Wolfson, H. (1965), "Immortality and Resurrection in the Philosophy of the Church Fathers", in *Immortality and Resurrection*, K. Stendahl (ed.). New York: Macmillan, pp 54-96 .

Ultimate Explanations: Comments on Tipler

Lawrence Sklar

University of Michigan

1. Looking for Ultimate Explanations

Prof. Tipler and his colleague John Barrow, applying the results and methods of contemporary physics, have made a number of suggestions for pushing beyond the limits of science as ordinarily understood, sometimes suggesting that ideas once thought archaic and worthy to be discarded can be revivified in the light of our new theoretical understanding of the world. The role of the observer in quantum mechanics, they suggest, makes at least plausible "idealist" claims of the dependence of matter on mind. Cosmology, they suggest, provides a framework in which the idea of God as evolutionary endpoint of man-in-nature, familiar from Teilhard de Chardin, can be made concrete and plausible at least as a possibility. All of this, admittedly speculative, structure is presented in something like the form of an inverted pyramid, resting at its base on the much less ambitious, if much more secure, methodological posits of the sort grouped together under the (somewhat misleading) rubric of the "Weak Anthropic Principle."

This "principle" merely says such things as that, given the fact that the world does contain us as observers, it can be inferred to be a world of the sort where creatures like us are possible. And it cautions us with such methodological warnings, familiar explicitly since Boltzmann and implicit in much earlier science, as the injunction to be careful in positing what we observe locally in our universe to be true on a cosmic scale, since the local phenomena might be rare in our universe, although expectable in any region where observers can be found, since they are the phenomena necessary for the existence of observers in that local region of the universe. And it warns us to avoid other biases due to our nature as observers conditioning the kinds of evidence we are likely to come upon, such as the condition of the universe during the periods of time in which we exist being taken to be always the case.

Recently Prof. Tipler has added to the earlier arguments of himself and Prof. Barrow other attempts to restore to life and respectability modes of dealing with the world most have thought archaic and indefensible. These are proposals to provide "ultimate" explanations for the existence and nature of the world. "Ultimate" in the sense that these accounts of the world would show its existence and nature to be matters of necessity. The explanations would be unconditioned. Whereas we are now accustomed to scientific explanations of the states of the world as requiring the positing of other, usually earlier, states of the world in the explanation basis, and where we are accustomed to thinking of

PSA 1988, Volume 2, pp. 49-55

scientific explanations of general features of the world as requiring the positing in the explanatory basis of some "deeper" generality to do the explaining, the ultimate explanations will posit no contingent particular facts or generalizations at all. Rather that the world is, and that it is as it is, will follow as matters of unconditioned necessity.

The place of man in the universe, an important feature of the "anthropic" speculations and arguments, plays, as we shall see, only a limited place in these new speculations. Some of the new arguments do, however, share with the older ones of Tipler and Barrow an important feature in that in some of the arguments it is alleged that one can resolve older "philosophical" questions only by a strategy which invokes at crucial points facts about the world revealed to us in contemporary theoretical physics. That such "ultimate" explanations would bring together "philosophical" reasoning with modern mathematical physics is, of course, not a new idea. Eddington's "Fundamental Theory" and von Weizäcker's proposed "transcendental deduction" of basic physical features of the world are noteworthy predecessors of this sort of argument. Other more recent arguments which brood about the conceivability of the universe as a quantum fluctuation out of nothing may also be members of this family of speculations which attempts to use physics to generate unconditioned explanations. But at least one of Tipler's arguments has curious features of its own worth exploring.

Other of Tipler's more recent arguments eschew the reference even to contemporary results of physics and seek the unconditioned, rather, in more purely "metaphysical" considerations.

2. Is the Nature of the Universe a Matter of Necessity?

Let us look first at an argument of Tipler's to the effect that, given the existence of the universe, its nature is not a matter of contingency but of necessity (Tipler,l987,sec.IV,pp.13-16). I am not sure that Tipler still holds to the soundness of this argument, but it was recently a part of his overall program, and its structure captures much of the flavor of the arguments from a combination of physics and philosophy which seek to avoid the contingency of explanation. The argument, as far as I understand it is this:

1) Philosophers frequently assert that we can imagine many alternative theories which would account for our observational data.

2) But an explanatory theory must be self-consistent.

3) Given the data we now have, it must be the case that the correct theory of the world must be something at least roughly like our current fundamental theory.

4) But physics has shown us that as the generality of data gets larger and larger, it becomes harder and harder to encompass it with a theory which is self-consistent.

5) So physics seems to be converging on a unique self-consistent TOE - Theory of Everything.

6) Well, it might be countered, couldn't a much simpler self-consistent world be imagined?

7) But existence of the world implies that the world is observed by something in it. And existence of an observer demands great complexity in the world.

8) Therefore, if the universe exists, it must be this one, since this one is the unique, self-consistent world complex enough for it to be observed by a portion of itself.

Needless to say this precis is only my understanding of Prof. Tipler's argument, and may be misrepresenting his ingenious if briefly presented sketch of its structure. While Prof. Tipler himself describes his argument as having "many gaps," I find my own doubts about it not resting on its incompleteness but, rather, on some serious puzzles about whether anything like it in structure can be expected to show what it sets out to show, viz. the necessary uniqueness of the world as a possible world.

(1) The argument that recent physics seems to indicate the possibility that only one self-consistent theory might be found which could do even rough justice to the torrent of data which we have accumulated, while extremely interesting from an epistemological perspective, seems irrelevant when the issue is the alleged necessary uniqueness of the world as a possible world. While the introduction of self-consistency would seem to be an interesting way of excluding from considerations the alleged "alternative but distinct" hypotheses which would also explain the data explained by the standard theory, so beloved by epistemological skeptics and conventionalists, or, rather of excluding at least some of those alleged alternatives, the allegation that other universes are "possible" and that it is only a contingent fact that this one exists presupposes that had one of those alternative possible worlds been the case, the data we now have wouldn't be had by the scientists, if there were any, in that alternative world. So the unique self-consistency of the only one theory even roughly empirically adequate relative to the data of this world is irrelevant to whether other possible worlds with other radically different totalities of empirical data could have existed.

(2) The main argument, that an existent world must of necessity be one which contains observers of itself within it; and that the existence of such observers demands of the world an extraordinary complexity; and that such complexity could only be encompassed by the uniquely self-consistent theory which, perhaps, does justice to the complexity of facts about this world, is much more relevant to showing what was intended to be shown. But I doubt that it is very persuasive.

(a) The claim that any imaginable possible world must contain observers within it is dubious. To be sure, were we not aware of the world we would have no concept of it and no concept of its existence. To be sure also, as empiricists, Kantians, pragmatists and others have insisted, our comprehension of being and its nature extends only as far as our awareness of it. But it is far from clear why any of the familiar arguments of that sort ought to lead us to conclude that the notion of a world devoid of observers of it internal to it is in any way, shape or form absurd or the notion of an impossible world in any sense of impossibility. Indeed, if one accepts some of Barrow and Tipler's arguments to the effect that there may be, in this universe, no observers but us, and that our own existence is quite improbable as well, ought not we to think , *a priori*, that if anything observerless universes are much more likely to be the case than those with observers?

(b) Be that as it may, let us restrict our attention to observer equipped worlds. Let us admit that such worlds would have to have a fair degree of complexity: in order to allow within them subsystems of sufficient complexity to have "internal maps" to somehow or other "represent" the world external to them; in order for these subsystems to interact in an appropriately complex causal way with the world outside them for them to respond to that world so as to become more-or-less reliable representers of it; in order for the world in question to have a sufficiently complex history for such representing-interacting observers to have spontaneously evolved in the world in question. But why on earth should we think that the world would have the kind of complex structure which could only be described by the unique self-consistent Theory of Everything that, if it exists, masters the complexity of this world?

I just don't see any plausibility to the claim that sufficient complexity to embody observers-as-representers in a world, even spontaneously evolved observers, could only be achieved by the kind of order of complexity we find in our world. I think that we find here a problem quite familiar from other "transcendental deductions" of the nature of the world. In the argument it is assumed that the world under discussion is sufficiently like this world so that features of it alleged to exist of necessity and knowable to be in it *a priori* (in this case, as is usual, the existence in the world of "knowers" of it) could only be achieved in a manner more-or-less like the way that state of being is achieved in this world.

(c) Even were we able to convince ourselves that only one possible lawlike structure could hold of a universe sufficiently complex to contain observers of itself, this would be insufficient to demonstrate the necessity of the world as we find it. Are we to abandon altogether the idea that within a given lawlike structure a vast variety of worlds can be found, differentiated by the particular facts, the "initial conditions" characterizing them? Is Tipler only claiming that one class of possible worlds, those described by the lawlike structure of this world, contains all possible worlds, or is the argument meant to convince us that only a single possible world, this one, is possible - inclusive of its specific particular circumstances? Would the claim then be that only one possible initial conditions could lead to the existence of observers? Why should we believe anything like that?

3. Is the Existence of the Universe a Matter of Necessity?

While Prof. Tipler may no longer hold to the soundness of his earlier alleged proof that the nature of the world is a matter of necessity, he does still claim confidence in another argument designed to convince us that the existence of the universe is a matter of necessity and not contingency. Indeed, combined with his arguments concerning the evolution of the universe to the "Omega Point," his and Barrow's cosmological instantiation of Chardin's evolutionary God-in-nature, he offers what he takes to be a modernization and reconstruction of the Ontological Argument's proof of the existence of God as a necessary being - a reconstruction which would, I am sure, horrify Anselm!

Let us look at some of the basic steps in the argument that the universe exists of necessity.

(1) If there were a simulation in a computer of a natural world, complete with internal observers to it, the simulated observers would take their simulated world as real, and would be as right in doing so as we are in taking our world as real. Indeed, for all we know we are components of such a simulation.

(2) The real being of a simulation is not a physical process on a computer, but the abstract mathematical structure which is the "program" considered as a formal, mathematical structure.

(3) For any description of the world, no matter how complete, the description will have a simulation in the form of a model of the assertions in the real of abstract mathematical objects.

(4) Since everything which is true of the world is true of the realm of abstracta, and vice versa, the two realms of being are, by the Principle of the Identity of Indiscernibles, one and the same world.

(5) Therefore the real world is identical to a world of mathematical abstracta.

(6) But the existence of mathematical structures is a matter of necessity and not contingency.

(7) Therefore the existence of the real world is a matter of necessity.

Let us ignore the dogmatic claim, borrowed from D. Dennett and D. Hofstader, that "persons" in a computer simulation would think themselves inhabitants of a world, indeed that they would be persons that thought at all. Instead let us focus on the argument from (2) through (7).

(1) The claim that mathematical abstracta exist as a matter of necessity is, of course, the dogmatic assertion of a Platonism which to many is far from fully intelligible, much less obviously true. Indeed, the claim of the existence of a body of necessary truth is itself problematic as we all know. But let us give Tipler (6) for the present.

(2) The crucial step of the argument, (4), is simply not plausible at all. There is not the slightest inference from the fact that one and the same logical form can be modelled in two different domains to the claim that the domains share all their properties in common, much less that they are identical. The fact that $\forall x(Ax \rightarrow Bx)$ can be instanced by all men who are mortal and all prime numbers greater than two which are odd hardly shows that men are prime numbers greater than two or that mortality is oddness. And this is true even if we use all the propositions describing reality. The only thing the existence of a model of all of those assertions in the realm of mathematical abstracta shows us is that the logical form of propositions instanced in the domain of the real world can be instanced in a body of propositions about the world of mathematical abstracta as well. In the new interpretation of the assertions which makes them truths over the domain of, say, sets, all of the meaningful predicates of the assertions have been reinterpreted. Nothing in the possibility of such a reinterpretation shows us that the predicates "mean the same thing" in the two interpretations or that the two domains of the two models are one and the same domain. The application of the Principle of Identity of Indiscernibles would be legitimate only if one already believed that the only thing characterizing the meaning of the terms of a theory was the purely logical structure of the theory, and that the only relevant properties of things were purely logical properties. But why should anyone believe any such thing?

(3) The kind of "global Platonism" advocated by Tipler here (and, to be sure, sometimes flirted with by Quine) ought to be contrasted with the more familiar claims that some region of discourse ought to be reinterpreted as being merely discourse over representational abstracta. Some functionalist accounts of some mentalistic language as "programming" language over behavior identifies cognitive states as mere representational abstracta. Some instrumentalistic-representationalistic views about higher theoretical entities in physics may try to do the same thing, taking, for example, spacetime points as mere ordered n-tuples of numbers and viewing reference to spacetime entities and structures as representational of what is real, that is of the relational spatio-temporal features of concrete material entities.

But in these cases the merely representational reference of some of the theoretical terms is being contrasted with the genuine reference to the non-Platonistic features of the world the theory is designed to account for, even if only in an instrumentalistic-representational way, be that overt physical behavior for the mental functionalist-representationalist or the spatio-temporal relations among material objects for the spacetime relationist-representationalist. No wholesale Platonism is intended, and the meaning fixing of the theoretical terms is fixed in something beyond the pure logical structure of the theory, i.e. in the association of the base level terms with the "real" features of the world to be accounted for by the theory.

I suppose a phenomenalist might take such a representationalist line with the whole of the material world, taking all reference to the physical as reference to abstracta used in a representational coding of the general relational features among sense-data. But even he has the sense-data, the red-patches-here-now, and they are not numbers or sets or abstracta of some other kind. And it is the sense-data to which the basic vocabulary is associated when meaning is given to the theoretical vocabulary.

It isn't all that clear, I suppose, what we are talking about, metaphysically, when we talk about the world. But I, for one, am pretty sure it isn't merely some arbitrary set of Platonistic entities having some appropriate structure sufficient to make it a model of some specified set of uninterpreted logical forms. The problem of the meaning and reference of our scientific vocabulary is a difficult one, but I think Tipler's hyper-Platonism can only amount to a *reductio ad absurdam* of one way of looking at reference, that is having it determined by the logical structure of a theory alone.

(4) As Prof. Tipler realizes, there is a conflict between his desire to prove the existence of the universe necessary by identifying the universe with an abstract structure, and his earlier desire to prove the nature of the universe necessary. For all manner of abstract structures, and hence all manner of universes, necessarily exist if any one of them does. Tipler's attempt to make the consequences of this argument seem more innocuous by referring to the plurality of worlds posited by the Many Worlds interpretation of quantum mechanics misses the point. For if the universe-as-abstraction thesis is true then all logically consistent universes exist, not just the limited subset that are possible worlds consistent with the actual laws of quantum theory and the Many Worlds interpretation of it. This point holds true even if one restricts one's attention to possible worlds with internal observers. Even if such worlds are hard to come by given the laws of the actual world, they are much too easy to come by if any possible abstract structure is allowed as a possible, and, hence, from Tipler's perspective, actual, universe.

Many if these universes will also fail to have "Omega Points" as well, making a mockery of the claim that God necessarily exists, for in some universes he will and in others he won't - even if all are actual in Tipler's sense. His attempt to argue that any actual world is indiscriminable to its inhabitants from one with an "Omega Point," is unconvincing. Even if convincing it still would not argue against the existence of lots of actual universes in which an "Omega Point" doesn't exist.

4. Are Ordinary Explanations Explanation Enough?

Should we even hope for explanations of the existence and nature of the universe which are unconditioned on any antecedent contingent posits? Well Aristotle and Leibniz certainly thought so, blaming our familiar resort to generalization from the particular observed to the apparently contingent general on our intellectual limitations. If only we were smart enough, they and others suggested, we could derive all from first principles metaphysically necessary and epistemically a priori, since self-evidently true. Goodness knows many have tried to find "ultimate explanations" behind the merely conditional explanations familiar from workaday science.

But many tried to find Eldorado and the Kingdom of Prester John as well. Alas, they didn't exist. I still find Kant's treatment of the issue in his *Prolegomenon to Any Future Metaphysic* persuasive. Conditioned by the ever increasing success of deeper and deeper explanations of phenomena, we hope that someday we will find the ultimate explanation, which, unlike all of its transient anticipators, requires no contingent posit at all. But such an unconditioned accounter of the conditioned is, according to Kant, an unobtainable ideal. The belief in its possibility is a useful illusion. Useful because the eternal hope that the final explanation will be found, like the carrot forever in front of the donkey, goads us ever onward in our search for the obtainable deeper and deeper relative explanations. Perhaps this is wrong, but I don't find anything either in the invocation of the difficulty of finding self-consistent supersymmetric string theories or in the invocation of the existence of abstract models isomorphic to the intended models of our theories to convince me that what Kant thought unobtainable is now within reach of being achieved.

I can't resist ending with a Sidney Morganbesser story. Heidegger in his *Introduction to Metaphysics* tells us that the one and only "authentic" question of metaphysics is: "Why is there something rather than nothing?" Morganbesser came upon a group of graduate students brooding over the Heideggerian ultimate question. After listening to them for a few minutes he finally said, "You guys! Even if there was nothing you still wouldn't be satisfied."

Well there is something. And most of us pretend to ourselves most of the time that we are satisfied with finding our way about in it by means of the conditioned relative explanations of ordinary science. Some, like Tipler, are not. But why shouldn't we believe, with Kant and the Rolling Stones, that this is just one more familiar case of not getting what you want but getting all you need?

References

Tipler, F. (1987), "The Omega Point Theory: A Model of an Evolving God," Paper delivered at the Vatican Observatory Conference on Science, Philosophy and Theology, September 1987.

The Hole Argument

John Norton

University of Pittsburgh

The "hole argument" shows that spacetime substantivalism leads to a radical form of indeterminism in a broad class of spacetime theories.[1] My purpose in this note is twofold. First I shall present a short and informal version of the argument in the form developed in Earman and Norton (1987). Second I will show how the argument can be extended to the case of "manifold plus further structure" substantivalism whenever that further structure admits certain common symmetries.

1. An Illustration of the Mathematical Devices Used

The hole argument depends on the possibility of displaying two models of some spacetime theory which agree everywhere but within a small neighborhood of the spacetime manifold. To illustrate how two such models are arrived at, I will describe the construction for the easy to visualize special case of a spatially homogeneous and isotropic expanding universe in general relativity. It will be clear how the construction can be extended to other spacetime theories.

A model of general relativity is a triple <M,g,T> which represents a physical situation deemed possible by the theory. M is a four dimensional differentiable manifold which is a set of point-events laid out in a continuum with neighborhoods that are four dimensional. For the above case of an expanding universe, the stress-energy tensor T represents the smoothed out matter of the galaxies. In Figure 1, this smoothed out matter is pictured by the world lines of the galaxies. The diverging of these world lines is a manifestation of the expansion of the universe. Each of these galaxies is in free fall. The metrical structure g of the spacetime determines which trajectories in the manifold are free fall trajectories as well as a large number of other properties related to gravitation and the metrical behavior of rods and clocks. This metrical structure is pictured in Figure 1 by the little light cones drawn at various places.

A diffeomorphism on the manifold M is just a map that assigns points in M to points in M in a smooth, invertible manner. We are interested in a special case which we call a "hole diffeomorphism". To define one, we choose any neighborhood of M we please and call the chosen neighborhood "The Hole".[2] A hole diffeomorphism h is just the identity map outside The Hole and comes smoothly to differ from the identity within The Hole. An example is shown in Figure 1. A diffeomorphism can "carry along" the structures g and T defined on the manifold. If the diffeomorphism h maps the point p to the point hp,

PSA 1988, Volume 2, pp. 56-64

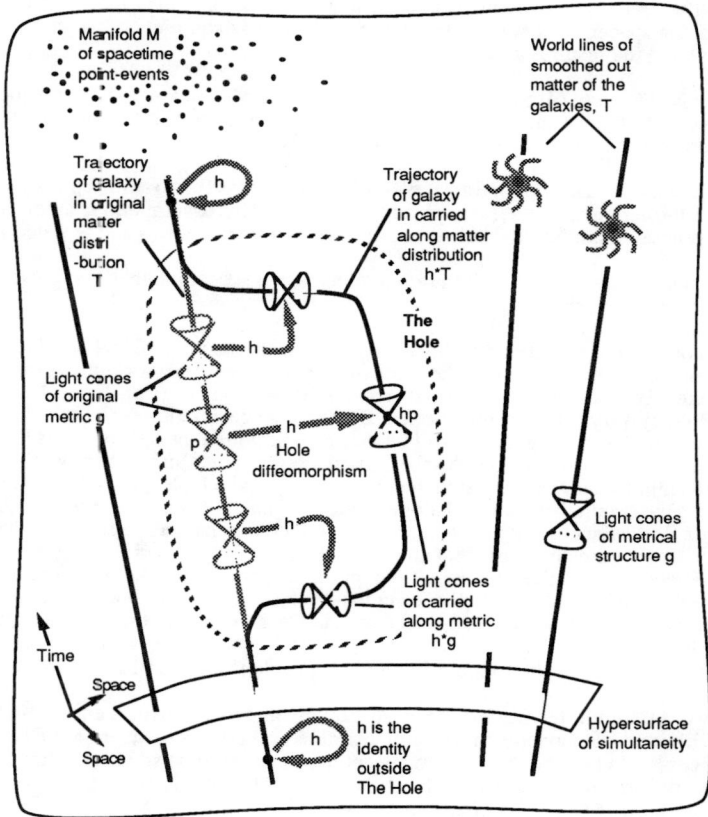

Figure 1   Hole Diffeomorphism Applied to an Expanding Universe in General Relativity

then a world line that passes through p is carried along to a world line that passes through hp. The carry along under hole diffeomorphism h of T is called h*T. It is shown in Figure 1. Notice that the carried along trajectories of the smoothed out matter of the galaxies are no longer free fall trajectories of the metric g. However we can also define the analogous carry along h*g of the metric g. The trajectories of the carried along galaxies will now be free fall trajectories of the carried along metric h*g.

We now have two triples: the original model <M,g,T> and a diffeomorphic copy <M,h*g,h*T>. The fact that the first triple is a model of the theory does not guarantee that the diffeomorphic copy is also a model, that is, also represents a physically possible situation according to the theory. A very important theorem in general relativity, however, assures us that diffeomorphic copies of models are themselves models. This theorem depends only on the fact that general relativity is what we call a *"local spacetime theory"*: the fields g and T of the theory are specified solely by tensorial differential equations. Most commonly discussed spacetime theories can be formulated as local spacetime theories in which all the fields of the models are determined by tensorial differential equations. The hole argument applies to all spacetime theories in their local formulations. Newtonian spacetime theory and special relativity, for example, both admit local formulations[3] and the result is given generally as:

Gauge Theorem (Active General Covariance): If Mod is any model of a local spacetime theory and h any diffeomorphism defined on its manifold then the diffeomorphic copy h*Mod is also a model of the theory.

The standard assumption in general relativity texts is that two diffeomorphic models such as these represent the same physical situation. This assumption is an instance of the more general assumption which applies to all spacetime theories and is stated as:

Leibniz Equivalence: Diffeomorphic models in a spacetime theory represent the same physical situation.

A common justification for this equivalence is the fact that diffeomorphic models agree on all observables under the standard physical interpretations of the mathematical structures. For example, the original and carried along galaxies of Figure 1 may traverse The Hole by very different trajectories. But the time each requires to traverse the hole, as measured by co-moving clocks will be the same provided the first is determined by the original metric and the second by the carried along metric. Similarly any other observable that we might select pertaining to the phenomena in The Hole will fail to distinguish between the original and carried along model. A general justification of this claim and Leibniz Equivalence is based on Einstein's "point-coincidence" argument. (See Norton, 1987.) This argument asserts that all observables are fully reducible to coincidences at point-events such as the collision of two particles or the coincidence of a pointer with a given mark on a scale. Since all such coincidences are preserved under the carry along, all observables must also be left unchanged in the transition to the carried along model.

2. Consequences of the Denial of Leibniz Equivalence: Radical Local Indeterminism

Two diffeomorphic models are in general distinct mathematical entities—they will have different components in the same coordinate chart. Thus we are not forced by logical necessity to assume that both represent the *same* physical situation, although, of course, this assumption is routinely made in general relativity texts. Let us pursue the consequences of denying Leibniz equivalence.

If we deny Leibniz equivalence, we conclude that diffeomorphic models represent distinct physical situations. However diffeomorphic models agree on all observables. So we must conclude that they represent distinct physical situations which cannot be distinguished by any observationally verifiable differences. In the heyday of logical positivism this conclusion alone would have been sufficient to terminate any further consideration of the denial of Leibniz equivalence.

For those undeterred by observationally unverifiable differences, there is a further undesirable consequence revealed by the hole argument: radical local indeterminism in all local spacetime theories. To see how this indeterminism arises, take the example of the local spacetime theory of general relativity and consider the two models of an expanding universe pictured in Figure 1. The two models agree exactly everywhere outside The Hole since the diffeomorphism that carries one into the other is the identity outside The Hole. But they come smoothly to differ within The Hole. Thus the fullest specification of all the fields outside The Hole will not enable the theory to determine how the model will develop into The Hole, *no matter how small The Hole is in spatial and temporal extent* . This indeterminism is of a very extreme form rarely encountered in non-quantum theories. Under it, the only way that one can determine the model over the entire manifold is by specifying it everywhere. If the specification omits any neighborhood no matter how small, the above construction shows us that the theory will fail to determine the model within that neighborhood.

It is important to see that this form of indeterminism is undesirable because of the special way that it arises. Since diffeomorphic models agree on all observables, the denial of Leibniz equivalence amounts to the assumption that there are physically significant properties of the models that transcend observational verification. The construction of the hole argument reaffirms the dubious nature of these extra properties. It shows that these extra properties are not only opaque to observation but are also opaque to the theory itself in this sense: the theory is unable to determine how these properties will develop into an arbitrarily small neighborhood of the manifold, given a full specification of the properties everywhere else.

3. Who Must Deny Leibniz Equivalence?

Spacetime substantivalists must deny Leibniz equivalence. Spacetime substantivalism is an extreme form of realism about certain structures in spacetime theories. It holds that spacetime is a substance in so far as it is something that has an existence independent of its contents. I do not know how to make the notion of "independent existence" precise here. But this much is clear of the substantivalist position. If the contents of spacetime are rearranged in some way in spacetime—for example everything is spatially translated three feet in some direction—then the substantivalist must say that we arrive at a physically distinct situation. For an important physical property has changed: the spatiotemporal locations of the contents.

This necessary commitment of substantivalists and the ensuing conclusion of the physical distinctness of observationally indiscernible states of the world is precisely what Leibniz exploited in his challenge to the Newtonian Clarke in his third letter of their famous correspondence (Alexander, 1956). Leibniz considered, for example, the case of the bodies of the world replaced in space in such a way that East and West are exchanged but all other relations preserved. The Newtonian space substantivalist must insist that a new world has been formed even though it is indiscernible from the old one.

In local spacetime theories, the mathematical entity which most naturally represents spacetime are the manifolds of the models. This view is defended in Earman and Norton (1987). It leads the substantivalist to what we call "manifold substantivalism". The manifold substantivalist must insist that the rearrangement of spacetime structures against the background of the spacetime manifold leads to a structure that represents a different physical situation. Since the carry along under diffeomorphism effects just such a rearrangement, the manifold substantivalist must insist that a model of a spacetime theory and a (non-identical) diffeomorphic copy of it represent different physical situations. Thus the manifold substantivalist must deny Leibniz equivalence and accept the undesirable consequences outlined in Section 2. In particular the extra physical properties introduced by this substantivalist must be opaque to both observation and the laws of the physical theory in the sense given above.

4. An Escape? Manifold plus Further Structure Substantivalism

The full force of the hole argument is directed against manifold substantivalism. One might think, therefore, that the substantivalist can escape the undesirable consequences of the hole argument by choosing not just the manifold, but the manifold plus some further structure to represent spacetime. For example one might choose the manifold plus metric in both special and general relativity. Or in Newtonian spacetime theory one might choose the manifold plus absolute time one-form, Euclidean spatial metric and the affine connection adapted to them. For concreteness below, I shall assume that manifold plus further structure ("mpfs") substantivalists do make these choices for the "further structure" in Newtonian theory, special and general relativity.

A simple intuitive consideration reveals roughly when this escape via mpfs substantivalism will succeed or fail. The hole argument goes through against the manifold substantivalist because the spacetime theories we consider provide no means of individuating physically the points of the manifold other than through the further structures defined on them. Thus one produces no changes in the observational consequences by rearranging the individuating structures over the manifold by means of a carry along. Moreover the laws of the theory seem indifferent to whether one carries out such rearrangement. Thus we would expect similar problems for mpfs substantivalism if the "further structure" exhibits symmetries. For loosely speaking, the presence of these symmetries represent a failure of the further structure to individuate fully the points of the manifold. These intuitions are made more precise in two claims.

Claim 1: Observational indistinguishability. If the "further structure" selected by the mpfs substantivalist admits any symmetry transformation at all, then the mpfs substantivalist is committed to the distinctness of observationally indistinguishable states of affairs in local spacetime theories.

Justification. Let the theory have models <M,S,C>, where M is a differentiable manifold, S represents the "further structure" and C represents the remaining "contents" of spacetime. Let S have the non-identical symmetry h, so that by definition h*S=S. Consider the two structures <M,S,C> and <M,S,h*C>. If one of them is a model of a local spacetime theory, the gauge theorem guarantees that both are models. The mpfs substantivalist must say that they represent physically distinct situations since C has been rearranged against the spacetime background of <M,S>. But we have that <M,S,h*C>=<M,h*S,h*C>. Thus the mpfs substantivalist must say that the two diffeomorphic models <M,S,C> and <M,h*S,h*C> represent physically distinct situations. However from earlier we know that such diffeomorphic models agree on all observables.

Theories that exhibit symmetries required for Claim 1 to hold include flat Newtonian spacetime theory, special relativity and general relativity applied to spatially homogeneous and isotropic cosmologies.[4]

Claim 2: Indeterminism. If the "further structure" includes certain common symmetries, such as spatial homogeneity and isotropy, then we can recover a form of indeterminism analogous to that arrived at in the hole argument in local spacetime theories.

Justification. The construction needed to establish this result is more complicated than that required for the hole argument. I will give it for the case of special relativistic electrodynamics. It will be clear that analogous constructions are possible for the other theories cited. Special relativistic electrodynamics has models <M,g,F>, where M is an R^4 differentiable manifold, g a Lorentz signature metric tensor and F the Maxwell field tensor. The model set of the theory contains just all such triples in which g satisfies the vanishing of the Riemann curvature tensor as its tensorial field equation and F satisfies Maxwell's equations. For simplicity we assume that F is fully inhomogeneous and anisotropic. Assume that the mpfs substantivalist selects <M,g> as representing spacetime. Let t be a translation by unit spatial distance in some direction. t is a symmetry of g so that we have

$$t*g=g.$$

It is possible to decompose the translation t into the composition of two parts as shown in Figure 2. To do so, we select any two parallel hypersurfaces of simultaneity, which divides the manifold into three regions. Call the the one between the hypersurfaces the "present" and the remaining two the "past" and the "future".

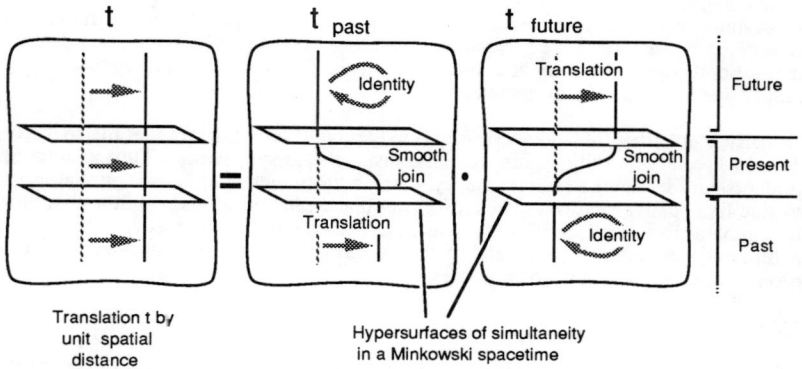

Figure 2   Decomposition of a Spatial Translation in a Minkowski Spacetime

Define t_{past} to be the identity map in the future, to coincide with t in the past and, in the present, to be a smooth interpolation between them. Similarly define t_{future} to be the identity in the past, to coincide with t in the future and, in the present, to be a smooth interpolation between them. The two interpolations in the present are to be chosen so that we have:

$$t = t_{past} \cdot t_{future}.$$

In accord with the earlier discussion, the mpfs substantivalism must hold that translating F to t*F across the spacetime background of <M,g> will produce a structure representing a physically different situation; the various parts of F are now located at different spatiotemporal events. That is, this substantivalist must conclude that the two models <M,g,F> and <M,g,t*F> represent different physical situations. Exploiting the fact that t*g=g we can say:

Model I, <M,g,F>, represents physical situation I.

Model II, <M,t*g,t*F>=<M,g,t*F>, represents physical situation II.

Now recall that we could proceed from model I to model II by a two step transformation: the application of t_{future} followed by the application of t_{past}. Figuratively:

$$<M,g,F> \; (t_{future}) \rightarrow \; <M,t_{future}*g,t_{future}*F> \; (tpast) \rightarrow \; <M,t*g,t*F>=<M,g,t*F>$$

Let us call the intermediate model <M,t_{future}*g,t_{future}*F> "model III". Note that it is diffeomorphic to both model I and II, so it represents a physical situation observationally indistinguishable from those represented by models I and II. Now ask what physical situation is represented by model III. There are three possible answers:[5]

(a) physical situation I
(b) physical situation I
(c) a physical situation other than I or II.

Any answer leads to indeterminism. For example, consider answers (b) or (c). The selection of either entails that the two diffeomorphic models, model I and model III, represent distinct physical situations. However by the construction of the diffeomorphism t_{future} that relates the two models, they agree in the past and come smoothly to disagree in the future—and the disagreement matters physically since the models represent different physical situations. This is a classic instance of future directed indeterminism. The full specification of the model in the past fails to determine the development of the model into the future. Correspondingly, if the answer selected is (a), one considers model III and model II which leads by analogous arguments, to past directed indeterminism. (The time reversibility of the theory enables us to convert this past directed indeterminism into a future directed indeterminism if we wish.)

The construction given is one of the simplest. It could be extended to give the indeterminism more of a "hole-like" character. For example, by decomposing the translation t into more parts we could give a version of the construction in which the underdetermined neighborhood consists of only an arbitrarily thin slice of the manifold bounded by two parallel hypersurfaces of simultaneity.

Finally, I note that the above construction exploited only the fact that the spacetime <M,g> of special relativity admits spatial translations as symmetries. Thus analogous arguments can be mounted in any theory which admits such symmetries or analogous ones like like spatial rotations. Thus it follows that an indeterminism akin to that of the hole argument faces mpfs substantivalists in at least the following theories: flat Newtonian spacetime theory, special relativity and general relativity applied to spatially homogeneous and isotropic cosmologies.

5. Conclusion

The analysis of spacetime substantivalism given here is not based on the belief that determinism is or ought to be true. Earman (1986) has catalogued admirably the many ways that determinism can fail even in classical theories. The force of the attack on spacetime substantivalism comes from the way that the indeterminism arises. The spacetime substantivalist is forced to introduce properties which must have physical significance, even though they remain inaccessible to observational verification and are opaque to the laws of the spacetime theory in so far as the theory cannot determine their development.

Addendum: Replies to the Criticism of Maudlin and Butterfield

Tim Maudlin's essentialism offers an escape for substantivalists from the hole argument which depends on the analysis of which properties are essential to spacetime. In so far as the escape reduces to the endorsement of manifold plus further structure substantivalism,[6] I have outlined in Section 4 above the circumstances under which the escape fails and succeeds. Whenever the spacetime admits no symmetries, then the mpfs substantivalist escape succeeds. Whenever the spacetime admits symmetries in the way indicated, then the mpfs substantivalist escape fails. Overall we might say that the escape enjoys partial success. For, through it, spacetime substantivalism need not *always* lead to the disastrous consequences—they only arise sometimes. Of course there is a surer escape: the denial of spacetime substantivalism.

Unfortunately I do not think that Jeremy Butterfield's most ingenious escape via counterpart theory even enjoys this type of partial success. Butterfield's escape depends on endorsing what he calls "One", which asserts that at most one of two diffeomorphic models of a spacetime theory can represent a physically possible world. If "model" means "mathematical structure which a theory selects as representing a physically possible situation" then "One" ought to assert that at most one of two diffeomorphic *structures* can represent a physically possible world according to a given spacetime theory. The first problem is that "One" directly contradicts the active general covariance of our local spacetime theories as expressed in the gauge theorem above. The models of a local spacetime theory are *all* structures of the appropriate type that satisfy the theory's tensorial field equations. John Earman and I stressed this "all" in our (1987, p.517) by explicitly introducing a "completeness condition". This completeness condition allows derivation of the gauge theorem in local spacetime theories. This theorem contradicts "One" by guaranteeing that a diffeomorphic copy of a model is itself a model and thus represents a physically possible situation.

Let us set this worry aside. We might choose, for example, to deviate in some way from the local formulation of the spacetime theory. I still do not think that the escape works. Given two diffeomorphic structures, at most one is a model of the theory according to "One". Take the case in which one of them is a model. How are we to distinguish the real model from the imposter? There must be some property which distinguishes them and the property must be physically significant in so far as it tells us which structure represents a physically possible world. Since the real model and the imposter are diffeomorphic, this property cannot have observational consequences. If we mistakenly choose the imposter as a model, we would interpret it as representing a physically possible world indistinguishable observationally from that represented by the real model. Similarly the property eludes the tensorial field equations of local spacetime theories. Take exactly the set up of the hole argument with the real model specified on the manifold everywhere outside The Hole. These field equations will be unable to distinguish between the development into The Hole of the real model or of one of its infinitely many diffeomorphic copy-imposters. Thus the counterpart theorist is in precisely as bad—or as good—a situation as the manifold spacetime substantivalist. Both introduce properties which must have physical significance, even though they remain inaccessible to observational verification and are opaque to the laws of local spacetime theories.

Notice finally that the problem of the real model and the imposter must face anyone who would seek an escape from the hole argument by avoiding the local spacetime formulation of spacetime theories.

Notes

[1]The hole argument was advanced by Einstein in 1913-1914 as an argument against the acceptability of generally covariant field equations in general relativity. Its clearest statement is Einstein (1914, pp.1066-67) and the role it played in Einstein's thought is discussed in Norton (1987). The non-triviality of the argument was revealed to modern readers by Stachel (1980). For a novel approach to the reading of Einstein's version of the argument, see Norton (forthcoming).

[2]The name was introduced by Einstein for his version of the argument which applied to a special case in which The Hole was a matter free neighborhood—a hole!—in a matter distribution.

[3]For special relativity, the theory has models <M,g>, where M is a four dimensional differentiable manifold and g a Lorentz signature metric. The field equation is simply the vanishing of the Riemann curvature tensor.

[4]In both Claim 1 and 2, for the case of general relativistic cosmologies, the constructions must use a matter distribution whose *stress-energy tensor* is spatially homogenous and isotropic, but which is spatially inhomogeneous or anisotropic in some other property. An example of such a matter distribution is a uniformly expanding cloud of non-interacting dust particles, each with the same mass, but not all of them identical particles.

[5]One might be tempted to add the fourth answer "no physical situation at all". To do so violates the gauge theorem stated above for local spacetime theories. If one persists nonetheless, one must then face the problem of the real model and the imposter described in my reply to Butterfield below.

[6]A closer reading of Maudlin's text reveals to me that he cannot advocate the type of mpfs substantivalism described above. He holds that a hole diffeomorphism fails to transform a model of a theory into another model, since the transformed mathematical structure does not represent a physically possible situation. Thus his escape fails in the same manner as Butterfield's: first, he violates active general covariance and, second, he must face the problem of the real model and the imposter. (I thank Jeremy Butterfield for this point.

References

Alexander, H.G. (1956) (ed.), *The Leibniz -Clarke Correspondence*. Manchester Univ. Press.

Earman, J. (1986), *A Primer on Determinism*. Dordrecht: Reidel.

Earman, J. and Norton, J. (1987), "What Price Spacetime Substantivalism? The Hole Story", *British Journal for the Philosophy of Science* 38: 515-525.

Einstein, A. (1914), "Die formale Grundlage der allgemeinen Relativitaetstheorie", *Preuss. Akad. der Wiss., Sitz.* pp.1013-1085.

Norton, J. (1987), "Einstein, the Hole Argument and the Reality of Space", in *Measurement, Realism and Objectivity*, J.Forge (ed.). Dordrecht: Reidel, pp.153-188.

_ _ _ _ _ . (forthcoming), "Coordinates and Covariance: Einstein's View of Spacetime and the Modern View", *Foundations of Physics*.

Stachel, J. (1980), "Einstein's Search for General Covariance". Paper read at the Ninth International Conference on General Relativity and Gravitation, Jena. Published in D.Howard and J.Stachel (eds.), *Einstein and the History of General Relativity: Einstein Studies*, Volume 1. Boston: Birkhauser, 1989 forthcoming.

Albert Einstein Meets David Lewis[1]

Jeremy Butterfield

Cambridge University

1. Introduction

With help from Einstein, John Norton and John Earman (1987; this volume) have invented a fascinating argument against spacetime substantivalism. It gives the philosophy of space and time a refreshing stimulus: all credit to them. The idea of the argument is that substantivalism rules out determinism, even of very weak kinds. Norton and Earman have no special brief to defend determinism. But they believe that if it fails, it should fail for reasons of physics that vary from case to case—it should not fail at a stroke for the sake of a philosophical doctrine such as substantivalism. They conclude that in the conflict between substantivalism and determinism, it is substantivalism that must go.

The conflict arises from 'hole diffeomorphisms'. These enable us, once we are given a model for a spacetime theory, to produce another model. A diffeomorphism is a permutation of the points that induces a permutation of their properties and relations, according to the rule: image-points are to have the properties and relations to one another that were possessed by their pre-images. A hole diffeomorphism is such a permutation of the points, that reduces to the identity map except on a small region, called the 'hole'. So the points in the hole get shuffled, and there is an induced shuffling of their properties and relations. Norton and Earman now claim: substantivalism must hold that these two models represent different physically possible worlds. After all, the points in the hole have different properties in the two models. (And in general all points will have different relations in the two models—for example different distances to a given point in the hole.) But this means that substantivalism rules out determinism, even of very weak kinds. For determinism requires that one physically possible world is singled out by the specification of the physical state on some region of spacetime; and the larger this region is, the weaker is the corresponding version of determinism. And Norton and Earman's claim means that substantivalism prevents such a singling out, no matter how large this region. For given any model and any region in it, no matter how large, we can apply a hole diffeomorphism that is the identity map on this region so as to produce another model, which according to substantivalism represents another physically possible world: it differs in the remaining region, the hole.

I am an aspiring scientific realist and so find substantivalism attractive. I think that, fortunately, the hole argument can be rebutted: substantivalism can be reconciled with determinism. Surely that is as one would expect: 'isms' like substantivalism are not pre-

PSA 1988, Volume 2, pp. 65-81

cise, and so are unlikely to be refuted by technical arguments. I have described this reconciliation elsewhere (1987; 1989). In this paper, I shall summarize the main ideas; but devote most of the paper to developing some further points, correcting an error in my (1987) and replying to some objections.

There are two aspects to such a reconciliation: technical (Section 2) and philosophical (Sections 3 and 4). (Readers who want to avoid technicalities can skip Section 2.) The technical aspect concerns the fact that Earman and Norton's argument does not use a precise definition of determinism. So perhaps there is a precise definition that is not violated by pairs of models that are hole diffeomorphs of each other. In fact, this is so: while one definition of determinism is violated by hole diffeomorphs, there is another definition that is not violated. The philosophical aspect concerns the question whether this second definition is as good an explication of the basic idea of determinism as is the first one: if it is not, the technical victory would be hollow. All participants in the debate agree that the basic idea is as above: one physically possible world is singled out by the specification of the physical state on some region of spacetime. So this question about explication leads into the issue of what is the relation between models of our theories and physically possible worlds.

I will argue that the answer to the question is Yes—the second definition is as good an explication of the basic idea. This answer will be based on denying Norton and Earman's claim about substantivalism: that it must hold that the two hold diffeomorphs represent two different physically possible worlds. More specifically, my answer Yes is based on the idea that at most one of a pair of hole diffeomorphs represents a physically possible world. This idea is common to both Maudlin (1989; this volume) and me. And accordingly, I like Maudlin's rebuttal of the hole argument, which is based on the idea that metrical relations are essential to spacetime points. But I think we can do better. In brief, I hold that non-isometric models cause trouble for metrical essentialism; and that the best way out is to deny that a spacetime point can be an inhabitant of two possible worlds—thus applying the modal metaphysics of David Lewis.

A disclaimer. I do not think that the position I expound here, or the disjunction of it with Maudlin's position, is the unique rational response to the hole argument, even among scientific realists. (Non-realists will respond that the argument, whatever its technical or historical interest, flogs a dead horse.) There surely are scientific realists whose other commitments will prompt them to respond to the hole argument by giving up one of the pair, substantivalism and determinism, rather than trying to reconcile them as Maudlin and I do. Thus realists who are less enthusiastic than I am about substantivalism may well respond by trying to rewrite spacetime theories without reference to spacetime points. That is a challenging project, for which some interesting and perhaps workable proposals have been made by Earman and Stachel. And realists who are less enthusiastic than I am about modal metaphysics and/or about the intuitive idea of determinism may well respond by taking determinism as just a technical feature of a class of models, which can be allowed to fail when it threatens substantivalism. (Recall Norton and Earman's claim that, although determinism can be allowed to fail, it should fail for reasons of physics—not at a stroke for reasons of philosophy. In effect, this position simply disagrees. And it can take heart from the fact that while one precise definition of determinism is violated at a stroke by hole diffeomorphs, another is not—so that it is hardly fair to say that substantivalism rules out at a stoke determinism simpliciter.)

2. Determinism in Terms of Models

In this section, I will show that while hole diffeomorphs always violate one precise definition of determinism, to be called Dm1, there is another definition, Dm2, that they do not violate. (Indeed, we shall also see that there is a third definition, Dm0, that they need not violate; but I will not emphasize Dm0, and will discuss it only in the course of motivating Dm1.) The distinction between the two definitions, Dm1 and Dm2, is closely related to whether a theory postulates metric structure(coded by metric fields and connections) once and for all, as a fixed framework the same in every model of the theory. (Or as Earman (1986, p. 24) puts it: as a fixed canvas on which the material contents of spacetime (coded by matter fields) get painted.) The alternative is for a theory to treat metric structure on a par with matter fields: to constrain metric fields and connections only by requiring them to satisfy field equations, just as for matter fields.

Only the second kind of theory can be subject to the hole argument. Thus recall the hole argument's assumption that applying a hole diffeomorphism to a model of the theory yields another model. This is a substantive assumption: since the metrical properties and relations of points get shuffled by the hole diffeomorphism, along with their material properties and relations, this assumption prohibits the theory from postulating a fixed framework of metrical structure the same in every model. In other words, this assumption restricts the hole argument to those theories that treat metric structure on a par with matter fields. This restriction involves no criticism of Norton and Earman, for two reasons. First, they are aware of it; they call the second kind of theory 'local' (Norton this volume, p. 56-64; Earman and Norton 1987, p. 517-518). Second, and more importantly, the substantivalist cannot sidestep the hole argument simply by considering only theories of the first kind. At least, she cannot do this if her substantivalism is based on scientific realism about our current best spacetime theory, namely general relativity. For in general relativity, the metric structure is affected by matter, so that there are non-isometric models: pairs of models for which no diffeomorphism (smooth invertible point-to-point map) between them drags the metric structure of the one into coincidence with the metric structure of the other. (Indeed, general relativity has non-diffeomorphic models; think of how the global topology can be different in different models, so that there is no diffeomorphism between them.) These non-isometric models prevent one from writing down general relativity as a theory of the first kind; and so general relativity is subject to the hole argument. (Of course, general relativity was historically the first theory of the second kind. Newtonian theory and special relativity, as traditionally presented, are of the first kind. But with general relativity in hand, we can with hindsight formulate them as theories of the second kind; and Earman and Norton recommend doing so: 1987, p. 518.)

For the first kind of theory, the first definition of determinism, Dm1, works. That is, there is no automatic violation of Dm1 via the hole argument. And Dm1's verdicts about whether determinism holds in these theories are intuitively right; (they are described in detail by Earman (1986, pp. 23-40, 58-61)). On the other hand, any theory of the second kind will trivially violate Dm1—whatever the details of the theory. And the reason is precisely the hole argument! But there is a second definition of determinism, Dm2, that works for theories of the second kind. That is hardly surprising: general relativity texts that discuss the initial value problem prove a 'uniqueness of solution' result; which suggests that determinism, according to some decent definition, holds good. And in fact, we can extract Dm2 from such discussions, by abstracting from the specific details of general relativity.

Motivating Dm1 is a little subtle. We need to lead up to Dm1 through two preliminary definitions. First I shall motivate a preliminary definition, Dm, which is easily seen to be too strong; (I thank John Norton for pointing out to me this undue strength, thus correcting my (1987: p. 14)). How should we weaken Dm? There are two ways to do so. Both weakenings work for the first kind of theory in the sense that they deliver intuitively right verdicts about whether determinism holds. But the first weakening, to be called Dm0, does not suit Norton and Earman's polemical purposes: it is not automatically violated by theories of the second kind. The second weakening does suit their purposes: it is thus violated. And accordingly, it is this second weakening that is Dm1. In short, I will lead up to Dm1, which is designed to suit Norton and Earman, through two preliminary definitions: Dm, which is too strong, and Dm0 which is weaker but does not suit their purposes.

We want determinism to mean that agreement on regions of a certain kind (typically sandwiches or slices) implies agreement elsewhere. But there is no meaning to a vector or tensor at a point in one manifold being the same as a vector or tensor at a point of another manifold. So we spell out agreement in terms of a diffeomorphism dragging the vectors, tensors etc. (the so-called geometric objects) in one model into coincidence with the corresponding geometric objects of the other. We would also like to cover the case of a theory (like general relativity) not all of whose models have manifolds that are diffeomorphic to one another. The natural tactic is to make the statement conditional on the existence of a global diffeomorphism between the models' manifolds, and on the manifolds containing regions of the right kind—and if the regions do not exist, determinism will be vacuously and harmlessly true. Thus suppose we have a theory whose models comprise a manifold and various geometric objects: $<M,0_i>$, $<M', 0_i'>$ etc. We spell out agreement of two such models in terms of a diffeomorphism d between their manifolds having an induced drag-along map d^* that drags the 0_i into coincidence with the $0_i'$. That is, we require that for all image-points $d(p)$, $d^*(0_i)(d(p)) = 0_i' (d(p))$. With this motivation, we might write down this definition:

> Dm A theory with models $<M,0_i>$ is S-deterministic, where S is a kind of region that occurs in manifolds of the kind occurring in the models, iff:
> given any two models $<M,0_i>$ and $<M',0_i'>$ and any diffeomorphism d from M onto M', and any region R of M, of kind S:-
> if d(R) is of kind S, and also for all i $d^*(0_i) = 0_i'$ on d(R), then:
> for all i, $d^*(0_i) = 0_i'$ throughout M'.

But this definition is too strong! Because its quantifier "any diffeomorphism d from M to M'" has such a large range, Dm is violated by paradigm deterministic theories such as electromagnetism in Minkowski spacetime. (And since these theories can be of the first kind, the violation is quite independent of the hole argument.) To see this, we first recall that for any theory, models $<M,0_i>$ and $<M',0_i'>$ are called isomorphic if there is a diffeomorphism d from M onto M' such that $d^*(0_i) = 0_i'$ throughout M'. Similarly two regions of two models can be called isomorphic. (Thus Dm states, roughly speaking, that isomorphism of regions implies isomorphism of models.) And a model $<M,0_i>$ is called symmetric if it is isomorphic to itself by a diffeomorphism other than the identity map. Similarly a region of a model is called symmetric if there is a isomorphism from the region to itself other than the identity map. (And if we consider only the dragging into coincidence of metric fields and connections, then we say models are isometric; and that a model has a metric symmetry.)

To violate Dm for our paradigm deterministic theory, we now take any non-symmetric model of the theory. In general there are many such models; we can even allow arbitrarily large regions of spacetime to be symmetric—we only require that there be no total, global symmetry of the model. Now take any region R of the model of the 'determining' kind; i.e. any region whose state we think of as determining the state throughout spacetime—e.g. a spacelike hyperplane, for electromagnetism in Minkowski spacetime. Now take any isomorphic copy of our model: there are many such. (Again: our models can be of the first kind, so that the two models have a fixed framework of metric structure in common between them.) Dm is violated! There are plenty of diffeomorphisms between these two models that yield agreement on the region R but not elsewhere. To yield agreement on R, a diffeomorphism need only have as its restriction to R, the given isomorphism. To yield non-agreement somewhere beyond R, it need only differ from the given isomorphism somewhere beyond R. For if it differed somewhere from the given isomorphism and yet yielded agreement throughout the manifolds, then the first model (and so both models) would be symmetric, contrary to our choice of the first model.

(To put the argument formally: let $<M,0_i>$ be the chosen model, so R is a region of M; let $<M',0_i'>$ be the isomorphic copy, with d the given isomorphism, so that the isomorphic copy is $<M',0_i'> = <d(M),d^*(0_i)>$. Let f be a diffeomorphism from M to M' that (i) restricts to d on R: $f(p) = d(p)$ for all points p in R; (ii) differs from d on some region Q. By (i) f satisfies the antecedent of Dm: it gives agreement between R and $f(R) = d(R)$. But if f satisfied the consequent of Dm, (i.e. if f were a global isomorphism), then the composite map f^{-1} o d, ' first d then the inverse of f', being a composition of isomorphisms, would be a symmetry of $<M, O_i>$ – contrary to assumption.)

In short, Dm is too strong because the large range of its quantifier "any diffeomorphism d from M to M'" makes it easy to have local but not global agreement. So we need to weaken Dm by shrinking the range of the quantifier. One suggestion is to write instead 'any isometry d from M to M'". Let us call this Dm0. That is, Dmo requires only that for any diffeomorphism that drags the metric(s) and connection of $<M,0_i>$ to those of $<M',0_{iV}>$, local agreement imply global agreement. In one important respect, DmO works very well; but it does not suit Norton and Earman's purposes. I shall treat these points in turn.

DmO works very well in that it delivers intuitively right verdicts about whether determinism holds, for some familiar theories of the first kind. Namely, theories using a classical spacetime with or without absolute rest, and theories using Minkowski spacetime; where the regions S are time-slices, or thin sandwiches, across the manifold. In such theories, the postulation of a fixed framework means that there are isometries between any two models. And more important, the restricted quantifier eliminates unwanted violations of determinism of the kind that plagued Dm. The main reason is that in all three spacetimes, there is a unique extension of an isometry on a sandwich to a global isometry. That is: an isometry between two models is determined by its restriction to a sandwich. To see this for classical spacetime without absolute rest, recall that the global isometries are the Galilean transformations (considered as active point-to-point maps), and that these cannot be identity on a sandwich (i.e. for a stretch of time) and differ from identity elsewhere. Thus with d and f global isometries: if d = f on a sandwich, then f^{-1} o d is a global isometry that is identity on the sandwich, and so everywhere; and so d = f everywhere. In classical spacetime with absolute rest, and in Minkowski spacetime, the point is strengthened; the global isometries cannot be identity even on an instantaneous slice and differ from identity elsewhere. (For details, cf. Earman (1986; 23-40; 58-61), or the summary in Butterfield (1987: 15-17).) The fact that global isometries are determined by their restrictions to a sandwich clearly implies that our argument above for violating Dm will fail, once we are restricted to isometries rather than arbitrary diffeomorphisms, and once we take the region R to be a sandwich. Given a global isomorphism (and so isometry) d, there is just no other isometry of the kind required by the argument, i.e. that agrees with d on the sandwich R but that differs from it elsewhere!

But DmO does not suit Norton and Earman's purposes in arguing against substantivalism. For they need a definition of determinism that is violated by all theories of the second kind, as a result of the hole argument; i.e. a definition that is violated by any two models that are hole diffeomorphs, say ($<M, O_i>$ and $<M,d*(O_i)>$), with d the hole diffeomorphism, and the determining region R taken as all of spacetime less the hole, so that d is identity on R. And DmO is not violated by any two such hole diffeomorphs! Indeed it is easy to show that the violation of DmO by a pair of hole diffeomorphs requires the existence of a metric symmetry. The argument for this will by now have a familiar ring: it turns on composing two isometries, in opposite directions between two models, so as to get a metric symmetry of one of the models. Thus suppose that DmO is violated by the hole diffeomorphs above, with d the hole diffeomorphism (so that d is an isomorphism between the models). Then there is a global isometry f that yields agreement between R and f (R), but not global agreement; so f is not a global isomorphism. So f differs from d and f-1 o d, being a composition of isometries, is a metric symmetry of $<M, O_i>$. Now, in general, a model of a theory of the second kind does not have a metric symmetry. So by contraposition of this argument, such a model and a hole diffeomorph of it cannot violate DmO. In short, DmO is not violated by any two hole diffeomorphs.

You might say that Norton and Earman can reply as follows. For the theories of the second kind that we usually consider, some of the models do have metric symmetries; and in order to violate DmO, it is of course enough to find one pair of models with a global isometry between them yielding agreement on R but not global agreement. So they can propose that the hole argument involves building a hole diffeomorph only of some metrically symmetric model, and then arguing that DmO is violated.

However, there are three reasons to think that this is not what Norton and Earman really have in mind: two negative and one positive. First, the fact that in general relativity the metric is affected by matter means that metrically symmetric models are unrealistic; so an argument that required them would be rather limited. Second, Norton and Earman make no mention of the idea that their argument is limited to theories (of the second kind) with such models: they take the hole argument to use a hole diffeomorph of an arbitrary model. Indeed, their discussion of metric symmetries goes rather in the opposite direction; they see metric symmetries not as something they need, but as trouble for a certain kind of substantivalist (Earman & Norton 1987: p. 522 fn. 2; Norton this volume, Section 4, p. 59; Section 4 below will deny the trouble).

The positive reason is that there is a second way to weaken Dm, by shrinking the range of its quantifier; this is Dm1. Like DmO, it delivers intuitively right verdicts about whether determinism holds, for some familiar theories of the first kind. And unlike DmO, it suits Norton and Earman's purposes for theories of the second kind: it is violated by any two models that are hole diffeomorphs. Dm1 gains these two merits by explicitly using the distinction between absolute geometric objects that are meant to be "the same" in every model and dynamical geometrical objects that "vary". We have in effect met this distinction already; apart from its not explicitly mentioning metric structure, it is tantamount to my distinction between theories of the first and second kind. Thus Dm1 weakens Dm by requiring that the diffeomorphism considered should drag any absolute objects there are in the first model into coincidence with the corresponding objects in the second model. That is, we rewrite the quantifier as "for any diffeomorphism d from M to M' that drags any absolute objects on M into coincidence with the corresponding absolute objects on M'": and the conditional statement that follows is then in effect only about the dragging into coincidence of dynamical objects (since the coincidence of absolute objects is already assumed). Thus we write:

Dm1 A theory with models $<M, O_i>$ is S-deterministic, where S is a kind of region that occurs in manifolds of the kind occurring in the models, iff: given any two models $<M,O_i>$, $<M',O_i'>$, and any diffeomorphism d from M onto M' that drags any absolute object among the O_i to the corresponding absolute object among the O_i' throughout M': and given any region R of M, of kind S:-
  if d(R) is of kind S, and also for all i $d*(O_i) = O_i'$ on d(R), then:
  for all i, $d*(O_i) = O_i'$ throughout M').

(Although Norton and Earman do not use Dm1 in their argument against substantivalism, Earman writes down essentially Dm1 at the start of his thorough discussion of determinism of theories of the first kind (1986; p. 24, fn. 1). This is further evidence that Dm1 is the definition that they have in mind.)

To apply Dm1 to a theory, one needs to know which if any of its models' geometric objects are absolute. But the usual judgments about this issue suffice for the two merits stated above. Thus if one judges that in the familiar theories of the first kind using classical and Minkowski spacetime the metric(s) and connection are absolute—the usual judgment—then Dm1's quantifier will reduce for such theories to DmO's "for any isometry". And so Dm1 will join DmO in giving the intuitively right verdicts about whether determinism holds in such theories. As for the second merit; if one judges that in all theories of the second kind there are no absolute objects, the quantifier will reduce to the "for any diffeomorphism from M to M'" of Dm—i.e. the large range again. We saw above that this large range makes for violation of Dm by non-symmetric models of theories of the first kind. But we can now check that it also makes for violation of Dm1 by any two hole diffeomorphs, and thus by any theory of the second kind. The identity map suffices for the check! Let $<M,O_i>$ and

$\langle M, d^*(0_i) \rangle$ be related by a hole diffeomorphism d which is identity on R—as extensive as you like. The identity map id on M is a diffeomorphism between the models with $id^*(0_i) = 0_i'$ on R; while $id^*(0_i) \neq 0_i'$ in the hole, M - R. And since there are no absolute objects that id is required to drag into coincidence, id provides a counterexample to Dm1.

In gaining these two merits as an interpretation of Norton and Earman, Dm1 does however pay price. It gains its merits by using the distinction between absolute and dynamical objects. And that is a slippery distinction. Admittedly, it is familiar;and in order to gain the merits, the judgments one needs to make about what is absolute and what dynamical, are familiar ones. Nevertheless, the distinction has notoriously resisted general definition (cf. Friedman 1983, p. 59, fn. 1). Perhaps it is no wonder that Norton and Earman steered clear of a general definition of determinism!

So much for Dm1. I turn to this Section's more positive task: providing a definition, Dm2, that is not violated by pairs of hole diffeomorphs. Dm2 will be hard to violate. It is not just that (unlike Dm1) not every pair of hole diffeomorphs will violate Dm2: DmO had that feature, since only metrically symmetric hole diffeomorphs could violate it. Rather, no pair of hole diffeomorphs violates Dm2. Starting from our current knowledge of general relativity, one has every reason to hope that there is some such definition. For some general relativity texts discuss the initial value problem for general relativity; and they prove a 'uniqueness up to isomorphism' result, suggesting that determinism in some decent sense holds good. And indeed, by abstracting from the details of general relativity, one can extract Dm2 from these texts. (For the details of the extraction, see Butterfield (1987: 17-19, 26-29). The definition is:-

Dm2 A theory with models $\langle M, 0_i \rangle$ is S-deterministic, where S is a kind of region that occurs in manifolds of the kind occurring in the models, if:
given any two models $\langle M, 0_i \rangle$, $\langle M', 0_i' \rangle$ containing regions R, R' of kind S respectively, and any diffeomorphism d whose domain of definition includes R and which maps R onto R':-
if $d^*(0_i) = 0_i'$ on $d(R) = R'$, then:
there is an isomorphism f from M onto M' that sends R to R', i.e. $f^*(0_i) = 0_i'$ throughout M' and $f(R) = R'$.

This differs from Dm1 in three ways. First, there is no use of the idea of an absolute object: all diffeomorphisms are considered. Second, the diffeomorphism d assumed to exist (i.e. given by the antecedent) need not be global; it need only be defined on R or some superset of R. Third, the (global) isomorphism f that the consequent asserts to exist need not extend d; that is, it need not agree with d on d's domain of definition. Thus even if the given diffeomorphism d is global, Dm2 does not reduce to Dm (or Dm1 in application to theories with no absolute objects), with its single diffeomorphism in antecedent and consequent. In this case, Dm2 is weaker than Dm (and Dm1), since its consequent replaces reference to d by an existential generalization. Indeed, we cannot require that f extend d on all of d's domain, on pain of having Dm2 violated by any pair of hole diffeomorphs. For given one model, the identity map id on it is global, so that if f extended id, then f would equal id, and Dm2 would reduce to Dm (and Dm1) and so be violated by a pair of hole diffeomorphs. I thus claim that Dm2 is the general definition of determinism implicit in modern presentations of general relativity's initial value problem, and more generally, is suited to theories of the second kind. (Note added in proof: David Malament and John Norton have pointed out to me that general relativity obeys not only Dm2 but also the stronger definition that conjoins to Dm2 the requirement that f equal d on R. But I still claim that Dm2 is a suitablel definition for theories of the second kind.)

I can clarify Dm2 and defend this claim by considering an objection; (I thank Roberto Torretti for correspondence). The objection uses the now familiar device of composing two diffeomorphisms in opposite direction between two models so as to give a symmetry. Thus

Dm2 entails that the restriction of f to R, call it f_R, is such that its drag-along f_R^* agrees with d^* on all the 0_i. That is, at all points $d(p)$ of R', $f_R^*(0_i)$ at $d(p) = d^*(0_i)$ at $d(p)$. For both must equal $0_i'$ at $d(p)$! (Unless f equals d on R, f_R^* doesn't equal d^* as a function, of course.) So if f and d are not equal on R, then this implies in turn that $f_R \circ d^{-1}$ is a symmetry of the sub-model $\langle R', 0_i' \rangle$. So if f does not equal d on R, then $\langle R', 0_i' \rangle$ has a symmetry; and since $\langle R, 0_i \rangle$ and $\langle R', 0_i' \rangle$ are isomorphic, this implies that $\langle R, 0_i \rangle$ has a symmetry. To put it contrapositively: if we are given that $\langle R, 0_i \rangle$ has no symmetries, then f agrees with d on R.

This point can lead one to suspect that for a theory of the second kind to violate Dm1 and satisfy Dm2, it must have models with regions R, with symmetries. Thus one might argue, first, that for such theories, with no absolute objects, the initial quantifier is in effect the same in the two definitions, viz. "for any global diffeomorphism d". The fact that in Dm2, d need only be locally defined should make no difference, since if two models are diffeomorphic as asserted by the consequent, then one can always extend a locally defined diffeomorphism to a global one. And so one might argue, second, that the difference between the definitions must turn on Dm2's existential quantification in its consequent; that is, on Dm2's introducing f which is allowed to differ from d. And so one might cite the previous paragraph to argue, third, that violating Dm1 and satisfying Dm2 requires models with regions R with symmetries. That is, one might argue for the claim: any pair of models $\langle M, 0_i \rangle$ and $\langle M', 0_i' \rangle$ with regions R and R', that satisfy the antecedents of Dm1 and of Dm2, and which satisfy the consequent of Dm2, but do not satisfy the consequent of Dm1, are such that $\langle R, 0_i \rangle$ has symmetries.

If this claim were true, then my dialectical position against Norton and Earman would be much weakened. For they can just give me a model $\langle M, 0_i \rangle$, with an R with no symmetries, produce another model by a hole diffeomorphism on a hole ahead of R; (in the new model R' is isomorphic to R, as well as identical to it as a point-set, and so has no symmetries); these two models violate Dm1 by our earlier argument, using the identity map. And so if the claim were true, Norton and Earman would have given two models which also violate Dm2 (by the contrapositive of the claim). In short, if the claim were true, Norton and Earman could easily produce two models which violate Dm2 no less than they violate Dm1.

But the claim is false. (Indeed the argument just given shows it must be, once we assume that general relativity satisfies Dm2: an assumption based on the fact that Dm2 was extracted from general relativity's initial value problem. For general relativity does have models with R lacking symmetries, and so pairs of models of the kind described). The argument for the claim goes wrong at the third step. To see the error, recall that the contrast between Dm1 and Dm2 lies in whether the map that gives the global match (M to M') must equal the map d that gives the local match (R to R'). Dm1 says Yes; (and so d is assumed globally defined, though assumed to give matching only locally). Dm2 says No: the idea is 'why commit yourself in seeking a global match, to global maps that equal d on d's domain of definition? Perhaps, d is good for a local match but no good as part of a global match.' Now we can see the error in the third step of the argument: the claim focuses this contrast on R; but the contrast is not just for R but for the entire domain of definition of the map d. That is, the claim proposes that for Dm1 and Dm2 to be respectively violated and satisfied by a pair of models, the global map f must differ from d on R (and so induce a symmetry in R' and so R). But f can differ from d elsewhere than on R; it can differ from d on other parts of d's domain; and by differing, hold out the prospect of Dm2 being satisfied while Dm1 is violated. This is exactly what happens with the two models that we imagined (in the previous paragraph) Norton and Earman giving to me. $\langle M, 0_i \rangle$ has an R with no symmetries. $\langle M', 0_i' \rangle$ is produced by a hole diffeomorphism. Since R has no symmetries, R' has no symmetries, and f cannot differ from d on R. But f can differ from d elsewhere; and so Dm2 can be satisfied while Dm1 is violated (with d as the identity map).

To sum up this Section: Dm2 provides a precise sense of determinism, in which determinism is not violated by hole diffeomorphs. And that completes the technical aspect of my reconciliation of substantivalism with determinism. But for philosophers, questions will remain: does Dm2 capture the basic idea of determinism? Or at least, does it do so as faithfully as Dm1? And how does Dm2 relate to substantivalism? Fortunately, we need not worry about getting consensus on what is the basic idea of determinism. For as mentioned in Section 1, all participants in this debate agree on it: a single physically possible world is specified by the physical state on a certain region of spacetime. But since this Section's definitions are cast in terms of models <M, O_i> rather than possible worlds, this consensus means that to tackle the philosophical questions I need to consider the relation between models and possible worlds: see Section 3.

3. Models and Worlds

This is not the place to defend the use of possible worlds, or to enter the debate about realism concerning them (cf. Lewis 1986). But I do need to address three questions. What is the relation between models and worlds? Does Dm2 explicate the basic idea of determinism at least as faithfully as Dm1? (If not, the technical victory of Section 2 will be hollow. Readers who skipped Section 2 need only know that Dm1 is violated by any pair of hole diffeomorphs, and Dm2 is not.) And can a substantivalist endorse Dm2? The discussion of each question will lead into the next. And the third question will lead into Section 4, where we will see that it is connected to what the substantivalist says about the transworld identity of spacetime points.

I need to distinguish between physically possible worlds, and models (tuples) that purport to represent them. This will not prejudge the realism debate: it may be that possible worlds are ersatz objects, it may be that they are set-theoretic tuples. But even if this is so, we can ask about the relation of representation between tuples of the kind, <M,O_i>, that occur in discussion of spacetime theories and physically possible worlds. In particular, in assessing the hole argument, we must consider whether this relation is one-one, or one-many (some tuple represents more than one world), or many-one (some world is represented by more than one tuple), or many-many (both one-many and many-one). But we need to set aside one issue which affects this relation, but is irrelevant to us: the issue whether the physics expressible in terms of the O_i of a given spacetime theory determines all the facts about a possible world that obeys the theory. To say Yes is to espouse a strong determinationist physicalism. To say No means that in general one tuple <M, O_i> represents different worlds, so that the relation is one-many (it may also be many-one). To simplify discussion, I shall write as if we say Yes. (Norton and Earman tacitly adopt the same simplifying policy.) Given that we say Yes, I see no other reason why the relation should be one-many: so that in effect the options are now, one-one or many-one.

Norton and Earman take the relation to be many-one. They endorse their 'Leibniz Equivalence' (1987: p. 522; Norton, this volume, p.58), which says that any two isomorphic models represent the same possible world. Thus all and only the models in an isomorphism equivalence class represent the same world. They also make two further claims. First, that the practice of physics is to endorse Leibniz equivalence. Second, that the substantivalist must deny it. Indeed, as we saw in presenting their argument in Section 1, they claim that the substantivalist must deny the special case of it, where the two models are hole diffeomorphs; (this is a special case since the two manifolds share the same base-set of points, which isomorphic models in general do not).

I agree with the two further claims. As to the first, physics texts do not usually distinguish models and physically possible worlds (as Maudlin emphasizes; this volume, p. 82). But the more careful modern texts of general relativity do so, and they endorse Leibniz equivalence (Hawking & Ellis 1973: 56, 227-8; Sach and Wu 1977: 27). As to the second, the substantivalist believes in points. And belief in points implies belief that a possible world fixes its population of points and their properties and relations; so models distributing such properties and relations differently simply cannot both represent the same world — even where the base-sets of points are the same. (So I admit that there is a clash between the relativity texts, and substantivalism; but I will hold that this clash is less sharp than one facing Norton and Earman.)

Yet I deny that the substantivalist (or that matter, any denier of Leibniz equivalence) is committed to radical indeterminism. And not just because they can resort to the technical definition Dm2. Even using the basic idea of determinism, denying Leibniz equivalence does not commit one to denying determinism! The reason is that one can deny Leibniz equivalence in two different ways. Indeed, one can deny the special case where the isomorphic models are hole diffeomorphs (and so have the same base-set of points), in two different ways. The first way, tacitly assumed by Norton and Earman, leads to indeterminism; but the second way does not.

To see how this works, it is best to keep things simple by concentrating on the special case relevant to the hole argument: the case of two isomorphic models with the same base-set of points. In denying that two such models represent the same world, one has two options. First, (Each): in general, *each* of the two models represents a different world. Second, (One): in all cases, *at most one* of the two models represents a world. Note that in (Each), 'in general' has its usual semi-technical meaning: 'in some, perhaps the typical, cases', not 'universally'. On the other hand, (One) says 'in all cases'. So (Each) and (One) are mutually exclusive and jointly exhaustive ways to deny that two such models represent the same world; and (One) is logically stronger than (Each).

If we say (Each), and some pairs of hole diffeomorphs are among the 'some cases' in which each represent, then such pairs clearly violate the basic idea of determinism. And this is of course the crux of Norton and Earman's argument: they tacitly assume that by denying Leibniz equivalence, a substantivalist is committed to taking each of any pair of hole diffeomorphs to represent different worlds. (Though it would be enough for them that each of some pairs do so.) But if we say (One), then no pair of hole diffeomorphs violates the basic idea of determinism. To put it more positively: any pair accords with determinism, because at most one member of the pair represents a world. Indeed, the bearing of hole diffeomorphs on the basic idea of determinism is exactly as it is for someone who endorses Leibniz equivalence: for both, there is no threat to determinism — there is at most one world at issue, and thus no prospect of two worlds disagreeing on a hole.

(One) may at first seem an unmotivated escape from Norton and Earman's argument; it is, after all, logically stronger than (Each), and so initially less plausible. But I claim that there is good reason for it, quite apart from securing the reconciliation of substantivalism with determinism. Section 4 will consider two separate arguments for it.

But before doing that, we can already partially assess Dm1 and Dm2 as explications of the basic idea of determinism. Since these definitions are cast in terms of models, assessing them requires us to ask whether two hole diffeomorphs represent the same world. It is easy to show that is we answer Yes to this question, or answer No and take option (One), then Dm2 is a better explication of the basic idea than Dm1; while if we answer No and take option (Each), then Dm1 is the better explication. Suppose we answer Yes; so that for hole diffeomorphs there is only one world at issue and thus no threat to the basic idea of determinism. Yet Dm1 is violated by hole diffeomorphs , and Dm2 is not. So this Yes answer should prefer Dm2 as explicating the basic idea. Similarly, if we answer No and take option (One): at most one world, no threat to the basic idea, Dm2 to be preferred. On the other hand, if we answer No and take option (Each), hole diffeomorphs violate the basic idea, and of course Dm1 but not Dm2; and so Dm1 is to be preferred. To sum up: Norton and Earman, and those who agree with me in taking option (One), will concur in preferring Dm2 to Dm1. (Champions of Dm1 might conclude 'So much the worse for answering Yes, and for taking option (One)'. But I doubt that there are such champions. Dm1, with its use of the imprecise idea of absolute objects, has no special claim to be the correct explication of the basic idea.

And as hinted at the end of Section 1, even substantivalists who say they are happy to rule out determinism at a stroke, since determinism is just a technical feature of theories, should surely take some solace from the existence and non-violation of Dm2.)

Before I justify (One), there are two final points about substantivalism having to answer No: having to deny the special case of Leibniz equivalence where the two isomorphic models have the same base-set of points. The first is short. I promised to argue that the clash between physics' practice in endorsing Leibniz equivalence and substantivalism is less sharp than another clash facing Norton and Earman. The point is this: Leibniz equivalence implies that each model in an isomorphism class has redundancy in the way it represents the possible world common to the class: the membership of the base-set, and the way the properties and relations are distributed among these members, are an artefact of the representation. So one faces a challenge: rewrite spacetime theories so as to eliminate the redundancy: that is, give a direct account of worlds, and show that the models in an isomorphism class arise as equivalent representations of a single world. Earman (1977) and Stachel (1985, sect. 6) have both risen to this challenge. Earman has explored a way of writing spacetime theories without reference to points. And Stachel has explored a way of retaining points, while eliminating the redundancy by incorporating the arbitrary choice, of which point is the locus of a given physical point-event, into the representation of the point-event. Thus Stachel takes each point-event to be a map sending each point of the manifold to a set of geometric objects at that point. I believe that both approaches may well be workable; but I cannot discuss them here — see my (1989,p. 14-15). Suffice it to say here that both approaches clash with the practice of the physics texts, with their explicit and frequent quantification over points. I submit that this clash is sharper that the clash between the texts' endorsement of Leibniz equivalence and substantivalism's denial of it.

The second point is that one might even suggest that substantivalism can endorse the special case of Leibniz equivalence where the two isomorphic models have the same base-set of points. (Mark Wilson suggests this in correspondence; my thanks to him. He gives a technical motivation which for lack of space I cannot discuss). Clearly, this suggestion promises the substantivalist a very rapid escape from the hole argument! Can it work? I will briefly argue that if it is filled out in what I think is the natural way, then it has two minor disadvantages. But I agree that it may well be workable — it certainly deserves more investigation.

The suggestion is: the substantivalist can hold that a single possible world is represented by an entire equivalence class of isomorphic models with the base-set — and there is no need to find a new account of worlds, which eliminates the apparent redundancy in the representation given by any single model. Filling this out, I think the suggestion must agree that for a substantivalist, a possible world fixes a population of points and a painting of properties and relations on them. (Without that, the substantivalist does not really believe in points). And of course it must agree that an object cannot have contradictory properties. So how does it answer the accusation that for it, the contrary paintings of geometric objects in different models represent the attribution of contradictory properties, in a single world, to spacetime points?

I think it can do so by saying that points' properties and relations are far more conjunctive and extrinsic than we usually think. To see how this works, consider the simple case with just one scalar field as the only geometric object considered by the theory. Then the suggestion is: there is only one property assigned to a point p by the entire equivalence class of models; viz. the property of having value x_1 in model m_1, i.e. when the point q has value y_1, r has z_1, etc., and of having value x_2 in model m_2, i.e. when the point q has value y_2, r has z_2, etc., — and so on through all the models. Similarly in cases with more geometric objects: the only properties and relations attributed are highly conjunctive ones, with the different conjuncts picking up the paintings (which we naively read as a complete property attribution!!) in different models of the equivalence class.

There is an interesting analogy here with Benacerraf's well-known discussion of different identifications of natural numbers with sets (1965). Recall Benacerraf's basic point: 0 can be identified with a set in many different equally good ways. Two well-known ways are: 0 is identified with { }, when 1 is identified with {{ }} etc.; 0 is identified with the set of all sets with no members, {{ }}, when 1 is identified with the set of all unit sets, etc. Benacerraf concludes that no identification of numbers with sets is correct, indeed that numbers are not objects. But the conclusion that is analogous to the suggestion above would be as follows. Don't conclude that 0 and 1 are not objects; or that our arithmetic needs reformulation to prevent singular terms standing for numbers, and to expunge other indications of ontic commitment to numbers as objects (cf. Earman and Stachel above!) Numbers are indeed objects; but they have highly conjunctive properties, a different conjunct for all the ways we naively think of them as being identified with sets. So 0 has no members, when 1 has a single member viz. { }, and 2 has a single member viz. {{ }} etc.; and 0 has one member vis. { }, when 1 has the host of unit sets of members, when 2 has the host of duos as members; and so on through all the possible identifications of numbers with sets. In short, the analogy is: an identification of the whole numbers with sets is like a spacetime theory's model.

This suggestion has two minor disadvantages. First, it pays a price for its very rapid escape from the hole argument. Recall two philosophical points. First: for most substantivalists, spacetime points are not only objects, but also the basic objects of the ontology of spacetime theories. Indeed, for many substantivalists, points and perhaps regions (which may harmlessly be regarded as mereological fusions of points) are the only ontology of these theories: the so-called geometric objects of these theories, i.e. the scalars vectors etc., are taken as properties of points, properties of properties of points etc.. Second: scientific realists typically think of theories as attributing to their objects basic theoretical properties that, though they might be esoteric (e.g. spin, charm), are intrinsic to the objects and have a relatively simple representation in the formalism . (Admittedly, what it is for a property to be intrinsic is obscure—I like the theory of Lewis (1983); and 'relatively simple' is vague. But I think the point holds good.) The suggestion's disadvantage is now clear. It makes substantivalism swim against the current; it makes substantivalism take the basic theoretical properties of points to be highly extrinsic, and to be represented in the formalism by very long (uncountably infinite!) conjunctions. (But I should mention that Wilson's technical motivation may overcome this disadvantage.)

The second disadvantage is that the suggestion seems to collapse into the leading idea in Stachel's (1985, Section 6) rewriting of spacetime theories; the idea that each point-event is a map sending each point of the manifold to a set of geometric objects at that point. And while I have agreed that Stachel's rewriting may well be workable, such a collapse robs the suggestion of its distinctiveness. Thus, according to the suggestion, each equivalence class of models attributes to each point a conjunction of properties, with a conjunct for each model in the class. Similarly in Stachel's rewriting, pick a member of some point-event map. We can think of maps in the usual way, as sets of ordered pairs; so this member is an ordered pair of a point and a set of geometric objects located at the point. And now pick such a member (ordered pair) for all point-events, with no two members (ordered pairs) having the same first member. This collection is a model. And it corresponds to the suggestion's conjunct. The class of all such pickings is equivalent in information to the set of all Stachel's point-event maps; and it corresponds to the set of all the suggestion's conjuncts, and thus to the equivalence class of models. In short, there seems little to choose between Stachel's idea and the suggestion.

4. Justifying (One): Essentialism and the Denial of Transworld Identity

We can at last attack the question: can substantivalism justify (One)—the claim that at most one of two hole diffeomorphs represents a world? For clarity, it is best to argue first for something simpler: that some models fail to represent a world. Admittedly, endorsing

Norton and Earman's Leibniz equivalence enables one to avoid this argument; but my aim from Section 3 onwards has not been to convince all-comers, but to show the viability of (One). In fact we will see that the only way to avoid this argument is to endorse Leibniz equivalence; so the argument is as strong as I want.

The argument is based on a point which is almost universally agreed, whatever one's views about the nature of possible worlds; namely, that some objects have essential properties. Agreed, which objects have which essential properties is in general obscure. But some cases are clear: Hubert Humphrey is essentially human, or at least essentially animate; any molecule of H_2O is essentially a molecule of H_2O, or at least essentially a molecule; etc. And such cases mean that some models of our spacetime theories fail to represent a world. For as many authors emphasize, the manifolds in these models require nothing about the intrinsic nature of their points; all that matters is the structure of the manifold, and the further structure of scalars vectors etc. that get painted onto the points. So these theories have models with manifolds with Hubert Humphrey as a member of the base-set of points; and with molecules of H_2O as members. But given that Humphrey could not be a point, and nor could a molecule of H_2O, such models do not represent a possible world.

You might object that physical objects like Hubert Humphrey are never members of the base-set of a manifold of a model. That is, you might hold: although informal presentations of our theories require nothing about the intrinsic nature of the points, in a more formal presentation we should require all the points to be purely mathematical objects—say quadruples of reals, perhaps themselves identified with certain sets in a set-theoretic reconstruction. A similar position can of course be made out for pure mathematics. There also, the informal practice is to take a set and posit some structure on it; and from then on the structure, not which objects we happened to take as the set, is the focus of interest. And so you might hold: in a more formal presentation we should require all the objects treated in pure mathematics to be themselves mathematical—so a set-theoretic formalization of pure mathematics should use pure set-theory.

But there are two replies to this position. First, why should we require such uniformity? Even if uniformity is elegant, who is to say that it outweighs the merit of liberality? For in both areas—physics and mathematics—a more formal presentation can perfectly well take the liberal alternative: allow physical objects in—so that a set-theoretic formalization would use an impure set theory, having physical objects and perhaps also some primitively given sets in the bottom rank of its hierarchy. (This liberal alternative, with its use of physical objects, need not run the risk of having only finite sets—and thus of having to follow Russell and Whitehead in postulating an axiom of infinity. It need only posit what they did not: an infinitely high rank. Even with only finitely many bottom-rank objects, such ranks contain infinite sets.) The second, and more important, reply is that this position fails to block the argument's conclusion that some models fail to represent a world! The argument still goes through if one believes that some pure mathematical objects occurring in some manifolds cannot be points. I in fact believe more: that each pure mathematical object cannot be a point. Points are strange objects, to be sure; but they are not pure mathematical! (Incidentally, note a cousin of the above position: that the manifolds of our models should be subsets of R^4. Though irrelevant to us, this position is historically significant; Norton (1989) argues that prior to the definition of a differentiable manifold it was a natural position for Einstein—and that his views on such matters as general covariance arose from his adopting it.)

This second reply shows that the only way to avoid the argument is to say that all models represent worlds—even when their base-sets contain Hubert Humphrey, the number 1, or whatever. There are then in principle two ways to go. One is simple: endorse Leibniz equivalence, so that regardless of their base-sets any two isomorphic models represent the same world (and thus face the challenge described in Section 3, to rewrite spacetime theories). The other is very complicated; deny Leibniz equivalence and so embark on the game of specifying which pairs of isomorphic models represent different worlds, and for such pairs, which two worlds they represent. This leads to countless hard questions. If the set of actual points is the base-set of one model, while the union of {Humphrey} and this set is the base-set of another, do the models represent the same world? What if two models are exactly the same, except for a transposition of points; i.e. they have the same base-set, and the same painting of properties on the base-set's members except that two members are transposed? And so on. This is clearly a mug's game: one does better not to play. Thus I claim that Leibniz equivalence is the only way to escape the conclusion that some models do not represent a world; and so for present purposes, the conclusion is established.

To justify (One), we of course need something stronger than this conclusion; we need at most one of any two hole diffeomorphs to represent a world—even when the common base-set of the two models consists entirely of points, so that Humphrey and his ilk are not to blame for the failure to represent. There are two ways to get what we need: Maudlin's essentialism (Maudlin 1989; this volume); and the way I will prefer—denying transworld identity.

We have already seen the basic idea of essentialism: some models fail to represent a world, because they attribute to points properties or relations that are impossible for them. To justify (One), essentialism must claim for points a collection of essential properties and relations, rich enough that a hole diffeomorphism applied to any representing model gives a non-representing one. There is no technical problem here: since a hole diffeomorphism shuffles all those properties and relations of points in the hole that are coded by geometric objects, essentialism can get a rich enough collection by choosing one or more geometric objects. Whichever it chooses, (One) follows: with the phrase 'at most' being operative—there will of course be countless pairs of hole-diffeomorphs (even with a common base-set entirely of points) neither of which represents a world since both attribute impossible properties! But what about the philosophical issue: which geometric objects can plausibly be held to code essential properties of points? Maudlin argues strongly in favor of the metric field(s) and connection. For a discussion of his position (and other varieties of essentialism), see my (1989, §5). Here I have space only for two tasks. First, giving what I take to be our joint replies to Norton's two arguments (this volume) that (1) essentialism about the metric, and any similar appeal by the substantivalist to structure beyond the diffential structure, will founder on symmetries: and (2) that (One) founders on the impossibility of knowing which of two hole diffeomorphs represents. Second, summarizing my disagreement with Maudlin, thus motivating my preferred justification of (One)—denying transworld identity.

Norton's first argument (this volume § 4, pp. 60 ; cf. also Earman and Norton 1987, pp. 519-520, p. 522 fn. 2) relates to the point we rehearsed at the start of Section 2: that the hole argument applies only to theories of the second kind (which he calls 'local'). That is: if all the models of the substantivalist's favored spacetime theory have some common structure, then the substantivalist can in general avoid the hole argument by writing his theory as positing a single manifold plus common structure, the same in every model. The classic case is where all the models are isometric; so that the substantivalist can posit a single manifold and metric structure as a fixed canvas on which matter fields get painted. But—says Norton's argument—if this common structure has symmetries, as it usually does, then the threat of indeterminism can be resurrected. The idea of the argument is: (i) take a model, and the model produced by dragging-along with the symmetry; (ii) argue that for the substantivalist they represent two different worlds; (iii) remark that the symmetry can be factorized into a product of two maps; (iv) consider the intermediate model produced from the original one by dragging-along with one of the factor maps, and argue that whatever world it represents (even a world other than the two mentioned in (ii) there is indeterminism. I reply: the substantivalist can and should say: the intermediate model represents no world at all—even if (ii) holds good so that the first two models each represent a world. This reply follows immediately from essentialism about the properties coded by the common structure which has the symmetry. For example, take the case of metrical essentialism, so that the symmetry is a metric symmetry (Norton himself chooses this case). Then the first two models agree exactly on each points' metrical properties and relations (so neither or both is faithful to points' essential metrical properties and relations). And both disagree in this respect with the intermediate model. So given that one of the first two models represents a world, as (ii) requires, the intermediate model does not. It no more represents a possible world than does Humphrey's being a point.

Norton's second argument (addendum, p. 62) is against (One). He says, first: (One) contradicts 'local' i.e. second-kind theories, which by definition take all structures of the appropriate type as models. And second: even if we are willing to abandon local theories, so as to give (One) a chance, how could we ever know which of two hole diffeomorphs represented a world and which did not—which is the 'real model' and which the 'impostor'? By now, after our exercises in modal metaphysics, the reply is clear. The first point involves no real contradiction; someone who endorses (One) can of course use local theories with their plethora of models, though they must then distinguish models and worlds—since not all models represent a world. But as argued at the start of this Section, that is nothing extraordinary; only endorsers of Leibniz equivalence can avoid it. As to the second point, I agree there is a deep general question about how we know, or have rational belief in, modal propositions, such as that Humphrey could not be a point. But there isn't a specific problem for the substantivalist trying to justify (One): she simply appeals to such modal beliefs so as infer that some models are non-representing. And as we have seen, she only needs one geometric object to code essential properties of points to be able to infer (One). (In this second argument, Norton writes as if only I, not Maudlin, endorse (One); but I take this reply to be common between Maudlin and me.)

I turn to my disagreement with Maudlin. Namely: his essentialism cannot handle a theory with non-isometric models. For his essentialism concerns only the actual world's points. He claims only that the actual points have their metrical properties and relations to one another essentially, so that a possible world containing the actual points must be isometric to be actual world. So Maudlin cannot save modern-day substantivalism based on our current best spacetime theory, general relativity, which has pairs of models that are non-isometric, indeed non-diffeomorphic. (The same problem afflicts other non-metrical essentialisms: whichever geometric objects are held to code essential properties, general relativity has pairs of models which are not isomorphic with respect to these objects.) Maudlin is of course aware of this; and he briefly suggests (1989, p. 37-38) that the substantivalist should handle non-isometric models, by denying transworld identity for the points concerned and instead using counterparts: a point in a world that is not isometric with the actual world is not identical with any actual point, but at best a counterpart of it in the sense of Lewis (1968; 1986, Chapter 4). I agree with Maudlin that this is the best response the essentialist can make; see my (1989, pp. 19-22) for arguments against other responses. But non-isometric models are endemic to general relativity, and there will be countless possible points that are not identical with any actual point. I think we do best to come clean, and make a more radical denial of transworld identity: any point is an inhabitant of just one possible world.

Thus I propose that any point is a part of just one possible world. (It is of course a set-theoretic constituent (member, or member of a member, or ....) of many base-sets, and so many manifolds, and so many models.) Similarly for mereological fusions of points, i.e. spacetime regions. This proposal is of course inspired by Lewis (ibid.), who holds that no object (point or otherwise) occurs in any two worlds. This proposal will clearly secure (One). My (1989, § 6) defends it in three stages, which I can only summarize here. First, I connect it to the notions which for Lewis take the place of transworld identity: counterparts and duplicates. The proposal turns out to use a hybrid of Lewis' notions a hybrid that (like counterparts, unlike duplicates) involves denial of transworld identity, and no commitment to Lewis' theory of natural properties; and that (like duplicates, unlike counterparts) is precise and matches objects between worlds in terms of their intrinsic properties. So I have a choice of terminology: I prefer to use 'counterpart'. Second, I cast the definition of spacetime regions agreeing on their physical state, and so the definition of determinism, in terms of spacetime regions being counterparts (understood as the hybrid notion). In short: a diffeomorphism d becomes a mode of comparison between regions so that regions R and d(R) are counterparts relative to d iff the dragged-along objects $d^*(O_i)$ coincide throughout the image-region with the given objects O_i'. It turns out that Lewis himself gives a definition of detterminism that is essentially equivalent to Dm2, the definition which Section 2 showed the substantivalist needs to endorse so as to avoid the hole argument: a happy agreement. Third, I argue that this proposal has two advantages over essentialism—or any doctrine of the transworld identity of points. (1) For any such doctrine, the definition Dm1 seems to have a merit that Dm2 lacks—so that any such doctrine has trouble accommodating the needed definition Dm2. (2) The proposal accommodates the intuition that in Leibniz's thought-experiment about translating the material contents of the Universe three feet East, we must also identity the points according to the matter that inhabits them, so that the translation does not produce another world. It accommodates this intuition by the fact that all the fields O_i count towards counterparthood.

So much for advertising my proposal. Let me end with one remark motivating the proposal; and a reply to an objection. First, note that the proposal does not depend on Lewis' controversial realism about possible worlds! This independence does not turn on the nature of points: for any kind of object, counterpart theory has significant advantages over doctrines of transworld identity—not only on Lewis' realistic conception of worlds (1986, p. 198f.) but also on the various *ersatz* conceptions of them (pp. 194-197). So I urge acceptance of counterpart theory on *ersatzers*.

Second, you might object that by taking all the fields O_i to count towards counterparthood, I cannot capture the idea of a dynamical metric (a metric affected by matter) in the natural way: namely by saying of some point within some world, that in a world with a different mass distribution, the counterpart of the point has a different curvature. For different curvature implies not being counterparts! Similarly for counterparthood among regions. (I thank John Earman for this objection.) My reply goes back to Lewis' distinction between counterparts and duplicates. He argues that modal discourse exploits various different counterparthood relations, which weigh respects of similarity differently from one another, and that the respects of similarity are in general extrinsic. (He contrasts this with the unique relation of duplicatehood, for which all and only intrinsic properties count.) Thus I reply that when someone describes a dynamical metric by saying 'if the mass distribution were different, then the (strictly: counterpart) point would have a different curvature', they are simply using a counterpart relation more extrinsic than the one used in the definition of determinism. The point with the different curvature is a counterpart of the given pont, despite the difference in curvature, because counterparthood is cueing into extrinsic properties like the nature of the surrounding points, and what objects are at them. The same reply works for regions. Indeed, we get exactly this situation in everyday examples where alteration of one variable lawlikely alters another. Here I am, brown-eyed (an intrinsic property of me). I have many counterparts that are brown-eyed (say, counterparts by being offspring of counterparts of my parents—thus showing extrinsicness of the counterpart relation at issue). It is also true that if the eye-color gene of both my parents were blue, then my (strictly: my counterpart's) eye-color would be blue. Thus the counterpart can have different intrinsic properties such as eye-color, and yet be a counterpart, when counterparthood cues into extrinsic properties. To sum up: suppose I had simply chosen the other terminology, and used 'duplicate' not 'counterpart' for the hybrid notion my proposal needs; then I could use 'counterpart' in the way the objector wants to, and so say just what he does.

Notes

[1] I would like to thank my co-symposiasts, Philip Catton, John Earman, David Lewis, David Malament, Michael Redhead, Paul Teller and especially, John Norton and Roberto Torretti, for discussions and correspondence.

References

Benacerraf, P. (1965). "What Numbers Could Not Be", *Philosophical Review* 74: 47-73.

Butterfield, J. (1987). "Substantivalism and Determinism", *International Studies in the Philosophy of Science* 2: 10-32.

_ _ _ _ _ _ _. (1989). "The Hole Truth", *British Journal for the Philosophy of Science* 40: 1-28.

Earman, J. (1977). "Leibnizean Spacetimes and Leibnizean Algebras", in *Historical and Philosophical Dimensions of Logic, Methodology and Philosophy of Science*, R. Butts & J. Hintikka (eds.). Dordrecht: Reidel, pp. 93-112.

_ _ _ _ _ _ , (1986). *A Primer on Determinism*. Dordrecht; Reidel

_ _ _ _ _ _ . and Norton. J. (1987). "What Price Spacetime Substantivalism? The Hole Story", *British Journal for the Philosophy of Science* 38: 515-525.

Friedman, M. (1983). *Foundations of Spacetime Theories*. Princeton: University Press.

Hawking, S. and Ellis, G. (1973). *The Large-scale Structure of Spacetime*, Cambridge: University Press.

Lewis, D. (1968). "Counterpart Theory & Quantified Modal Logic", *Journal of Philosophy* 65: 113-126.

_ _ _ _ _ _. (1983). "New Work for a Theory of Universals", *Australasian Journal of Philosophy* 61: 343-377.

_ _ _ _ _ _. (1986). *On the Plurality of Worlds*. Oxford: Blackwells.

Maudlin, T. (1989). "Substances and Space-Time: What Aristotle Would Have Said to Einstein", forthcoming in *Studies in the History and Philosophy of Science..*

_ _ _ _ _ _ _. (1989). "The Essence of Space-Time", *PSA 1988,* Vol 2: 82-91.

Norton J. (1989), "Coordinates and Covariance: Einstein's View of Spacetime and the Modern View", forthcoming in *Foundations of Physics*.

_ _ _ _ _ _. (1989), "The Hole Argument", *PSA 1988*, Vol. 2: 56-64.

Sachs, R. and Wu, H. (1977). *General Relativity for Mathematicians*, London: Springer.

Stachel, J. (1985), "What a Physicist can Learn from the Discovery of General Relativity", *Proceedings, 4the Marcel Grossmann Meeting on Recent Developments in General Relativity*, Rome, Italy, 17-21 June.

The Essence of Space-Time

Tim Maudlin

Rutgers University

The debate over the ontological status of space has been waged over the millennia on many battlefields. In ancient times the major focus was on the possibility of a vacuum, an entity that was both something and yet, in another way, nothing. Newton shifted the scene of action to dynamics, to the question of the ontological commitments needed to give an adequate explanation of observable phenomena. But along with the dynamical effects of Newtonian absolute space came also the spectre of motions unobservable in principle, a notion attacked from Leibniz onward as conceptually absurd or meaningless. So the theater of operation was moved again, especially by verificationists, this time into theories of meaning. Now John Norton and John Earman have opened an entirely new line of assault on the question. They argue that considerations about the possibility of determinism have direct implications for the substantivalist-relationist controversy. The linkage to determinism is made via the hole diffeomorphism argument which Prof. Norton has just outlined. Before the substantivalists abandon their trenches, I would like to analyze the tactics of this new incursion and probe for its weaknesses.

The effectiveness of the hole argument depends upon the concession that the diffeomorphs represent, under some interpretation, ontologically distinct, possible states of affairs. For they then would depict contrasting possible world histories that are entirely identical up to some moment yet diverge afterward. According to Norton and Earman, substantivalism about space-time commits one to endorsing an interpretation under which the each model represents a distinct possible state, and so leads inevitably to indeterminism. Relationists, in contrast, may regard the diffeomorphs merely as different representations of the same situation, thus avoiding the dilemma. Before we can evaluate this claim we must consider how it is that mathematical objects represent the world in the first place.

In contemporary physics, the nature of the relationship between mathematical entities and the physical realities they represent is easily overlooked due to a systematic ambiguity of usage. For example, one might read at one time that Brans and Dicke postulate an energy-carrying scalar field and at another that a scalar field is a map from space-time points into the reals. This hardly makes Brans and Dicke into Neo-Pythagoreans who believe that mathematical maps can carry energy. Rather the gap between the mathematics and the physical objects has been submerged by a usage of "is" which really means "is represented by". The depths of the submersion are exposed when we consider exactly which use of "is" is elliptical. Is a scalar field really a physical entity which has one degree of freedom that exists at every spatio-temporal point, so constituted as to be best

PSA 1988, Volume 2, pp. 82-91

represented by a map from the points into, e.g., the reals? Or is a scalar field the mathematical map itself, properly constructed to represent certain sorts of physical entity? Obscurities like this pervade physicist's locutions.

As an illustration, consider this example of the blurring between physical objects and mathematical representations which can be found in Robert Geroch's *General Relativity from A to B* (Geroch 1987, p. 8). Introducing the notion of space-time, Geroch writes:

> Here and hereafter, we shall denote by M the set of all possible events in our universe: all those events that have occurred in the past, all those occurring now, and all that will occur in the future; those in this room, in our solar system, in other galaxies. This one enormous set M will be called space-time.
>
> A point of M, then, represents an event...

Having first insisted that the points of M *are* events, Geroch immediately speaks of them as *representing* events, eliding the link that connects the mathematical order to the physical.

That link becomes critical for the problem at hand. For if we fail to distinguish the representation from the objects represented we might simply accept without further examination that the diffeomorphs, which are distinct *mathematical* objects, must each correspond to a different state of affairs. Or at least we might infer that the substantivalist, being committed to the existence of physical space-time points, must regard the two models as presenting distinct possibilities. Yet explicit acceptance of space-time points into one's ontology need not entail such a commitment, as the following case illustrates.

Imagine someone who is a substantivalist about space-time in virtue of these theses. First, she asserts that there are physical space-time points and allows, in logical reconstructions, variables to range over them. According to Quine's dictum, this puts the points into her ontology. Further, she explicitly states that the event locations (space-time points) are substances, capable of existing independently of other physical entities, subjects of predication, substrata for properties, &c. By whatever direct manner you please, she places the event locations in her ousiology, into the category of substance. However, she also maintains that mathematical points in solutions to the field equations *represent* these physical event locations by functioning as existentially quantified bound variables. She interprets the mathematical objects by Ramsifying out reference to specific event locations. So the metric tensor being flat at some point in the mathematical model only asserts of the physical world that *some event location or other* has no gravitational field. If a neighborhood of points in the model has a constant curvature tensor, that implies only that some neighborhood of physical points is constantly curved. Since she allows the variables to range over physical space-time points, she will regard the mathematical model as either being true or false. Of course, if the physical universe has any symmetries, a true mathematical representation will have many interpretations that satisfy it.

I have invented this hypothetical substantivalist to demonstrate that substantivalism *per se* does not guarantee the applicability of the hole dilemma. For under this interpretation the diffeomorphs do not represent distinct possible states of affairs. Although the diffeomorphs are separate *mathematical* objects, assigning different tensors to individual mathematical points, the differences wash out when we give the physical interpretation. Since the mathematical points are replaced by variables, and since the diffeomorphs can be generated from one another simply by renaming (re-coordinatizing) points, the two mathematical structures have the same physical content. So our substantivalist escapes the implication that the diffeomorphs represent distinct, possible physical situations by evading the requirement of distinctness. On this view, there is only a passive and no active interpretation of the diffeomorphism. Since the two models do not present alternative possibilities, determinism survives.

A little reflection shows, however, that this brand of substantivalism buys its solution to the hole dilemma at too high a price. For the representational capacity of the physics so construed is too impoverished to accommodate the central case in the substantivalist-relationist debate, viz. the Leibniz shift. Take as a paradigm Leibniz shift the displacement of all physical objects in a Neo-Newtonian or Minkowski space-time 3 meters to the North. According to the substantivalist such a shift is metaphysically possible and would result in an ontologically distinct state of affairs.[1] Further, due to the space-time symmetries of the physical laws, such a shifted situation is physically possible if the unshifted is. Now it is not clear whether there is any universal analogue to the Leibniz-shift operation in the General Theory of Relativity, where the space-time usually does not admit of symmetries. But perhaps in certain situations in the GTR, and certainly in the Newtonian and Special Relativistic regimes, the substantivalist must contend that the Leibniz shift is a metaphysical possibility.

Unfortunately, the Ramsifying gambit washes out differences in physical content not only between diffeomorphs, but also between Leibniz-shifted models. If we move all of the mathematical structures that represent physical objects some specified fixed amount in the mathematical model and then Ramsify the result we get the same physical content as the original. The two models will either both have interpretations that make them true or neither will, so they fail to depict alternative possibilities. Our Ramsifying substantivalist cannot maintain a difference in ontological content when she needs it.

The problem is that in specifying the Leibniz shift we must refer to physical event locations not via bound variables but by name. We want to say that in the shifted situation objects that are *here* (or objects qualitatively identical to them) would be *there*. If one really believes in event locations, believes that there is a deep ontological fact about at which space-time point a particular event occurred, then one ought to be able to discuss the possibility of that event (or one qualitatively identical) occurring somewhere else.

From this perspective, the Ramsifying solution to the hole dilemma seems highly artificial. After all, if event locations are fully in the ontology, why should we not be able to refer to them as specific individuals? The restriction to bound variables simply has no reasonable justification within the substantivalist program. So we should allow the mathematical points in our solutions of the field equations to act as *names* of physical space-time points. We now recover the ontological distinguishability of the Leibniz-shift case, for under the shift particular mathematical points come to be assigned new tensors, where the tensors represent physical fields. If the mathematical point names a physical location, then the original model and the Leibniz-shifted one make different physical claims.

But in recovering the Leibniz-shift case we have also impaled ourselves again on the horns of the hole dilemma. For the hole diffeomorphism also changes the field tensors assigned to particular mathematical points. If the point acts as a name of an event location then the diffeomorphs describe incompatible physical states and indeterminism threatens. Have we made any progress?

So far, our main progress has come in seeing how the hole dilemma turns on a particular picture of the way mathematical points represent physical ones. The difficulties arise exactly when we regard the former as names for the latter. But now we are in a position also to recognize that the hole argument has little to do with the GTR *per se*. Rather, it is an instance of a problem that arises for *any* theory that posits namable substances. On the surface, it appears that *any* such theory will have problems with determinism.

To take a very general example, consider any theory according to which both I and the Eiffel Tower are named substances. There will be some maximal description of the universe as it is, a description given in vocabulary of the theory. If the theory is true, this description will constitute a model of the theory. Now consider the description that results from this one if we exchange the names "Tim Maudlin" and "Eiffel Tower" wherever they occur. The new description contains such sentences as "T. M. is constructed in 1889, scandalizing Parisians." and "Eiffel Tower addresses a group of philosophers in Evanston on October 28, 1988". There is a "passive interpretation" of this new description which construes it as a case of merely *renaming* the Eiffel Tower and me, applying different conventional linguistic tags to those objects. Under the passive interpretation the two descriptions have the same content. But we are interested instead in the active interpretation, according to which the new description is just plain false. In the active interpretation "Eiffel Tower" refers to the very same object in both descriptions. And since the Eiffel Tower is not 5' 8", does not wear glasses, &c., the description we get by interchanging names is not equivalent to the original, and is not true.

So under the active interpretation, our new description depicts a distinct ontological state of affairs from the old. Further, unless the laws of our theory make explicit reference to the Tower or me by name, the new description will also be a model of the theory. For the theory will be blind to the referents of the linguistic tags, the names that we apply to substances. Without information about every named substance directly built in, how is the theory to judge one description to be an acceptable state of affairs and the other not?

But we now seem to be able to demonstrate that the theory must exhibit radical indeterminism. For under the active interpretation the two descriptions present situations that are a) ontologically distinct b) consistent with the theory and c) in exact agreement on all of the facts about the universe up to the construction of the Eiffel Tower. Apparently our theory cannot predict on the basis of the history of the world up until 1889 whether it is I or the Eiffel Tower which is to be built for the Paris exposition. Perhaps all models that agree on the facts up to 1889 also agree that some substance made of girders appears on the banks of the Seine in that year, but just *which* substance so appears is left open. Evidently one must either renounce determinism or abandon substantivalism about anything at all.

The parable of the Eiffel Tower suggests that the difficulty Norton and Earman have discovered arises not from features peculiar to the GTR but rather from a very broad metaphysical picture. Roughly, the problem arises as soon as one introduces names for substances and further assumes that the substances have all of their properties only accidentally. For a world made up of bare particulars and accidents automatically falls prey to a kind of permutation argument: permute some of the bare particulars, have them exchange all of the properties with which they have been clothed, and the result is a distinct yet observationally identical state of affairs. Verificationists would object to the very coherency of distinct but observationally identical states. But verificationism is no longer in favor, so such a metaphysical possibility is not automatically foreclosed. What Norton and Earman have discovered is is that the possibility of certain permutations also conflicts with determinism: if the permutation effects no substances that existed before time t_o, then the permuted model will represent a world that agrees with the original in *all* facts (both observable and unobservable) until t_o yet diverges after.

If the hole dilemma is generated by such a broad metaphysical picture of substances and their attributes, it should be countered in equally general terms. And one traditional metaphysical response immediately presents itself to oppose the doctrine of bare particulars and accidents, viz. essentialism.

The essentialist would have little difficulty defeating the Eiffel-Tower-swap argument for indeterminism. For although under the active interpretation the swap engenders a description of a *distinct* state of affairs, it is equally clear that the new state of affairs is not a *possible* one. It is not among the vicissitudes that might have occurred to me that I be made of metal girders, nor among the possibilities open the the Tower that it fly Eastern to O'Hare. I might have been somewhat fatter or thinner, tidier or messier than I actually am, but there are limits to my metaphysical plasticity. Those bounds are set by my *essential* features, the properties whose loss I could not survive.

So too in the arena of space-time we must ask: what are the *essential* properties of space-time points regarded as substances? Which of their properties might possibly be stripped from them and which not? Perhaps upon answering this question we may find that the hole diffeomorphism generates, as does the Eiffel-Tower-swap, a description of a distinct but *impossible* state of affairs.

One holds little hope of having such metaphysical questions as this addressed in the physics literature. But the arch-substantivalist, Newton, tackles precisely this issue in several places. In the Scholium on space and time he writes:

> As the order of the parts of time is immutable, so also is the order of the parts of space. Suppose these parts to be moved out of their places, and they will be moved (if the expression may be allowed) out of themselves. For times and places are, as it were, the places of themselves as of all other things. It is from their essence or nature that they are places; and that the primary places of things should be movable, is absurd. (Newton 1966, p. 8)

And even more clearly in de *Gravitatione et Aequipondio Fluidorum*

> For just as the parts of duration derive their individuality from their order, so that (for example) if yesterday could change places with today and become the later of the two, it would lose its individuality and would no longer be yesterday, but today; so the parts of space derive their character from their positions, so that if any two could change their positions, they would change their character at the same time and each would be converted numerically into the other. The parts of duration and space are only understood to be the same as they really are because of their mutual order and position, nor do they have any hint of individuality apart from that order and position which consequently cannot be altered. (Newton 1962, p. 136)

Newton evidently saw that the parts of space and time, being intrinsically identical to one another, had to be differentiated by their mutual relations of position. Parts of space bear their metrical relations essentially.

Newton's doctrine on absolute space and time must be updated once we adopt the 4-dimensional picture, but the translation is not difficult. The order and position of space-time locations one to another are determined by the metric. Hence the general relativistic version of Newton's doctrine would hold that space-time is an essentially metrical object and that the points of space-time bear their metrical relations essentially. We might imagine a point of space-time supporting different material fields or containing different particles than it actually does, just as Newton would affirm that absolute space might have been differently populated than it is. But to suppose that a pair of space-like separated events might have instead been time-like separated, or that a region in the absolute past of an event might have been instead in its future, is to posit the absurd.

If space-time is an essentially metrical structure then the hole diffeomorphism employed by Norton and Earman, when actively interpreted, does not generate a description of a *possible* situation. For not only are the material particles and fields reassigned under the mapping, but the metric itself is "moved", altering the metrical relations between points. The essentialist objects to this as a metaphysical absurdity. So whereas the Ramsifying interpretation attempted to avoid the dilemma by denying the *distinctness* of the two represented situations, the essentialist insists rather that the second requirement, viz. that the represented situations both be possible, has not been met. If one is to use mathematical points as names for physical points- a condition necessary for generating the hole dilemma- then the essential features of the named individuals must be acknowledged and respected when describing real possibilities.

Earman and Norton do not suppose that the substantivalist must conceive of event locations as entirely bare particulars. They attribute to the substantivalist the view that space-time is essentially only a differentiable manifold and hence of itself only has topological structure. Since the diffeomorphism respects topology, such a view would succumb to the hole dilemma. The question is: Why should any serious substantivalist settle on manifold substantivalism? What would recommend that view?

Prima facie it seems like a peculiar position to hold. The substantivalist is interested in ascribing full ontological status to space-time. But just *qua* differentiable manifold, abstracting from the metrical (and affine) structure, space-time has none of the paradigm spatio-temporal properties. The light-cone structure is not defined; past and future cannot be distinguished; distance relations do not exist. Spatio-temporal structure is metrical structure, and the substantivalist will certainly insist that space-time have spatio-temporal structure.

Norton and Earman provide several considerations designed to promote the bare manifold as the most plausible candidate for substantival space-time. The leading arguments are: a) In the GTR the metric field is just like any other field, both mathematically and physically. It is represented by a tensor similar to, e.g., the electromagnetic field; it is governed by local differential equations; it carries energy and momentum. Hence it should not be essential to space-time (Earman & Norton 1987, p. 519). b) In the GTR the metric becomes a dynamical object. Hence it should not be considered essential to space-time (Earman forthcoming, chapter 10). In both cases the substantivalist can question the cogency of the inference.

The similarity of the metric to other fields at first seems to lend aid and comfort to the metrical essentialist. Space-time becomes a fully causal object, interacting via tidal forces and gravitational waves with other more traditional physical entities. Yet if the metric tensor is really no different from the electromagnetic tensor, why should metrical features be deemed essential to space-time points and electromagnetic features not?

Granting certain similarities of the metric field to other physical fields, still the metric plays a unique role in any relativistic theory. For the metric is indispensable. There is a theory of what the universe would be like without electromagnetism or without strong gauge fields. One can describe universes that are complete vacua, with stress-energy tensors identically zero. But there is no theory of space-time with electromagnetism but with no metrical structure. Without affine structure and light-cone structure and covariant differentiation the laws governing other fields could not be written. It is the physics itself, not just philosophical fancy, that singles out the metric as peculiarly intertwined with space-time.

Another possible worry to be raised against the substantivalist is that space-time might be artificially inflated by a theorist intent upon reducing other physical features to purely metrical ones. Can one tell by inspection of a theory just which physical effects really arise from the metrical structure and which don't? Paradigmatic of the approach which expands space-time in order to subsume new phenomena is the Kaluza-Klein theory, which sought to account for electromagnetism via the addition to 4-dimensional space-time of a fifth closed spatial dimension. But a closer look reveals that general covariance was sacrificed in Kaluza-Klein theory: the 5-dimensional cylinder naturally foliates into rings which the point transformations have to respect. In fact, Kaluza-Klein actually introduces a fiber-bundle over a 4-dimensional space-time and misleadingly describes it as a 5-dimensional space-time. The fiber-bundle, which associates some distinct internal degrees of freedom with each space-time point, is the natural structure for representing physical entities that reside *at* space-time points and should not be construed as adding additional *spatio-temporal* dimensions.

This is not to say that space-time might not have more than four dimensions. String theory legitimately postulates such a possibility. But general covariance is a minimal criterion that such theories must satisfy, and there is no reason to believe that the effects due to any non-spatio-temporal degree of freedom can be reproduced by artificially expanding the space-time.

The second line of argument for metrical structure being accidental to space-time contends that since the metric is dynamic, varying from place to place within a model and between models, since it is determined in part by the distribution of energy via the field equations, it should naturally fall on the accidental side of the essential/accidental dichotomy. According to this approach only absolute features of space-time, features shared by all of the models of the theory, are suited to be considered essential to space-time.

If we still operated with a Newtonian picture of space as an entity enduring through time the dynamism of the metric would indeed create difficulties. Since metrical structures differ at different times this would require the same point of space to have now one curvature, now another. Metrical features could not be essential to such parts of space.

But in the 4-dimensional scheme the ultimate subjects are perfectly evanescent, each event unenduring and hence unchanging. Nothing prevents these punctate entities from bearing their metrical features essentially.

The sense in which the metric is dynamic is not that event locations ever change their metrics. Rather, it is dynamic because it nomologically depends upon the distribution of matter and the boundary conditions, and hence varies from place to place and model to model. If space-time points have their metrical features essentially, then we must also say that space-time points are contingent entities: particular points may be represented in one model but not in another. But there is no reason why contingent, dynamically produced entities cannot have essential properties. Essentiality implies a sort of necessity but not such as to make items which have essences into necessary existents. My humanity is essential to me although I am a contingent being. It is necessary only in that wherever I appear or could appear, there too humanity must follow. The suggestive equation of absolute with necessary and hence with essential and of dynamical with contingent and hence with accidental pivots on an equivocation. The metrical properties of space-time points may be essential to them although neither they, nor they points themselves, are absolutely necessary entities. The necessary/contingent or absolute/dynamical distinction cuts at different joints than the essential/accidental does.

Further, if the only acceptable essential features of space-time were its non-dynamical ones then the substantivalist could not retain even the bare topological manifold. For topology is in this sense dynamical too: it varies from model to model and is determined in part by the mass distribution via the field equations. So the substantivalist would be left with only a collection of unrelated points, and the diffeomorphism could be replaced with a direct permutation a la Eiffel Tower swap. Norton and Earman do not think that the substantivalist must be reduced to such a view, but if all the dynamical properties are stripped from space-time nothing but bare points seems to remain.

Essentialism is not a popular metaphysical doctrine. Modal versions of Theseus's ship quickly convince us that separation of the essential from the accidental properties of ordinary objects is a hopeless, and probably meaningless, task. Quite unpredictably, the hole argument has provided us with one of the strongest arguments for essentialism ever produced. For if this diagnosis has been correct, the hole argument and its analogs reveal that any ontology which contains namable substances must either ascribe essential properties to those substances or else fall prey to radical indeterminism. If you want to have substances at all, you better have essences as well.

One extreme response to Theseus's ship is not to abandon essences but to revel in them, making every single property of a substance essential to it. Arguably, this is

Leibniz's position. It would be preferable to be able to draw a non-arbitrary line between essential and accidental features of the basic entities, but the grounds for such a distinction have been elusive.

I believe that Norton and Earman have unintentionally presented us with a tremendously powerful argument in favor of substantivalism about space-time. For if ultimately the only true substances are space-time points, if our ontology finally resolves into event locations and their properties, then we can make a principled and motivated distinction between essential and accidental properties. The hole argument has shown us that metrical structure plays a unique role in grounding the possibility of determinism, and that it must do so in any theory, regardless of its ontology. So those who have space-time points in their ontology, those who embrace the most natural recipients of metrical structure, will be in the best position to exorcise the unholy spectre of indeterminism that Norton and Earman have raised.

Addendum- Reflections and Responses

In light of the presentations and remarks of Professors Norton and Butterfield, I would like to record some comments that I would have made in Chicago had time permitted.

First I should note that the brand of substantivalism I explore in this paper- metrical essentialism- is not the only form of that doctrine which can elude the hole dilemma. The Ramsifying substantivalist portrayed above can do so without any commitment to essentialism, albeit she must forego the ability to name space-time points. Butterfield's counterpart theory also provides a means of escape. However, neither the Ramsifier nor Butterfield can admit that the classical Leibniz shift (viz. moving all material bodies 3 meters North in Neo-Newtonian or Minkowski space-time) describes a metaphysically coherent operation that would generate a distinct state of affairs. The Ramsifying case has already been treated; Butterfield would get the same result because, roughly, if one adopts a Lewisian account of possible worlds as spatio-temporally *disjoint* entities, the only diffeomorphisms at issue are from one manifold onto *another*, rather than from one onto itself. Strictly speaking, a *hole* diffeomorphism cannot even be defined if the map is between two distinct manifolds, for the map outside the hole has to be the identity (and hence the identification of the points in the domain with those in the range must already be settled). If the diffeomorphism is between two manifolds, identity of domain and range can never be at issue- only counterparthood. And Butterfield reasonably conjectures that if a diffeomorphism exists between two manifolds, it is the diffeomorphism itself which will provide the most reasonable pairing of counterpart points. But this holds also in the case of the Leibniz shift, so it too cannot be construed as describing an ontologically distinct situation.

In a way, Butterfield and I recapitulate the the argument between Lewis and Kripke on the nature of modality. He sees modal claims about individuals as being underwritten by a counterpart relation between distinct worlds or models, and then seeks the principle by which such a relation is to be established. I believe that in many situations- the Leibniz shift being a paradigm- we can meaningfully discuss what might have happened at some event location just by stipulating that we are talking about that event location. This requires names or name-like terms for the points, but no counterpart relation. However, on this picture our linguistic reach exceeds our metaphysical grasp. Just as we can say "If Nixon were a ham sandwich...." even though Nixon could not have been one, so we can generate hole diffeomorphisms that don't represent metaphysically possible situations. There is no problem of identifying counterpart points here. We know perfectly well *which* event locations we are talking about (if we use the mathematical point as names), but we describe them as having properties that they could not possibly have.

This is not to say that we *only* consider possibilities that are described using names. As Butterfield points out, on my view no non-isometric models can represent possible states of this very same space-time. In particular, no model not isometric to the actual world can represent how *this* space-time might have been. But we can still consider non-isometric possibilities: they are just different possible space-times, not different possible states of this space-time.

The mathematician who writes down a solution to the field equations- the Schwarzschild solution for example- does produce a mathematical object that represents a real physical possibility. But here the mathematical object is to be interpreted more along the Ramsifying line: the claim is not, of some particular actual event locations, that they might have had this metric, the claim is rather that there might have been some event locations that had this metric. The mathematical points are clearly not names for physical event locations. Imagine asking the mathematician, for example, exactly *which* point in the model represents the top of the Eiffel tower at 11:34 AM October 28, 1988...

Finally, on my view *certain* counterfactuals will have to be given a counterpart-theoretic explanations. If we say that this region of space-time would have been more curved had the Earth been heavier, we cannot hold that *this very region* might have been more curved. That is metaphysically impossible. But we often have no trouble using counterfactuals that involve situations that are not metaphysically possible. Had Nixon been a ham sandwich he would not have been kosher. Or, more dramatically, if you could trisect an angle with ruler and compass then you could square the circle. This last is *provable* , and demonstrates conclusively, to my mind, that counterfactuals do not provide automatic insight into possibility. So in explicating counterfactuals about what would have happened to this space-time under various circumstances, we can use both counterpart relations to non-isometric models and even antecedents that are not metaphysical possibilities. Metrical essentialism can countenance both of these ploys.

Metrical essentialism is not the only possible substantivalism. It may not even be the most plausible. But it does avoid the hole dilemma and, unlike Butterfield's solution, it still endorses the coherence of the classical Leibniz shift. Norton and Earman claim that acceptance of the Leibniz shift is an "acid test" of substantivalism. I leave it to them to defend that claim. But since it was *Clarke* who introduced the "Leibniz" shift- introduced it as an argument for substantivalism- retaining the possibility of the shift in a sufficiently symmetric space-time is a desideratum for any view that purports to be an elaboration of *classical* substantivalism. Further, I hope to have shown that metrical essentialism is not just an *ad hoc* reaction to Norton and Earman, but can be clearly found in Newton's writings. Whether for all that it is the best substantivalism, I leave to others to judge.

A Final Note

The metrical essentialist would respond to Norton's t_{future} trilemma as follows: $\langle M, t_{future}*g, t_{future}*F \rangle$ represents a *non*-physical (indeed not *metaphysically* possible) situation distinct from both Model I *and* Model II. Since t_{future} changes, e.g., the affine structure of M it cannot represent a possible state of the same space-time that is represented by $\langle M, g \rangle$.

Notes

[1]The substantivalist *need not* accede that the shifted objects would be the *very same* objects as the unshifted. It is surely possible on the substantivalist picture that a planet qualitatively identical to Earth have existed 3 meters to the North of the Earth's present location, although it may not be metaphysically possible that the *Earth* have done so. Some further comments on the substantivalist's account of the Leibniz shift may be found in the addendum.

References

Earman, J. (forthcoming), *World Enough and Space-Time: Absolute vs. Relational Theories of Space and Time.*

Earman, J. & Norton, J. (1987), "What Price Spacetime Substantivalism? The Hole Story", *British Journal for the Philosophy of Science* 38: 515-525.

Geroch, R. (1978), *General Relativity from A to B.* Chicago: University of Chicago Press.

Newton, I. (1962), *Unpublished Scientific Papers of Isaac Newton* , A. Rupert Hall & Marie Boas Hall (eds.). Cambridge: Cambridge University Press.

_ _ _ _ _ _. (1966), *Principia* . Berkeley: University of California Press.

Part IV

PHYSICS

A Philosopher Looks at String Theory

Robert Weingard

Rutgers University

Before we, as philosophers, take a look at string theory I want to mention that more than one person has suggested to me that it is still too early for philosophical and foundational studies of string theory. Indeed, the suggestion emphasizes, since string theory is still in the process of development, and its physical and mathematical principles are not completely formulated, there is, in a sense, no theory for the philosopher to analyze. And I must admit that I think there is something to this suggestion. In a sense I hope I will make clear, there does not yet exist a precise mathematical formulation for string theory as there is Hilbert space for (elementary) quantum theory, and Riemann spacetime for general relativity. Because these latter formulations exist, we can ask precise questions, and prove precise theorems about their interpretation. The Kochen-Specker theorem about noncontextualist hidden variable theories, the Fine-Brown proof of the insolvability of the quantum measurement problem, and the current determinism-hole argument debate are some examples. Without a clearly formulated mathematical structure, I don't think we can expect to get analogous distinctly stringy results.

This suggests a related worry. String theory, at least in the first quantized theory, is a relativistic quantum theory of strings (one dimensional extended objects). One may well agree that of course all of the standard philosophical and foundational issues of quantum theory and relativity are still there, but be skeptical about whether string theory will either shed any light on these old problems, or give rise to stringy problems. Some philosophers have expressed analogous doubts about quantum field theory.

Again there is something to this worry. But only something.I think it is true, as I have already said, that string theory is not ready for certain kinds of foundational studies. And I don't think string theory (or quantum field theory) will shed light on those favorite topics of philosophers of physics, Bells theorem and realism-hidden variable theories in quantum mechanics. But there is more to physics than nonrelativistic quantum mechanics.

For the past fifteen years, quantum gauge field theories have been our best theories of the physical universe. If we wanted to understand the basic ontology of the universe, at least according to the best source of information available, then we needed to understand these theories. And we needed to answer such questions as, What is the ontological significance of a quantum field?, Do Fermi fields have a different ontological status than Bose fields?, What is the geometrical significance of the gauge connection and its associated fiber bundle?, among many others.

PSA 1988, Volume 2, pp. 95-106

These questions still need answering. But now many physicists believe that string theory may (will?) replace quantum field theory as our best fundamental theory of physics. And in the process it is claimed that it will give us a unified theory of all the fundamental interactions, a consistent quantum theory of gravity (and thus of the metric field), and an explanation of the family structure of elementary particles, to name a few. If this is correct, then we will have to ask of string theory the kind of questions we should have been asking of quantum field theory.

Therefore I would like to talk about the following. First, I want to describe some simple, but interesting, string theory to give you some idea of what this theory is all about. Then I want to discuss string field theory and suggest that even though we do not have a complete mathematical formulation, we can get an idea of some of its ontological implications. This is one of the philosophically exciting aspects of string theory. However, I also want to ask whether there is, in fact, (good) reason to think that string theory may (or will) emerge to replace quantum field theory. Unfortunately, this may dampen our excitement somewhat.

It is time, then, to turn to string theory. Strings-one dimensional extended objects-come in two kinds, open and closed.

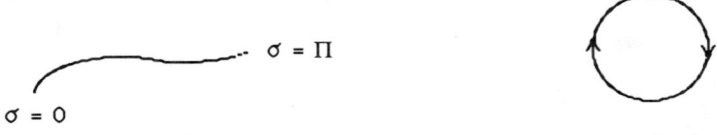

$\sigma = \Pi$

$\sigma = 0$

open string   closed string

Classically, strings can have translational as well as vibrational motion. The vibrational motion can be decomposed into left and right moving normal modes. For the open string, boundary conditions require that these left and right movers be identified, while for the closed string they are independent. When we quantize, these different normal modes become states of different mass and spin. The difference between the open and closed string is that the Hilbert space of states for the closed string is a product space $H = H_R \times H_L$, of the left and right movers-roughly speaking, its a product of two open string Hilbert spaces.

Of special interest are the massless modes. String theory is fundamentally a theory of the very early universe, when the mass scale was the planck mass, m_p. This is the natural mass scale of the theory, so that the massive modes of the string will have masses that are, roughly, multiples of m_p. Thus, only the massless modes of the string will correspond to the particles we see at the relatively low energy scale of the laboratory-the masses of observed particles will instead come from symmetry breaking.

The massless states of the open string form a spacetime vector, so the massless states of the closed string form a second rank spacetime tensor. The significance of this emerges when we try to make string theory into a unified theory of the fundamental interactions of physics. Since the strong and electroweak interactions are mediated by spin one vector particles, while gravity is carried by a spin two graviton-the quanta of the second rank metric field g_{uv}, a stringy unification requires at least that the gauge vector particles and the graviton are massless states of a single string. Our above remarks, however, suggest that the graviton must come from the massless modes of the closed string, while the gauge bosons must be massless states of the open string. As we will see, string theory has the resources to overcome this problem.

So far we have been talking about the bosonic string-the degrees of freedom are the spacetime coordinates, x^u, of the string, and its excitations have integral spin. To add fermions to the theory we put a spinor at each point of the string world sheet (the surface the string sweeps out in space time). Lets draw a picture of this situation,

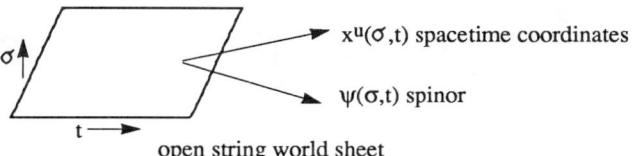

open string world sheet

Amazingly, ψ can be either a world sheet spinor (that is, a spinor with respect to transformations of the worldsheet coordinates σ,t) or an explicitly spacetime spinor (the Green Schwartz string). When done properly, in either case we get a theory with the same spectrum of spacetime bosons and fermions. In particular, we can note two things. First, the purely bosonic string contains a tachyon in its spectrum (a state with spacelike four momentum). By requiring the fermionic string to have spacetime supersymmetry the tachyon is eliminated. Second, it turns out that the (interacting) bosonic string can be consistently formulated only in twenty six dimensional spacetime. When we add fermions, the number of required spacetime dimensions is reduced to ten (at least for the standard formulations).

Since in everyday life, only four dimensions of spacetime are apparent, in either case all but four of the dimensions must be compactified. That is, each of them is rolled up into "cylinders" of very small radius, explaining why we normally do not "see" them. Compactification may seem like an *ad hoc* move designed merely to remove the embarrassment of more than four spacetime dimensions. But in fact(at least in the standard version of the theory) it does much more than hide the extra dimensions. It is meant to explain how the spectrum of particles and interactions we see at our energy scale come from the "pure" high energy string theory. We can get an idea of how this is supposed to work by looking at the heterotic string.

But first, a preliminary point. Consider a vector A^u in a D dimensional spacetime, in which i=1....n dimensions are not compactified, the N=n+1....D dimensions are compactified. The A^i are thus the components of A^u in the n uncompactified dimensions and A^N the components in the compactified dimensions. Then from the point of view of the uncompactified dimensions, since N is not a spacetime index, only the A^i form a vector, while the A^N are D-n scaler fields, which transform into each other according to what the i dimensions regard as an internal symmmetry group. In fact, this is a geometrical symmetry of the compactified dimension.

We turn next to the Heterotic string. This is a theory of closed strings, in which the left moving modes are those of a twenty six dimensional bosonic string, while the right movers are those of a ten dimensional fermionic string. This is possible because, as we mentioned , the left and right movers of the closed string are independent. Spacetime therefore has twenty six dimensions, but the right movers have nonzero components in only ten of them.

We can think of the compactification as occurring in two stages. In stage one we compactify the sixteen purely bosonic dimensions, and then in stage two, six more spacelike dimensions are compactified to give an effectively four dimensional theory. Lets consider stage one. A vector excitation of the bosonic left movers will be of the form, $\alpha^u_{-n}|0>$, where α^u_{-n} is a creation operator with n labeling the mode created. This will break up into $\alpha^u_{-n}|0> = (\alpha^i_{-n}|0>, \alpha^N_{-n}|0>)$. The i'th components form a vector in the ten (so far) uncompactified dimensions while, as we saw above, the N'th components will be sixteen scaler fields from the ten dimensional point of view, with the internal symmetry index N.

Similarly, a right moving vector will be of the form $\beta^i_{-m}|0>$, where the β^i_{-m} are suitable creation operators. And note that here we only have the i index since the right movers have only ten components. The closed string states, as we saw, will be products of the left and right movers, so here of the form, $\alpha^u_{-n}|0> \times \beta^i_{-m}|0> = (\alpha^j_{-n}|0> \times \beta^i_{-m}|0>, \alpha^N_{-n}|0> \times \beta^i_{-m}|0>)$.

The case of interest to us is n = m = 1, the massless vectors. Then there are two important points. First, since the $\alpha^N_1|0>$ are ten dimensional scalers, the products $\alpha^N_{-1}|0> \times \beta^i_{1}|0>$ are still (massless) ten dimensional vectors. Second, these vectors form a multiplet indexed by N, which transforms according to a symmetry determined by the compactification. Thus, we have solved two problems.

We have obtained massless vectors from the closed string, and we have introduced an internal symmetry. (The open string has two distinguished points, the two end points, on which we can put charges. But each point of the closed string is "the same", so there is no place to put a charge without breaking the symmetry of the closed string). It turns out that given the proper compactification, we get an N = 1 Yang-Mills supermultiplet, whose internal (gauge) symmetry is either $E_8 \times E_8$ or SO(32). In addition, the $\alpha^j_n|0> \times \beta^j_1 |0>$ form a second rank tensor in ten dimensions and, as you would expect, this contains the graviton.

We have obtained the graviton and the Yang-Mills vector bosons, but it has been accomplished in ten, rather than four dimensions. To achieve a realistic theory, we need to complete stage two and compactify from ten down to four dimensions. Here the $E_8 \times E_8$ gauge group offers a possibility. Namely, that the compactification breaks one of the E_8's to an E_6 symmetry in four dimensions. Unlike E_8, E_6 has complex representations(which means left and right handed particles transform differently under the group), and ones large enough to contain a complete family of the known fermions. The other E_8 would remain unbroken and could be the gauge group of the so called "shadow" matter, matter that would interact only gravitationally with normal matter. It would be a candidate, then, for the missing mass of the universe.

The heterotic string provides a way of getting massless vector bosons from the closed string. It turns out that conversely, the graviton is contained in the open string. Not as an excitation of the free string, however, but as an intermediate state in the interaction of open strings. To see this, consider the one loop contribution to an amplitude involving four external strings. There are three kinds of loops that will contribute to this; a planer loop, a nonplaner loop(a planer loop with an even number of twists), and a nonorientable loop (a planer loop with an odd number of twists). We can picture these as,

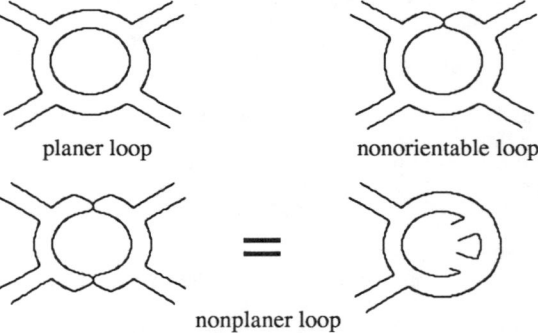

planer loop nonorientable loop

nonplaner loop

Because the theory is conformally invariant, the nonplaner loop is equivalent to the world sheet pictured below,

which shows two open strings intracting thru a closed string intermediate state! And indeed, exact calculation shows that in D = 26, the nonplaner loop amplitude contains a closed string pole. Interestingly, here is a place where the necessity of twenty six dimensions for the bosonic string can be seen. In $D \neq 26$ this simple pole becomes a branch point which destroys the unitarity of the one loop amplitude.

We see here an important difference between string theory and quantum field theory. We can do quantum field theory without having to have a (quantum) theory of gravity. For example, we can do quantum electrodynamics in flat spacetime without the involvement of gravity because the photon field A^u, and the Dirac fields $\psi(\bar\psi)$ create only photons and positrons (electrons) respectively. This is fortunate because Einsteins gravity is nonrenormalizable as a quantum field theory.

However, since the free closed string already contains the graviton in its spectrum, only the classical open string is consistent without gravity. The minute we quantize, that is, go beyond tree diagrams to the one loop level, then we have to include gravity to get a consistent theory. Given the nonrenormalizability of Einsteins gravity this might have spelled disaster for string theory. But there is a lot of confidence that the superstring is not just renormalizable but finite. That is, that the different divergences that arise cancel each other, and do not have to be removed by counterterms. Unfortunately, I cannot tell whether this confidence is justified.

Even if we do not have to worry about divergences, however, gravity still posses a problem for string theory. The reason is that at present, string theory provides only a perturbative theory of gravity. Given a background spacetime in which the string propagates, one can calculate graviton scattering to any order in perturbation theory, i.e., up to any number of loops. But the metric of the background spacetime is itself, presumably, a condensate (or coherent state) of some of the massless modes of the strings propagating in it, since, macroscopically, the metric is the gravitational field. At present the background spacetime has to be assumed, but clearly this is at best incomplete. In a deeper and more complete theory, we would expect an explanation of how the metric of spacetime emerges from the string dynamics.

Another aspect of this problem emerges when we realize that so far we have been speaking of first quantized strings. The string is given as the basic object, and when we quantize we get an Hilbert space of states of the string. But this theory can be reformulated as a string field theory, where the fields are functionals of the string coordinate functions, $x^u(\sigma,t)$, as opposed to being functions of spacetime position as in point particle field theory. But this field theory has two draw backs. It too is defined only in perturbation theory, and it has resisted a covariant formulation. It must be done in a particular coordinate system, such as light cone coordinates. Clearly, this to is unsatisfactory.

On the one hand, such nonpertubative objects as monopoles and instantons emphasize that the content of a field theory is not exhausted by its perturbation theory. On the other hand, a covariant formulation of a complete string field theory would (presumably) reveal symmetries of the theory that are hidden by the noncovariant formulation. Therefore, we can see at least two possible benefits from having such a covariant string field theory.

First, we should get an explanation of the fundamental interactions of the theory. For example, the gauge invariance of qed explains the origin of the basic trilinear coupling,

$$= \psi \gamma^u A_u \psi$$

between electrons and photons. And the nonabelian gauge symmetry of qcd explains why there are four gluon and three gluon self interaction terms. Unlike the abelian field strength, F_{uv},

$$F_{uv} = \partial_u A_u - \partial_v A_u,$$

whose lagrangian, $F^{uv} F_{uv}$, contains only kinetic terms like $\partial_u A_v \partial^u A^v$ and no self interaction terms, the non abelian field strength, F_{uv}, is

$$F_{uv}^a = \partial_u A_v^a - \partial_v A_u^a + gf^{abc} A_u^b A_v^c.$$

This has the extra piece proportional to $[A_u, A_v]$, so that it transforms correctly under the gauge group. And it is this extra piece which gives rise to the new trilinear and quartic self interaction terms in the lagrangian, $F_{uv}^a F^{auv}$.

Second, we might get an explanation of the metric in terms of some kind of symmetry breaking of the field theory. We can illustrate both of these points, in a stringy context, by looking at a proposal of Wittens[1] for a covariant field theory of open strings.

Wittens proposal is a generalization of Yang-Mills theory, involving a noncommutative algebra of string forms. If we write Yang-Mills theory in the language of forms, then the gauge invariant action, S_{YM} is,

$$S_{YM} = \int tr \, F \wedge \hat{F},$$

where F is the gauge field strength 2-form, $F = dA + A \wedge A$, A is the gauge field 1-form, and $\hat{F}$ is the dual field strength. As a stringy generalization of F Witten writes,

$$G = QB + B*B,$$

where G and B are the stringy analogues of F and A, and * generalizes the wedge product ∧. It is defined by $(B*C) = (B_L, C_R) \, \delta \, (B_R - C_L)$, that is, the left half of B joins to the right hand of C. By singling out the midpoint we get an associative product, $(A*B)*C = A*(B*C)$, for paramaterized strings, but one that is not commutative, even up to a sign like the wedge product.

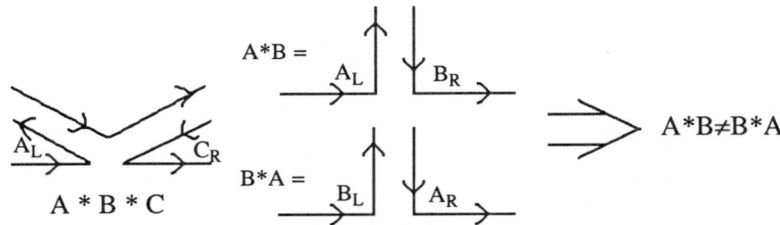

Finally, Q plays the role of the exterior derivative in string form space. Witten identifies it with the BRST charge.[2]

There is no analogue of the dual of a form (we can't raise and lower indices) in this formalism, so the closest analogue of S_{YM} for strings would be, $\int G*G$ (with $\int$ combining integration and taking the trace). But an arbitrary variation of this vanishes identically, and thus cannot yield any equations of motion. To get an alternative gauge invariant action, Witten turns to the analogue of the Chern Simmons three form w, which in Yang-Mills theory obeys, $dw = tr \, F \wedge F$. He postulates the action, $S = \int w$,

$$S = \int B*QB + 2/3 B*B*B ,$$

which is invariant under the infinitesimal gauge transformation $\delta B = Q\Lambda + B*\Lambda - \Lambda*B$ (Λ is a zero form gauge function), and under an arbitrary variation yields the equation of motion $G = 0$. Unlike Yang-Mills theory, therefore, we have only a cubic interaction term $2/3 B*B*B$.

But perhaps of more interest to us is the fact[3] that we can derive Wittens action from an action containing no kinetic term, by expanding around a background. We postulate an action S_c, containing only the cubic interaction of S,

$$S_c = \int B*B*B,$$

where S_c is invariant under the infinitesimal gauge transformation $\delta_c B = B*\Lambda - \Lambda*B$, and $\delta S_c = 0$ implies the equation of motion $B*B = 0$.

To obtain Wittens form of the action we need the fact that each 1-form B determines a derivation D_B given by,

$$D_B C = B*C - (-1)^{n_c} C*B,$$

where nc is an integer. If we then split B into a fluctuation $\tilde{B}$, and background B_0,

$$B = B_0 + \tilde{B},$$

and require that B_0 satisfies the equation of motion, $B_0 * B_0 = 0$, we get,

$$\int B*B*B = 3/2 \int (\tilde{B} *D_{B_0} \tilde{B} + 2/3 \, \tilde{B} * \tilde{B} * \tilde{B}).$$

If $\delta_c B_0 = 0$, then B transforms according to δ under a gauge transformation and for suitable B_0, $D_{B_0} = Q$. Thus, the recovery of Wittens action is complete.

The important point for us is that Q depends on the metric. But if we can define $\int$ and $*$ so that they do not depend on a metric, as in the case of differential forms, then with the simple cubic action we do not have to assume a background gravity field or metric. Instead, the metric of spacetime comes from how the fundamental field B is split (by God?) into a background B_0 and fluctuation $\tilde{B}$.

Earlier, when I said that a precise mathematical formulation of string theory did not exist yet, this is exactly what I had in mind. Wittens theory is suggestive, but a completely satisfactory covariant string field theory does not exist yet. None the less, as I have tried to show, I think we can get some idea of the ontological implications of such a theory, namely, that the metric of spacetime is not a fundamental field.

I want to return now to the topics of supersymmetry and extra dimensions, to amplify the brief remarks I made earlier. After all, we are discussing the superstring. Supersymmetry is a symmetry of a theory (of the action) under a transformation which changes bosons to fermions, and fermions to bosons. The supersymmetry transformations are closely connected to spacetime transformations. In particular, the result of two supersymmetry transformations is a spacetime translation. In four dimensional spacetime this is expressed by

$$\{Q_\alpha, Q_\beta\} = (\sigma_u)_{\alpha\beta} P^u,$$

where P^u is the four momentum operator (the generator of translations), Q_α the supersymmetry generators and σ_u the Pauli matrices. This is an important point so let me try to make it plausible, and illustrate supersymmetry generally, with a simple example from quantum mechanics.

Consider a system consisting of an harmonic oscillator and a spin 1/2 fermion. Then we know that we can represent the harmonic oscillator by annihilation and creation operators, a and a^+, such that $[a, a^+] = 1$, and $[a, a] = [a, a^+] = 0$. The hamiltonian is then, $H_a = N_a + 1/2$, where $N_a = a^+ a$ is the number operator. If $|0>$ is the ground state, then

$$a|0> = 0, \qquad\qquad N_a \, |n> = n|n>,$$
$$\frac{(a^+)^n \, |0>}{(n!)^{1/2}} = |n> .$$

For the fermion we introduce the operators b and b^+, which obey the anti commutators $\{b, b^+\} = 1$, $\{b, b\} = \{b^+, b^+\} = 0$. Then if $b|0> = 0$, it follows that,

	$|\alpha>$	$|\beta>$
b	0	$|\alpha>$
b^+	$|\beta>$	0

,

and b and b^+ are annihilation and creation operators for the states$|\alpha>$ and$|\beta>$. For our analogy we will regard$|\beta>$ as a spin 1/2 state, while $|\alpha>$ is the ground state. The states $|n>|\alpha>$ are then regarded as bosonic, and the $|n>|\beta>$ as fermionic.

It follows that for this system, $Q = a^+b$ changes fermions to bosons, while $Q^+ = b^+a$ changes bosons to fermions,

$$Q|n>|\beta> = (n+1)^{1/2} \, |n+1>|\alpha>,$$
$$Q^+|n>|\alpha> = (n)^{1/2}|n-1>|\beta> ,$$

and,

$$\{Q, Q^+\} = (a^+b \, b^+a + b^+a \, a^+b) = a^+a + b^+b,$$
by $[a^+, a] = \{b, b^+\} = 1$, and $[a,a]=[a^+, a^+]=\{b,b\}=\{b^+, b^+\}=0$. So
$$\{Q, Q^+\} = N_a + N_b = H .$$

The hamiltonian H is the generator of time translations and we have the analogue of the relation $\{Q_a, \overline{Q}_\beta\} = (\sigma^u)_{ab} P_u$ (note that we are taking $H_b = N_b - \frac{1}{2}$).

That the supersymmetry generators form a nontrivial algebra with spacetime generators suggest that we can interpret a supersymmetry transformation as a transformation in some sort of an extension or generalization of spacetime. And this is indeed the case. This extension, called superspace, is obtained by adding anticommuting spinoral coordinates ϑ^α to the usual spacetime coordinates x^u and then the supersymmetry generator Q^α can be expressed in terms of both the spacetime and spinorial derivitives, $\partial/\partial x^u$ and $\partial/\partial\Theta^\alpha$.

Earlier we obtained the fermionic string by putting a spinor at each point of the string world sheet. But this was unmotivated. We had no understanding of why there should be a fermionic string. But if superspace is the fundamental manifold, then the coordinates of the string will be the superspace coordinates $z^m = (x^u(\sigma,\tau), \vartheta^\alpha(\sigma,\tau))$. Both x^u and ϑ^α will be degrees of freedom of the string, and quantizing the string will mean quantizing both of them.

Since supersymmetry transformations change bosons and fermion into each other, bosons and fermions occur in the same multiplets (representations) of the supersymmetry transformations. In our simple example the multiplets have the form, ($|n>|\alpha>$, $|n-1>|\beta>$), each state in the multiplet having energy n. According to superstring theory, before compactification the fermionic string in ten dimensions has supersymmetry. This implies that every bosonic state (particle) has a fermionic partner of the same mass. But the compactification to four dimensions must break the supersymmerty because observed particles do not form supersymmetry multiplets. Since the supersymmetry generators change the spin by 1/2, there should be a scaler electron, a massless spin 1/2 partner of the photon (this can't be the neutrino which interacts weakly, unlike the photon), and a spin 3/2 partner of the graviton. None of these super partners have been observed.

None the less, at a fundamental level, there is supersymmetry, and the question arises as to the ontological significance of this fact. Does the fact that (before symmetry breaking) a boson can be transformed into a fermion by the supersymmetry transformations mean that the two are, at a deeper level, a single particle that can appear as a boson or a fermion? Or does it just mean that physically the two can change into each other?

An often expressed point is that matter is composed of fermions (quarks and leptons), while the forces which hold matter together are bosonic, that is, the interactions are mediated by gauge bosons. The division is not absolutely sharp because bosonic gluons can self interact and form glue balls. But that is a minor exception. If nature is supersymmetric then force and matter appear in the same multiplets. But is this really ontological unification?

The answer appears to hinge on how realistically we can (or should) interpret super-space.[4] The $2n + 1$ z components of a particle with spin n (in four dimensions) form a representation of the three dimensional rotation group. Therefore, for example, the difference between a spin $z = 1/2$ and $z = -1/2$ electron is that of the orientation of a single particle. And this is so even if the symmetry is broken by a magnetic field and the two have different energies. But such an interpretation does not work when the symmetry transformations are merely "rotations" in an internal space like isospin space. Because such an internal space is, apparently, just a mathematical space, the symmetry transformations just represent the fact that the particles can be transformed into each other. In the case of supersymmetry, the appropriate space, as we have seen, is superspace. Whether super-space can (or should) be interpreted realistically as an extension of spacetime, ontologically on a par with extending spacetime by adding more (regular) spacetime dimensions, is a question I find it hard to deal with clearly. In as much as I can, the anticommutator connecting Q_α and P_u argues that it can. Perhaps another relevant point is the fact, emphasized by Green and Schwartz,[5] that the superstring field is not a finite polynomial of the ϑ^α as in the case of ordinary (point particle) super fields.

This question about superspace is similar to the question I raised at the beginning of the talk about the geometrical significance of the gauge connection. These are difficult interpretive questions but I think they are important.

Finally, I want to turn to the question I raised earlier, about whether there is any reason to think string theory may (or will) emerge to replace quantum field theory as the framework for fundamental physics. It is often pointed out, even by the most enthusiastic string theorists, that there is absolutely no experimental evidence for string theory. Partly, no one knows how to calculate precisely anything observable at conceivable laboratory energies. But also, there is not even qualitative evidence, such as jets provide for qcd.

But I do not want to dwell on this here. It seems to me that one could be optimistic about a developing theory, even without such evidence, if there were at least some clues in either theory or experiment to suggest that the new theory was on the right track. Unfortunately, even this is lacking in string theory. Let me give some examples of what I mean here by a "clue" that the theory is on the right track.

I want to briefly mention two examples, general relativity and the SU(2) x U(1) elec-troweak theory. In both cases we have a problem which gives a clue to its solution. For the case of general relativity, a clue emerges when we consider the problem of trying to fit gravity into flat spacetime (into special relativity). The natural move would be to con-sider the gravitational field to be a flat spacetime tensor field on analogy with the electr-magnetic field. But then, gravity would not affect the null cone structure of spacetime, and either light would not be affected by gravity (since light world lines would remain null geodesics in the presence of gravity), or light would be affected by gravity and light worldlines would be timelike in the presence of gravity.

However, if we think the connection between special relativity and local inertial frames is correct, then it would follow both that light is affected by gravity (the equivalence principle), and that light travels null geodesics. These can be reconciled only if we give up flat spacetime and postulate that spacetime in the presence of a gravitational field is curved but locally flat. Even if we did not have the ability to determine whether or not light is affected by gravity, we would have a clue, or suggestion to an important part of general relativity.

In the case of the weak interactions, the problem was that the V-A theory of Feynman and Gell-Mann is nonrenormalizable. According to this theory, the weak interaction has the current-current form $J_{\bar{u}} J^{+u}$, an example of which is pictured below,

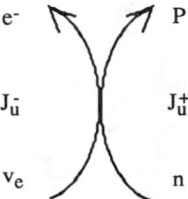

The current $J_{\bar{u}}$ is an (electric) charge lowering operator, here changing the neutral electron neutrino (v_e) into a negatively charge electron (e⁻), while J_u^+ is a charge raising operator , here changing the neutral neutron (n) to the positive proton(p). And in each case the currents change the electric charge by one unit (the charge on the electron).

Now the fact that the weak currents are electric charge raising and lowering operators certainly suggests a connection between the weak and electromagnetic interactions. But the electromagnetic interaction is not a current-current point interaction. Instead, qed is a renormalizable theory in which a gauge vector field mediates the interaction. So given that there is a connection between the two interactions, a natural thought is to try to gauge the weak interaction as well, hopefully getting a renormalizable theory. But how do we do this? What is an appropriate gauge group, and how do we fit the relationship between the two interactions into this theory? Here is where the weak currents provide another clue.

We know that any continuous group G, with n generators, which is a symmetry of the weak interactions (of the action) will generate n conserved currents, $J_{\bar{u}}^a$, a = 1,...,n. And the spatial integral of the time component of $J_{\bar{u}}^a$ will be a generator, Q^a of the group G,

$$Q^a = \int J_0^a \, d_x^3.$$

If G is a Lie group, the Q^a will satisfy the the Lie algebra of the group. Of relevance here, we know that SU(2) has the generators L^a, where $L^{+(-)} = L^1 \pm iL^2$ are raising (+) and lowering (-) operators with respect to L^3. That is, if the charge of ψ is m, so that $L^3\psi = m\psi$, the charge of $L^{+(-)}\psi$ is m±1.

This suggests that SU(2) is a symmetry of the weak interaction, with

$$J_u^1 = 1/2(J_u^+ + J_{\bar{u}}), \quad J_u^2 = 1/2i(J_u^+ - J_{\bar{u}}),$$

forming two of the conserved currents of that symmetry. And there should be a third conserved neutral current whose time component J_0^3 gives us the charge operator, $L^3 = \int J_0^3 d_x^3$. We would like this to be the electromagnetic current. However, when we use the known particles to form the currents, the generators do not obey the SU(2) commutation relations.

There are at least two moves we can make, while staying with the basic idea of SU(2) symmetry. We can postulate the appropriate additional particles so that the generators close under commutation to form SU(2), as in the theory of Glashow and Georgi.[5] Or we can try a minimal enlargement of the symmetry group by adding a U(1) factor, which gives a weak neutral current as well as the electromagnetic neutral current. The detection of the weak neutral current in 1973 therefor ruled out the pure SU(2) theory.

This latter example is interesting for us because two things that are often cited in support of string theory are that it provides (hopefully) a finite theory of quantum gravity, and that it unifies the fundamental interactions. And if correct these do make string theory tremendously interesting. But one thing we have learned from the many attempts at unified field theories in this century, I would argue, is that unification, in itself, is not a guide to a successful theory.

Consider the attempts by Einstein, Weyl, Schrodinger, Misner and Wheeler, and others to fashion a unified geometrical theory of gravity and electromagnetism. All of these were essentially mathematical investigations, because unlike the two cases we have just discussed, there were no empirical or theoretical clues as to how such a unification should take place. Indeed, there was no indication, classically, that such a unification was needed. After all, electromagnetism, conceived as an antisymmetric second rank tensor field in curved spacetime is perfectly compatible with Einsteins gravity.

True, Einsteins gravity is nonrenormalizable, while string theory promises to be a finite theory of quantum gravity. But again, there is not even a hint in what we know about physics, that this is the right way to get a workable quantum theory of gravity. Unless we have some reasons for thinking this, I don't think either of our reasons-providing a unified theory of the fundamental forces, and a finite theory of gravity-can, by themselves, give us (much) reason for thinking string theory may (or will) emerge as the fundamental framework for physics.

Undoubtedly, this is because, as I mentioned earlier, string theory at the fundamental level is a theory of very high energies, many orders of magnitude greater than are obtainable in the Laboratory.And this raises an interesting question. If there are no clues in our low energy world to the high energy world of strings, how was string theory discovered? Why did anyone think of it in the first place? A two paragraph history follows.

In the late 1960's it was proposed that the strong interactions obeyed duality. Duality involves the ideas of s-channel and t-channel processes, which are pictured,

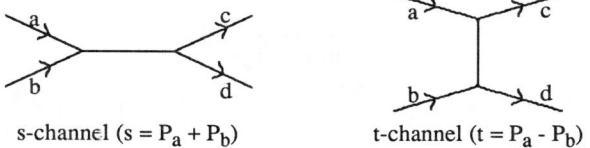

s-channel ($s = P_a + P_b$)   t-channel ($t = P_a - P_b$)

In the s-channel the incoming particles combine to form an intermediary particle, while in the t-channel, the incoming particles exchange an intermediate particle. If M(s,t) is the amplitude for $a + b \rightarrow c + d$, then duality says that we can regard M as a sum of s-channel amplitudes or a sum of t-channel amplitudes. The two are just different ways of describing the same process.(Compare this with quantum field theory where we have to include both types of diagrams to get the correct amplitude.)

In 1968 G. Veneziano wrote down a formula for M(s,t) that actually exhibited duality. From the analysis of this formula and its generalizations developed the theory of dual models. Then in the early 1970's, it was shown that all the properties of the model could be derived from an underlying lagrangian theory-that of the relativistic string! However, remember that this was supposed to be a theory of the strong intractions.By then qcd was emerging and it became clear that dual models do not give a correct account of the strong interaction. But in 1974,Scherk and Schwartz showed that the scattering amplitudes of massless spin two states of the closed string equals the graiton-graviton scattering amplitudes of quantum gravity(in the lowest order of approximation). They therefore proposed reinterpreting string theory as a theory of all interactions, but at the mass scale of the Planck length, rather than that of the strong interaction.

Thus was string theory born. To my mind, this is the most amazing thing about string theory, that it exists at all!

Notes

[1]Witten (1986).

[2]Using the BRST charge operator Q we can define a ghost free spectrum (i.e., no negative norm states). Namely. $Q|\psi> = 0$ implies $< \psi|\psi \geq 0$. However, since $Q^2 = 0$, there are zero norm states. For example, $Q|\phi >$ has zero norm if $|\phi >$ is not a physical state (so $|\phi >$ is not annihilated by Q).

[3]Horowitz (1986).

[4]This has been discussed by Redhead (1983) and Weingard (1984).

[5]See Green and Schwartz (1984).

[6]Georgi and Glashow (1972).

References

Georgi, H.,and Glashow, S.L. (1972), "Unified Weak and Electromagnetic Interactions without Neutral Currents", *Physical Review Letters* 28: 1494-1497.

Green, M.B. and Schwartz, J.H. (1984), "Superstring Field Theory", *Nuclear Physics* B 243: 475-536.

Horowitz, G.T., Lykken, J., Rohm, R., and Strominger, A. (1986), "A Purely Cubic Action for String Field Theory", NSF-ITP-86-61.

Redhead, M.L.G. (1983), "Quantum Field Theory for Philosophers", in *PSA 1982* vol.2, P.D. Asquith and R.N. Giere(eds.). East Lansing: Philosophy of Science Association, pp. 57-99.

Weingard, R. (1984), "Grand Unified Gauge Theories and the Number of Elementary Particles", *Philosophy of Science* 51: 150-155.

Witten, E. (1986), "Non-Commutative Geometry and String Field Theory", *Nuclear Physics* B 268: 253-294.

String Theory, Quantum Gravity and Locality

Cyrus C. Taylor

Case Western Reserve University

In this talk, I'd like to explain a little bit about what string theory is, why theoretical physicists are so excited about it, and why I think that it will have a rather profound impact on some of our ideas about the structure of the physical world.

Let me begin by reviewing the way we think about quantum mechanical particles in a relativistic setting. The classical dynamics of the theory is specified by assuming that the action for a given particle trajectory is proportional to the relativistic interval traversed by the particle. If one sets up a canonical formalism with the variable parametrizing the particle trajectory playing the role of time, then one finds that the four-momentum of the particle is constrained, with the constraint just being the relativistic relation between the mass, energy, and three-momentum of a particle. The particle thus lives on a subspace of the full phase space, and one finds that on this subspace, the Hamiltonian for the system vanishes. The quantization of this system is staightforward. Let us adopt the Schrodinger picture. Then the Schrodinger equation just tells us that the wave function doesn't depend on how we parametrize the particles trajectory. (This is a consequence of the fact that the Hamiltonian vanishes). However, we have to implement our constraint. Demanding that it annihilate the wave function when we treat it as a quantum mechanical operator, we find that we have derived the Klein-Gordan equation. For a review of the general procedure, see for example (Scherk 1975).

Quantization of the string proceeds in an analogous fashion. One postulates that the action for a given string trajectory is proportional to the area of the world sheet, sets up a canonical formalism, and again finds that the system is constrained. In this case, however, one finds that there are an infinite number of constraints, with the Hamiltonian again vanishing on the constraint surface. These constraints, which have non-vanishing Poisson brackets with each other, can be thought of as generating gauge transformations. Because of normal ordering ambiguities in the expressions for the constraints, the quantization requires more care than in the case of the particle. (This is perhaps not terribly surprising, since we can think of the phase space of the string as being infinite dimensional, corresponding to the Fourier modes of the position of the string). For details of the procedure, see, for example. (Green, Schwarz and Witten 1987). The upshot is that in twenty-six dimensions, (or ten in the case of the superstring), one can construct a unitary quantum theory. The theory now has an infinite number of states, including states that can be identified with gauge bosons (open string) and with the graviton (closed string). Further, the theory is again invariant under reparametrizations of the string world sheet. This is quite important once we consider interacting strings.

In the case of a theory of interacting particles, the picture of interactions is not terribly nice from a geometric viewpoint: the process in which, say, three particles interact at a local vertex prohibits us from thinking of the worldlines of the particles as a manifold. Three-particle interactions are distinct from four-particle interactions, and so on. In the context of a second-quantized theory, this becomes quite important, and is one way of understanding problems associated with renormalizability. Briefly, one specifies the theory by assuming some finite set of kinds of particle interactions. For example, in quantum electrodynamics, one has a single kind of three-particle interaction, corresponding to the interaction of a photon with, say, an incoming and outgoing electron. When one calculates amplitudes for the scattering of various particles, one finds typically that they diverge. If these infinities can be removed by a systematic scheme for redefining the coupling constants associated with the original interaction vertices, then the theory is renormalizable. (See Itzykson and Zuber 1980 for a textbook account of this procedure.) If not, then new vertices must be introduced to cancel the infinities. These typically require the introduction of still more vertices. Typically, these will in turn result in new infinities, and one is left with a theory in which one would need to specify an infinite amount of information in order to calculate anything. Such theories are said to be non-renormalizable, and since they have essentially no predictive power, are avoided like the plague.

Einstein's theory of general relativity, as presently understood, is non-renormalizable in four space-time dimensions. These problems are particularly evident when matter is included. See (Goroff and Sagnotti 1986), and references therein.

In the context of string theory, the picture is much rosier. Strings don't join at a point, but rather (in the case of closed strings, for example) through a process which can be depicted as a plumber's T-junction. In twenty six dimensions, the theory is reparametrization invariant, so that we can think about Feynman diagrams as being built out of T-junctions and lengths of pipe, all of which are made out of rubber. That is, we identify diagrams if they can be continuously deformed into one another. The upshot is that there really is only one kind of interaction, and so there appears to be no mechanism by which the theory can be non-renormalizable. This doesn't constitute a proof of renormalizability, and indeed, in the case of superstrings, there appear to be some technical obstructions to carrying out an actual proof; nevertheless, the basic picture is very appealing and widely accepted as being (provisionally) true. For a review of the current status of string perturbation theory, see (D'Hoker and Phong 1988).

Thus far, we have discussed string propagation in a flat twenty-six (or ten) dimensional spacetime. The first question one asks as a physicist is: where is the real world? Candelas, et al., (1985) proposed that, as a first step, we assume that the ten dimensional spacetime has the structure of a product of flat four-dimensional Minkowksi space with a six-dimensional compact manifold. The rough idea is that if the size of the internal space is small compared with the distances which we can currently probe, then we are only seeing the low energy effective string theory, which will be four dimensional. Candelas, et. al, (1985) showed that the heterotic string consistently propagates on such a manifold if it has the structure of a Calabi-Yau manifold. Further, they showed that such compactifications can lead to realistic low energy theories. The number of such Calabi-Yau compactifications is quite large, but believed to be finite (Yau 1985). Among these is a manifold yielding a low energy theory with only three generations of fermions. This may be an essentially unique three generation model (Rusjan and Senjanovic 1988; Mohapatra 1988).

Constructions of this sort are in the realm of model building, and seem rather far from deductions from some "fundamental theory of everything". Nevertheless, there are a number of beautiful aspect to the Calabi-Yau approach. The number of generations of fermions is determined by the Euler number of the internal manifold. The role of the Higgs field is played by expectation values of path-ordered exponentials of the gauge fields about non-contractible loops. Yukawa couplings are, in principle, calculable. (For a general review, see Green, Schwarz and Witten, 1987).

It is fascinating to note that it appears that some of the Calbi-Yau models can be constructed in a radically different fashion, directly in four dimensions, by adding additional degrees of freedom to the string world sheet which can be used to define conformal field theories. (See, for example, Gepner 1988/89). This raises questions about the meaning of the "dimensionality" of space in a particularly acute form. In the same context, it is interesting to note that it is possible to embed 10-dimensional superstring in 26 dimensional bosonic string theories (Casher, et al., 1985).

At this point, it should be clear that there are some rather intriguing questions of principle associated with string theory as we currently understand it. In the first place, it is apparently a theory of general relativity, in that it has a dynamical graviton, yet we have to specify some background metric in order to formulate the theory. The origin of general covariance seems terribly obscure, and the process of constructing the theory seems rather circular. Second, rather than having a "fundamental theory of everything", we appear to have at least several thousand consistent string theories, with rather obscure interconnections between them. A serious student of string theory can't help but suspect that there is some deeper theory lurking behind what we currently know of string theory. This hidden theory presumably stands in much the same relation to the string perturbation theory that we know as a full quantum field theory stands to its perturbation theory about some set of extrema of the classical action.

Perhaps the most pedestrian approach to constructing such a non-perturbative string theory is string field theory. Here, the basic idea is to define a theory which stands in the same relation to strings as an interacting scalar field theory does to particles. A variety of proposals have been made, of which the most successful thus far has been an open string field theory proposed by Witten (1986). Space does not permit a detailed discussion of the theory, but I would like to comment on a few aspects of it. The theory is formulated in terms of a field defined on the infinite dimensional configuration space of the open string, the BRS operator of the open string, and a definition of what it means to multiply two strings together. The theory looks very much like a gauge field theory defined on an infinite dimensional space, and bears many formal similarities to gauge theories in 2+1 dimensions in which the action is just a Chern-Simons term. The operators in the theory have been explicitly constructed (Gross and Jevicki 1987), and the theory apparently reproduces what is known of string theory when treated perturbatively (Giddings, Martinec and Witten 1986). Further, it has an intriguing non-perturbative structure, in which it appears that the dependance on any particular background metric can be thought of as the expansion of a more fundamental action about a particular solution of the classical equations of motion (Horowitz, et al. 1986). However, because of the essential non-locality of the string interaction, the canonical structure of the theory is not well-understood outside of the context of perturbation theory, and there are reasons to suspect that a correct canonical quantization may not reproduce the known S-matrix (Eliezer and Woodard 1989).

Finally, I would like to comment briefly on some very recent developments. Witten has been pursuing a program of study of topological quantum field theories; that is, quantum field theories in which the physical states are associated with a non-trivial topological structure. Because there is no dependence on the metric, there is no notion of locality. Particularly interesting examples are provided by the 2+1 dimensional Chern-Simons theories I mentioned above (Witten 1988a). It is particularly intriguing to note that the Einstein-Hilbert action in 2+1 dimensions can be interpreted as such a theory: general covariance is apparently realized in what Witten terms an "unbroken" phase since the expectation value of the metric vanishes and there is no notion of locality (Witten 1988b). That these theories are related to string theories comes from the observation that the whole of current algebra theory can apparently be reconstructed from such three dimensional theories.

In conclusion, it is apparent that a variety of intriguing ideas about the structure of space-time are being rapidly injected into theoretical physics. It would seem that these ideas should be of some interest to philosophers of science.

References

Candelas, P., Horowitz, G., Strominger, A. and Witten, E. (1985), "Vacuum Configurations for Superstrings", *Nucl. Phys.* B258, 46.

Casher, A., *et al.*, "Consistent Superstrings as Solutions of the D=26 Bosonic String Theory", *Phys. Lett.* 162B, 121.

D'Hoker, E. and Phong, D. H. (1988), "The Geometry of String Perturbation Theory", *Rev. Mod. Phys.* 60, 917.

Eliezer, D. A., and Woodard, R. P. (1989), "The Problem of Nonlocality in String Theory", *Brown* HET-693.

Gepner, D. (1988/89), "Yukawa Couplings for Calabi-Yau String Compactifications", *Nucl. Phys.* B311, 191.

Giddings, S., Martinec, E. and Witten, E. (1986), *Phys. Lett.* 176B, 362.

Goroff, M. and Sagnotti, A. (1986), "The Ultraviolet Behavior of Einstein Gravity", *Nucl. Phys.* B266, 709.

Green, M. B., Schwarz, J. S., and Witten, E. (1987), *Superstring Theory.* (Two volumes). New York: Cambridge University Press.

Gross, D. and Jevicki, A. (1987), "Operator Formulation of Interacting String Field Theory", *Nucl. Phys.* B283, 1.

Horowitz, G., Lykken, J., Rohm, R. and Strominger, A. (1986), "A Purely Cubic Action for String Field Theory", *Phys. Rev. Lett.* 57, 283.

Itzykson, C. and Zuber, J.-B. (1980), *Quantum Field Theory.* New York: McGraw Hill.

Mohapatra, P. (1988), "Comments on Possible New Three Generation Calabi Yau Manifolds", *Phys. Lett.* 214B, 199.

Rusjan, E. and Senjanovic, G. (1988), "Honest Symmetries and Complex Structures of the Three-Generation Superstring Model", *Phys. Lett.* 214B, 193.

Scherk, J. (1975), "An Introduction to the Theory of Dual Models and Strings", *Rev. Mod. Phys.* 47, 123.

Witten, E. (1986), "Non-Commutative Geometry and String Field Theory", *Nucl. Phys.* B268, 253.

_____. (1988a), "Quantum Field Theory and the Jones Polynomial", to appear in *Comm. Math. Phys.*

_____. (1988b), "2–1 Dimensional Gravity as an Exactly Soluble System", to appear in *Nucl. Phys.* B.

Yau, S.-T. (1985), "Compact Three Dimensional Kahler Manifolds with Zero Ricci Curvature", in *Symposium on Anomalies, Geometry, Topology*, W. Bardeen and A. White (eds.). Singapore: World Scientific, p. 395.

Lorentz Invariant State Reduction, and Localization

Gordon N. Fleming

Penn State University

1. Introduction

In this paper I will present conceptions of state reduction and particle and/or system localization which render these subjects fully compatible with the general requirements of a relativistic, i.e. Lorentz invariant, quantum theory. In the case of state reduction, the concept presented has, in it's main features, been advanced by several investigators in recent years, (Giovannini 1983), (Aharonov and Albert 1984), (Malin 1984), (Dieks 1985). My own contribution to this topic has been to elaborate the concept in question and examine, in some detail, the often counterintuitive but never causally anomalous or internally inconsistent consequences that emerge from taking the concept seriously (Fleming 1985, 1986). In the case of particle and system localization, the concepts are more nearly my own and have been pursued by me for some time (Fleming 1964, 1965, 1966) as a way of freeing the subject of localization in relativistic quantum systems from the conceptual problems pointed out by Newton and Wigner (1949) myself (Fleming 1965) and extensively studied by Hegerfeldt et. al. (1974, 1980) and Ruijsenaars (1981). The connection between the two topics of state reduction and localization lies in the fact that, as presented here, both issues are clarified and formulated in a Lorentz covariant way by the same approach. That approach consists of the systematic generalization of the following concepts: initial data assignment at definite times, initiation and completion of measurement at definite times, and dynamical evolution as time dependence; to the concepts of initial data assignment on arbitrary space-like hyperplanes, initiation and completion of measurement on arbitrary space-like hyperplanes, and dynamical evolution as space-like hyperplane dependence, respectively.

These opening comments concerning state reduction should suggest to the reader that I take state reduction as representing a real physical process. I am of the persuasion, which I will not defend here, that state vectors represent ontological, physically real situations, and not, in any sense, our knowledge of those situations. By contrast, density operators for so-called mixed states, (when they are not the density operators of subsystems of composite systems which, latter, are in pure states — the relation between the two uses of density operators being a very interesting and, as yet, obscure subject) represent a hybrid combination of ontological and epistemic situations. Consequently, when the correct state vector for describing a physical system or an ensemble of physical systems must change as a result of a measurement act this change represents an ontological change, the occurrence of a physical process. I, furthermore, am of the opinion that state vector reduction is not confined to measurement events in the laboratory. With Abner

PSA 1988, Volume 2, pp. 112-126
Copyright © 1989 by the Philosophy of Science Association

Shimony (1986) I believe that "We are just not that important in the scheme of things." to be the only agents of state reduction. If, with John Wheeler, (1983) I must regard these reductions as bringing into existence the properties of the macroscopic world as we know it, I do not concur that it is all our doing. The Universe is fast at work bringing it's own macroscopic properties into existence, thank you! We have, of course, very little understanding of these stochastic dynamical processes at present. We must turn to the model builders of state reducing dynamics (Zurek 1981), (Joos and Zeh 1985), (Ghirardi *et al.* 1986), (Karolahazy *et al.* 1986), (Pearle 1986), and others, for what understanding has been garnered. But we do know that if quantum theory is correct then the accumulation of state reductions will not yield Peirce's prediction (1891) "At any time, however, an element of pure chance survives and will remain until the world becomes an absolutely perfect, rational, and symmetrical system, in which mind is at last crystallized in the infinitely distant future." More than an element of pure chance must always remain since with every "actualization of a potentiality" (Riezler 1940), (Heisenberg 1958), (Shimony 1986) quantum theory requires a corresponding potentialization of actualities. This being an inevitable consequence of the existence of incompatible observables.

For my part the best verbal description of the ontological situation represented by a state vector occurs in the writings of Abner Shimony, whose language I have already employed. He refers to "networks of evolving potentialities" (Shimony 1986) some of which are "actualized" (ibid.) under those circumstances in which state vectors are reduced. Consequently, what we normally call properties of physical systems are brought into existence by these actualizations and, with the exception of those cases of conserved properties, do not remain actualized for long but are replaced again by potentialities. To this surreal vision one hears the response "What, then, are the enduring properties of quantum systems? Do they have no properties independent of these actualizations? And if not, what are we talking about when we refer to them?" They do indeed have enduring properties independently of the occurrence of actualizations, although the actualizations tend to alter the enduring properties when they occur. It is just that the enduring properties are very unlike classical properties of physical systems. The enduring properties are just the evolving potentialities! They evolve deterministically in the absence of actualizations and stochastically in the presence of them. An electron is a system with a particular manifold of possible potentialities. A proton is a system with a different manifold of possible potentialities. The manifold for an electron-proton system is different again and not just the union of the manifolds for the previous systems. Thus the potentialities are in no sense less real than the results of actualizations. Rather potentialities and actualities are, again in Shimony's words "modalities of reality" (1986) enjoying equal status.

My initial task here is to comment on the issue of state vector assignments prior to and anterior to potentiality-actualizing-state-reducing-processes and the changes forced upon us, in those assignments by the special theory of relativity. This will be the subject of section 2. In section 3 the subject of localization will be taken up and in section 4. we will discuss briefly the superluminal propagation that emerges from the localization study and the manner in which causal anomalies are nevertheless avoided.

2. Lorentz Invariant State Reduction

A basic assumption of the theory of relativity is that if an hypothetical physical phenomena is describable from the standpoint of one inertial frame of reference then it must be describable from the standpoint of any inertial frame of reference. Furthermore, the rules for translating the description of such a phenomena from one inertial frame to the other must be capable of formulation without regard to the possibility or impossibility of the phenomena, as a consequence of the laws of dynamical evolution, but only with regard to the relation between the inertial frames in question and the nature of the mathematical structures required for the description of the phenomena in the initial frame. This assumption is rarely explicitly stated, but is nevertheless universally employed, even in

114

the process of elucidating the basic consequences of the theory of relativity itself, which process often employs the frames of reference in which the description of the phenomena of interest is simplest, whether those frames of reference include the initial one or not. An illuminating discussion of this assumption is to be found in Wigner (1956). The considerations of this section are motivated by the observation that a naive application of the concept of state reduction, as a process beginning and ending at definite times, to the relativistic domain, entails the violation of the aforesaid assumption (Bloch 1967), (Aharonov and Albert 1981). If the assumption is to be retained the concept of state reduction must be modified, for we will see that there are circumstances in which state reductions occurring in definite time intervals in one inertial frame have no translation at all to state reductions within definite time intervals in most other inertial frames.

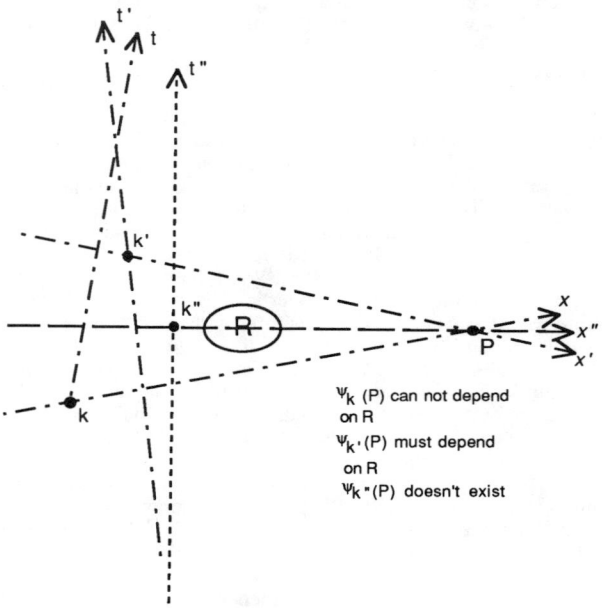

Figure 1. Bloch's paradox

The first indication that definite time state reductions were conceptually inadequate in the relativistic context was provided by Bloch (1967). A simplified version of his argument goes as follows: Consider a bounded space-time region in which a state reducing interaction will take place. Call the region R (Fig. 1.). Suppose the system subject to the interaction consists of a single particle with position observable, x , and that the state reducing interaction leads to a state with a finite probability that the system remains a single particle of the same type. Now consider a space-time point, P, separated from R by space-like intervals exclusively. In some frames of reference, call them K, this point will be earlier than any point in R and so the position representation wave functions for the particle, at the point in question, will have values depending only on the initial state of the system and completely independent of the state resulting from the reduction, which, in the frame of interest, has not occurred yet. In some other frames of reference, call them K', the space-time point will be later than any point in R and so the position wave functions for the particle at that point will have values that do depend on the result of the reduction. Consequently there can not be any translation rule for inferring the position representation wave function at the point of interest in any of the K' frames from the value of the wave function at the point of interest, or, for that matter, from the values of the wave function at all points of space at the time of the point of interest, in any of the K frames. The reason is that the change is, in part at least, a dynamical one and the translation rule we are considering can not depend on the conditions of dynamical evolution. Furthermore, there is a third set of frames, call them K", in which the situation is even worse. For these frames in which the space-time point of interest has a time coordinate simultaneous with points lying inside R, the state reducing interaction is still in progress and so, at the point P, no definite state function for the particle exists at all! For these frames there is nothing to translate to.

The example we have just considered employed the notion of a position representation wave function and, since that notion is itself a matter of uncertain status, for relativistic particles, and will be taken up at length in the next section of this paper, the reader may question the significance of my conclusion on that account. The next example does not rely on any such notion, but obtains the requisite consideration of space-like separated regions by employing two state reducing interaction regions. It is related to but distinct from several ingenious examples presented by Aharonov and Albert (1980, 1984) and was presented, in altered form, by Shimony (1986).

We consider an idealized EPR experimental apparatus lying on an optical bench (Fig. 2a). The detectors at either end of the optical bench are equidistant from the correlated photon emitter, at the center of the bench. Consequently, after the photons emerge from the emitter in the two photon entangled state

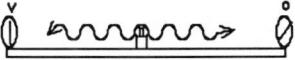

(2a) No intermediate state.
Bench or flatcar frame.

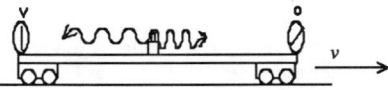

(2b) Intermediate state, $|v'>_L |h'>_R$.
Ground frame.

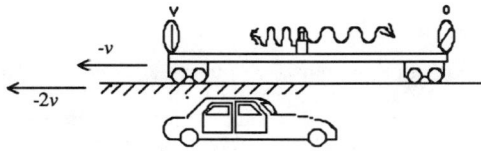

(2c) Intermediate state, $|\underline{o}">_L |o">_R$.
Car frame

Figure 2. EPR from three frames.

$$|A> = (1/2)^{1/2}\{ \; |h>_L|v>_R + |v>_L|h>_R \} = (1/2)^{1/2}\{ \; |o>_L|\underline{o}>_R + |\underline{o}>_L|o>_R \}$$

(where $|h>$, $|v>$, $|o>$ and $|\underline{o}>$ represent horizontal, vertical, and two mutually perpendicular oblique linear polarization states for single photons, respectively) they will each encounter their respective detectors simultaneously. If, as is customary, the detectors are sensitive to the linear polarization of the photons, and if the left detector lets vertically polarized photons through while the right detector lets photons through with polarization in the oblique direction designated by "o", then there are four possible distinct outcomes of the photon detector encounters; neither photon gets through, only the photon moving to the right gets through, only the photon moving to the left gets through, or both photons get through. For the purposes of this discussion I will assume the last possibility occurs but any of the possibilities would yield a similar analysis. For the last possibility then, both photons getting through the detectors, which occurs with a probability depending on the angle between the polarization axes of the two detectors, the final state of the system, after the simultaneous detection encounters, is

$$|v>_L|o>_R.$$

We now recognize that our optical bench is on a rolling flatcar (Fig. 2b) moving to the right with uniform velocity relative to the ground. From the standpoint of an observer at rest on the ground the detection events are not simultaneous. Instead the left moving photon encounters the leftmost detector first. When this happens, given our hypothesis that both photons get through their detectors in the instance considered, the state of the system changes to

$$|v'>_L|h'>_R$$

as a consequence of the correlation between the photons in the original entangled state. The two photon state retains this form until the right moving photon encounters the right most detector, whereupon the state changes to

$$|v'>_L|o'>_R$$

(The primes refer to the fact that these factor vectors are the unitary Lorentz transforms of the formally corresponding unprimed vectors from the flatcar frame.) The problem with this description is that the intermediate state in the ground frame has no translation in the flatcar frame as a state! There is no *time* in the flatcar frame at which the Lorentz transform of the ground frame intermediate state is the correct flatcar frame state. Finally, if we consider an observer in a car moving to the right over the ground faster than the flatcar is (Fig. 2c), then for such an observer the flatcar is moving to the left and it is the right moving photon that will encounter its detector first. Again there will be an intermediate state for the car frame time interval between the two detections. The intermediate state will be,

$$|\underline{o}''>_L|o''>_R,$$

Again this state will have no translation to any state actually occurring, at *any time*, in the flatcar frame. Furthermore, the intermediate states in the ground and car frames are clearly not the Lorentz transforms of each other since the intermediate state in the ground frame depends on the outcome of the left detection event while the intermediate state in the car frame depends on the outcome of the right detection event. The translation requirement of the theory of relativity is clearly violated here! As was the case for the previous two frames, translatability is restored for the final state, since, after the leftmost detection event, the state in the car frame is,

$$|v''>_L|o''>_R$$

How are we to understand this apparent breakdown of the translatability requirement? The answer that follows focuses upon the spatially extended character of the photon state vectors in the example and contains no dependence on the degree of that extension. At first glance this may seem like an extreme measure when one contemplates application to spatially local states such as position eigenstates, but, as we shall see, the solution is compatible with what is needed to implement a consistent relativistic quantum theory of localized position eigenstates. To proceed with the solution, then, we note that in all the frames considered the states referred to definite time intervals, and may be regarded as referring to all the instants of time lying in those intervals, but they did not refer to definite intervals of space; in a position representation they would be indefinitely spatially extended. If we now ask for a space-time geometrical construct which corresponds to a definite time during the intermediate period in either the ground or car frame of reference we recognize that without specified spatial limitations associated with those times we must identify the construct with the space-like hyperplane which is instantaneous in the reference frame of interest and has the desired time coordinate in that frame. This is the invariant geometrical construct on which we can expect to find the translation of the intermediate state of interest when we transform to any other frame of reference. In particular, in the flatcar frame, there are two sets of non-instantaneous, i.e. "tilted" space-like hyperplanes, the members of which have definite intermediate time values in the ground and car frames respectively (Fig. 3). The unitary Lorentz transforms of the intermediate ground and car frame states are to be associated with these "tilted" hyperplanes in the flatcar frame. Similarly, the states associated with definite times in the flatcar frame translate into states associated with the same invariant hyperplanes, which now appear non-instantaneous, in the relatively moving ground and car frames. The reason we had no trouble translating the flatcar states to the ground and car frames before mentioning "tilted" hyperplanes is that the instantaneous hyperplanes of the flatcar frame prior to or anterior to both detection events can be continuously moved into ground or car frame instantaneous hyperplanes prior to or anterior to both detection events, respectively, without crossing either of the detection events. Since we are working in the Heisenberg picture the state vectors associated with various hyperplanes in a single reference frame do not change if the dynamical evolution

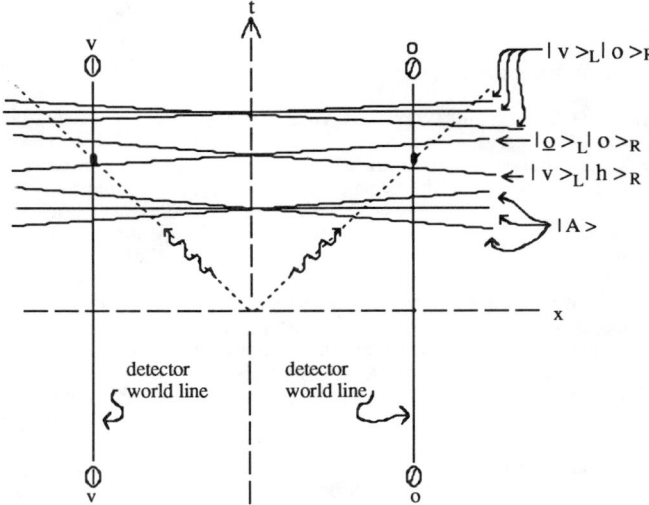

Figure 3. State vector assignments to hyperplanes in the flatcar frame.

from one hyperplane to another is purely unitary. In fact, in any reference frame, the Heisenberg picture state vector assignment changes only when we pass from one hyperplane to another which lies on the opposite side of one state reducing interaction region or another. For any hyperplane not intersecting any state reducing interaction region, state vectors can be assigned to it in any reference frame and the assignments will always be the unitary Lorentz transforms of one another. In this way the translation problem is completely resolved. But we must now take seriously the notion of dynamical evolution as the evolution of dynamical variables and, if state reducing interaction regions are present, state vectors from space-like hyperplanes to space-like hyperplanes.

In terms of the resolution offered here to the translation problem for the relativistic aspects of the EPR experiment, we can begin to see how the Bloch paradox, discussed earlier, can be resolved. The novelty is that we must take into consideration the hyperplane on which the state function is being evaluated and not only the point at which it is being evaluated!

In the first set of reference frames considered, in which the space-time point of interest lies earlier than any point in the state reducing interaction region, R, the state function is being evaluated on hyperplanes, passing through the point of interest, that lie prior to R. On such hyperplanes the state function can not depend on the outcome of the state reduction. In the second set of reference frames considered the space-time point lies later than R and the state function is being evaluated on hyperplanes, passing through the point of interest, that lie anterior to R. On such hyperplanes the state function must depend on the outcome of the reduction. Finally, in the third set of reference frames, one is asking for the value of the state function on hyperplanes passing through R itself. But no state function can be associated with such hyperplanes in *any* reference frame. In any reference frame in which a hyperplane, to which a state function can be assigned, does not appear instantaneous, that hyperplane nevertheless carries the unitary transformation of the state function originally assigned in the frame in which the hyperplane was instantaneous. In the next section we will see that the novel, and perhaps disturbing, feature of hyperplane dependence of position representation state function assignments in any single frame of reference is, in fact, required by a Lorentz invariant concept of localization, even in the absence of state reducing interaction regions.

3. Lorentz Invariant Localization

In 1949 Newton and Wigner (N-W) identified those non-normalizable state functions in the positive energy subspaces of the Klein-Gordon and Dirac formalisms which corresponded to precisely spatially localized, positive energy, Klein-Gordon or Dirac particles. The need for an explicit identification arose from the fact that the spatial coordinates of the Klein-Gordon or Dirac wave functions can not be identified as the eigenvalues of any self adjoint operator defined within the positive energy subspaces of those formalisms. This, in turn, is a consequence of the facts that in the Klein-Gordon formalism the state function and it's first time derivative, and in the Dirac formalism the four component state function, can not be proportional to Dirac delta functions of position at any given time unless they contain both positive and negative energy contributions. The focus on the positive energy subspaces was, and is motivated by the earlier reinterpretation of the negative energy state functions as anti-particle positive energy state functions. Some results of the N-W identification seemed troubling; for having uniquely determined the localized position eigenstates by reasonable assumptions concerning their transformation properties under an active application of the members of the Euclidean group of spatial translations and rotations, Newton and Wigner found that, for example, a pure Lorentz transformation, or boost, of a position eigenstate localized at the spatial origin of coordinates at the time, $t=0$, resulted in a state that was not localized at all, anywhere, at any time! Subsequently others (Fleming 1965), (Hegerfeldt 1974) found that the N-W position eigenstates evolved superluminally as regards the spatial distribution of the position probability density after the initial localization. These features undermined confidence in the use of these state functions for their originally intended purpose. In assessing this situation, however, it must be kept in mind that these localized states were not created anew. Rather, they were identified as already existing structures within a widely used and accepted formalism. If we can not interpret them as localized states, then we are confronted with the problem of how they are to be interpreted.

Before showing how the N-W states can be regarded as members of a larger class of states, which, taken together, generate a coherent concept of Lorentz invariant localization, there is a simple and instructive way to see that the concept of *localization at a definite time* is, by itself, too restrictive for a relativistic quantum theory. Working in the Heisenberg picture, and in one spatial dimension for simplicity, let $X(t)$ be some particle position operator for the time, t. The expectation value of this operator in the state, $|A>$, is; $<A|X(t)|A>$. Under a passive Lorentz transformation to a new frame moving with velocity, v, relative to the original frame, we naively expect the transformed state vector, $|A'>$, to yield the transformed expectation value,

$$<A'|X(t')|A'> = (<A|X(t)|A> - vt)/(1 - v^2/c^2)^{\frac{1}{2}}$$

where

$$t' = (t - (v/c^2)<A|X(t)|A>)/(1 - v^2/c^2)^{\frac{1}{2}}$$

But this transformation scheme, in which the time variable of the operator after the transformation depends on the state used in calculating the expectation value, makes it impossible to extract a transformation rule for the position operator itself, and consequently for it's eigenvectors, the localized states!

To resolve these problems of the N-W localized states, which are fundamentally problems of the conceptual interpretation of a formalism, we turn our focus from the position operator and it's eigenvectors themselves, to the projection operators for locating the physical system, particle or otherwise, within some spatially extended but bounded region, S, at a definite time, t. The eigenvalues and most general eigenvectors of this projection operator, refer to the possible outcome of a search, over the region S, at the time t. If we now ask what this search or it's most general possible outcomes look like from the standpoint of a Lorentz transformed reference frame, we see immediately that the resulting description would not refer to any definite time at all in the new frame. Instead, because of the spatial extension of S in the original frame, we have to recognize region S as belonging to a space-like hyperplane, which was instantaneous at the time t in the original frame, but which is not instantaneous, but rather "tilted" in the new frame. Accordingly, the projection operator, in the new frame, is not associated with any definite time at all, but rather with the "tilted" hyperplane. The hyperplane in question is, of course, *the same one* in both frames of reference; it just has a different appearance and description in the two reference frames.

Continuing this line of reasoning we see that the general eigenvectors of the projection operator will, in the new frame, also be associated with the "tilted" hyperplane, and not with any definite time. We do start to balk here because we want to argue that this reasoning is all well and good for those eigenvectors which are only localized within S but not at any point of S. But what of the latter? Surely they can be associated with a definite time in the new frame as well as the old one. Of course a trivial association of the "tilted" hyperplane localized states with a definite time can be made since the point of localization in S has a definite time coordinate. The physically significant question, however, is the relationship of these "tilted" hyperplane localized states to the states which are the eigenstates of the position operator at a definite time. The experience of Newton and Wigner implies that the two kinds of states are very different. This is confirmed by constructing the position operator for localization on the "tilted" hyperplane out of the "tilted" hyperplane projection operators, in the same way that the position operator at a definite time is constructed out of the projection operators for regional localization at a definite time. Two interesting results emerge from this construction. First, the "tilted" hyperplane position operator turns out to satisfy a unitary, manifestly covariant, Lorentz transformation rule, both with respect to the original definite time position operator from which the projection operators were originally obtained, and under arbitrary inhomogeneous Lorentz transformations as soon as one extends the concept of a hyperplane dependent position operator to arbitrary space-like hyperplanes. There is no residue of the original transformation problem entangling the transformed time coordinate with the state dependent expectation value of the position. Second, the position eigenvectors of our constructed, "tilted" hyperplane position operator, or any generalization of it to an arbitrary hyperplane, are very specific to the hyperplane with which they are associated. For a given space-time point and any two distinct space-like hyperplanes containing that point, the localized states associated with the two distinct hyperplanes, although localized at the same space-time point, are different states! In particular each of them is orthogonal to all of the other localized states belonging to the same hyperplane but localized at different points on it, while at the same time neither of them are orthogonal to any of the localized states belonging to the other hyperplane! Can we make sense of these features?

The manifestly covariant transformation properties of the position operators and their eigenvectors are very easy to understand. So we will look at that first. In any inertial reference frame employing Minkowski coordinates the coordinates of the space-time points comprising a space-like hyperplane satisfy an inhomogeneous linear equation of the form

$$\eta^\mu x_\mu = \tau$$

where, η^μ, is a time-like unit four vector with, by convention, a positive time component. The hyperplane as a whole, then, can be represented by the ordered pair, (η, τ). In any other inertial frame set of Minkowski coordinates, $x^\mu{}'$, related to the first set by the inhomogeneous Lorentz transformation

$$x^\mu{}' = \Lambda^\mu{}_\nu x^\nu + a^\mu$$

the same hyperplane will be represented by the ordered pair, (η', τ'), where

$$\eta^\mu{}' = \Lambda^\mu{}_\nu \eta^\nu$$

and

$$\tau' = \tau + a_\mu \Lambda^\mu{}_\nu \eta^\nu$$

Now if the state vector in the first inertial frame (and, in the presence of state reducing interaction regions, for the given hyperplane) is, $| \psi >$, and the transformed state vector for the second inertial frame is

$$| \psi' > = \cup(\Lambda, a) | \psi >$$

where, $\cup(\Lambda, a)$, is the unitary operator effecting the inhomogeneous Lorentz transformation in the state space, then the transformation of position vector expectation values that we intuitively require is

$$<\psi' | x^\mu (\eta', \tau') | \psi' > = \Lambda^\mu{}_\nu <\psi | x^\nu (\eta, \tau) | \psi > + a^\mu$$

This already has a manifestly covariant form and we extract the transformation equation for the operator itself by simply writing, $| \psi' >$, in terms of, $| \psi >$, on the left hand side and then removing the arbitrary state vector, $| \psi >$. This yields

$$\cup(\Lambda,a)^* \, x^\mu \, (\eta',\tau') \, \cup(\Lambda,a) = \Lambda^\mu_{\ \nu} \, x^\nu \, (\eta,\tau) + a$$

It should be borne in mind that the operator valued 4-vector, $x^\mu \, (\eta,\tau)$, does not have four *independent* operator components since all the expectation values and eigenvalues of it must lie on the (η,τ) hyperplane. This requires the operator equation

$$\eta^\mu x_\mu \, (\eta,\tau) \, = \, \tau$$

to be satisfied. In that frame of reference in which the hyperplane is instantaneous this constraint equation tells us that the non-trivial operator content is confined to the spatial components of the position operator and, in that case, the position operator is just the N-W position operator.

The hyperplane dependence of the eigenvectors of the position operators, while following formally immediately from the hyperplane dependence of the position operators themselves, requires more in the way of interpretive deliberation to understand. I think that here we must employ Bohr's famous emphasis on the dependence of the outcome of measurement on the conditions of measurement. Because of our inclination to detach state reduction from measurement as its exclusive arena of occurrence we should rephrase Bohr's dictum to assert the dependence of the outcome of state reduction on the conditions of state reduction. In the present instance the specification of the conditions of state reduction include the identification of the hyperplane over which, in effect, the particle or the position eigenvalue is searched for. From the formalism one finds that the position operators and the associated projection operators for intersecting distinct hyperplanes do not commute, and, in accordance with the general principles of quantum theory this means that the corresponding measurements are incompatible. No wonder, then, formally or intuitively, that the position eigenstates for coincident localization on distinct intersecting hyperplanes are quite different.

Furthermore, there is a classical analogue for this feature (Fleming 1965). Suppose we have a classical stress-energy-momentum tensor field which is everywhere locally conserved and, on any space-like hyperplane has a spatially extended but spatially bounded support. In any inertial frame we can define a center of energy position and trace out its world line as a function of time. For most stress-energy-momentum fields, one can show that the world lines we get for the centers of energy from relatively moving inertial frames will all be different. In any one frame all of these different position vectors look like generalizations of the center of energy to arbitrary hyperplanes, and the different world lines indicate the non-trivial hyperplane dependence of the generalized position vectors. Now if, at the classical level, the world lines are hyperplane dependent, then, at the quantum level, localization at a given point on a given hyperplane must result in a different state than localization at the same point but on a different hyperplane!

To get this classical analogue we have had to resort to the notion of spatially extended systems. Since, in fact, quantum mechanical position operators do behave in the manner we are focusing on, I take this as prima-facia evidence that we should regard all relativistic quantum systems, including single particles, as spatially extended, quite independently of the presence or nature of any dynamical interactions!

4. Superluminal Propagation and Causal Anomalies

It remains for us to consider the dynamical evolution of the position operators and the associated localized eigenstates that our considerations have led us to introduce. Alternatively, we may consider the evolution of state functions in the position representation defined by these localized eigenstates. In either case we see the hyperplane dependent version of the space-like or superluminal propagation seen long ago in the original Newton-Wigner time dependent representation. Nevertheless, as a consequence of the way in which the superluminal propagation is interwoven with the hyperplane dependence, the former is not internally inconsistent, just highly counterintuitive! My task here is to make this last statement plausible.

First let us be clear about what is meant by the claim of superluminal propagation. If | x, η > and | x ', η' > are two position eigenvectors for localization on hyperplanes of orientation η^μ and $\eta^{\mu'}$ respectively and at the Minkowski points x^μ and $x^{\mu'}$ respectively on such hyperplanes, then if x and x' are space-like separated, i.e.

$$(x - x')^2 = (x^0 - x^{0\prime})^2 - (\mathbf{x} - \mathbf{x}')^2 < 0$$

then

$$< x, \ \eta \mid x', \ \eta' \ > \ =/= 0$$

unless $\eta = \eta'$ and $(x - x')\eta = 0$.

In other words, since the inner product in question is the conditional probability amplitude for finding the position observable to acquire the value x' on a hyperplane of orientation η,' given that it definitely had the value x on a hyperplane of orientation η, we conclude that this conditional probability is not zero for space-like separation of the points of localization unless the two distinct points are localized on the same hyperplane. In terms of position representation state functions,

$$\psi(x, \ \eta) = < x, \ \eta \mid \psi >,$$

we conclude from this that if such a state function vanishes outside of a bounded region on one hyperplane, it will nevertheless be found to have non-zero values at some points lying outside the envelope of the forward light cones of the bounded region on other hyperplanes. In fact, detailed calculation shows that such non-zero values will be the case almost everywhere outside the envelope (Hegerfeldt and Ruijsenaars 1980). This means that if a state reduction resulted in definite confinement of the position observable to such a bounded region, there would be a non-zero probability of a "subsequent" state reduction confining the observable to a region that is wholly space-like with respect to the original region, on a different hyperplane.

From the purely formal standpoint, the internal consistency of the dynamical evolution I have described is guaranteed if one can show that the differential forms of the integrability conditions for η and τ dependence are satisfied. These integrability conditions take the form of commutation relations among the η and τ derivatives. They are given by

$$(\delta/\delta\tau)(\delta/\delta\eta^\mu) - (\delta/\delta\eta^\mu)(\delta/\delta\tau) = 0$$

and

$$(\delta/\delta\eta^\mu)(\delta/\delta\eta^\nu) - (\delta/\delta\eta^\nu)(\delta/\delta\eta^\mu) = \eta^\mu(\delta/\delta\eta^\nu) - \eta^\nu(\delta/\delta\eta^\mu)$$

and they are always satisfied in the formalism. Their being satisfied assures that the value of a dynamical variable or an eigenvector or a state function on a given hyperplane is unambiguous in the sense of being independent of the sequence of hyperplanes used to propagate the value from the hyperplane on which initial conditions were prescribed.

Nevertheless, it is still difficult to see, intuitively, how the hyperplane dependence can avoid the kind of causal paradoxes that are familiar to every student of special relativity and which are used to compel assent to the proscription against superluminal propagation. Let us consider such an apparent paradox. We consider a particle confined to a box at rest in some inertial frame, K, say until the time, t, at which the box is opened and the particle escapes. The state function for the escaping particle, evaluated at definite times in the frame K, disperses superluminally, although with an exponentially damped amplitude outside the forward light-cone envelope of the box opening (Fig. 4). Now imagine a particle detector moving relatively to $\overline{K}$ and receding from the box at the time, in K, that the particle escapes. There is a finite probability that this detector will detect the escaped particle at a space-like interval from the box opening "event". Such a detection, however, itself consists of localizing the particle again, and, in the frame, K', of the receding detector the subsequent space-like propagation of the particle can take it to points spatially near to the box but preceding the opening. If, now, we presume another detector attached to the outside of the box and set to lock the box forever if it is excited, we have the makings of our paradox: for upon opening the box, there is a finite probability that the escaping particle will trigger a sequence of two detections that will prohibit the original opening of the box!

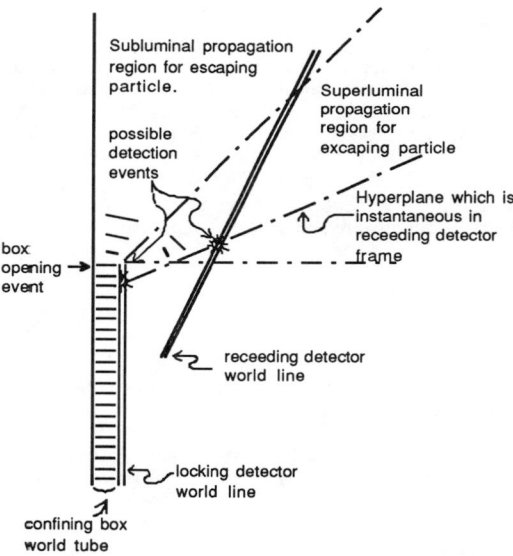

Subluminal propagation region for escaping particle.

possible detection events

Superluminal propagation region for excaping particle

Hyperplane which is instantaneous in receding detector frame

box opening event

receding detector world line

locking detector world line

confining box world tube

Figure 4. Confining box "paradox".

The resolution of this paradox depends crucially upon careful attention being paid to the specification of the orientations of the hyperplanes on which the original confinement and the subsequent detections occur! For a free particle it is easy to show that confinement within a bounded region of one hyperplane is incompatible with such confinement on any other hyperplane whatsoever. That similar incompatibilities exist in the presence of interactions is not so clear as a universal result but it has been shown to be the case for an interesting family of interactions (Fleming and Bennett 1989) and we accept the generality of the result for our present purposes. This being presumed, let us first suppose that confinement in the box is on those hyperplanes that are instantaneous in the rest frame of the box and that detection-localization occurs on those hyperplanes that are instantaneous in the rest frames of the detectors. In this case, the escaped particle, relocalized by the receding detector, can superluminally propagate into the local past of the box opening event only on hyperplanes on which the locking detector is not sensitive. This is a consequence of our general prescription that the effect of a state reduction is manifested only on those hyperplanes lying anterior to the state reducing region. Therefore, if, as supposed, the receding

124

detector localization takes place after the box opening, in the box frame, then the paradoxical locking detection can not occur. On the other hand, since the particle is never confined on the hyperplanes that are instantaneous in the frame of the receding detector, that detector could register the presence of the particle on such hyperplanes, and at box frame times that are earlier than the opening of the box. In that case the superluminal propagation of the particle could put the particle just outside the box on box frame instantaneous hyperplanes prior to the opening of the box and the locking detector could be triggered. This is consistent, however, since the particle did not get out of the box by opening it but rather by being detected early on a hyperplane within which the particle is never confined. There are clearly two distinct ways of getting out of the so-called confining box!

The problem is simpler to analyze if the box confines and the detectors all detect on the same set of parallel hyperplanes. In that case the detectors can never be excited by the particle unless the box is opened, and once opened the subsequent superluminal propagation will never take the particle to a point prior to the opening on a locking detector sensitive hyperplane. These arguments certainly do not constitute a proof of the internal consistency of the hyperplane dependent formalism and its associated conceptual scheme. They do, however, make the internal consistency more plausible than seems the case at first glance. With the identification of hyperplane dependence we see the need to specify at least some kinds of initial data in a hyperplane specific way. The hyperplane dependent dynamical evolution gives rise to paradoxes if we ignore this hyperplane specific character of certain kinds of initial data. Considerations of space prevent me from elaborating this point here and I must content myself with these plausibility arguments. A careful treatment will be presented elsewhere.

References

Aharonov, Y. and Albert, D. Z. (1980), "States and Observables in Relativistic Quantum Field Theories." *Physical Review*, D21: 3316.

_____. (1981), "Can We Make Sense out of the Measurement Process in Relativistic Quantum Mechanics?" *Physical Review*, D24: 359.

_____. (1984), "Is the Usual Notion of Time Evolution Adequate for Quantum Mechanical Systems? II. Relativistic considerations." *Physical Review*, D29: 228.

Bloch, J. (1967), "Some Relativistic Oddities in the Quantum Theory of Observation." *Physical Review*, 156: 1377.

Dieks, D. (1985), "On the Covariant Description of Wavefunction Collapse." *Physics Letters*, 108A: 379.

Fleming, G. N. (1964), "Covariant Position Operators, Spin and Locality." *Physical Review*, 137: B188.

_____. (1965), "Nonlocal Properties of Stable Particles." *Physical Review*, 139B: 963.

_____. (1966), "A Manifestly Covariant Description of Dynamical Variables in Relativistic Quantum Theory." *Journal of Mathematical Physics*, 7: 1959.

_____. (1985), "Towards a Lorentz Invariant Quantum Theory of Measurement." (Penn State preprint), unpublished.

_____. (1986), "On a Lorentz invariant quantum theory of Measurement" in *New Techniques and Ideas in Quantum Measurement Theory.* D. Greenberger (ed.) New York Academy of Sciences, no. 480.

_____ and Bennett, H. (1989), "Hyperplane Dependence in Relativistic Quantum Mechanics." *Foundations of Physics*, 19: 231.

Giovannini, N. (1983), "Relativistic Kinematics and Dynamics: a New Group Theoretical Approach." *Helvetica Physica Acta*, 56: 1002.

Hegerfeldt, G. C. (1974), "Remark on causality and Particle Localization." *Physical Review*, D10: 3320.

_____ and Ruijsenaars, S. N. M. (1980), "Remarks on Causality, Localization, and Spreading of Wave Packets." *Physical Review*, D22: 377.

Heisenberg, W. (1958), *Physics and Philosophy.* Harper and Row, New York.

Ghirardi, G. C., Rimini, A. and Weber, T. (1986), "Unified Dynamics for Microscopic and Macroscopic Systems". *Physical Review*, D34: 470.

Joos, E. and Zeh, H. D. (1985), "The Emergence of Classical Properties Through Interactions with the Environment." Zeitschrift f. *Physik,* B59: 223.

Ka'rolyha'zy, F., Frenkel, A. and Luka'cs, B. (1986), "On the Possible Role of Gravity in the Reduction of the Wave Function" in *Quantum Concepts in Space and Time,* R. Penrose and C. Isham (ed.) Oxford University Press.

Malin, S. (1984), "The Meaning and Significance of Quantum States", *Foundations of Physics*, 14, 1083.

Newton, T. D., and Wigner, E. P. (1949), "Localized States for Elementary Systems". *Reviews of Modern Physics*, 21: 400.

Pearle, P. (1986), "Models for Reduction" in *Quantum Concepts in Space and Time* R. (ibid.)

Peirce, C. S. (1891), "The Architecture of Theories". *The Monist*, 1, 161; reprinted in *Collected Papers of Charles Sanders Peirce*, VI, 6.33 C. Hartshorne and P. Weiss (ed.) Harvard University Press (1935).

Redhead, M. L. G. (1983), "Nonlocality and Peaceful Coexistence", in *Space, Time and Causality*, R. Swinburne (ed.) Dordrecht: Reidel.

Riezler, K. (1940), "Physics and Reality". Yale University Press.

Ruijsenaars, S. N. M. (1981), "On Newton-Wigner Localization and Superluminal Propagation Speeds". *Annals of Physics*, 137: 33.

Shimony, A. (1986), 'Events and Processes in the Quantum World", in *Quantum Concepts in Space and Time* (ibid.)

Wigner, E. P. (1956), "Relativistic Invariance in Quantum Mechanics". *Nuovo Cimento*, III, 517.

Wheeler, J. A. (1983), "Law Without Law", in *Quantum Theory and Measurement*, J. A. Wheeler and W. H. Zurek (ed.) Princeton University Press.

Zurek, W. (1981), "Pointer Basis of Quantum Mechanics; into What Mixture Does the Wave Packet Collapse?" *Physical Review*, D24, 1516.

On the Possibility that the Present Quantum State of the Universe is theVacuum[1]

David Z. Albert

Columbia University

This note will consist of a single rather elementary remark about the overall Quantum state of the world.[2]

Let me say something at the outset about what this remark is supposed to be good for. It will not support and it will also not raise any doubts about any particular cosmological model.[3] The problems which this remark will turn out to have a bearing on are problems which those models aren't intended to address, problems (about which, however, there has lately been a good deal of speculation[4]) connected with the 'ultimate beginnings' of the world; and the remark will merely suggest that those problems need not be worried about; that perhaps, after, there simply *aren't* genuine problems like that.

The remark is that if any linear Quantum-Mechanical equations of motion happen to be the true and universal equations of motion of the world, then it is consistent with everything we observe, and it is consistent with everything we are even *likely* to observe, to suppose that the present (and past and future) state of the physical world is the vacuum state.

I want to discuss this remark at first a bit loosely and hand-wavingly (in section one) so as to make the discussion less cumbersome and more accessible at the outset. Then (in section two) I'll describe a certain recent re-interpretation of the Quantal formalism[5] which, as it turns out, will accommodate a far clearer and more precise formulation of the remark, one that will suffice to rigorously fill in the gaps in the discussion of section one.

1. The Argument

Suppose that every single physical system in the whole world is a quantum-mechanical system, and that the universe itself has some quantum state, and that the equations of motion of the universe are linear quantum mechanical equations of motion; and (consequently) that all the 'conventional' accounts of quantum mechanics, the accounts that talk about a 'collapse' of the wave-function, the accounts which make use, at one point or another, of an un-quantum-mechanical measuring apparatus or an unquantum-mechanical 'observer', are false.[6] These suppositions are called the 'many-words' interpretation of quantum mechanics. The many 'worlds' of that name refer to the many terms, the many *branches,* of the wave-function.

PSA 1988, Volume 2, pp. 127-133

Now, it happens to be a property of these 'many worlds' (that is, it is a property of any solution of the quantum-mechanical equations of motion) that an observer whose direct experience constitutes some single one of those worlds, some single 'branch' of the wave-function, can deduce hardly anything by observation about how many *other* such branches may exist, or what particular branches they may be; he can deduce (to put it another way) hardly anything about the overall quantum state of the universe. The reason is that the equations of motion are linear; a simple example (in which, for the sake of definiteness, a quantum-mechanical automation will play the role of the observer) will remind us of how that works:

Imagine that a certain particle is prepared in a superposition of two position eigenstates:

$$| \alpha >_p = \frac{1}{\sqrt{2}} [|x_1 >_p + | x_2 >_p]$$

and that a quantum-mechanical automaton (equipped with the necessary sensory apparatus) is instructed to measure and record (in one of its memory registers; memory register A, say) the position of that particle. The state of the composite system consisting of the particle and the automaton, when that procedure is complete, will be:

$$| \beta_1 >_{p+A} = \frac{1}{\sqrt{2}} [|x_1 >_A |x >_p + | x_2 >_A | x_2 >_p]$$

where $| x_1 >_A$ is a state of the automaton (or, more particularly of memory register A) wherein it is recorded in memory register A that the position of the *particle* is x_1 (and similarly for $| x_2 >_A$). Now, suppose that we were to inquire of this automaton whether the present state of the world is an eigenstate of the position of the particle (the correct answer, of course, is 'no'); and suppose that the automaton were programmed so as to answer that question *correctly* in the event that the particle had initially been prepared in any eigenstate of the position.[7] The linearity of the equations of motion will require that such an automaton,[8] in the state $| \beta_1 >_{p+A}$ of equation (2), answer, mistakenly, 'yes'; and a little reflection will show that no amount of cleverness in the construction or the programming of such an automaton can possibly suffice to avoid mistakes like that.[9] So what is it that an observer (whose experience constitutes some single branch of the wave-function) *can* ascertain by experiment about the overall state of the world?: Merely that that branch which he sees constitutes some *part* of that overall state; merely that that overall state isn't orthogonal (that it isn't *perfectly* orthogonal, and that is very little information indeed) to the branch he sees.[10]

Consider, for example, whether it is possible to establish by experiment that the overall state of the world is not the vacuum. In the non-relativistic case, it is possible to establish that: I see a table, say, across the room, and I reflect that any state wherein a table exists is orthogonal (and *perfectly* orthogonal) to the non-relativistic vacuum state, and I conclude, with certainty (albeit I can conclude very little about the overall state of the world), that the overall state of the world is not the vacuum. But what if we were to entertain such considerations in the context of relativistic quantum field theories? The *non*-relativistic vacuum is an eigenstate of an infinity of *local* observables: one can say with certainty, in that state, that there are zero tables at point x and zero tables at point y. The relativistic vacua aren't like that. Those vacua are eigenstates only of *global* observables (observables like the total energy, the total charge, and so on), and *not* of any local ones.[11]

States which entail, say, that there is a table across the room, and states wherein the world appears roughly as it appears to us (full of approximately localized objects, full of systems which are changing with time), are *not* necessarily orthogonal to *those* vacua.

The properties of a quantum state which can suffice to establish that that state is orthogonal to relativistic vacua are global properties; the measurement of those properties requires measurements that (since the measuring apparatus is a part of the world, too) are at best stupendously difficult, and at worst, even in principle, impossible.[12] That (in view of what has been said above) is a curious fact: it means that observers such as ourselves cannot by any practical means (nor, perhaps, by any means whatever) establish that the overall state of the world (at present and for all past and future times) is not the vacuum; observers such as ourselves cannot establish, by any practical means, that our experience is not merely a constituent, merely a *branch,* of that vacuum.

Now we are ready to come to the point of this note. From time to time, throughout the long history of cosmic speculation, and (more recently) in the physical literature about the quantum state of the universe, something called 'the problem of the creation of the world' comes up (Tryon's paper, ref. 3, is a well-known example from the literature of quantum field theory; and there are others). The problem of the creation of the world might be posed like this: 'How can the world have arisen, at the outset, from the vacuum?; or, better, like this: 'Why *is* there a world; why isn't there rather, *nothing*?'. Consider what an odd light our remarks here shed on 'problems' such as that.[13] Perhaps, given those remarks, (and *this* is the point of this note) there *are* no problems such as that! Perhaps there is simply nothing here to wonder about. Perhaps it is the case that nothing ever arose from the vacuum save the vacuum itself.[14]

Consider this: What if the Creator, the Selector of Initial Conditions. had decided *not* to create; to create *nothing*, to create the vacuum?[15] That vacuum would already have contained us and what we see around us. The option *not* to create some world like ours, given the physics relativistic quantum field theory (and this is perhaps a better way to make the point) is not a logical possibility.[16]

2. The Many-Minds Interpretation

We have been supposing here that some linear quantum-mechanical equations of motion will ultimately turn out to be the true and universal equations of motion of the whole world; and that supposition; as I mentioned before, is called the many-worlds interpretation of quantum mechanics. But too much has as yet been left unsaid about precisely what that supposition amounts to. The many-worlds interpretation (to paraphrase Richard Healey)[17] is itself badly in need of an interpretation!

The languages in which the many-worlds interpretation is conventionally discussed, the languages wherein one actually talks (in one way or another) about the *multiplication of worlds*, are too vague and too problematic and too obscure to be of any use in the sharpening of the present discussion. There are problems about the consistency of any such talk with the quantal equations of motion (which, of course, are precisely the equations which that talks seeks to *interpret*),[18] and there are problems about the consistency of any such talk with the invariance of the theory under changes of basis,[19] and there is a mystery, when one speaks in those languages, about what can possibly be signified by the term 'probability'.[20]

The present discussion, if it is to be carried any further, will require a clearer, more accurate, more intelligible, more consistent account of the story wherein the entire physical universe is a purely quantum mechanical system. I want to describe an account like that (the *only* one that has thus far come to light, I think) in this section, an account which is no less intuitively implausible than the more conventional talk about the multiplication of worlds, but wherein the story can at least be told with absolute explicitness from the beginning to the end; and then I want to apply that account, that language, to the discussion of the possibility that the present state of the universe, is the vacuum.

The new account is called the 'many-minds' interpretation of quantum mechanics.[21] It goes like this:

1) Every single physical system there is (sub-atomic particles, tables, measuring instruments, people, stars, the universe) is a quantum mechanical system, and evolves entirely in accordance with linear quantum mechanical equations of motion.

2) Every sentient physical system (every *observer.* for example) is associated not with a single mind, but rather with *continuous infinity of minds.* The individual minds (unlike the *brains* with which they are associated) *aren't* quantum-mechanical objects; they're *never* in superpositions; those minds are always necessarily thinking some particular thought. The *measure* of that infinite subset of minds which happen to be thinking some specified thought at some specified time is equal to the square of the absolute value of the coefficient of the brain state, associated with that specified thought, in the universal wave-function, at that specified time (the measure of those minds associated with some particular sentient object which, for example, believe that X, is equal to the square of the absolute value of the coefficient of the physical state of that object which is associated with the belief that X).

Suppose, for example, that the states $| x_1 >_A$ and $| x_2 >_A$ in equation (2) represent belief states of a sentient observer, rather than a simple automaton. Then, according to the many-minds interpretation, when the state $|\beta_1 >_{p+A}$ obtains, half of that observer's continuous infinity of minds believe that the position of the particle P is X_1, and the other half of her minds believe the position of P is X_2.

The time-evolution of each individual mind is probablistic; but, since there are always a continuous infinity of minds (or else no minds whatever) in any particular mental state, the evolution of the minds *as a set* is always deterministic. Consider, for example, the observer described in equation (2). In the course of the observation which produces the state $|\beta_1 >_{p+A}$, each of the observer's individual minds had a probability of ½ of developing the belief that the position of P is X_1, and each has a probability of ½ of developing the belief that the position of P is X_2.the other hand, since there are a continuous infinity of such minds, it will be true with *certainty,* when the observation in question is complete, that the *distribution* of mental states will be precisely in accordance with the prescription described above (i.e. the measure of the subset of minds with one of the above beliefs will be precisely equal to the measure of the subset with the *other* of the above beliefs).

This language of many minds is free of the various logical diseases of the languages of multiple worlds. This language, as advertised, is the sort of language we need here. It is consistent with the claim that the linear quantum-mechanical equations of motion are the true and universal equations of motion of the whole physical world, and it is consistent with the invariance of the physical theory under changes of basis,[22] and there is no mystery whatever here about what 'probabilities' refer to (in the many-minds picture, probabilities refer to nothing at all in the physical universe, which is, after all, absolutely deterministic; rather, probabilities refer here invariably and exclusively to sequences of *mental* states). It is a language in which every story about the world can be told explicitly. I want to make use of it now to elucidate the story about the possibility that the present quantum state of the universe is the vacuum.

The discussion of section one (which the reader can now easily translate into the language of many minds) left something unsettled. That discussion established (given its premises) that what we observe of the universe is consistent with the possibility that the present quantum state of the universe is the vacuum. But nothing was concluded there about the *probability* that, given what we observe, the state of the universe is the vacuum.

That sort of *inverse* probablistic inference, even more so than *direct* inference (inference, say, about the likely result of some future experiment), is an immensely foggy and ambiguous business within the conventional languages of the many-words interpretation.[23] It is a trivial business, on the other hand, within the language of many minds. The fact (within that new language) is that if the state of the universe is not *perfectly* orthogonal to states such as we observe, then, with *certainty,* there must exist a continuous infinity of minds with mental states precisely like our own, a continuous infinity of minds which seem to themselves to be observing what we seem to ourselves to be observing (the *measure* of that set of minds is equal to the absolute square of the inner product of the state we observe with the true overall state of the universe). If the present state of the universe were the vacuum (or if, more generally, the present state of the universe were not precisely orthogonal to the state we observe), then the probability that we, our minds, our experiences, our observations, our thoughts, would exist, is exactly equal to *one.*

The many-minds account, then, entails that probabilities refer exclusively to *sequences* of mental states, and not to the existence or nonexistence of such states at a given time, which is invariably a matter of certainty. The existence of some particular mind in some particular mental state (the fact, say, that my mind exists, and that I have the experiences that I do) entails nothing whatever, nothing certain and nothing probablistic, about the overall state of the world (nothing, that is, other than that that overall state is not precisely orthogonal to states wherein such minds exist). Given what we observe, and given the premises of the many-minds picture, it is not impossible, and it is not improbable, that the present and past and future state of the universe is the vacuum state.

It would be good to know of a less implausible account of what it means to suppose that the whole world is a quantum-mechanical system, and it would be good to know how the present discussion might look in the context of an account like that. The argument of section one of this note could clearly be translated into the language of any such account, and so any such account would certainly entail that it isn't *impossible* that the present state of the universe is the vacuum; but questions about the *probability* that the present state of the universe is the vacuum seem more difficult. All that can be said about those latter questions, I suppose, is that unless and until such an account many happen to emerge. the answers given here are the only ones we shall be able to imagine how to give.

Notes

[1] I am thankful for many helpful conversations with Yakir Aharonov and Aaron Casher.

[2] Several authors have made remarks which are *related* to the remark to be made here. DeWitt (Phys. Rev. 160, 1113 (1967)) considered the possibility that the overall state of the world might be very different from what we see; and Page and Wootters (Phys. Rev. D, 27, 2885, 1983) have written an interesting paper on the possibility that the quantum state of the world might be some eigenstate (other than the vacuum) of a non-relativistic Hamiltonian.

[3] It will, however, put those models in something of a novel context, but of that more later (see, in particular, footnote 13).

[4] See, for example, E.D. Tryon, Nature 246, 5433 (1973) and A. Vilenkin, Phys. Lett. B117, 1-2 (1982).

[5] This re-interpretation, which will be briefly described in section II, is due to Barry Loewer and the present author. A detailed account of this re-interpretation is now being prepared for publication.

[6]Many authors have argued that all of this must necessarily be supposed if one is to talk sensibly about quantum cosomology at all. See, for example, J. Bell in *Quantum Gravity* (edited by Isham, Penrose, and Sciama; Oxford, 1981).

[7]That is, the automaton is programmed like this: if the state of the world is $|x_1 > _A | x_1 > _p$, the automaton says 'yes, the present state of the world is an eigenstate of the position of the particle', and if the state of the world is $| x_2 > _A | x_2 > _p$, the automaton say 'yes, the present state of the world is an eigenstate of the position of the particle.'

[8]One programmed as described in footnote (7).

[9]Any automaton which has been programmed to answer such questions correctly in the event that the particle had initially been prepared in an eigenstate of some particular observable, can always be 'fooled' by preparing the particle initially in an eigenstate of some other, incompatible, observable.

[10]Perfectly really means perfectly. What an observer can ascertain is merely that his branch *exists,* but the *measure* of that branch within the overall wave-function of the world remains entirely unknown to him. That measure, indeed, might even be zero (just as there are sets of points on the real line, say, which albeit they are nonempty, are of measure zero).

[11]These uncertainties in local observables, which are a very general property of relativistic vacua, are sometimes called vacuum fluctuations.

[12]A measurement of the total charge might, in principle, be possible. One would need to use an infinite array of local measuring devices whose own charges are known to be zero. On the other hand, there is no way whatever, even in principle, to isolate the contributions of working measuring devices to the total *energy* of the world.

[13]It sheds no light at all, as we mentioned at the outset, on any of the questions (other than these questions of *ultimate* beginnings) which usually concern us in cosmology. The sort of cosmology we usually concern ourselves with, after all, is the cosmology of *our branch* of the wave-function.

[14]Perhaps, that is, it will turn out that the present state of the world is the vacuum.

[15]Something ought to be said about the meaning of the word 'nothing'. Vilenkin (in reference 3) uses the word 'nothing' to denote not the vacuum but, rather, a particular point of superspace, a particular space-time geometry. But Vilenkin's 'nothing', unlike the vacuum, has non-trivial dynamical properties: it *evolves,* it *changes;* 'nothing' shouldn't do that!

[16]That Creator could, of course, have created a world perfectly orthogonal to ours; but what would surely amount to *creating* something; that would be a *world,* a world with *things* in it.

[17]'How Many Worlds?' *Nous* 18 (1984) p. 591.

[18]This problem is described in considerable detail in reference 16.

[19]This problem is described in the paper cited in footnote 6.

[20]The mystery is this: if *all* of the worlds, *all* of the possible results of any given experiment, come into being with *certainty,* what sense does it make to speak of the *probability* that a certain experiment will have a certain particular result?

[21]This is the account referred to in footnote 5.

[22]The basis which is in some sense 'preferred' in the many-minds picture (the basis of brain states associated with definite *mental* states) isn't given *a priori*, rather, it follows from the particular *physical structures* of the brains involved. Moreover, this 'preference' is really nothing more than a calculational convenience; unlike in the language of splitting worlds (wherein the preferred basis determines the lines along which the world splits!), this preference, here, has no physical effects whatever. *The physical content of the many-minds picture is absolutely invariant under changes of basis.*

[23]The reason, of course, is the fundamental fogginess about direct probabilistic inference in those languages (see footnote 20), which is simply intensified in attempting to think about inverse inference.

References

Bell, J. (1981), "Quantum Theory for Cosmologists", in *Quantum Gravity*, Isham, Penrose, and Sciama, editors: Oxford University Press.

DeWitt, B. (1967), "Quantum Theory of Gravity 1" *Physical Review* 160: 1113-1195.

Healy, R. (1984), "How Many Worlds? *Nous* 18: 591-620. Page and Wooters (1983) 'Evolution Without Evolution: Dynamics Described by Stationary Observables" *Physical Review* D27: 28852891.

Tryon, E.D. (1973), "Is the Universe a Quantum Fluctuation?" *Nature* 246: 5433-5440.

Vilenkin, A. (1982), "Creation of Universes from Nothing" *Physics Letters* B117: 1-2: 25-28.

From Micro to Macro: A Solution to the Measurement Problem of Quantum Mechanics

Jeffrey Bub

University of Maryland

1. Introduction

Philosophical debate on the measurement problem of quantum mechanics has, for the most part, been confined to the non-relativistic version of the theory. Quantizing quantum field theory, or making quantum mechanics relativistic, yields a conceptual framework capable of dealing with the creation and annihilation of an indefinite number of particles in interaction with fields, i.e. quantum systems with an infinite number of degrees of freedom. I want to show that a solution to the standard measurement problem is available if we exploit the properties of the infinite quantum models available in this broader conceptual framework.

There is a *qualitative difference* between quantum systems with an infinite number of degrees of freedom and quantum systems with a finite number of degrees of freedom. In the infinite case, there exist many unitarily inequivalent irreducible Hilbert space representations of the algebra of observables of the system. In the finite case, there is only one such representation.[1] The Hilbert space of an infinite quantum system[2] decomposes into a direct sum of Hilbert spaces, each of which is a representation space for the algebra of observables. This means that there exist non-maximal observables that take the same eigenvalue for all the vectors in any given Hilbert space in the direct sum. Each such observable Q commutes with all observables, so the algebra of observables of an infinite quantum system has a non-trivial center. Q defines a superselection rule, a restriction on the superposition principle that arises naturally in the transition from non-relativistic to relativistic quantum mechanics. The different Hilbert spaces in the direct sum, which correspond to different eigenvalues of observables in the center of the algebra, are referred to as superselection sectors. There is no interference between states belonging to different superselection sectors (different irreducible representations), so states represented by superpositions of vectors belonging to different superselection sectors represent mixtures.[3]

I want to argue that this restriction on the superposition principle for quantum systems with an infinite number of degrees of freedom provides a solution to the standard measurement problem of non-relativistic quantum mechanics.

Consider a measurement interaction in quantum mechanics as an interaction between a finite quantum system S and an infinite quantum system M, where the 'pointer reading' observable of M lies in the center of the algebra of observables of M and commutes with all other obervables. (In the case of Schrodinger's cat, the 'pointer reading' observable takes

PSA 1988, Volume 2, pp. 134-144

two values: 'alive', and 'dead.') It can be shown (see §3 and (Bub 1988)) that there exists a measurement interaction between S and M that, in a *finite* time, yields a state vector in the Hilbert space of S+M representing a mixture (with the required probabilities) over different determinate alternatives: a particular measured value for the S-observable coupled with a particular 'pointer reading' for M. So, if S is initially in the state $\phi = \Sigma c_i \alpha_i$, where the α_i are eigenvectors of A for the eigenvalues a_i, the probabilities $|c_i|^2$ are to be interpreted as referring to 'pointer reading' transitions of a suitable infinite quantum system M in an appropriate interaction between S and M, i.e. in accordance with the Copenhagen interpretation $|c_i|^2$ is indeed the probability of finding the value a_i in a measurement of A.

The picture I propose is this: The observables of a finite quantum system like S are all indeterminate (except for observables like charge, mass, etc., for which superselection rules apply in virtue of a full field theoretic description of S). Only infinite quantum systems have determinate properties (those corresponding to observables in the center of the algebra). The determinateness of quantum properties is a collective phenomenon in quantum mechanics, emerging in the description of systems with an infinite number of degrees of freedom. The probabilities defined by states of a finite quantum system S refer to transitions between determinate properties of a suitable infinite quantum system M in a suitable interaction between S and M.

Why is this a solution to the measurement problem? The problem arises because quantum mechanics provides a set of 'statistical states' but no 'property states' for the description of physical systems. By a property state I mean a certain assignment of values (or ranges of values) to the physical magnitudes of a system, or a 'list' of properties of the system. A statistical state assigns probabilities to values, or ranges of values, or properties of the system. Formally, such a property state might be represented by an ultrafilter in an algebra, or the atom generating the ultrafilter. For example, the algebra of physical magnitudes of a classical mechanical system is a commutative algebra of real valued functions on the position-momentum phase space of the system. The subalgebra of idempotent magnitudes is a Boolean algebra, isomorphic to the Boolean algebra of Borel subsets of phase space. The property state of the system is represented by a phase point (which corresponds to an atom in the Boolean algebra), or collection of sets to which the point belongs (this is the ultrafilter of propositions generated by the point), or a listing of all the properties characterizing the system at a particular time. The statistical state is represented by a probability measure (in the standard Kolmogorov sense) over these atoms, or ultrafilters, or phase points, or lists of properties. The 'no go' hidden variable theorems of Kochen and Specker (1967) and Bell (1964) show that the probabilities generated by the statistical states of quantum mechanics cannot be generated by measures over property states, without violating certain well-motivated restrictions on these states. So, there are good grounds for supposing that there are no property states for quantum systems. The question is then: What do these statistical states mean if there are no property states, no lists of properties characterising quantum systems at a particular time?

The point is that we must have determinateness *somewhere* in the theoretical scheme for the probabilities to make sense. What we want is to relate the probabilities to a generalized 'counting' (in the measure-theoretic sense) over determinate possibilities, whether we interpret the probabilities epistemically, or as propensities, or what have you. Measurement results are determinate, so it is not surprising that many physicists feel comfortable with an interpretation of quantum mechanics that takes the probabilities as referring to the results of measurement. But our most fundamental physical theory is not simply about measurements - it is about the behavior of physical systems. The problem of interpretation is to make sense of the picture presented by the theory of these physical systems and their properties as they change in time through mechanical interactions.

The measurement problem is one facet of this general problem of interpretation. In the form of the problem of Schrodinger's cat, it arises in part from a specific proposal for introducing determinatess into the theory, i.e. from a specific proposal for defining property states for quantum systems. I shall call this the 'orthodox' property state assumption. The orthodox property state assumption states that when the state of a system S is ϕ, the property state of S is the 'list' of properties represented by Hilbert space subspaces containing ϕ (i.e. all quantum propositions represented by subspaces that contain ϕ are true, all those orthogonal to ϕ are false, and the rest are neither true nor false). It follows from this assumption and the linearity of quantum mechanics that if S interacts with a cat (system M) in such a way as to set up a correlation between S-eigenstates and the two M-states 'alive' and 'dead,' then after the interaction the cat is neither alive nor dead.

On the picture presented here, a quantum system is associated with a noncommutative algebra of physical magnitudes (observables), or a non-Boolean possibility structure or logical space (given by the subalgebra of idempotent observables). Only quantum systems with an infinite number of degrees of freedom, i.e. macrosystems or fields, have determinate properties associated with the Borel subsets of values of these magnitudes. The possibility of determinate proerties at the macrolevel depends on the existence of superselection rules, in effect a restriction on the superposition principle, reflecting the decomposition of the Hilbert space representation of the algebra into a direct sum of Hilbert spaces corresponding to the inequivalent irreducible representations of the algebra, which is a feature of the infinite case. The probabilities specified by the quantum state of a finite system (understood as a statistical state) refer to measures over the determinate alternatives associated with the inequivalent irreducible representations of the algebra characterizing a suitable system-instrument composite system, where the instrument is a macrosystem. Quantum mechanical systems have determinate properties, or determinate (ranges of) values for observables, only to the extent that these properties or observables do not exhibit interference effects, and this means that only infinite quantum systems have determinate properties. The notion of 'measurement' applied to a finite quantum system S refers to a special kind of interactive coupling between S and an appropriate macroscopic system M, idealized as an infinite quantum system in an appropriate initial state (the 'zero pointer reading state'), in virtue of which the system M makes a transition (with a certain probability, depending on the initial quantum state of the finite system) to a new macrostate, the 'i'th pointer reading state,' which is coupled to the i'th eigenstate of the observable 'measured' by the interaction. What is objective before the measurement interaction is the possibility structure (given by the algebra of idempotent observables) of the finite quantum system S and the initial quantum state (statistical state) of S, the possibility structure of the measuring instrument M and the initial quantum state (statistical state) of M, and the determinate macroproperties of M as specified by the macrostate (property state) of M. The final position of the pointer depends (stochastically) on the initial quantum states, the nature of the interaction, and the underlying possibility structure or logical space that determines the totality of possibilities and their interrelationships open to the system.

In the following, I shall take a closer look at how this works out and consider certain difficulties for the position.

2. Strategy

Consider two finite quantum systems S and M. Let $\phi = \Sigma c_i \alpha_i$ be the initial state of S and ψ the initial state of M. In a measurement of S by M, the closest we can get by a unitary transformation to the mixture

$$W = \Sigma |c_i|^2 P\alpha_i \otimes P\psi_i$$

that we apparently want as a final post-measurement state is the pure state

$$\Psi = \Sigma c_i \alpha_i \otimes \psi_i.$$

The problem, of course, is that $P_\psi \neq W$, i.e. these two statistical operators do not generate equivalent statistics.

In the following, the superscript S denotes an operator or observable of the measured system S, and similarly the superscript M refers to the measuring instrument M.

Lemma:

Write:

$$PS_i = P\alpha_i \otimes I^M, PM_i = I^S \otimes P\psi_i$$

Then

$$W = \Sigma_i PS_i P_\psi PS_i = \Sigma PM_i P_\psi PM_i$$

It follows immediately that

$$Tr(WA^S) = Tr(\Sigma_i PM_i P_\psi PM_i A^S) = Tr(P_\psi A^S)$$

Similarly,

$$Tr(WA^M) = Tr(\Sigma_i PS_i P_\psi PS_i A^M) = Tr(P_\psi A^M)$$

Also, if A^M commutes with $P\psi_i$ (or if A^S commutes with $P\alpha_i$) then

$$Tr(WA^S \otimes A^M) = Tr(P_\psi A^S \otimes A^M)$$

This result[4] suggests two pseudo-solutions to the measurement problem that have been proposed in the literature: that W and P_ψ are effectively equivalent because they yield the same statistics for all S-observables (Margenau 1963), and that W and P_ψ yield the same statistics if $\{A^M\}$ is abelian, which we are entitled to assume if M is a macroscopic (hence, presumably, classical) system (Jauch 1968). In order to use the indistinguishability of P_ψ and W as the basis for a solution to the measurement problem, we need some *principled* way to restrict the set of observables of M so that

$$Tr(WA^S \otimes A^M) = Tr(P_\psi A^S \otimes A^M)$$

for all observables A^M, and a transition of the form

$$(\Sigma c_i \alpha_i) \otimes \psi \rightarrow \Sigma c_i a_i \otimes \psi_i$$

can describe a measurement process without requiring the 'collapse' sanctioned by a separate projection postulate. We need to show, on the basis of a purely quantum mechanical analysis, that there exists a class of systems - 'macroscopic measuring instruments' - with a restricted set of observables (in some sense). That is, we need to show that within quantum mechanics there exists a class of systems which play a role analogous to that of Bohr's classically described measuring instruments. The quantum mechanical probabilities could then be taken as referring to the probabilities of the outcomes of appropriate measurement interactions between other quantum systems and these systems.

3. From Micro to Macro

Consider a spin-1/2 system. The spin observables satisfy the commutation relations

$$[S_x, S_y] = ih/2\pi S_z, \quad [S_y, S_z] = ih/2\pi S_x, \quad [S_z, S_x] = ih/2\pi S_y$$

and the condition $S_x{}^2 + S_y{}^2 + S_z{}^2 = 3/4(h/2\pi)^2 I$. The Hilbert space representation of the spin algebra is given by the spin operators $S = h/4\pi\sigma$ on a 2-dimensional Hilbert space $\mathcal{H}$, where the components of σ are the Pauli spin operators. This representation is irreducible (i.e. $\mathcal{H}$ contains no proper subspace $\mathcal{K}$ whose vectors remain in $\mathcal{K}$ after being acted upon by σ). If σ' defines another irreducible representation of the spin algebra on a Hilbert space $\mathcal{H}'$, then there exists a unitary transformation $U: \mathcal{H} \to \mathcal{H}'$ such that $\sigma' = U\sigma U^{-1}$. This is von Neumann's representation theorem (1931) for the spin algebra: all irreducible representations are unitarily equivalent to the Pauli representation.

In the case of infinite quantum systems, this uniqueness theorem no longer holds. To see this, consider an infinite 1-dimensional array of spin-1/2 systems, located at sites n= 0, ± 1, ± 2, ...[5] We seek representations of the spins S_n which satisfy the conditions:

$$[S_{nx}, S_{ny}] = ih/2\pi S_{nz}, [S_{ny}, S_{nz}] = ih/2\pi S_{nx}, [S_{nz}, S_{nx}] = ih/2\pi S_{ny}$$

$$[S_n, S_m] = 0 \text{ if } n \neq m$$

$$S_n \cdot S_n = S^2{}_{nx} + S^2{}_{ny} + S^2{}_{nz} = 3/4(h/2\pi)^2 I$$

Define a z-configuration to be a sequence $s = \{s_{nz}: n = 0, \pm 1, \pm 2, ...\}$ with each $s_{nz} = \pm 1$ representing the eigenvalue of S_{nz} in units of $h/4\pi$ (i.e., s_{nz} represents the eigenvalue of σ_{nz}). A configuration is essentially positive if all but a finite number of its components s_{nz} are equal to +1. Associate each essentially positive z-configuration s with a unit vector $\Psi_s(z+)$, such that

$$\Psi_s(z+) \perp \Psi_{s'}(z+) \text{ if } s \neq s'$$

Define $\mathcal{H}_{z+}$ as the ∞-dimensional Hilbert space spanned by the vectors $\Psi_s(z+)$ for all essentially positive z-configurations. It can be shown that there exists an irreducible representation of the spins S_n in terms of the operators $S^{(z+)}{}_n = h/4\pi\, \sigma^{(z+)}{}_n$ on $\mathcal{H}_{z+}$.

We can also construct another representation $\mathcal{H}_{z-}$ based on z-configurations which are *essentially negative* (all but a finite number of components s_{nz} are equal to -1). These two representations of the spins are unitarily inequivalent, since the spin density of the system is $+h/4\pi$ in the z-direction for all quantum states in the $\mathcal{H}_{z+}$ representation, and $-h/4\pi$ in the $\mathcal{H}_{z-}$ representation.

The spin density in $\mathcal{H}_{z+}$ is

$$\overline{S^{(z+)}} = \lim_{n \to \infty} 1/(2n+1) \Sigma_n h/4\pi\, \sigma^{(z+)}{}_n$$

The direction z is arbitrary, so similar representations can be obtained for any orientation, leading to an infinity of inequivalent, irreducible representations of the spin algebra.

These considerations provide a framework for a quantum theory of macrosystems, idealized as infinite quantum systems. The Hilbert space of such a system decomposes into a direct sum of ∞-dimensional Hilbert spaces, each of which provides a different irreducible representation of the algebra of observables. These irreducible representations or superselection sectors correspond to different macrostates of the system. So a macrostate is identified with an equivalence class of microstates. A macro-observable takes the same value for every microstate in a given superselection sector.

For example, the spin density of the infinite 1-dimensional array of spin-1/2 systems defines a macro-observable - the polarization of the idealized macrosystem - that takes different values for different superselection sectors, different equivalence classes of microstates. If $\mathcal{H} = \oplus_k \mathcal{H}_k$, where $\oplus_k$ denotes the direct sum over the ∞-dimensional subspaces corresponding to the different macrostates, and P_k is the projection operator onto the subspace $\mathcal{H}_k$, then (following Wan (1983)) a macro-observable is represented by an operator of the form $Q = \Sigma q_k P_k$ and a general (bounded) observable by an operator of the form $A = \Sigma_k P_k A_k P_k$, where A_k is a bounded self-adjoint operator in $\mathcal{H}_k$ (the restriction of A to $\mathcal{H}_k$).

The measurement process is idealized as an interaction between a microsystem and a macrosystem represented as a quantum system with an infinite number of degrees of freedom. Consider a measurement of spin in the z-direction on a spin-1/2 system S. Suppose S is initially in the pure state

$\phi = c_+\phi(z+) + c_-\phi(z-)$

The measuring instrument M is a macrosystem idealized as the infinite 1-dimensional array of spin-1/2 systems considered above. Suppose M is initially in a macrostate of polarization 'up in the x-direction,' i.e. the microstate of M is initially some pure state $\Psi(x+)$ in the superselection sector of microstates of the $\mathcal{H}_{x+}$ representation.

An interaction Hamiltonian can be constructed along the following lines: In a 2-dimensional Hilbert space, the self-adjoint operators

$$L_+ = 1/\sqrt{2}(\sigma_x + \sigma_z), \quad L_- = 1/\sqrt{2}(\sigma_x - \sigma_z)$$

have the properties

(i) $L_+\psi(x) = \psi(z+)$

(ii) $L_-\psi(x) = \psi(z-)$

(iii) $(L_\pm)^n = I$ if n is even, $(L_\pm)^n = L_\pm$ if n is odd.

Analogously, in the infinite tensor product Hilbert space $\mathcal{H}^M = \otimes_n \mathcal{H}_n$ of the measuring instrument, the operators

$$LM_+ = \otimes_n 1/\sqrt{2}(\sigma_{nx} + \sigma_{nz}), \quad LM_- = \otimes_n 1/\sqrt{2}(\sigma_{nx} - \sigma_{nz})$$

have the properties

(i)' $LM_+\Psi_s(x+) = \Psi_s(z+)$

(ii)' $LM_-\Psi_s(x+) = \Psi_s(z-)$

(iii)' $(LM_\pm)^2 = I^M$ if n is even, $(LM_\pm)^2 = LM_\pm$ if n is odd.

Following Wan (1986), define H_I as

$$H_I = \lambda(PS_+\otimes LM_+ + PS_-\otimes LM_-)$$

where PS_+, PS_- are the projection operators onto $\phi(z+)$, $\phi(z-)$ in $\mathcal{H}^S$, and

$\lambda = \pi h/4$ if $t = (0,T)$, $\lambda = 0$ if $t \notin (0,T)$.

Then the temporal evolution of the state of S+M induced by H_I during the measurement interaction is governed by the unitary operator

$$U(t) = \exp\left(\int_0^t -2\pi i/h\, H_I(t')dt'\right)$$

This has the effect of 'moving the pointer' from the superselection sector corresponding to 'up in the x-direction' to the superselection sector corresponding to 'up in the z-direction' or 'down in the z-direction,' depending on the state of S. (See Bub (1988) for details.)

Suppose the interaction is switched on for a time T. If the initial state of the system S+M is:

$$\chi(0) = (c_+\phi(z+) + c_-\phi(z-))\otimes\Psi(x+)$$

then the interaction yields a final state after a time T:

$$\chi(T) = c_+\phi(z+) \otimes\Psi(z+) + c_-\phi(z-)\otimes\Psi(z-)$$

Since $\Psi(z+)$, $\Psi(z-)$ belong to different macrostates of the measuring instrument M, this vector state represents a mixture of composite system states $\phi(z+)\otimes\Psi(z+)$, $\phi(z-)\otimes\Psi(z-)$ with weights $|c_+|^2$, $|c_-|^2$, respectively. For every observable of the the composite system $A^S\otimes A^M$ we have

$$\mathrm{Tr}(P_{\chi(T)}A^S\otimes A^M) = (\chi(T), A^S\otimes A^M \chi(T)) =$$

$$|c+|^2(\phi(z+),A^S\phi(z+))(\Psi(z+),A^M\Psi(z+))+|c-|^2(\phi(z-),A^S\phi(z-))(\Psi(z-),A^M\Psi(z-))$$

$$= \mathrm{Tr}(WA^S\otimes A^M)$$

where $W = |c_+|^2 PS_{\phi(z+)}\otimes PM_{\Psi(z+)} + |c_-|^2 PS_{\phi(z-)}\otimes PM_{\Psi(z-)}$

4. Difficulties

There is, firstly, the problem pointed out by Bell. Commenting on Hepp's theory of measurement (Hepp, 1972), Bell writes (1975, p. 51):

> The continuing dispute about quantum measurement theory is not between people who disagree on the results of simple mathematical manipulations. Nor is it between people with different ideas about the actual practicality of measuring arbitrarily complicated observables. It is between people who view with different degrees of concern or complacency the following fact: so long as the wave packet reduction is an essential component, and so long as we do not know exactly when and how it takes over from the Schrodinger equation, we do not have an exact and unambiguous formulation of our most fundamental theory.

Hepp was perhaps the first to propose a solution to the measurement problem along the lines sketched here. (Recent work includes a series of papers by Machida and Namiki (1980, 1984), Araki (1980, 1986, 1987), and Yanase (1982).) He develops a number of detailed models of the measurement process in the C*-algebra formulation of quantum mechanics, but the essential idea is that contained in the model of a macroscopic measuring instrument as an infinite linear array of spin-1/2 particles.

In Hepp's analysis, the array M is a semi-infinite array, fixed at positions $n = 1, 2, ...$, and the effect of the measurement interaction between M and a single spin-1/2 particle S is to flip the spins consecutively from $n = 1$. This means that for any time t, however large, there will always be a value of n for which the spins beyond n have not yet been flipped. But the measurement process is only complete when there is no interference between different 'pointer readings' (different macrostates of the array), i.e. when the spin flipping induced by the interaction Hamiltonian has been completed. Hepp shows that, in the limit as $t \to \infty$ and $n \to \infty$, there will be no interference between the two macrostates of M correlated with the spin eigenstates of the system S by the measurement interaction for any local observable. A *local* observable is an observable defined on a finite segment of the array. (A 'classical' observable for Hepp is a function of an infinite set of spin observables.) As Bell points out, for any time t, there always exists a local observable that will show interference between the two states of the array that end up correlated with the spin eigenstates of S. Roughly, the idea is to define a local observable for a value of n large enough to outstrip the interaction. Bell comments (1978, pp. 48, 49):

> While for any given observable one can find a time for which the unwanted interference is as small as you like, for any given time one can find an observable for which it is as big as you do *not* like.

As I see it, there are two difficulties with Hepp's formulation, associated with the two different kinds of infinity in the theory: (i) measurement interactions are only completed at t = ∞, and (ii) measuring instruments are idealized as quantum mechanical systems with a *literally infinite* number of degrees of freedom. As far as the first problem is concerned, it is clear that in order to have a measurement completed in a finite time the Hamiltonian that generates the time evolution in the measurement process cannot be an observable (roughly, because an operator representing an observable cannot carry a vector from one superselection sector to another). In the model I sketched above, the interaction Hamiltonian is not an observable (it is not of the form $A = \Sigma P_k A_k P_k$).[6] It is, of course, a self-adjoint operator and the time evolution of S+M during the measurement process is unitary. I see no reason in principle to reject such interactions as characteristic of the measurement process in quantum mechanics in a formal sense.

As far as the second problem is concerned, the idealization of the macroscopic measuring instrument as a quantum mechanical system with a literally infinite number of degrees of freedom cannot be avoided. For von Neumann's representation theorem does not fail for a system with a very large by finite number of degrees of freedom, and the superposition principle still applies to such systems. But even in classical mechanics we idealize a macrosystem, which we think of informally as composed of a very large but finite number of particles, as an infinite system in order to provide a description of collective phenomena at the macrolevel (such as phase transitions), that are relatively insensitive to the number of constituent particles and their precise motions at the microlevel and involve macro-concepts (such as 'gas,' 'liquid,' 'solid,' 'temperature,' etc.) that do not apply to single particles. For example, Boltzmann's derivation of the Maxwell velocity distribution law for a gas at equilibrium at a given temperature applies classical mechanics and statistical assumptions to a model of N particles with total kinetic energy E divided into J discrete units e with E = Je. A microstate of the system is defined by assigning J_i units to each particle i, so that $\Sigma J_i = J$. The Maxwell distribution is obtained by taking the limit J → ∞, e → 0, N → ∞, with E/N fixed. The model is an asymptotic idealization sanctioned by classical mechanics that yields properties characterizing the collective behavior of large ensembles of particles that are qualitatively different from the properties of individual systems in the collective. As such, the asymptotic character of the idealization is essential and cannot be dispensed with, if we want to describe the thermodynamic features of the system *within the framework of classical mechanics*.

I want to suggest that the situation in quantum mechanics is analogous, with one crucial difference. In a classical world, thermodynamic properties emerge as collective phenomena in a many-body system that has determinate property states at the microlevel. In a quantum world, *the very determinateness of properties is itself a collective or cooperative phenomenon*. Schrodinger's cat is presumably composed of 10^{23} particles, give or take a few, but if we want to describe features of this many-body system that are *essentially collective* in nature, an asymptotic idealization is required that is essential to the description and not merely pragmatic. The determinateness of properties (via the failure of von Neumann's representation theorem) follows automatically in such a model, but this is achieved through the formal vehicle of an idealization that is not exclusive to quantum mechanics but characterizes the description of genuinely collective phenomena in both classical and quantum theories. The necessary use of such idealizations in the classical or quantum mechanical characterization of macrosystems might well be taken as an inadequacy of these theories as mechanical theories - a non-linear mechanical theory would no doubt treat collective phenomena differently. But this inadequacy does not vitiate the use of asymptotic idealizations to resolve the measurement problem, insofar as this is a purely *theoretical* or foundational problem internal to quantum mechanics.

There are two further points I want to make. One concerns the possibility that linear superpositions of macroscopically distinguishable states actually exist in nature. (See Leggett (1986) for a review of current evidence.) The other concerns the relationship of the account of measurement given here to the Daneri-Loinger-Prosperi theory (1962). The underlying idea of the proposal outlined above is to exploit the *qualitative difference* between finite and infinite systems: Superselection rules, i.e. restrictions on the superposition principle, arise naturally in the algebraic structure of the quantum mechanical description for infinite models. The claim is that if we idealize a macrosystem via an infinite model, then the quantum mechanical description of the macrosystem will introduce a set of non-trivial observables that commute with all other observables and are in this sense 'classical.' This provides a level of determinate properties to which the probabilities of the theory can refer. It is not required for this solution to the measurement problem that the superposition principle breaks down completely at the macrolevel so that interference effects are never possible between macroscopically distinguishable states. Insofar as the determinateness of quantum properties precludes the existence of interference effects between quantum states for these properties, and hence requires a restriction on the superposition principle, only infinite quantum systems can have determinate properties. Insofar as the superposition principle holds for a particular macrosystem, an infinite model of the sort sketched above is not an appropriate model for the system. The claim is only that the theory has the resources - infinite models - to introduce determinateness for some properties of some quantum systems, thereby providing a reference for the probabilities.

The Daneri-Loinger-Prosperi theory is a quantum ergodic theory of measurement. It is supposed to follow from this theory that a state of the form Ψ, resulting from a measurement interaction between a microsystem S and a macro-instrument M, will undergo an ergodic motion after the interaction has ceased, so that eventually the composite system S+M will be 'macroscopically characterized' by a mixture over (time evolved) eigenstates of the S-observable A^S correlated with macrostates associated with the different 'pointer readings' of M, with the appropriate weights $|c_i|^2$. I have criticized this theory elsewhere (Bub, 1968, 1989). What Daneri, Loinger, and Prosperi actually establish is this: On the assumption that measurements on the macroscopic instrument are restricted to observables compatible with a certain fixed set of non-maximal observables (the 'macroscopic observables' of M), structural features of M as a quantum mechanical macrosystem ensure that Ψ undergoes an ergodic motion which, after a sufficiently long time, yields a time evolved state that can be 'macroscopically characterized' by a statistical operator representing a mixture over alternatives corresponding to the (time evolved) eigenstates of the observable measured correlated with equilibrium ('pointer reading') macrostates of M. The qualification 'macroscopically characterized' refers to the experimental indistinguishability after a sufficiently long time between the actual pure state of the composite system and the mixture, with respect to the restricted set of non-maximal observables.

Now, it does not follow from any principles of the quantum theory in the Daneri-Loinger-Prosperi analysis that, for sufficiently large systems, there exists a unique set of non-maximal observables which plays the privileged role required by the ergodic theory. Nor does it follow from their analysis that other observables, incompatible with these, cannot be measured on the system. In order to use the indistinguishability of Ψ and W as the basis of a solution to the measurement problem, we need some *principled* way to restrict the set of observables of M. We need to show, on the basis of a purely quantum mechanical analysis, that there exists a class of systems - 'macroscopic measuring instruments' - with a restricted set of observables (in some sense). That is, we need to show that within quantum mechanics there exists a class of systems which play a role analogous to that of Bohr's classically described measuring instruments. The Daneri-Loinger-Prosperi theory does not show this. Bell's comments on Hepp's theory apply equally well to the Daneri-Loinger-Prosperi theory. For all t < ∞, the pure state is not equivalent to the mixture, i.e. interference is not destroyed. A little bit of interference, however little, vitiates the solution. The problem is not to show that, for all practical purposes, the probabilities are 'classical.' Rather, for a solution to the measurement problem along these lines, we have to show, on the basis of the quantum mechanical structure of the measuring instrument that, after a *finite* time t, there is *no interference* between the different alternatives associated (via the measurement interaction) with the initial linear superposition over eigenstates of the observable measured.

The solution to the measurement problem sketched in §3 yields a unified, realist interpretation of quantum mechanics as a fundamental theory of physics. The underlying problem is to provide an interpretation of the theory that allows the probabilities to refer to appropriate determinate alternatives, without going beyond the resources of the theory. The solution is to see that there are perfectly good theoretical models - systems with an infinite number of degrees of freedom - that provide the appropriate determinate referents for the probabilities. Whether we believe that this theory can adequately describe real macrosystems, or the functioning of real macroscopic measuring devices, is another question. If we have grounds for believing that this is not the case, then we have grounds for modifying the theory.

Notes

[1] Von Neumann's representation theorem (1931).

[2] For brevity, I shall refer to a quantum mechanical system with an infinite number of degrees of freedom as an 'infinite quantum system.'

[3] Thus, for infinite quantum systems, some vector states represent mixtures, while other mixtures - mixtures of coherent states - are not represented by vector states.

[4] First formulated by Furry (1936). See also Ghirardi, Rimini, and Weber (1988).

[5] The following example is taken from Sewell (1986).

[6] The precise relation between the set of observables as defined here and Hepp's 'local' observables, and between the set of macro-observables and Hepp's 'classical' observables, is not entirely clear. It seems fairly obvious that a macro-observable is a 'classical' observable and that a 'local' observable is an observable.

References

Araki, H. (1980), *Progress of Theoretical Physics* 64: 719.

_ _ _ _ _. (1986), "A Continuous Superselection Rule as a Model of Classical Measuring Apparatus in Quantum Mechanics", in V. Gorini, A. Frigerio (eds.), *Fundamental Aspects of Quantum Theory*. New York: Plenum Press.

144

_____. (1987), "On Superselection Rules", in M. Namiki et al. (eds.), *Proceedings of the 2nd International Symposium on Foundations of Quantum Mechanics*. Tokyo: Physical Society of Japan.

Bell, J.S. (1975), "On Wave Packet Reduction in the Coleman-Hepp Model", *Helvetia Physica Acta* 48: 93-98. Reprinted in J.S. Bell, *Speakable and Unspeakable in Quantum Mechanics*. Cambridge: Cambridge University Press.

_____. (1964), *Physics* 1: 195-200.

Bub, J. (1968), *Nuovo Cimento* 57B: 503-520.

_____. (1988), *Foundations of Physics* 18: 701-722.

_____. (1989), "On the Measurement Problem of Quantum Mechanics", in M. Kafatos (ed.), *Bell's Theorem, Quantum Theory, and Conceptions of the Universe*. Boston: Kluwer Academic Press.

Daneri, A., Loinger, A., and Prosperi, G.M. (1962), *Nuclear Physics* 33: 297-319.

Furry, W.H. (1936), *Physical Review* 49: 393.

Ghirardi, G.C., Rimini, A., and Weber, T. (1988), *Foundations of Physics* 18: 1-27.

Hepp, K. (1972), *Helvetia Physica Acta* 45: 237.

Jauch, J.M. (1968), *Foundations of Quantum Mechanics*. Reading, Mass.: Addison-Wesley.

Kochen, S. and Specker, E.P. (1967), *Journal of Mathematics and Mechanics* 17: 59-87.

Leggett, A.J. (1986), "The Current Status of Quantum Mechanics at the Macroscopic Level", in *Proceedings of the 2nd International Symposium on Foundations of Quantum Mechanics*, M. Namiki et al. (eds.). Tokyo: Physical Society of Japan.

Machida, S. and Namiki, M. (1980), *Progress of Theoretical Physics* 63,:1457, 1833.

_____. (1984), "Critical Review of the Theory of Measurement in Quantum Mechanics", and "Macroscopic Nature of Detecting Apparatus and Reduction of Wave Packet", in S. Kamefuchi et al. (eds.), *Foundations of Quantum Mechanics in the Light of New Technology*. Tokyo: Physical Society of Japan.

Margenau, H. (1963), *Philosophy of Science* 30: 1, 138.

Sewell, G.L. (1986), *Quantum Theory of Collective Phenomena*. Oxford: Clarendon Press.

Von Neumann, J. (1931), *Mathematische Annalen* 104: 570.

Wan, K.-K. (1983), *Canadian Journal of Physics* 58: 976.

Yanase, M.M. (1982), *Annals of the Japanese Association for the Philosophy of Science* 6: 83.

Part V

EXPERIMENT

Philosophers of Experiment

Ian Hacking

University of Toronto

The Neglect of Experiment: that is the title of Alan Franklin's (1986). He did not mean to imply that scientists were neglecting experiments, spinning well financed cobwebs of theories while laboratories decayed for lack of funds. He meant that historians and philosophers neglected the experimental side of science. That was true, and is no longer so. Although his title was fine when he was writing, the times have passed it by.

A decade before there had been almost no reflective philosophy of experiment. What little had been published was not seen as writing about experiment—that was not something to write about—but as discussion of the theory/observation distinction, or the impossibility of eliminating a theory by crucial experiment, etc. The even-handed *Dictionary of Scientific Biography* discreetly cut articles on experimenters and expanded those on theorists. Thaddeus Trenn's (1977) on the experimental discovery of isotopes was poorly received. The principle of not opening old wounds prevents me from quoting here remarks in conversation made by some of our most distinguished historians, *a propos* of that 'tedious recounting of test tubes and jottings'.

The contrast with the past 3 or 4 years is extreme. There have been historico-philosophical international conferences devoted to experiment. They have dutifully produced volumes of collected papers, just as if experiment were a legitimate subdiscipline (Batens and van Bendegem, 1988, Gooding et. al. 1988). The practice is so new that I think that it began with the three 'experimental' papers by Peter Galison, J.S. Rigden and Roger Steuwer in Achinstein and Hannaway (1985).

There has been a growing number of books. Often, as in the case of Galison (1987), they present a rich tapestry woven from incidents in the history of science, but glowing with philosophical colours. As I write, the current issue of Isis (79, 1988, no. 3) is dedicated to our topic. Several of the papers contributed to *PSA 1988* are about the philosophy of experiment (Baird 1988, Stump 1988). Nor should we become fixated on such local events; we must also turn to fundamental studies of experiment and technology that find some of their roots in the work of Habermas, *e.g.* Radder (1988).

So intense and continuing has been this activity that I shall present a highly selective retrospective exhibit of ten years of collective thinking about the laboratory sciences. What has been the interest of experiment for philosophers? Part of the answer is that we have been addressing old questions in new ways: fact, fiction, forecast, rationality, justifi-

PSA 1988, Volume 2, pp. 147-156

cation, irrealism, demarcation, Duhem's thesis, and so forth. But I shall also develop new themes, and show how thinking about experiment bears on philosophical issues in ways that have not yet been much noticed.

In no way do I want to make experiment more important than theory. One of my messages is the richness, complexity and variety of the scientific life. In particular, theory is not one thing but many, and experiment not one thing but many. Philosophy of science has been impoverished not only by its obsession with theory, but also with its complacent doctrine that there are one or at most two kinds of theories (*e.g.* real theories and bridge principles). If we may for a moment speak of the experimental liberation movement, one of its aims has been not just to elaborate the life of experiment, but also to improve the quality of life for theories—along with making the theory/experiment distinction not obsolete but multifaceted.

In presenting the work of others I should mention the obvious, that there is a division between sceptics and admirers of experiment. That is nothing new. Philosophy of science has always been riven by that difference in instincts. The adulation of science was characteristic of Popper and Carnap, however heated their superficial differences of opinion. Feyerabend was sceptic, Lakatos admirer, and Kuhn, seen as sceptic, probably isn't. It is a misfortune of this session that both Galison and I respect science too much. The chairman, Andrew Pickering, provides some valuable counterweight.

In the late 1970's historians, philosophers, sociologists, anthropologists, and retiring scientists began to write, in a reflective mode, about experiments. On the surface, their endeavours were largely unrelated, although this shift from theory to experiment as object of enquiry does betoken a larger movement in contemporary culture. Once people did begin to think about experiment, those conducting social studies of science got there first, starting with Ravetz (1971). The sceptics among them riled admirers of science and provoked heated discussions of evidence and rationality.

What data, what experiments, furnish significant evidence? Gilbert and Mulkay (1984) interviewed investigators in bioenergetics to see what results were regarded, not as 'crucial', but at least as 'key' to the recent development of their field. They reported on the basis of interviews that the "selection of certain experiments as key did not depend on any clearly identifiable qualities of the experiments themselves, nor even on the reception reportedly given to the experiments by contemporary researchers" (p.122). They doubted that there is "any kind of data that can be used to provide a firm bedrock for historical description and analysis" (p.124). They were writing for historians, but the message for philosophers was manifest. This was not the old rationalism of Popper or Lakatos, teaching that there are no crucial experiments, on the ground that theory is paramount. It is the far more radical claim that as a matter of brute fact, there are no criteria for distinguishing key data. There is a blooming confusion. It may end in consensus, but not because the community is constrained by the evidence.

This sceptical stance begged for counter-attack. According to Joseph Robinson (1986), Gilbert and Mulkay lacked "an understanding of the particular scientific issues involved in the specific cases" (p.52). To follow the discussion seriously, we have to look up *e.g.* a 1966 National Academy of Science paper "ATP Formation Caused by Acid-Bath Transition of Spinach Chloroplasts." These debates can be conducted only by bringing out what is specifically at issue in the laboratory and its environs.

This tiny fracas illustrates literally dozens of rather detailed controversies between sceptics and admirers of experimental science. Both parties use a detailed case, but want the reader to infer, "that's how it is, everywhere".

Such debates have shifted philosophical questions about evidence. Probabalists like Keynes and Harold Jeffreys, and later Carnap and to some extent the students of subjective, personal or judgmental probability, began with a clear problem. What is the relation between the evidence (or, the available evidence) and an hypothesis of interest? Carnap's symbol 'c(h,e)' epitomized the conception of inductive rationality and evidence. Later, as the programme degenerated, the question became, how does new evidence modify my structure of beliefs or judgements? Social studies of science, in the vein of Mulkay and many others, put the very idea of "the" evidence (or, "the available evidence") in question.

The doctrine of the strong programme in the sociology of knowledge is better known than individual disputes about spinach chloroplasts or whatever. The explanation of the acceptance of a proposition or a practice shall not include "and the proposition is true and supported by evidence, that's why people believed it" or "and the practice works and attains its ends, that's why people adopted it." Above all, don't say that a belief was accepted because it was reasonable to do so! That is the strongest of attacks on a timeless concept of rationality (like the one epitomized as 'c(h,e)'. The idea is that what is rational, and what counts as working, is determined in an historical and social setting. There is no such thing as a good reason, tout court.

I find spinoffs of the strong programme more fascinating than its original tenets. The programme has nothing in particular to do with experiments, but inevitably it has application to them. I think of the exceptionally original work of Schaffer and Shapin (1986). Their "Hobbes, Boyle and the Experimental Life" is the most important contribution to discussion of what A.C. Crombie (1981) called styles of scientific reasoning in the European tradition. The book is about the introduction of a new style of reasoning, the laboratory demonstration and the probing of nature by instruments. It is an epic whose protagonists are Hobbes and Boyle, with the latter as victor.

Ernan McMullin (1988a) discusses the shaping of scientific rationality, of "what counts as a good reason in a scientific argument," of "second-level disagreements" (p.3). McMullin runs through a familiar gamut of theory-obsessed contributors such as Kuhn and Lakatos. The lecture on which his paper is based was given just as the book by Shapin and Schaffer was published. The book, with its imaginative and controversial discussion of how an experimental style of reasoning was put in place, transforms the level of debate. A new generation is speaking, one informed by careful reflection on experiment, rather than theory. We can now consider in detail what it is for one style of reasoning to replace another (as the sceptic would put it) or how styles of reasoning evolve (as Crombie would put it.) This, I suggest, is how people in the future will address McMullin's theme, "the shaping of scientific rationality." Notice how we have moved on from the strident tones of strong programme confrontations. The issue is less, "what non-rational elements determined belief?" than, "how is it that we came to call these procedures the reasonable ones?" That this is my own preference is evident from the use of styles of reasoning in Hacking (1982).

The tenets of the strong programme are, in my opinion, a minor aspect of a broader movement, the "social construction of scientific facts" school. Bruno Latour calls it constructionism; philosophers will recognize it as close kin to Nelson Goodman's "skeptical, analytic, constructionalist orientation" (Goodman, 1978, p.1). Latour's ideas are elaborated in his (1987), but here I attend more to the *Laboratory Life* of Latour and Woolgar (1978). It is one of several recent books with "construction" in its subtitle, not to mention Pickering's movement of the word "constructing" to the title of *Constructing Quarks* (1984). The great and long forgotten pioneer of this genre was of course Ludwig Fleck (1935), writing about the construction of the Wasserman test, and also, incidentally, writing about a style of reasoning, or *Denkstil*.

Construction has nothing special to do with experiment, as witnessed by another construction-subtitle book, Donald MacKenzie's (1981) on British statistics. It has, nevertheless, a strong overlay of scepticism about experiment. Facts settled by experiment are (it is urged) facts only after they have been settled. Hence the idea of experiment constraining theory by revealing facts is made to collapse. R.B. Braithwaite (1953) asked in 1946, "Is science invention or discovery?" His answer was classic: "Man proposes a system of hypotheses. Nature disposes of its truth or falsity" (pp. 367-8). The second clause has been constantly whittled away at this past half century. The social constructionists make out that people, not nature, dispose of the truth or falsity of the proposed hypotheses. The issue has been well joined. Robinson v. Gilbert & Mulkay, already cited, is an instance of the debate. More important are the matters at issue between Galison and Pickering. Their debate is, in my opinion, one good model (among others) of how to carry on. On the one hand they can draw on instrumentation and experimentation that they know almost as well as the agents who did the work in the first place (weak neutral currents, the bubble chamber). On the other, they turn this into an epistemological and metaphysical parable, telling the way it is in a certain kind of science, in general.

There is, incidentally, a problem for those who want to follow their example. Do we have to learn about spinach chloroplasts and bubble chambers, and thousands of cunning instruments? The brief but wrong reply is: philosophers of theory have to learn evolutionary biology or quantum mechanics, so what's new? The truth is that there are more kinds of instruments even than there are subatomic particles or species. So many experiments, so many details! The consequence is one you would not at first think of: the philosopher of experiment must be a better writer, a better artist, have more highly developed literary skills, than the philosopher of theory.

As an outsider sees it from a distance, Pickering says that there are very few constraints on how experimental work will proceed, whereas Galison sees that there are a great many. The outsider will notice Pickering attending to how the goals of a research programme are negotiated between the players, so that success is defined by the consequence of social interactions (Pickering, forthcoming). The same outsider will crassly say that Galison must have some truth on his side, because most experiments won't work: that's the unnegotiable constraint. But Galison has something subtler in mind, based on the way in which instrumental technique, tradition and availability determines what it is possible to do in an experimental situation, largely independently of the theoretical structures in the background.

Pickering, on the other hand, in current work, increasingly makes plain that he does not think that anything goes. Like all of us, he proceeds by delving deeper than the bland labels "experiment," "observation," of old. He has three levels: theory, phenomenology of the apparatus, and the material instrumentation and objects being investigated. He calls these three plastic resources. In the course of an experimental investigation you may change your account of how the apparatus works, your account of the world that you are trying to find out about, or may modify your instruments. The final product—in the unusual event that there is a final product—is a moulding of the three of these together.

Philosophers will see that Pickering is extending Duhem's thesis about auxiliary hypotheses. Duhem's thesis has long been decked out in Quinery and hence been largely irrelevant to real science. It is an important achievement to return it to its proper station. I combine it first with Cartwright (1983, ch. 6) on phenomenological laws and approximation; secondly, with Ackermann (1985), a much more abstract and less example-oriented vision of a dialectic involving data, instruments and theory. Cartwright has a good deal to teach about how phenomenological laws of the apparatus are modified, Ackermann about how data are on the one hand material objects given regardless of theory, and on the other hand used as signs and regularly reinterpreted.

Duhem's thesis is widely regarded as an indeterminacy thesis—and so it is, when restricted to theories and hypotheses. In his simplest and astronomical example, he thought about changing our theory of the stars, and our changing the theory of the telescope; he did not think about changing the way we make telescopes, or about how the dialectic between our theories (auxiliary hypotheses) about telescopes and our telescopes, each of which we modify to try to keep them in some kind of harmony with each other. I read Pickering's extension of Duhem's thesis as implying more determinacy than indeterminacy. I do not mean that the world pre-ordains what shall be our theory of the stars, our theory of our apparatus, and our apparatus. I do think that only a few such combinations persist, and that the plasticity noted by Pickering turns into a sort of glue that keeps so much of our science stable. Patrick Heelan and I will discuss that later this year. (Heelan 1988, Hacking 1988).

Now I shall turn to realism. Latour and Woolgar described a discovery that won a Nobel prize. The first readers of the book read it in the anti-rationality mode: the research reached a successful conclusion not because the competing investigators produced compelling evidence, but because they negotiated with the larger community of endocrinologists and compelled acceptance of their analysis and synthesis of a particular tripeptide. That is indeed a sceptical theme of their book, and admirers of the scientific achievement naturally insist that there were far more constraints on the laboratories—constraints imposed by reason and nature—than Latour and Woolgar were willing to countenance.

Yet the subtitle about constructing facts makes plain that their iconoclasm is directed elsewhere. Their book is the most powerful work of scientific anti-realism to have emerged in the past decade. It is entirely different from instrumentalist anti-realism. Van Fraassen's constructive empiricism takes for granted that a given theory either is, or is not, empirically adequate to the phenomena. He is an admirer of science. He has no scepticism about phenomena, and never considers whether facts are constructed before theories can be adequate to them. Conversely, Latour and Woolgar have in principle no anti-realist instincts about unobservable (theoretical) entities. They claim only that they don't exist until they are constructed. It is part of the rhetoric of science, they say, to erase all memory of the construction, so that we speak of discovering phenomena and of discovering the ("unobservable") structure of a tripeptide. Once again, I urge comparison with Nelson Goodman. To distinguish this radical position from the merely verbal anti-realist science-admiring perspective of people like van Fraassen, we should use Goodman's self-appellation, and call Latour an irrealist.

As an admirer of science I restrain my enthusiasm for this kind of irrealism. Hacking (1988b) sketches the development of TRH—the substance that is the topic of Latour's book—over the past ten years. It furnishes additional considerations in favour of Latour's story. My purpose, however, was not merely to welcome the book into the fold of more conservative philosophy. It was rather to show that the "constructionist" story can be retold in an entirely non-constructionist way, so long as you do not think that there is one unique description of the real world that is the ideal endproduct of inquiry. You can understand the negotiations so highlighted by Latour and Woolgar as negotiations aiming at settling one possible description, at agreement on one set of criteria for judging the specifics of some endocrinological experiments. The description, says the conservative thinker, was always true of the world, and not made true. It excludes other possible descriptions, more on grounds of incommensurability than inconsistency. There is no uniquely right description, but that is just a pleasant meta-fact about the world and its describers. To say this is not to become subjective. It is to become pluralistic in one's meta-physics. Most descriptions won't wash (denial of subjectivity). There is no reason to think that only one will (pluralism). The world is so complex that we cannot compose the one true complete story about it. There is no one exhaustive true story: the idea does not make sense, as P.F. Strawson (1959, p. 128) wisely noted long ago, speaking of Leibniz.

Is there then nothing at issue between Latour's irrealism and my wishy-washy pluralistic realism? On the contrary. I think that there is some truth in the notion of natural kinds (although I don't think there are canonical, uniquely natural, kinds). The classifications of the human sciences are different from natural kinds; I call them human kinds. Latour doubts that there is any important difference. (Hacking 1988a) Here we are again, discussing an old question, the identity or disparity between the natural and human sciences. But we are doing it on largely new and chiefly experimental ground.

This thought naturally leads on to the traditional question of demarcation of science from pseudoscience. That was a significant practical problem for late 1920's Vienna and Berlin. It is commonly said that the whole issue is dead. Yet, as Steve Fuller (1988, ch. 7) urges, other problems of demarcation surface at present. We can witness some in an ongoing spat in which Richard Rorty (1988, p 54) attacks Bernard Williams (1985, p.139). But those two men debate solely at the level of theory and "realism", running around the usual squirrel cage of theory-obsessed philosophy. They barely acknowledge interaction with the world, and when they do so, do it is in terms of that jaded hack, "information and control." That is a phrase that conceals the complexities of the ways in which we interact with the world, and implies that it is the master, theory, that does the informing and the controlling.

In my opinion one of the fundamental ways in which laboratory science and ethics separate is that the former engages in what I have called the creation of phenomena: the purifying and maintaining of phenomena that do not exist in a pure state anywhere in the universe. (Hacking 1983, ch. 12). It is those phenomena to which theory answers. Phenomena may be maintained in the laboratory, brought back when interesting, or transformed into off-the-shelf transportable technology. This notion of purifying, creating and regularizing phenomena (and hence the world that we inhabit) certainly involves thinking and theorizing about the material world, but it also involves interaction with the world, and, in an unmetaphorical sense of the words, remaking it.

I have most often used examples of creation taken from physics, so it is important to insist that this is by no means necessary. As I wrote this paper, the two most publicized reports in Nature (13 September 1988) and Science (22 September 1988) were from immunology. Researchers have successfully implanted major parts of the human immune system into mice, giving them a small working model that can be used in testing drugs and vaccines, and, more importantly, enabling them to investigate this mysterious thing we call an immune system without doing experiments of human subjects. Both groups used a special strain of mice that have a genetic defect: they lack any immune system. So they are shortlived, and indeed persist only because experimenters maintain the race. Using different techniques two different groups have implanted human immune systems into such mice. This phenomenon, of the human immune system living in genetically defective mice, is new to the universe. It is one of many kinds of example of what I call the creation of phenomena.

I do not quarrel immediately with the constructionist who says that the "result", of a human immune system living in defective mice, is counted as result only in consequence of a lot of negotiations. Nor do I discount the obvious truth that the widespread reporting of these results is due to AIDS. I say only that whatever mixture of experimental ingenuity, the proclivity of nature, and the negotiation of competitive research groups brought the phenomena into being, a phenomenon was brought into being that did not exist outside the laboratory. We did not find it lying about, hidden like some lost island shrouded in mist.

The world with which we interact does not fix just what discoveries or inventions shall be made, nor does our past knowledge determine what we shall count as a discovery. It is not preordained which instruments we shall devise, nor which phenomena we shall be able to stabilize or purify. A science can develop among many possible paths, bringing into being different phenomena. Phenomena that we create on one possible historical path might not be created on another historical path, say because we have invented neither the instruments nor the theory of the instruments through which we could recognize or control them. Moreover any path that we do follow has its own momentum. Experimental techniques and instruments, when they produce what are taken to be stable results, themselves suggest further steps to take by analogy. Had we not started out on a path, we would not have created later on what we do in fact create. Thus this vision of experimental science is far more open than that of the philosopher of theory, who typically imagines that we are aiming at the one truth about a subject matter.

I speak of possibilities in the future. Many are open. But our knowledge also closes off possibilities. A picture is this: we set up benchmarks as we choose our paths through possible things to establish. Our procedures come to fix things—and these may be procedures as direct as the invention and perfection of modes of instrumentation, which then define questions for us and ranges of possible answers.

Instrumentation leads me to another topic, one that used to occur at each of our biennial meetings as a special symposium: the unity of science. I believe in the disunity of science, largely for the reasons Patrick Suppes (1979) urged at our meeting a decade ago. But science does have a unity quite different from that of the GUTs. It is kept together by a motley of disunified unifiers. Most important is one we now ignore, namely mathematics—Galileo's language of the author of nature, the differential and integral calculus, the Langrangians and Hamiltonians of the nineteenth century. We have some rather new unifiers, such as endemic statistical techniques, and brand new ones, such as fast computation that transforms the articulation of theory, the processing of data, and the simulation of synthetics (be they molecules or metals). There is another great unifier central to my present theme: instruments and apparatus.

Throughout most of the twentieth century, regimes and practices of experimentation and instrumentation have been a more powerful source of unity than grand unified theories. Instruments were speedily transferred from one discipline to another, not according to theoretical principles but in order to interface with and participate in the creation of phenomena. An extreme version of this 'instrumentalist' thesis: it is not high level theory that has stopped the innumerable branches of science from flying off in all directions and becoming different cultures. It is rather the pervasiveness of a widely shared family of devices. Nuclear magnetic resonance spectrometers, once on the frontier of experimental physics, are now the pedestrian stuff of biochemical assays. Not because the biochemist thinks that the physics used to make the device is right, but because the instrument is there, the one to use.

The unity of science is worked by a motley of diverse unifiers. Theory-oriented philosophy gave us a picture of nature as not only some hegemonic unified totality but also as passive and inert. It made us think that we discover her properties, we reveal her secrets. I saw a striking image of this the other day. 1905 has been called the *annus mirabilis*, the year in which Einstein put forth not only the special theory of relativity, but also the photon account of light, and the full understanding of Brownian motion in statistical mechanics. In November of that year the king of Portugal, a patron of science and literature, made a state visit to Paris. He was presented with a statue by Brassais over a metre high: *La nature se decouvrent devant la science*. A handsome young woman, Science, is lifting a cape high over her head, undressing herself to reveal her exposed bosom and sightly body.

The sculpture means many things. Nature is something to be discovered, revealed, a sexual object who can be got to undress before the inquisitive male, who will then dominate and possess her. She is passive. All that must be done is to take off her clothes. The given is the woman, nature, who gives herself, exposes her one truth. And the science that possesses her is theoretical science, which discovers this passive being.

That is an image entirely at odds with the experimental attitude. But our experimental predecessors were equally forthright in their sexual imagery of science and nature. We had a notable experimental philosophy of science long before the present one that is emerging. It was the philosophy of Bacon, or at any rate that named Baconian. It inaugurated the aggressive master-slave picture of the male master, who did not so much uncover as interfere with Mistress Nature. This, as writers in the women's movement have amply documented, has been incorporated into our accounts of how nature herself works, with a conscious attempt to find controlling forces, triggering mechanisms, targets and the like, a picture of nature in the image of macho militarism (Merchant 1980, Keller 1985).

With what images will a new mode of philosophizing about experiment reflect its underlying and implicit attitudes? Self-consciousness about the sexual imagery of the old philosophies, both theoretical and Baconian, has emerged at just the same time as the new thinking about experimentation itself. I hope that the image of the future will be one of the experimenter collaborating with nature rather than mastering it.

This tendency will assuredly be augmented by the fact that the most imaginative sciences of the present decade are biological and astrophysical, one a life science and one scarcely a laboratory science at all. It may be that the self conception of the experimental life sciences will displace the role model set by physics. Physics has long lived by the following picture of an experiment: there is a target, some apparatus used to interfere with the target, and a detector used to determine what is the effect of the interference. Those are the words and idea of James Clerk Maxwell. If one resists the military overtones of targets, recall Rutherford's imagery of splitting the first atom; he compared his alpha particles to shells from a 19 inch gun, at that time the noblest achievement of the Royal Navy. The more biology liberates itself from a desire to emulate physics, the more the physicists' conception of their work may in turn be modified.

That would involve a change in our idea of how we relate to nature. The Baconian image of the man-scientist interfering with the woman-nature was projected on to nature itself. The overriding theme was one of central causal structures that dominated everything that happened. One can speculate that if an image of experimenter as 'biological' collaborator were to take hold, we would come to think of the autonomous and independent activities of nature in a different light.

Science has from time to time served as a model for all culture, most notably in the Enlightenment, but also during the nineteenth century when the confrontation between religion and science so troubled reflective Europeans. At present it does not seem a model, but that is because science itself is so ill understood. The common image of science remains a modified version of Enlightenment science. The humanities have clung to the Enlightenment image of science as a grand unifying intellectual adventure, one that strives to find the ultimate theory of everything. Our civilization now values accommodation, variety, choice. It denies foundations but yearns for a stability that ensures coexistence of a multitude of interests. It wants toleration and respect, not unified hegemony. As the human sciences have become more and more diversified, as writing and composing and dancing and designing have become more varied, science had been cast as a stereotyped monolith. But science has become as multifloriate as the humanities. It has become a domain in which there can be stability without foundations, sharing without commensurability. It is a domain that favours realism about the material world, with a maximum of variety but a minimum of subjectivism. It has become a domain in which there can be coherent action within a thoroughly disunified world picture. I have been speaking for a modest scientific humanism, which we might call an experimental humanism. It will be rooted, perhaps, more in the life sciences, than in physics, that old bastion of theory and of unity.

References

Achinstein, P. and Hannaway, O. (eds). (1985), *Observation, Experiment and Hypothesis in Modern Physical Science*. Cambridge, Mass.: M.I.T. Press.

Ackermann, R. (1985), *Data, Instruments, and Theory: A Dialectical Approach to Understanding Science*. Princeton: Princeton University Press.

Baird, D. (1988), "Five Theses on Instrumental Realism", *PSA 1988*, 1: 165-173.

Batens, D. and van Bendegem, J. P. (eds.). (1988), *Theory and Experiment: Recent Insights and New Perspectives on their Relation*. Dordrecht: Reidel.

Braithwaite, R.B. (1953), *Scientific Explanation* (1946 Tarner Lectures). Cambridge: Cambridge University Press.

Cartwright, N. (1983), *How the Laws of Physics Lie*. Oxford: Clarendon Press.

Crombie, A.C. (1981), "Philosophical Presuppositions and Shifting Interpretations of Galileo", in *Probabalistic Thinking, Thermodynamics and the Interaction of the History and Philosophy of Science*, J. Hintikka et. al. (eds.). Dordrecht: Reidel.

Fleck, L. (1935), *Entstehung und Entwicklung einer wissenschaftlicher Tatsache. Einfuehrung in die Lehre vom Denkstil und Denkcollectiv*. Basel: Schwabe. F. Bradley and T. Trenn (trans.) (1979), *Genesis and Development of a Scientific Fact*. Chicago: Chicago University Press.

Franklin, A. (1986), *The Neglect of Experiment*. Cambridge: Cambridge University Press.

Fuller, S. (1988), *Social Epistemology*. Bloomington, Ind: Indiana University Press.

Gilbert, G.N. and Mulkay, M. (1984), "Experiments are the Key: Participants' Histories and Historians' Histories of Science", *Isis* 75: 105-25.

Gooding, D., Schaffer, S., and Pinch, T. (eds.). (1989), *The Uses of Experiment: Studies of Experimentation in the Natural Sciences*. Cambridge: Cambridge University Press.

Goodman, N. (1978), *Ways of Worldmaking*. Indianapolis: Hackett.

Hacking, I. (1982), "Language, Truth and Reason", in *Rationality and Relativism*, M. Hollis and S. Lukes (eds.). Oxford, Blackwell: pp. 48-66. Enlarged in "Styles of Scientific Reasoning", in *Post-Analytic Philosophy*, J. Rajchmann and C. West (eds.) New York: Columbia University Press, 1985, pp. 146-165.

_____. (1983), *Representing and Intervening*. Cambridge: Cambridge University Press.

_____. (1988a), "The Sociology of Knowledge about Child Abuse", *Nous* 22: 53-63, followed by comment by B. Latour, 64-7.

_____. (1988b), "The Participant Irrealist at Large in the Laboratory", *British Journal for the Philosophy of Science*, 39: 277-294.

_____. (1988c), "On the Stability of Laboratory Science", *The Journal of Philosophy*, 85:507-14.

Heelan, P. (1988), "After Experiment: Realism and Research", *The Journal of Philosophy*, 85: 515-524.

156

Keller, E. F. (1985), *Reflections on Gender and Science.* New Haven: Yale
   University Press.
Latour, B. and Woolgar, S. (1979), *Laboratory Life: The Social Construction of
   Scientific Facts,* Sage, Beverly Hills and London. Second unaltered edition
   with additional preface and the word "Social" stricken from the subtitle on
   the grounds that it is redundant, Princeton: Princeton University Press, 1986.

Latour, B. (1987), *Science in Action.* Harvard University Press, Cambridge, Mass.

MacKenzie, D. (1981), *Statistics in Britain 1865-1930: The Social Construction
   of Scientific Knowledge.* Edinburgh: Edinburgh University Press.

McMullin, E. (1988a), "The Shaping of Scientific Rationality: Construction and
   Constraint", in McMullin 1988b, pp. 1-48.

_ _ _ _ _ _ _ _. (ed.) (1988b), *Construction and Constraint: The Shaping of
   Scientific Rationality.* Notre Dame, Ind.: University of Notre Dame Press.

Merchant, C. (1980), *The Death of Nature.* San Francisco: Harper and Row.

Pickering, A. (1984), *Constructing Quarks: A Sociological History of Particle
   Physics.* Edinburgh: Edinburgh University Press.

_ _ _ _ _ _ _. (forthcoming), "Big Science as a Form of Life:, in *The
   Restructuring of the Physical Sciences in Europe and the United States,
   1945-1960,* M. de Maria and M. Grilli (eds.). World Scientific
   Publishing, Singapore. de Maria and Grilli.

Radder, H. (1988), *The Material Realization of Science.* Assen/Maastrict: van
   Gorcum.

Ravetz, J.R. (1971), *Scientific Knowledge and its Social Problems.* Oxford:
   Clarendon Press.

Robinson, J.D. (1986), "Appreciating Key Experiments", *British Journal for the
   History of Science* 19: 51-56.

Rorty, R. (1988), "Is Science a Natural Kind?" in E. McMullin (1988b). 49-74.

Shapin, S. and Schaffer S. (1986), *Leviathan and the Air Pump: Hobbes, Boyle
   and the Experimental Life.* Princeton: Princeton University Press.

Stump, D. (1988), "The Role of Skill in Experimentation", *PSA 1988,* 1: 302-311.

Suppes, P. (1979), "The Plurality of Science", *PSA 1978,* 2: 3-16.

Trenn, T. (1977), *The Self-Spitting Atom: A Study of the Rutherford-Soddy
   Collaboration.* Taylor and Francis: London.

Williams, B. (1985), *Ethics and The Limits of Philosophy.* Cambridge,
   Mass.:Harvard University Press.

Multiple Constraints, Simultaneous Solutions[1]

Peter Galison

Stanford University

Historians and philosophers of science have, of late, found a new common ground. During the 1960s the two fields worked together to answer the general question: "how does one theory supersede another?" — followed immediately by a second query: "Is the process rational?" Historians provided examples from the chronicles of science, and philosophers tested these cases against universal models of change. Together the two helped ground an important avenue of inquiry, and today the best of that work — by Feyerabend, Kuhn, Lakatos, Popper, and others — forms a staple of the philosophy of science.

But despite the best of intentions, in the 1970s the two fields began to pull apart centrifugally. On one side, the philosophers found their concerns concentrated ever more on fundamental issues about meaning and reference, and these seemed to demand idealized histories with simpler structures. On the other side, the historians had programmatic interests in the exploration of national scientific styles, relations between science and the state, and the formation of scientific institutions. By the mid-1970s something of a backlash had developed in each camp — the philosophers judged the particularism of the historians unhelpful to the formulation of their more general concerns about meaning and theory change, and the historians found the philosophers' attention to universal patterns of scientific reasoning uninformative about the evolution of disciplines under specific cultural circumstances.

The case for a joint historical-philosophical enterprise has recently been reopened. Above all, several historians and philosophers, with Hacking prominent among them, have discovered a common locus of concern in a set of questions that are intermediate in scope between the atemporal orientation of the search for universal procedures of science, and the more local orientation of the historian. They take as their target the historical development of standards of demonstration, both in theory and in experiment.

1. The Persuasiveness of Theoretical Arguments

Since the focus both of my own work and of Hacking's has been on experiment, I would like to emphasize the importance of investigations into theory, where the guiding question is: "What makes a theoretical argument compelling and how have these standards changed?" Let me give some examples of the ways in which such questions get worked out as the result of a joint philosophical and historical investigation. In a sense, all these examples undermine the dichotomy of justification and discovery by showing how the conditions of origin help shape the criteria of justification.

PSA 1988, Volume 2, pp. 157-163

Unification is frequently invoked as a theoretical virtue that lends force to a theory's persuasiveness. But Margaret Morrison (1987) argues that unification is highly context-dependent; unification has force just insofar as practitioners previously considered the components to have been genuinely distinct (e.g. the terrestrial and celestial before Newton). The historical categorization of phenomena is therefore crucial to any argument to be made in which consilience or unification is to serve as evidence for the truth of a theory or for the existence of the entities it employs.

Questions of concept formation and category application enter in a different way in the historical-philosophical work of Hacking (1986a) and Arnold Davidson (1986). Both have been concerned with the specification, primarily in the human sciences, of categories that are prerequisite to the meaningfulness of certain utterances. Davidson points out, for example, that the phrase "X is a homosexual even though he never acted upon his inclinations" would have been meaningless before the advent of psychiatric reasoning. To use Hacking's (1986b) apt slogan, the phrase is not even "a candidate for true-or-false" until a psychiatric concept of homosexuality has been provided. Considerations like these lead Hacking to conclude in his "Philosophers of Experiment" that while there may be something to the idea of natural kinds, "human kinds" are created in large part through our system of classification. At least in the human sciences there is no separating the philosophical and historical components of the inquiry into how such ways of talking came to be accepted, and how the group being classified reacted. The historical investigation is, in its result, a philosophical excursion into the formation and implementation of a world-shaping concept.

The historicity of background conditions forms an essential part of Nancy Cartwright's recent (1989) study of causality in econometric theory. Her goal is to challenge the "symmetry" interpretation of equations, according to which effects and causes can never be distinguished — only correlated. Such a quasi-Humean critique underestimates the sophisticated view of equations held by the twentieth-century originators of econometrics in which background knowledge broke the symmetry, and allowed a causal interpretation of structural equations. But, Cartwright insists, the only way to make sense of the econometricians' claims about causality is to unravel, through historical investigation, the nature of this background knowledge. Causality thus plays a role in the establishment of what counts as an adequate explanation, but its role will remain forever hidden from us if we look only at equations, not at the full historical context in which they were developed and interpreted.

As in our view of causal explanations, we lose a great deal of our understanding of statistical inference if we strip the use of statistics from its historical context. In the late twentieth century it is taken for granted that statistics form a part of the armory of arguments upon which science draws. But as both historians and philosophers have pointed out, statistics entered scientific discourse in halting steps, and its certification as legitimate was intimately connected to its original social and juridical application. (Porter 1986; Daston 1988; Krüger et al., 1986). Two centuries later, debate continues about the meaning and right role of statistics in science, and nowhere more forcefully than in the interpretation of quantum mechanics. Here too philosophical questions have been recast by Fine's (1986) historical-philosophical study of Einstein's views on statistics and realism.

2. The Force of Experimental Demonstrations

On the experimental side, there is a question that parallels the one regarding theoretical arguments. It is the one with which I have been most concerned: "What makes an experimental demonstration persuasive?" Here the focus is not just how certain forms of speech become reasonable within historical contexts, but in addition how certain laboratory moves are certified. Under what conditions do instruments become acceptable means of acquiring knowledge? What are the proper uses of these instruments? (Galison 1985, 1987, 1989) Since Hacking has asked me to comment on the long-standing, if amicable, disagreement I have with Andy Pickering (cf. Lenoir 1988), since Hacking has today adopted some of Pickering's language of plasticity, and since Pickering himself is here, let me try to characterize the essential differences in our approaches to laboratory knowledge.

Pickering's approach seems to me to be driven by two intuitions. The first is that laboratory practice is essentially *plastic* — underdetermined by data, and therefore highly malleable in the hands of the scientist. For example, "It is unproblematic that scientists produce accounts of the world that they find comprehensible: given their cultural resources, only singular incompetence could have prevented [high-energy physicists from] producing an understandable version of reality at any point in history." (Pickering 1984, p. 413) Elsewhere (1981, p. 236), he refers to this malleability when he says that "physicists 'tune in' on phenomena consistent with [their] commitments."

In part because the practices of physics are so pliable, laboratory practice is, to a significant extent, subject to *external* forces of career interests and high theory. Pickering (1984, p. 409 emphasis added): "To choose between the theories of the different eras *required* a simultaneous choice between the different interpretive practices in neutrino physics which *determined* the existence or nonexistence of neutral currents." This gets to the heart of one level of our disagreement. For Pickering, factors entirely external to laboratory practice ensured that the neutral currents be accepted; these external factors range from the micropolitics of the Academy, to the imposition on the experimenters of a dominant theory. In my account, theories — and indeed the broader culture — enter the laboratory in various ways. Experimenters draw on and are constrained by their understanding and use of available technology, theories, and ideologies of their time. Constraints may include the acceptance of particular uses of computer simulations, properties of material properties, electronic detectors, or calculational methods. But nothing *guarantees* that the theory (e.g. gauge theory) and the experiments would come into accord (Galison, 1987). The scientific community is not assured in advance of a harmonious outcome to the linkage between experiment and theory — as many a grand and dead theoretical program mutely testifies.

Not accidentally — since interest-theorists like Pickering draw heavily on it — this picture of an externally-driven and malleable laboratory practice resembles Paul Forman's (1971) account of the origin of quantum mechanics in interwar Weimar. According to Forman, the demands of theory and experiment were insufficient to explain the acceptance of quantum uncertainty. Instead, Forman locates the acceptance of indeterminism in the confrontation between physicists and the hostile audience of German academic mandarins. His fascinating account goes like this. Through deception and self-deception academics in Germany expected a victory in World War I. When the Kaiser's army collapsed, this broad academic culture was shocked. A powerful constituency of humanistic scholars, doctors and lawyers angrily turned their wrath against the physicists whom they saw as the incarnation of materialism, determinism, and rationality in a world that more than ever seemed to be fueled by the spiritual, the unpredictable, and the irrational. Faced with this anger, the physicists *capitulated* (Forman's term) and introduced indeterminism into their subject. Norton Wise (forthcoming) criticizes the Forman thesis for its reliance on an external force (Weimer pessimism) to drive internal changes within the practice of physics (indeterminacy). Rightly, I think, Wise argues that the quantum mechanicians ought to be considered part of their culture not as the pawns of outside powers.

Indeed, Forman's "capitulation," Pickering's "interests," and some reductionist materialist accounts are all instances of externalism. In each case the practice of physics itself is taken to be unconstraining in some essential way and fixed only from outside. To be fair, one should remember what they took to be the opposing view: a naive picture of scientists, isolated from their culture, producing theories that were dictated by unambiguous, "raw" experimental data. Even if this picture of experiment as an assemblage of logic and protocol reports fails to capture the sophistication of the major logical positivists, it may portray the textbook version of it. No doubt both as philosophers and as historians we have to sympathize with the externalists' rejection of such brute inductivism.

3. Multiple Constraints, Simultaneous Solutions

My own inclination, however, is quite orthogonal to the one painted by the plastic-externalists. On the one hand, the evolution of physics strikes me not so much as plastic, but rather as immensely constrained. Undoubtedly, Pickering sees these constraints as the experimentalists' realist illusion — a "false consciousness" papering over the real (external) forces guiding their hands. Both historically and historiographically this kind of false consciousness account lacks force. On historical grounds, it makes physicists out to be much more naive than they are about the constraints under which they labor. On historiographical grounds, it seems to me (in general) to be a poor strategy to depict a group in such a way that one has to dismiss its members' own experience of their situation. Instead of viewing laboratory scientists as manipulated by outside forces, let us view them as embedded in their culture, where the phrase "culture" includes both the dynamics of the research endeavor and the larger environment in which it is situated. I would like to suggest that one can capture both the constrained and embedded aspects of the enterprise by thinking of laboratory moves as satisfying simultaneously a variety of constraints with highly diverse origins.[2]

Take a real example. In the mid-1980s, the particle physics community was working hard to decide how to build the massive detectors for the Supercollider. Servicing a research team of five or six hundred physicists, the detector's design would shape their work for decades — and cost hundreds of millions of dollars. Numerous factors conditioned the proposed design — as a result physicists would never have said that the design was *caused* by this or that consideration. In making their decisions they used the language of constraints and consistently grouped them together as if they were a set of homogeneous "external" and "internal" limits to action. Here is the judgment of one detector designer: "Factors which favor decreasing the radius are the minimizing of pion and kaon decays and detector cost." Or elsewhere, about the decision to move the magnet outside the calorimeter, the designers put down a list of reasons from which I abstract three: 1. "No deterioration of calorimeter performance with respect to energy resolution, shower spreading or, especially, hermeticity." These are constraints dictated by the physics goals of the detector, in particular the search for heavy quark decays. 2. "Reduction of uranium mass and calorimeter cost." These are practical constraints that issue from the status of the uranium supply within the whole network of military and nuclear industrial concerns and with the politics of science budgeting. 3. "Simplification of calorimeter cryogenics." This is a technical constraint on the construction of very low-temperature components for device itself (Feldman et al., 1984). A technological determinist (externalist) account of the instrument would inaccurately render the physics constraints secondary; an "internalist" account would wrongly narrate solely in terms of quarks, gluon, and leptons.

As we move outward from the design of a specific piece of equipment to a whole experimental program, we see constraints operating not just within, but *between* different fields of science. When C.T.R. Wilson began experimenting in the late 19th century, he wanted (like a good Victorian naturalist) to reproduce optical wonders of the sky in the laboratory, like coronae. But at the same time he was immersed in the culture of the Cavendish laboratory that focussed above all on the ionic constitution of matter. He therefore designed his apparatus, on electrical principles, to precipitate water around ions in his (literal) "cloud chamber." Meteorologists understood his creation of clouds as contributions to the physical origin of rain. Physicists took Wilson's cloud work as a means of making the ion visible, and therefore of discovering the dynamics of the ion. A reductionist account would say either that Wilson's physics was driven externally by its practical applications; or else they might claim that his meteorology was an application of the Thomsonian program of fundamental research on ions. Neither captures the spirit of the investigation: Wilson created a field simultaneously constrained by the natural historical tradition (which I have called "mimetic" experimentation) on the one hand, and the Cavendish tradition of electrical experiments on ions on the other. His condensation physics was a *simultaneous solution* to these *multiple constraints* (Galison and Assmus, 1989).

Let me give an example from a very different time and place. In their important recent book, *Leviathan and the Air Pump*, Shapin and Schaffer (1985) show how the Boyle versus Hobbes debate over the nature of experimentation was simultaneously a clash over how to handle disputes in natural philosophy, and how to bring order following the Restoration settlement. Boyle, they contend, wanted to separate phenomena (fact) from beliefs (causes) and so create an arena of experimental facts in which disputes could be settled, even when the disputants came from radically different traditions such as Cartesianism or Alchemy. In politics Boyle wanted the same: a circumscribed domain of political dialogue, protected from any single tyrannical force, and where dispute could be resolved without coercion. Hobbes, by contrast, was steadfastly opposed to the separation of cause and fact as a guide to experimentation, and as a political philosophy. Indeed, Hobbes expected civil war to issue from any program lacking absolute compulsion. Just as civil society creates Leviathan to maintain order in the social sphere, so the community of natural philosophers needed to elevate geometry and logic to a compulsive status in the sphere of natural philosophy.

Some readers have taken Shapin and Schaffer to mean that the Hobbes-Boyle dispute over the vacuum and experimentation *reduces* to that of an underlying political dispute. In the language I am using here, I would describe the story as one of a conflict over a means of resolving disputes that is at one and the same time a solution to problems in both natural and political philosophy. Politics, on this reading, are not explained by air pumps any more than air pumps are explained by politics. And the participation of Boyle, for example, in the political arena does not univocally determine in advance either the form or outcome of his experiments on the spring of air.

The language of multiple constraints, then, serves to free our descriptions from the choice that seemed so crucial a generation ago: internal accounts that made science immune to the cultures in which it grew, and external accounts that discounted the manifold strictures on moves within theoretical and laboratory practice. Partially autonomous constraint networks[3] separate the practices of experiment from those of theory. Hacking rightly and eloquently protests against the notion of a science unified under "the master, theory, that does the informing and the controlling." But the idea of multiple constraints carries with it a second significance. Sometimes the imposition of overlapping constraints *allows no solution.* So it is in the coordination between experiment and theory, and it is here, ultimately, that the notion of plasticity seems to fail most dramatically. For the authors of the first of the grand unified gauge theories, there was every theoretical reason to believe in their SU(5) model. It embraced the aesthetic constraint of being contained in a simple group, it reproduced the Glashow-Weinberg-Salam model and all its phenomenology at low energies, and it fixed the most salient of the free parameters contained in the electroweak theory. Most strikingly it predicted the decay of the proton at a rate that would be experimentally observable. Experimentalists rushed to build deep-mine detectors; the race was on to detect the decaying core of matter. Without any doubt, all the key players had a professional interest in finding the dying proton; the tickets to Stockholm seemed ready and waiting.

But experiment and theory were not reconciled. There was no signal that satisfied the technical constraints on the muon detectors and scintillators, the theoretical constraints on conservation laws, symmetries, and dynamics, and the experimental constraints on statistical significance. Of course there were desultory attempts to rescue the meeting of experiment and theory by modifying bits of theory or apparatus. But they persuaded no one — least of all the heads of the experiments or the authors of the theory. Sometimes the theorists cannot "stabilize" the experimentalists' practices into conformity. It is precisely because of that difficulty that I do not take the embeddedness of physics in culture and in the multitude of human concerns to mean that physics is automatically self-authenticating. We need an understanding of experimental and theoretical standards that is at once conceptual *and* contextual. That seems to me the hallmark of the new history and philosophy of science.

Notes

[1]I have benefitted greatly from conversations with N. Cartwright, A. Davidson, C.A. Jones, T. Lenoir, M. Morrison, and N. Wise.

[2]The term "embedded" is used differently by other authors cf. Wise (1988); an important example of what I mean by "multiple constraints" in the biological realm can be found in Lenoir (forthcoming).

[3]Both Hacking and I contend that the multiple traditions within science lead to disunity; we also both argue (in different ways) that science achieves stability even without the positivistic unity. See Hacking, 1988 APA talk, "The Stability of the Laboratory Sciences," and (Galison 1988).

References

Cartwright, N. (1989), *Nature's Capacities and their Measurement.* (Oxford University Press.

Daston, L. *Classical Probability in the Enlightenment.* Princeton: Princeton University Press, 1988).

Davidson, A. (1987), "Sex and the Emergence of Sexuality," *Critical Inquiry* 14: 16-48.

Feldman, G. J.; Gilchriese, M. G. D.; and Kirkby, J., "4π Detectors" in *Design and Utilization of the SSC*, 1984 Snowmass Proceedings, p. 630.

Fine, A. (1986), *The Shaky Game.* Chicago: Chicago Univ. Press, 1986.

Forman, P., "Weimar Culture, Causality, and Quantum Theory, 1918-1927," *Historical Studies in the Physical Sciences* 3: 1-115.

Galison, P. (1985), "Bubble Chambers and the Experimental Workplace," in *Observation, Experiment, and Hypothesis in Modern Physical Science*, P. Achinstein and O. Hannaway (eds.). Cambridge: Bradford-MIT.

_ _ _ _ _ _. (1987), *How Experiments End.* Chicago: Chicago University Press.

_ _ _ _ _ _. (1988), "History, Philosophy and the Central Metaphor," in *Science in Context* 2: 97-112.

_ _ _ _ _ _, and Assmus, A., (1989), "Artificial Clouds, Real Particles," in *The Uses of Experiment*, D. Gooding, (ed.). Cambridge: Cambridge Univ. Press.

Hacking, I. (1975), *The Emergence of Probability,* Cambridge: Cambridge Univ. Press.

_ _ _ _ _ _. (1986a), in *Reconstructing Individualism*, T. Heller, M. Sosna, and D. Wellbery (eds.). Stanford: Stanford University Press.

_ _ _ _ _ _. (1986b), "Language, Truth and Reason," in *Rationality and Relativism*, M. Hollis and S. Lukes (eds.). Cambridge: MIT Press.

Krüger, L.,Daston, L., and Heidelberger, M., (1987) *The Probabilistic Revolution,* Vol. 1; and Krüger, L., Gigerenzer, G., and Morgan, M. (1986), *op. cit.*, Vol. 2, Cambridge Bradfort – MIT.

Lenoir, T. (1988), "Practice, Reason, Context: The Dialogue Between Theory and Experiment," in *Science in Context* 2: 3-22.

_ _ _ _ _ . (forthcoming), "The Eye as Measuring Device", unpublished typescript.

Morrison, M. (1987), "Dimensions of Theory Acceptance: Methodology and Experiments," unpublished PhD dissertation University of Western Ontario.

Pickering, A. (1984), *Constructing Quarks,* Chicago: Chicago University Press.

_ _ _ _ _ _ _. (1981), "The Hunting the Quark," *Isis* 72: 216-36.

Shapin, S. and Schaffer, S. (1985), *Leviathan and the Air-Pump,* Princeton: Princeton University Press.

Porter, T. (1986), *The Rise of Statistical Thinking: 1820-1900*. Princeton: Princeton University Press.

Wise, N. (1987), "How Do Sums Count? On the Cultural Origins of Statistical Causality". In Krûger, Daston, and Heidelberger (1987).

_ _ _ _ _. (1988), "Mediating Machines", *Science In Context* 2: 77-113.

_ _ _ _ _. (forthcoming), "Forman Reformed", to appear in S. Schweber and L. Daston, *Science in Context* 3.

Part VI

REALISM

Scientific Realism in Real Science[1]

Roger Jones

University of Tennessee

1. Introduction

Scientific realism is a doctrine about the relationship of our ideas on the nature of things to the nature of things itself. Part of the doctrine is that there is a nature of things itself. With regard to the rest, Jarrett Leplin has said, "Like the Equal Rights Movement, scientific realism is a majority position whose advocates are so divided as to appear a minority" (Leplin 1984, p. 1). Still, it can be said that what realists would like is that the account of the nature of things provided by science be true and that those things really exist.

Characterized in this way, realism would seem to be a majority position indeed. As Ernst Mach has said of the doctrine,

It has arisen in the process of immeasurable time without the intentional assistance of man. It is a product of nature, and is preserved by nature. Everything that philosophy has accomplished...is, as compared with it, but an insignificant and ephemeral product of art. The fact is, every thinker, every philosopher, the moment he is forced to abandon his one-sided intellectual occupation..., immediately returns [to realism]. (Mach 1959, p. 37, quoted in Fine 1986, p. 134)

Pre-analytically, we are all realists. We would all *like* to be realists. But analytically, in the course of their one-sided intellectual occupations, philosophers have offered a number of objections to realism.[2] These range from highly abstract and general objections, to objections based on analysis of specific historical circumstances in science.

Another academic community troubled by issues related to scientific realism is that of contemporary physicists. Their difficulties certainly have nothing to do with the general and sweeping objections of philosophers. But neither do such difficulties seem to have been examined in philosophers' analyses of historical cases. So what I want to do here is to provide three illustrations of just what these difficulties with realism are for contemporary physicists. These illustrations come from the basic pedagogical tradition in classical mechanics and from the interpretive traditions of non-relativistic quantum mechanics and of general relativity. I think philosophers ought to be aware of these real-science difficulties with scientific realism, and I will provide some advice about how they ought to respond at the end of the paper.

PSA 1988, Volume 2, pp. 167-178

2. Realism

I have said that beyond a commitment to a "nature of things itself," advocates of realism are severely divided. But I have also said that they share the general hope that the scientific enterprise has the capacity to provide accounts of this nature-of-things-itself that are true, accounts involving just the things that are there themselves.

In what is more or less the "classical" realist position, this hope is elevated to a belief. Indeed, such classical realists are willing to go out on a limb and claim that theories in the "mature" areas of science should already be judged as "approximately true," and that more recent theories in these areas are closer to the truth than older theories. They see the more recent theories encompassing the older ones as limiting cases, and accounting for such success as they had. These claims are all closely linked to the claim that the language of entities and processes — both "observational" and "theoretical" ones — in terms of which these theories characterize the-nature-of-things-itself genuinely refers. That is, there are entities and processes that are part of the nature-of-things-itself that correspond to the ontologies of these theories.

The way in which this reference is fixed, and thus the nature of this correspondence are topics of intense current debate even among the classical realists who follow the position this far. All I want to point out however, is what a hearty and confident doctrine this "classical" realism is. It envisions mature science as populating the world with a clearly defined and described set of objects, properties, and processes, and progressing by steady refinement of the descriptions and consequent clarification of the referential taxonomy to a full-blown correspondence with the natural order. It is surely a grand ideal. Let me illustrate now how it fares in real contexts.

3. Classical Mechanics

In the beginning of an undergraduate education in physics comes Newtonian dynamics. That is, physics is presented as beginning with Newton's three laws, and they are typically introduced as simply formal generalizations of directly observable particle behavior. (See, e.g., Symon 1960, Ch. 1.) Thus a "particle" is introduced as the unit of matter, a gritty bit whose size, shape, and internal structure are regarded as negligible. Particles have positions at various times, and quantitative measures of positions and times lead to (continuous) functional relationships between them. Properties of this kind of functional relation are the velocity and acceleration of a particle at each time. The properties of mass and force are initially introduced quite operationally, as simply other aspects of the functional behavior of particles in space over time. Mass (or the ratio of two masses) is asserted as a constant of proportionality in the relative accelerations of two particles involved in an interaction. Force is introduced as simply the name of the mass-acceleration product, automatically identical, though with opposite sign, for pairs of interacting particles. However operationally they are introduced, however, these two properties of force and mass are soon identified with particle-nature. Force in particular is able to be "exerted" on other particles, and is asserted to be the source of their accelerations.

An early application of this Newtonian dynamical apparatus is to planetary motion. In the style in which Newton himself asserted it, it is asserted that each body/particle in the universe exerts at each instant on each other body/particle an attractive force directly proportional to the product of their masses and inversely proportional to the square of the distance between them. In accord with the previously established dynamical properties of particles, each body reacts to this force (as to any force) by accelerating in the direction of the force in amount directly proportional to the magnitude of the force and inversely proportional to the particle's mass.

Somewhere toward the end of a good first-year course in physics, it is pointed out that this approach to gravitational interaction is only convenient so long as bodies can be treated as massive particles, i.e., so long as unique centers of gravity can be defined for the interacting bodies. For extended bodies in general, a different approach must be taken. In this approach the gravitational interaction is described by means of a new kind of (generally un-analyzed) entity — a field in space, the gravitational potential. This potential varies from point to point, and a massive object placed at a point in space experiences a force in the direction in which the potential gradient is greatest, in magnitude proportional to the potential gradient. This last statement, plus a statement characterizing the variation of the potential from point to infinitesimally nearby point, provides a mathematical route to deriving the law of universal gravitation in precisely the form of the original approach. Introduced to gravitational potentials late in the first-year course, a young physicist-to-be is generally trained to work fluently with them in a second course in mechanics.

Newton's force law and the laws of planetary motion can be derived from an approach more general yet, that based on "minimum principles." In this context, such a principle may be taken to assert that if a massive particle is to proceed from one point to another in some fixed time, then of all possible fixed paths between these two points, that path is physically realized on which some quantity associated with the motion is a minimum. Practically speaking, the generality of the approach stems from the fact that the path need not be described in the familiar coordinates of Euclidean space. The method of characterization of the motion of a particle in terms of more "generalized" coordinates opens up to dynamical scrutiny the behavior of systems of many particles coupled by various interactions, systems virtually untreatable using previous approaches. One may, for instance, speak of the "configuration" of a many-particle system as characterized by a single generalized coordinate location, and characterize the evolution of the system wholly in terms of a trajectory in "configuration space." But again, for simple systems, results mathematically identical to those derivable from the approaches above emerge. This approach to mechanics, the Lagrangian and Hamiltonian approach, is the substance of the first-year graduate course in mechanics still traditional in most physics departments.

Finally, in light of the most recent reformulation of the laws of planetary motion, one may consider the space of classical Newtonian theory to be curved by the presence of matter. In this most modern theory of "Newtonian gravitation," the gravitational potential field of the approach above is absorbed into the structure of space itself, though time remains an autonomous parameter (thus distinguishing this theory from relativity theory). There is no "gravitational force" in this approach. Massive bodies move in "straight lines" in the curved space (unless some non-gravitational, e.g., electromagnetic, interaction intervenes). This approach to particle interactions, particularly gravitational ones, would be introduced as part of a good upper-level graduate course in general relativity.

These then are the stages in the development of competence in classical mechanics, as reflected in the treatment of planetary motion. At each stage the new approach is introduced as a generalization of the old, as capable of handling a class of problem inaccessible, or not conveniently accessible, to the old. The power, breadth, and elegance of the new treatment is extolled; but it is never suggested that any "new physics" has been introduced. As one widely-used textbook describes the Lagrangian and Hamiltonian approach,

They are not new physical theories, for they may be derived from Newton's laws, but they are different ways of expressing the same physical theory. They use more advanced mathematical concepts, they are in some respects more elegant than Newton's formulation, and they are in some cases more powerful in that they allow the solutions of some problems whose solution based directly on Newton's laws would be very difficult. The more ways we know how to formulate a physical theory, the better chance we have of learning how to modify it to fit new kinds of phenomena as they are discovered. (Symon 1960, p. 3)

The point is, all the approaches to planetary motion described above, as viewed from the standpoint of the education of a young physicist, are somehow on a par, are "different ways of expressing the same physical theory." It is important that they all be part of the mathematico-conceptual repertoire of a young physicist, the better for him or her to be able to handle "new kinds of phenomena as they are discovered."[3] In an image provided by the physicist Richard Feynman to characterize his attitude toward such a multiplicity of approaches to a particular fundamental law, "It is like a bridge with lots of members, and it is over-connected: if pieces have dropped out you can reconnect it another way" (Feynman 1965, p. 47; his discussion of Newton's law appears on pp. 50-53).

Actually, though this over-connected bridge serves the young physicist best as one of both mathematical and conceptual structure, the mathematical structure is considerably more over-connected than the conceptual. Though they are rarely analyzed in standard textbook treatments, the explanatory accounts provided of planetary motion in the approaches above are in fact radically different. In the first approach, for instance, the approach attributed to Newton, the law of universal gravitation is usually taken to describe the properties of a fundamental gravitational force which has about it a renowned kind of dual non-locality: the gravitational force is associated with the instantaneous positions and masses of massive bodies in empty space, bodies which respond to this force instantaneously and as they move about instantaneously change it.[4] The second approach described above does characterize the gravitational interaction in purely local terms. But it does so by means of the introduction of a new physical entity into the explanatory picture — the potential field. This field is eliminated if physical space — heretofore treated (implicitly) as a flat, Euclidean space — is assumed to be curved by the presence of matter, as in the fourth account. But then a kind of causal efficacy is associated with the structure of space itself. Finally, the approach in terms of a minimum principle seems to have no connotations of causality at all. The instantaneous motion of a massive body in this approach is determined by a property associated only with a complete path between two points in space.

Of course these sketches of the explanatory structures associated with the diverse mathematical approaches to classical mechanics ignore a whole wealth of difficulties. It might be argued that none of the explanatory frameworks will bear detailed scrutiny, that none provides the kind of thorough causal account which may be considered an important part of any fully satisfactory explanation. Certainly Newton was never satisfied with his understanding of the nature of the gravitational force. (See Note 4.) Associating the approach in terms of a minimum principle with a causal account is difficult even on the face of it. (See, e.g, Yourgrau and Mandelstam 1968, Ch. 14.) The field concept itself has its own problems (again, for a recent reference, see Nersessian 1984), as does an account of the causal efficacy of spatial structure (see Friedman 1983, pp. 67ff).

So what is "the account of planetary motion provided by classical physics?"[5] I hope it is clear that in the mind of a young physicist, there is likely to be no univocal, canonical account of the above description. All the approaches described above lie within the vastly over-connected structure of concepts labelled "classical physics," even "Newtonian dynamics." They in some sense "save the same phenomena," but with very different explanatory frameworks, very different ontological commitments. Even if a young physicist is a non-critical realist, he or she will have trouble when asked to articulate the fundamental (theoretical) furniture of the Newtonian universe. He or she doesn't know, in some canonical sense, what to be a realist about.

4. Quantum Mechanics

In classical mechanics the alternative explanatory frameworks (with their alternative ontologies) were associated with different mathematical approaches. But even taking the most straightforward approach to quantum mechanics — ignoring the stepping stones of the original Bohr-Sommerfeld theory, Schrodinger's wave mechanics, and Heisenberg's matrix mechanics; ignoring contemporary reformulations such as those due to Feynman and Schwinger — considering only the non-relativistic theory in its standard formulation, the same kinds of difficulties for applied realism are present: there is a variety of interpretations of the mathematical formalism of the theory and no single interpretation has really survived detailed scrutiny.

The existence of such a multiplicity of interpretations for this formal apparatus of quantum mechanics is so well known that I will merely point to it here.[6] Physicists readily characterize themselves as holding a version of the Copenhagen interpretation, or the statistical interpretation, sometimes even as accepting an approach to the theory in terms of quantum logic. More recently, many physicists express themselves particularly interested in the so-called "many worlds" interpretation due to Hugh Everett (Everett 1957, reprinted in DeWitt and Graham 1973).

It is important to underscore three aspects of this interpretive multiplicity. First is the disparateness of these various interpretations. The worlds that they picture are utterly different. Their ontological commitments are different; their imputations of causal structure are different; their focus on the mathematical structure is different. In some cases it is difficult even to translate from the role of a certain element of the mathematical structure in one interpretation to that in another. A second aspect of this interpretive multiplicity is the failure of any interpretation to provide an "explanatorily satisfactory" link between the mathematical formalism and the world of laboratory experience. The unsatisfactory elements themselves vary among the interpretations, but there are difficulties with every one. This certainly is part of the reason for the persistence of multiplicity here, for the failure of any one interpretation to emerge as "standard." Finally, it is important to appreciate the vividness of these views for their adherents, and the fervor with which they identify the ontology of the theory thereby. This point can be fully appreciated only by observing physicists in their moments of speculative candor — in late Friday afternoon conversations with graduate students, in dinner-table controversies during conferences. I can but heartily recommend these moments.

But then what is "the account of microworld behavior provided by quantum physics?" The interpretive state of quantum physics — even in terms of this single theoretical framework — simply does not allow a univocal, canonical such account to be identified. The general approach of one interpretation may suit a physicist more than the general approach of others, and he or she may spend some time adapting it to issues that he or she thinks particularly important and developing arguments as to why its lacunae are not devastating for its coherence. But every physicist will admit that such allegiance is to some degree a matter of taste. No physicist is unaware of competing interpretations, and none expects there to be decisive evidence, or argument, for one against the others. Physicists don't know what deep explanatory structure of the microworld to be realists about.

5. General Relativity

Mathematically, the modern, standard approach to relativity theory begins with the postulation of a four-dimensional, differentiable manifold, generally called "space-time." Associated with this manifold, purely on the basis of its mathematical structure, is a host of derived structures, most notably tensor fields of various sorts. The "space-time framework," as an interpretive stance, postulates various of these fields as "physically distinguished," that is, as in some sense physically interpretable. What this amounts to generally is that one may identify various of these fields as having the attributes, or some of the attributes, by which we have traditionally recognized certain physical concepts. One may identify, for instance, fields taken to characterize the spatio-temporal history of material particles, their mass and charge distribution, their energy. More particularly, the general theory of relativity postulates a certain tensor field as "the metric field" of space-time, and postulates certain relationships among this field and other physically distinguished fields.

In textbook presentations, this metric tensor field is almost universally regarded as ontologically fundamental. (See, e.g., Misner, Thorne, and Wheeler (hereafter MTW) 1973, Chapter 13.) It is part of the essential nature of space-time, its basic geometricality, one of whose aspects is curvature. This curvature is linked by the field equations of the theory to the density of matter and energy, expressed in a tensor field usually characterized as the "stress-energy field." There is some ontological ambiguity even at this stage of typical textbook presentations, however. Does the stress-energy field share the fundamental status of the metric field? Or should this status be reserved for the fields from which it is computed — of which it is in some sense composed — for example the mass and charge fields just mentioned? What about the fields in terms of which it is locally measured by variously moving observers?

These kinds of interpretive issues — the status of composed versus composing fields, that of locally measurable properties of space-time and the intrinsic structures they arise as manifestations of — are not my particular concern here however. What I want to concentrate on here are difficulties which arise when one considers particular solutions of the field equations of the general theory, as one must in order to apply the theory in the concrete world.

One of the most discussed solutions of the field equations is that due to Karl Schwarzschild, interpreted as describing an isolated, spherically symmetric, massive body. All applications of the general theory to solar system phenomena usually begin with the Schwarzschild solution, so it is very important. To focus on an example, consider measurements of the deflection of starlight in passing close to the sun.[7] In these measurements apparent stellar positions are compared in the presence and absence of the sun. The Schwarzschild solution provides light and particle trajectories in the neighborhood of the massive body, and these trajectories can be compared to the results of measuring these positions.[8] More importantly for the ontological prospector, however, the solution provides the very notions of "isolated," "spherically symmetric," and "massive," as well as other important components of our intuitive conceptual vocabulary, e.g., the notions of global space and time.

The Schwarzschild solution is extremely unusual in this respect, for it is a static solution. From the four-dimensional space-time structure of such a solution global three-dimensional and one-dimensional sub-structures can be projected. But this possibility is not uniquely a property of a static solution. What is unique is that one can interpret the four-dimensional structure in this case to have been factored into a global three-dimensional "space" and a one-dimensional "time": the spatial distances between the points of the three-dimensional space can consistently be interpreted as the space-time distances between points judged, in terms of the one-dimensional time, to have simultaneous positions (Earman 1970, p. 260).

Another special feature of the Schwarzschild solution is its "asymptotically flat" character: the curvature of space-time vanishes as one approaches "infinity" purely spatially, and along the trajectories of particles and light rays. It is only for such a metric structure that the notion of an "isolated body" makes sense at all, and only for certain kinds of such structure that a property such as "mass" can be ascribed to the isolated body (Geroch 1977; MTW 1976, Ch. 19; Wald 1984, Ch. 11).

Thus, the availability of this whole classical vocabulary for the starlight deflection experiments hinges on the status of the Schwarzschild solution as a description appropriate for phenomena within the solar system. At the present time, there seems to be general consensus among working astrophysicists that it is quantitatively adequate for the analysis of starlight deflection data. (See, e.g., Shapiro 1980, p. 122.)[9] But the qualitative divergences of the idealization from what is known independently about solar system behavior are obvious. After all, the sun is neither static, nor spherically symmetric, nor isolated. It is rotating, oblate, and accompanied by some planets of considerable mass.

The problem in all this for the ontological prospector is that a progression to a less idealized description that takes the form of giving up the Schwarzschild solution also requires re-evaluation of the status of the traditional conceptual vocabulary in this case — the description in terms of space, time, and mass. Even treating the sun as a rotating body, for instance, obviously a less idealized approach than assuming it not to be so, and one accomplished analytically by employing the Kerr solution to the field equations (MTW 1976, Ch. 33), involves interpretive problems with "space" and "time": the delicate relationship between distances in the three-dimensional subspace and space-time distances between points judged simultaneous in terms of the one-dimensional subspace that permitted us to identify global "space" and "time" is lost.

Such problems with idealized treatments of complex systems are not unique to the applications of relativity theory. How does the typical physicist respond? One familiar response is to argue that, locally, the failure of the mathematical conditions of definability would be operationally negligible, that the gravitational field of the sun by itself is sufficiently weak and its rotation sufficiently slow that this local region might well encompass the solar system, and that the concepts of space and time are on these grounds applicable in these circumstances. While possible for the starlight deflection experiments, such a line cannot be taken even on experimental grounds in the analysis of planetary motion. Relativistic effects due to the sun's rotation play a distinct role in the interpretation of measurements of Mercury's orbit (Shapiro 1980, p. 127). In any case, the conditions for global definability of space and time clearly do fail analytically when the rotation of the sun is taken into account. Their existence thus becomes approximate, subject to detailed analysis of these particular circumstances of application of the Kerr solution.

One would expect a similar problem with the notion of mass were the sun and Jupiter, say, to be treated analytically as a genuine two-body system. No general solution corresponding to such a system is known. But certainly the picture of "two interacting massive bodies" would be compromised in a fully relativistic treatment, and more so in the case of an orbiting, close binary star system (such as the binary pulsar, Shapiro 1980, p. 127; see also Ehlers 1980, p. 287).[10]

What I hope to have illustrated in this extended example is that the ontological commitments of the general theory are far from straightforward. One might say of space and time that they "exist only locally," in several senses. "Total mass," even of something like the sun, treated in the Schwarzschild solution, may "exist only at infinity." (Specifying the location of, i.e., the conditions for the definability of, infinity itself is an interesting problem, one that has been largely solved, albeit with different strategies and results for approaches to it along spacelike and lightlike trajectories. See Geroch 1977.) In fact specifying the conditions under which properties such as "total mass of a system" may be defined in general, that is, the conditions on a solution of the field equations such that some derived structure with some of the attributes by which we recognize "mass" can be identified, is an ongoing research problem among relativity theorists.[11]

Let me recapitulate the points of this section to make the implications for ontological commitment concise.

General relativity postulates a four-dimensional manifold, ordinarily called space-time; it postulates a physical field, called the metric field, on this manifold. This field is coupled to other physical fields on the manifold by Einstein's field equations, which relate the curvature at each point of space-time to the density of matter and energy there. To apply the theory to the concrete world one must assume a solution of the field equations. But a solution describing all details of the known cosmos is unknown and almost surely unknowable, and all known solutions are enormous idealizations. For some solutions structures may be defined with attributes we have associated with such concepts as space, time, mass, energy, charge, momentum. Sometimes these structures may be defined conveniently; other times the definitions are extremely cumbersome; often they are not available. All such structures are tied together by the elegant and powerful mathematics of the theory as aspects of "the gravitational field," interwoven with the omnipresent metric tensor field of space-time with its dual geometrical and material significance. One will not, in this sort of theoretical and interpretive framework, I think, ever settle on an intuitively appealing canonical set of entities and their properties to characterize the relativistic universe. A contemporary relativity physicist, no matter how pre-analytically inclined to realism, will not be able to determine what to be a realist about.

6. Ontology in Contemporary Physics

I have tried in these three examples to capture at least the spirit of the ontological scene in contemporary physics. The examples have come from the pedagogical tradition in classical mechanics to which all physicists are exposed, and from two of the major theoretical frameworks which dominate contemporary research. The focus in the examples has been chosen in an attempt to make them broadly representative of one locus of the sources of ontological ambiguity in contemporary physics. Thus, the example of classical mechanics features a variety of interpretive frameworks each associated with a distinct mathematical formulation. In quantum mechanics, the focus is on the variety of interpretive frameworks arising from a single mathematical formulation. And in general relativity, the ontological ambiguity is presented as arising from within a single interpretive framework associated with a single mathematical formulation, as a result of the inevitable idealizations necessary to apply the theory in concrete contexts. The focus in each case seems to me appropriate to the way in which the area of physics is understood by most contemporary physicists.

But there are additional aspects of the ontological scene that need to be mentioned even to begin to fill out the account. In the first place, the interpretive situation in nearly all areas of contemporary fundamental physics is a good deal more complex than I have indicated. There are always alternative mathematical formulations for the fundamental equations in any of these areas of physics. There is always a multiplicity of interpretations, in each inevitably some problems of articulation, for any mathematical formulation of any fundamental equations. There is always a host of theoretical and physical idealizations variously appropriate to applying any fundamental equations in concrete circumstances. Moreover, the panorama of formulations, interpretations, and idealizations is always changing: new ones are being introduced and in some cases proving fertile; the potential of others is being exhausted and their role in active research is declining. So the characterizations of the areas of contemporary physics I have provided give only glimpses of what is a much more elaborate and constantly changing environment of ontological ambiguity within the community of physicists.

But there is yet another element of structure in this ontological environment. I have focussed on what physicists themselves would say when asked to provide an explanatorily satisfactory account of some class of phenomena. The additional element of structure comes from the fact that physicists think and talk about a theory in different ways depending on what they are doing with it. Physicists write textbooks about theories and teach the theories out of textbooks at basic and advanced levels; they engage in medium-scale theoretical speculation about theories; they articulate them theoretically, seeking solutions to fundamental equations; they do calculations seeking precise values for in-principle measurable quantities; and they do calculations aimed at a particular measurement of a particular quantity associated with a particular system by means of a particular piece of apparatus. It has been my experience that physicists' ontological focus (at least) depends strongly on which of these activities they are engaged in. Thus, a relativity physicist engaged with the theory in high and mid-level theoretical articulation — either in teaching or research — may speak exclusively in terms of relatively abstract manifold structure, at best of metric and stress-energy fields. Whereas in contemplating light trajectories in the solar system, the talk may be all of distances, velocities, and the mass of the system. Unchallenged, physicists do invoke these various ontologies in a realist way; so this observation adds only quantitative richness to the picture of ontological ambiguity I have developed so far. But sometimes even the nature of physicists' ontological commitment seems to vary with their focus on the theory.

In the first place, a physicist talking like a realist while engaged with a theory in one way, or concentrating on one level in its articulation and elaboration, may take a less than realist posture with respect to ontological commitments particularly associated with another level. Then again, it has been my experience that as one passes down the ladder of activities mentioned above, from fundamentally theoretical to more applied ones, one is more apt to hear talk of a more instrumentalist flavor. That is, physicists may talk like realists when engaged in theoretical speculation and articulation within the theory — seeking new solutions to fundamental equations, proving theorems about the general properties of solutions, elaborating the interpretations of solutions and such. They may talk like realists when they begin producing "numbers" from the theory. But somewhere down the road toward calculations aimed at particular measurement situations, a certain instrumentalist tendency often creeps in. Such instrumentalist hedging is most obvious when physicists are made self-conscious about the large number of assumptions associated with such calculations — the drastic idealizations necessary to generate solutions to fundamental equations, the vagaries of the search for data to instantiate the solutions, the equally drastic simplifying assumptions required to isolate a physical system likely to provide measurable comparisons with the resulting predictions, even the theories of instrumentation required actually to generate numbers from the system and provide measures of goodness of fit.[12] Still, such instrumentalist caution is by no means systematic, and one sometimes hears the most incautious ontological invocation at this most applied level.

Thus, the ontological scene in significant areas of contemporary physics seems to me to have an enormous amount of structure. There is the structure arising from the constantly shifting panorama of formulations, interpretations, and idealizations. Then there is the additional structure arising from the different kinds of activities which physicists engage in, the different levels of articulation, elaboration, and application of theories in which they are involved. And there is the structure arising from the shifting character of their ontological commitment as they move among these activities.

7. Advice to Philosophers

The lesson in all this for philosophers seems to me this: it is extremely difficult, if not impossible, to read off a clear, and clearly described ontology for much of contemporary fundamental physics from listening, however carefully, to what physicists say about it. In fact, the more carefully you listen, and the more physicists you listen to, the more jumbled it all becomes.

Thus I think it is extremely difficult, if not impossible, to use contemporary fundamental physics — after all the usual paradigm of "mature" science — to support the classical realist position I mentioned at the beginning. This position, remember, envisions mature science as populating the world with a clearly defined and described set of objects, properties, and processes, and progressing by a steady refinement of descriptions and consequent clarification of the referential taxonomy. One simply does not find in fundamental physics today the requisite unambiguous populations and clear descriptions. But it's not just realism that fails to draw support from listening carefully to contemporary physicists. I doubt that any global "ism," any large-scale attempt to provide an abstract account of what science aims and arrives at, can find consistent support in the shifting sands of ontological commitment I have described here.

It does seem to me, though, that there is yet good work to do for hopeful realists among philosophers, certainly in science generally, even in fundamental physics. This work involves going in with an open mind to see what it is that actually gets established in this kind of science — firmly, stably, and robustly established — whether entity, law, effect, phenomenon, or whatever. And it involves paying close attention to how it gets established, on what basis scientists become convinced that something is established. I am convinced that there is much more rationality and continuity in science than some critics have claimed. But putting one's finger on the nature of that continuity, particularly in fundamental physics, is not easy.

In any case, one can point to recent examples of this good work in fundamental physics. Nancy Cartwright (1983) has certainly done some of it; Ian Hacking (1983) has as well. And more recently, Peter Galison has provided an unprecedentedly clear picture of how things get established in high energy physics, in his book, *How Experiments End* (Galison 1987). I am very enthusiastic about this good work, because it seems to me that in such work lies our best hope for getting a firmer grip on what can be established — firmly, stably, and robustly established — in philosophy of science.

Notes

[1]The research on which this paper is based was partially supported by National Science Foundation Grant SES-86-18758. I would like to thank Paul Teller, Arthur Fine, and Ernan McMullin for comments on an earlier version.

[2]Jarret Leplin 1984, p. 3 provides a nice summary of philosophical arguments against realism.

[3]This is a good deal more than a promissory note to a young physicist. Each of these mathematico-conceptual approaches — save perhaps the action-at-a-distance approach — has been very important in the formulation of more contemporary theories. The field-theoretical approach is vital in the formulation of electrodynamics and later field theories; the Hamiltonian- Lagrangian approach is central to the formulation of quantum theory, and a formulation of relativity theory as well; the approach in terms of curved space is the heart of the general relativistic framework. These approaches continue to co-exist in the mind of a mature physicist who contemplates an overview of contemporary theories.

[4]As I point out below, Newton was never satisfied with his understanding of the nature of the gravitational force. But he did steadfastly reject the view that particles act at a distance across empty space. For a recent reference, see McMullin 1978.

[5]An "account" in a mature science is generally judged to consist of an explanatory model, the relationships between whose fundamental explanatory concepts are expressed in terms of an underlying mathematical theory, which makes possible quantitatively precise and qualitatively novel predictions.

[6]See, for instance, Jammer 1974. More recently, Pagels (1984) presents an engaging vision of this multiplicity in his chapter on "The Reality Market Place."

[7]These measurements constitute one of the three "classical" tests of relativity theory, along with measurements of the precession of the perihelion of Mercury's orbit and the red-shift of radiation emitted in a gravitational field. For a philosophical analysis of their status as "tests" of the theory see Glymour 1980.

[8]An interesting philosophical treatment of this experimental circumstance is presented in Laymon 1984.

[9]In fact, actual deflection experiments are analyzed under more idealized assumptions yet. These "weak-field, slow-motion" assumptions are manifested in a treatment of the metric structure of space-time as merely a perturbation of a flat, Minskowskian metric, a "post-Newtonian correction" to a flat, empty space (MTW 1976, Ch. 39). A full analysis of this experimental application would have to investigate possible differences in ontological commitment between this treatment and the Schwarzschild treatment.

[10]On the general problem of a description in general relativity of isolated, interacting bodies, see the contributions by Dixon and D'Eath in Ehlers 1979. A brief general discussion of the relationship between properties defined "exactly" and those associated with approximative methods appears on p. 165.

[11]For recent references, see the papers in Flaherty 1984. The following question and answer, from the paper by R. Penrose in this volume, gives some of the flavor of the current discussion:

Q: What properties of "mass" are you trying to capture in your formulation of quasi-local mass? A: I honestly don't have a complete answer to that question. We want agreement with the weak field limit. We want agreement on I(+) (future light-like infinity) and i(o) (spatial infinity). We would like positivity. Physical intuition is important, but it must be malleable, allowing certain principles to be dropped, if necessary. (Penrose 1984, p. 30)

[12]See Laymon 1984 for a bit more detail about these assumptions in the stellar deflection experiments mentioned in section 5.

References

Cartwright, N. (1983). *How the Laws of Physics Lie*. New York: Clarendon Press.

DeWitt, B. and Graham, N. (eds.) (1973), *The Many-Worlds Interpretation of Quantum Mechanics*. Princeton: Princeton University Press.

Earman, J. (1970), "Space-Time, or How to Solve Philosophical Problems and Dissolve Philosophical Muddles without Really Trying," *Journal of Philosophy* 67: 259-277.

Ehlers, J. (ed.) (1979), *Isolated Gravitating Systems in General Relativity* (Proceedings of the International School of Physics "Enrico Fermi," Course 67). Amsterdam: North Holland.

_ _ _ _ _. (1980), "Isolated Systems in General Relativity," in *Ninth Texas Symposium on Relativistic Astrophysics*, J. Ehlers, J. Perry, and M. Walker (eds.). New York: New York Academy of Sciences.

Feynman, R. P. (1965), *The Character of Physical Law*. Cambridge: MIT Press.

Fine, A. (1986), *The Shaky Game*. Chicago: University of Chicago Press.

Flaherty, F. J. (ed.) (1984), *Asymptotic Behavior of Mass and Spacetime Geometry*. Berlin: Springer-Verlag.

Friedman, M. (1983), *Foundations of Space-Time Theories*. Princeton: Princeton University Press.

178

Galison, P. (1987), *How Experiments End*. Chicago: University of Chicago Press.

Geroch, R. (1977), "Asymptotic Structure of Space-Time," in *Asymptotic Structure of Space-Time*, F. R. Esposito and L. Witten (eds.). New York: Plenum Press, pp. 1-105.

Glymour, C. (1980), *Theory and Evidence*. Princeton: Princeton University Press.

Hacking, I. (1983), *Representing and Intervening*. Cambridge: Cambridge University Press.

Jammer, M. (1974), *The Philosophy of Quantum Mechanics*. New York: Wiley.

Laymon, R. (1984), "The Path from Data to Theory," in Leplin (1984), pp. 108-23.

Leplin, J. (ed.) (1984), *Scientific Realism*. Berkeley: University of California Press.

Mach, E. (1959), *The Analysis of Sensations*. New York: Dover.

McMullin, E. (1978), *Newton on Matter and Activity*. Notre Dame: University of Notre Dame Press.

Misner, C., Thorne, K., and Wheeler, J. (1973), *Gravitation*. San Francisco: W. H. Freeman.

Nersessian, N. J. (1984), *Faraday to Einstein: Constructing Meaning in Scientific Theories*. Dordrecht: M. Nijhoff.

Pagels, H. (1982), *The Cosmic Code*. New York: Simon and Schuster.

Penrose, R. (1984), "Mass and Angular Momentum at the Quasi-Local Level in GR," in Flaherty (1984), pp. 23-30.

Shapiro, I. (1980), "Experimental Challenges Posed by the General Theory of Relativity," in *Some Strangeness in the Proportion: A Centennial Symposium to Celebrate the Achievements of Albert Einstein*, H. Woolf (ed.). Reading: Addison Wesley, pp. 115-136.

Symon, K. R. (1960), *Mechanics*. Reading: Addison-Wesley.

Wald, R. (1984), *General Relativity*. Chicago: University of Chicago Press.

Yourgrau, W. and Mandelstam, S. (1968), *Variational Principles in Dynamics and Quantum Theory*. Philadelphia: W. B. Saunders.

Modest Realism

William Newton-Smith

Oxford University

It was once fashionable to treat realism as an explanatory theory of science. This realism treats the sentences of science, both observational and theoretical, literally: such sentences are true or false as the case may be in some correspondence sense of truth. To this was added the epistemological claims that it was in principle possible to discover whether any given sentence was in fact true or false and that science has been progressive. In view of the fact that theories seemed to have a bad track-record - from the point of view of truth they all seem sooner or later come unstuck - the realist offered a more modest goal for science; producing more approximately true theories. The strategy for defending realism was explanatory. The realist claimed that inference to the best explanation (hereafter cited as IBE) was a standard epistemic tool of the scientist and ought to be available to the philosopher. The realist deployed IBE in the face of the fact that the history of the mature sciences was constituted by a sequence of theories which were ever more successful at the observational level. That success needed explanation. What was it about Einstein's theories that made them better able to make successful predictions than Newton's? It must be that those theories had captured more truth about the world. It would have been a total mystery, it was said, if theories should deliver the goods without providing more approximately true accounts of the underlying theoretical structures in the world (Newton-Smith 1981). To mark the fact that support for this style of realism has largely faded away it will be called faded realism.

Faded realism failed on many fronts, some of which are noted below. But the alternatives offered by those who diagnosed its failures are not convincing. And in this paper I explore the possibility of arguing for a more viable form of realism by shifting the content of realism and by deploying a non-explanatory defensive strategy. What is offered has close affinities with Fine's NOA. And that should not come as a surprise. For what Fine offers as an alternative to realism is very close to what many realists thought they were offering all along. However, the appearance is somewhat misleading. For NOA is too minimalist and needs strengthening.

Some of the criticisms of faded realism are fatal. For instance, Fine has drawn attention to the fact that the strategy of the faded realist in deploying IBE is question begging in confrontation with the instrumentalist. The faded realist purported to legitimize this use of IBE by reference to the fact that it was standardly deployed in science. The time had come, it was said, to apply this technique to science itself, giving a realism that was an empirical, explanatory theory of science. But the instrumentalist disputes the claim

PSA 1988, Volume 2, pp. 179-189

that it is standard practice to infer the truth or even the probability of a sufficient degree of truth of a hypothesis from its explanatory potential. An old-fashioned instrumentalist might deny that science seeks explanations. An sophisticated contemporary instrumentalist will deny that explanation is linked with truth in the required fashion. This instrumentalist holds that scientists attach pragmatic value to the explanatory power of a theory without regarding it as a sign (however fallible) of truth. To use IBE in a debate with an instrumentalist simply begs the question.

Not only is this explanatory strategy question-begging, there is not an appropriate analogy between the use of IBE in science and in philosophy by the faded realist which could rationalize the transference of IBE. Suppose for the sake of argument that scientists do treat IBE as having epistemic status. A standard putative illustration of IBE is Thomson's work on the electron. Thomson noted scintillations on the phosphorescent end of an evacuated tube when a charge was applied to a cathode at the opposite end. The best explanation available was that negatively charged particles, electrons, emitted from the cathode travelled with rectilinear motion to the screen. This inferential move was not the end of the matter for Thomson. It was the beginning of a programme which had subsequent success at the observational level. IBE gave initial plausibility to a posit which was then deployed. It suggested experiments to perform, the outcome of which gave increased credibility to the posit. If this is a typical deployment of IBE in science, IBE in faded realism is quite different from IBE in science. For the faded realist, the inference to realism is the end of the matter and not the beginning of anything. The faded realist did not use the posit of increasing truth at the theoretical level to derive novel results for which corroboration could be sought. Having made the inference, the faded realist regarded the instrumentalist with an unduly satisfied expression. The deployment of IBE in faded realism is sufficiently unlike the alleged employment of IBE in science that appeals to the acceptance of the latter cannot be deployed to legitimate the former.

No doubt some faded realists will claim that there is in fact an analogy. Having tentatively offered realism as an explanatory hypothesis, it might be said, one does put the theory of realism to work to give it further credibility. For certain features of scientific praxis can be rendered more intelligible on a realist rather than an instrumentalistic perspective. If this worked it might establish an analogy. Unfortunately it does not. To illustrate, consider the familiar claim that faded realism can make better sense of the drive to unification in science than can instrumentalism. A theory could not be true unless it can be consistently extended as it stands to become a total theory of nature. But if the aim were mere empirical success at the observational level there is no reason not to rest content with a plurality of different theories, treating them as tools with different domains of application. But this argument is not convincing, as van Fraassen has pointed out. For the instrumentalist simply notes that whenever we have encountered this situation, theoretical unification has tended to produce better empirical results. That being so the instrumentalist has every reason to seek unification. Van Fraassen's move generalizes. Let F be any feature of scientific praxis which the realist purports to explain or justify. The realist has to show that F is a feature operative in the history of science which has brought empirical success (even if we do achieve something more than empirical success in science, it remains the only sign we have of the something). But the instrumentalist can simply assert that the history of science provides inductive support for continuing to use F. Thus there is a justification, a justification which involves less contentious assumptions than the realist one for the continued deployment of F. Scientific praxis underdetermines the choice between realism and instrumentalism.

A successful argument for realism will have to take a different form. The strategy of this paper is to seek a common ground with the instrumentalist and to move from that to a modest realism. It is not so much a matter of arguing positively for realism but rather of removing apparent impediments. The common ground has three important landmarks. First, the parties to the debate agree that current theories have impressive observational successes. Guided by these theories we have built a technology, the fruits of which are there for all to see. Everyone can see Feyerabend live on television. Secondly, all parties agree that often we have reasonable beliefs about the existence and observable properties of medium size dry goods of the everyday world. This agreed starting point will be called common sense realism or CSR. The third landmark is the use of IBE in this domain as an epistemic principle. In my house in Wales I sometimes hear scrabbling noises in the night and find bits of food missing. That there is a mouse in the house is the best explanation I can think of (my neighbor, the practical joker, is too lazy to have gone to the trouble) and I infer that it is likely to be true. In taking use of IBE to be uncontentious at the this level we make common cause with van Fraassen.

This consensus about the ontology and epistemology of mice breaks down once we turn to the non-observable, to say, electrons. Those who unhesitatingly affirm the reasonableness of belief in the existence and the ravenous appetite of my mouse that no one has ever or may ever see, falter when we turn to electrons. Notwithstanding the fact that theories containing the term "electron" are much more impressive and successful than my theory about my nocturnal visitors, there is a hesitation or an opposition to saying that we are straightforwardly justified in believing that electrons exist with such and such properties. One who would make this affirmation is surrounded by a philosophical chorus of those who would say "no", "yes" (while winking), "yes, but truth ...", and so forth. This is a philosophical chorus, for it does not matter from the point of the praxis of science to which member of the chorus one listens. My strategy is to interrogate each member of the chorus, defusing their blocking moves and thereby removing the impediments to extending the stance of CSR to scientific discourse.

First there is the trivializer whose stance to scientific discourse is the analogue of Berkeley's stance to everyday discourse about material objects. Berkeley who advised speaking with the vulgar while thinking with the philosopher, affirmed the existence of mountains and streams with a wink. For, philosophically speaking, the mountains and streams were but collections of ideas in the mind of someone or other. The trivializer affirms the existence of electrons with a wink. Electrons turn out to be but collections of observable states. For the trivializer, the theoretical discourse of the scientist can be translated in a meaning-preserving fashion into the observational discourse. Such was the programme Russell set himself in *Our Knowledge of the External World*. It can be easily dismissed. No one has produced the translations in question. And, in any event, we would not actually welcome its success. For the point of theoretical discourse is to provide explanations of observational phenomena and if it were really just disguise observational discourse it could not provide this.

A more substantial member of the chorus is the reconstruer who belongs to the "yes but" school of philosophy. The reconstruer holds that there are electrons and that there are truths about electrons which it is reasonable to believe but truth is to be construed as a special kind of belief. This pragmatic move, fashionably called internal realism, is found in Ellis, Putnam and Jardine. Ellis (Churchland and Hooker 1985) talks of truth as being by definition what we would believe if our belief system were perfected, as if that state would be easier to recognize than any alleged state of correspondence between our beliefs and the world! Others (Putnam 1982) are simply vague about how truth is to be explicated. By contrast Jardine has offered a relatively precise version of the Peircean pragmatist account of truth, making truth what we would believe if we followed certain idealized modes of inquiry. However, there are insurmountable difficulties in his account which affect the other versions as well. The can only be indicated within the confines of this paper, for elaboration see Newton-Smith (forthcoming).

The problem is that internal realism is simply implausible at the level of CSR. For instance, internal realism is bound to make the truth now of particular singular judgments about the past a matter of the truth of predictions about the future. The CSRist takes it that many sentences about the past (about acorns falling in some remote forest yesterday, tiny icicles melting in the arctic last summer, about the number of Tyranosauri Rex that roamed the earth) have truth values now whether or not these values can ever be settled in the future. Certainly the instrumentalist can be expected to agree that unobserved instruments could have registered readings now lost to history, but not so lost that they do not give truth values now to sentences about those readings. And Jardine agrees that sentences of the sort cited above now have truth-values. According to him, this assumption has too robust a position within commonsense to be rejected. And so what is one to do? For Jardine the solution lies in the possibility that one day we might enlist time-travellers within our community of inquirers. The fruits of their labours could settle these truth-values. But in any consistent time travel stories, the future time travellers to our past have already been there. And that means that our friends will have already seen the acorn fall and have counted the dinosaurs. These creatures will be as Newton imagined God: everywhere and every when. And when there they would have had to have already noted all the observable states if all sentences about those states have truth-values now. Even if there are these beings and even if they agree to help us give truth-values to things, there is no guarantee that they will (or rather that they have already) cover the ground.

Internal realists castigate traditional scientific realists as metaphysical realists. It is said that they hark for some unexplicated relation of correspondence between theories and an utterly inaccessible reality. But whatever the situation for the scientific realist, the internal realist either has to fly in the face of CSR or has to adopt a metaphysics of inquiry inflated with observationally omniscient time-travellers in comparison with which the correspondence theory of truth looks positively down-to-earth. The internal realist should never have been invited onto the stage of current discussions in philosophy of science. Unless the internal realist can produce a convincing account of truth at the level of CSR, there is no point in considering the extension of that account to scientific discourse. To produce conviction at the level of CSR he or she must either convince us that the dinosaurs were a population without number or convince us that we should envisage our inquiries being assisted by God-like creatures (and, in that case, why not God himself?). We thus silence the reconstruer. Both the realist and the instrumentalist agree that with the proponent of CSR that there are sentences which have a truth-value which will never be determined no matter what time and energy is devoted to them and no matter what help we will in fact obtain in seeking to settle them.

Among our chorus are skeptics united by the slogan that seeing is believing. We are not entitled, says this skeptic, to claim justified beliefs about that which we cannot perceive. A familiar development of this line leads to the conclusion that the only epistemologically acceptable mode of seeing is real seeing, where really seeing is incorrigible. This skepticism restricts claims of knowledge to a world of sense data and we can set it aside as failing to meet the constraints of CSR. Someone who was not a skeptic at the level of CSR might nonetheless find their courage failing as the discourse moves away from familiar items. That sort of skepticism does not produce an interesting opposition to realism in the context of science. For it simply reveals the cautious nature of someone who has set their standards of credibility high with a bias in favour of relatively direct sensory evidence.

Skepticism would provide an interesting opposition to realism if it rejects the possibility of ever having rational beliefs about what is unobservable. Van Fraassen has offered such a skepticism based on a biological conception of the observable. Van Fraassen is a realist about items which beings of our biological kind could observe if they were in the right place at the right time, about other items he is a skeptic. But why should biology play this role in epistemology? Our theories may tell us about the observable states of the universe in spacetime regions which biological creatures like ourselves cannot visit. Why should it be reasonable to form beliefs about such states while it is not reasonable to form beliefs about the states of things around us which are too small or have too little energy for us to observe? Particularly in light of the fact that our theories explain why some things cannot be seen by us (though they might be perceived by creatures with different perceptual modalities). The provincialness of refusing to regard beliefs as well-grounded if they deal with the unobservable just because it is unobservable is epistemologically unreasonable (for further elaboration of this argument see Churchland 1985). If the skeptic is to remain on the stage, something must be done to rationalize this special treatment of our particular observational capacities. It maybe that the next member of the chorus, the science fiction writer, can help in this regard.

The science fiction writer, hereafter cited as SFW, listens with bemusement as the realist infers the existence of electrons in the face of the empirical successes of the theory of the electron. The SFW believes that there is always another story to tell. For the theory of the electron, there is also the theory of the nortcele (or nort for short). That theory is empirically equivalent to our best theory of the electron; it is logically incompatible with it (norts are hard and banana shaped) and it fares equally well on any epistemically viable principles of theory choice. It would perhaps be best to assume that when offered a theory of the electron, the SFW does not simply read off the rival theory. That might incline us to think that the two theories were mere notational variants of one another even if we could not see how to render the theories logically equivalent by reconstruing the predicates of one in terms of the other. Perhaps the SFW has something he calls a laboratory where he labours in what strike us as bizarre ways on unfamiliar apparatus before producing his logically incompatible, empirically equivalent rival theories. The SFW does not play tricks on us, taking our favorite theory and produce two rivals by adding in the one case the hypothesis that God is perfect and in the other case the hypothesis that God is really very good but slightly less than perfect. Some will argue that there could not be such a SFW. But as I have argued elsewhere there are no a priori grounds for denying this possibility (Newton-Smith 1980). Equally, no one has produced any convincing arguments that we need actually fear encountering a SFW in our world. At best we have evidence that some particular features of our world (notably features relating to the topology of space and time) may suffer from underdetermination. The fantasy of the SFW is no real threat to the realist. The best that has been offered to date are arguments which render underdetermined some small features of our theories relating to space and time. And the realist can well embrace these as isolated exceptions to his general stance. And he can even welcome any argument to support the coherence of the supposition of the SFW. For faded realism was claimed to be an empirical theory (broadly construed) and the existence of a SFW of global scope would utterly falsify his doctrine.

Contrary to the above remarks, it may seem that the SFW is already to hand. Suppose that T is our best theory of the electron. Let T* be the theory that everything is at the observational level as if T were true but there are no electrons. T and T* are logically incompatible and empirically equivalent by construction. But the invocation of T* is a question-begging move against the realist. For the realist, the explanatory success of T is a good reason to assume that there are electrons. Consequently, for the realist T* can be dismissed ab initio as evidentially deficient. T* does not make difficulties for the realist. It simply amounts to a denial of realism. What the instrumentalist needs to produce is a rival to theory T which on the realist's own principles is a genuine rival to T which cannot be dismissed straightaway. T* cannot play that role.

The SFW writer is at least offering the right kind of argument. We are considering possible impediments to the extension of the stance of CSR to scientific discourse. In principle underdetermination might be an impediment. But to block the use of IBE in the case of non-observational discourse we need some characterization of this distinction which rationalizes the difference in stance. A biologically based distinction did not provide a convincing rationale. On what characterization of the observable is it the case that passing from the observable to the non-observable co-incides with the passage from beliefs that are not subject to underdetermination to beliefs that are so subject? Suppose for the moment we use van Fraassen's biological characterization. And suppose we agree that there is underdetermination and that underdetermination (to the extent if obtains) undermines realism. But why should we assume that underdetermination arises just at the point where we pass from the biologically observable to the biologically non-observable? Might it not be the case that at the level of electrons there is just one theory? Might it not be the case that it is only when we turn to the level of ,say, the quark that underdetermination arises? If it is underdetermination that creates the problem for the realist, van Fraassen has to concede that there could be reason to believe in the existence of non-observables. Underdetermination, if actual, is threatening. But no reason has been given for thinking that that threat arises at just the point at which one enters the realm of the non-observable.

Another member of the chorus, Plain Person, is so emershed in common sense realism that he or she, finds talk of items that are not like the familiar items of common sense problematic for that very reason. Van Fraassen evinces something of the flavour of Plain Person in his talk of the "unimaginable otherness" of items in contemporary physical theories:

Do the concepts of the Trinity, the soul, haecceity, universals, prime matter, and potentiality baffle you? They pale besides the unimaginable otherness of closed space-times, event-horizons, EPR correlations, and bootstrap models. Let realists and antirealists alike bracket their epistemic and ontic commitments and contribute to the understanding of these conceptual enigmas. But, thereafter, how could anyone who does not say credo ut intelligam be baffled by a desire to limit belief to what can at least in principle be disclosed in experience? Or, more to the point, by the idea that acceptance in science does not require belief in truth beyond those limits? (Churchland and Hooker 1985, p.258)

Perhaps some argument not yet supplied will establish an epistemological principle which favours the introduction of items which are like items given in experience over items not like those given in experience. That might make a realist hesitant to embrace all that contemporary physics offers. But a realist can respect this hesitancy without abandoning his position. On its own it will not preclude his believing in non-observable items if those items are in someway analogous to familiar items. And it is hard to see why we should preclude the introduction of quite unfamiliar types of items if no items modeled on the familiar will suffice.

The reference to conceptual enigmas suggests a more serious impediment for the realist. This again is a reason for caution and not a reason to abandon the realist programme. It is a reason for caution for a realist should not infer the existence of an item which is conceptually problematic. If the theory which purports to introduce the items is incoherent, vacuous or conceptually absurd, we should hesitate believe the theory even if it has some empirical successes. But unless one had a proof that any item which was non-observable was bound to be conceptually problematic, this is just a reason for caution; a reason which any decent realist can and should respect. And, as van Fraassen suggests, this is a common problem for the realist and the anti-realist who should join forces in attempting to resolve the conceptual enigmas. Admittedly an instrumentalist can simply throw his hands up, remarking that the theory works and conceptual enigmas are quite irrelevant. But the history of science suggests that the attempts to resolve conceptual enigmas are productive of scientific progress at the empirical level. Plain Person rightly reminds us of the virtue of caution. We should not immediately give credibility to every new physical theory.

One of the finer members of the chorus, NOA, remains to be questioned once the conclusions to this juncture have been summarized. The modest realist takes the same ontological and epistemological stance to the theoretical discourse of science as to the discourse of common sense concerning the medium size dry goods of the everyday world. The members of the chorus offered putative impediments to this transfer. The most forceful members of that chorus were the skeptic and the SFW. In the case of the skeptic, no convincing grounds has been provided for thinking that it is illegitimate to extent IBE to the non-observable. To think otherwise is to indulge in an unconvincing epistemological provincialness. In the case of the science fiction writer, it was conceded that to the extent to which stories can be provided there are grounds for not extending the stance of CSR to the underdetermined parts of theories. But we have no reason to think that the SFW can in fact do this except in a marginal way.

My strategy for defending a modest realism is simple. Some will think that its very simplicity makes it suspect (see below). The simple defense rests on the claim that there is no viable conception of the observable which shows inferences going beyond the observable to be suspect for that very reason. There is no rationale for treating the observable and the non-observable as being significantly different in kind with respect to ontology or epistemology. No doubt one should be modest and cautious in moving from the relatively observable to the relatively non-observable. For there is more that can go wrong both conceptually and epistemologically and so one should be prudent. But nothing has been done to show that scientific discourse is so different kind from ordinary discourse that we need a different stances for the two discourses. There is a seamless web. When we move from inferring the existence of the mouse in the house to the existence of the mountains on the moon or to the existence of electrons in the cathode ray tube, we move in a direction along which prudence dictates care. We do not encounter a rupture beyond which all is utterly different and inaccessible. Modest realism is simply the extension of the stance of CSR to all scientific discourse. Truth is at stake in science and we have on occasion good reasons to believe in the truth or likely truth or likely approximation to the truth of some scientific claims. Consequently we have reason to believe in the existence of those theoretical entities which would have to exist were our reasonable beliefs to be true.

Some will feel that this route to realism is suspiciously easy. The ease arises our very strong starting point. CSR. The really difficult philosophical problems may well lie in the justification, if justification be needed or possible, for that starting point. The real challenge may be to show that we are not after all brains in a vat. But once we have come to believe that we are not, that CSR is to be our starting point, the extension of that stance to science is not such a great step. And so the argument for making the step is not of great moment. It is largely a matter of removing apparent impediments, impediments which are the legacy of an objectionably narrow empiricism.

In another sense this defense of realism is not at all easy. It says that some claims about the existence and properties of unobservable items can be justified. After this point it ceases to be easy. For it requires us to leave the ample armchairs of philosophy and examine in detail the content of, and evidence for, current physical theories. At this task philosophers may not be especially adept. The task is something that the scientific community is doing for itself. And the realist with his pro-attitude to science and tendency to progressivism, is bound to think that it is being done rather well. If one thinks, as I do, that there is not much point in trying to draw a distinction between philosophy and science, one might allow the philosopher with a serious interest in science to assist in these matters by providing epistemological caution and conceptual clarity. Scientists, speculating about the origins of the universe, can be a bit credulous. And they may find their mathematical creations so intriguing that the conceptual anomalies seem exciting rather than disturbing.

The realism to which the argument leads is called MOR for modest realism (and also to indicate that MOR seeks to say more than NOA). MOR extends the stance of CSR to all scientific discourse. MOR takes it to be legitimate to talk of truth and belief in the scientific course in the same sense that one talks of truth and belief in the context of CSR. MOR abandons the project of a general explanation of the success of science. MOR permits a plurality of particular attitudes to particular theories at particular moments of time. On occasion we may have sufficient evidence to justify belief in the theory. On other occasions, a theory may provides some useful results without being worthy of belief. NOA can be expected to urge that what has been defended is not what realism was intended to be by the realists; that in so far as it is defensible what has been defended is in fact NOA. To this juncture that is so. But following a brief, partial characterization NOA, I explore the viability of adding more to MOR under three headings; progress, philosophy and foot-stamping.

Fine diagnoses the realist as adding some interpretation to the discourse of the scientist. Accepting the legitimacy of talk of truth and belief does not give realism. Realism adds a special construal of truth: Truth is Correspondence with the World. This realist is accused of being hermeneutical. It is as if the realist has to give significance to science by making it Be About the World. NOA, on the other hand, "counsels us to resist the impulse to ask 'What does it all mean?'" (Fine 1985b, p.72). Science is a form of life that involves talk of truth and belief. No special construal of truth is called for. Truth is to be handed in terms of the familiar Tarski/Davidson referential semantics. The form of life is all right as it is. There is no need for philosophers to provide any justification or explanation of science.

MOR accepts these remarks about truth with one historical caveat. It is not clear that all those who styled themselves realists were invoking some metaphysical notion of Correspondence. It is not surprising that they were misunderstood for typically they said that truth was to be understood in terms of "cleaned-up" version of correspondence without providing any non-metaphysical elucidation of that notion. The mistake did not lie in assuming some notion of correspondence richer than what can be obtained from the Tarski/Davidson approach. It arose from assuming that truth in this minimal sense could be invoked as an explanation of anything. One way or the other the faded realist was mistaken: either in using a thick and unexplained notion of truth or in using a thin and explicated notion which failed to bear the weight placed on it.

Fine attacks the realist for taking a hermeneutical stance towards science. The realist is castigated for thinking that he, the philosopher, is needed to give the activity a point by positing for science a grand goal of Truth or Verisimilitude. Fine even accuses the realist of a simple fallacy of shifting quantifiers, passing from "each scientist has a goal" to "there is a goal that each scientist has". Fine's remarks have force up to a point. Science has the significance it has quite independently of any philosophical activities. And it is quite unhelpful to talk about the goal of science as if this was an active goal that each member of the community is consciously and continuously seeking. But there remains the matter of saying what the significance of science is for us.

It is at this juncture that MOR parts company with NOA. NOA, Fine tells us, is "not committed to the progressivism that seems inherent in realism" (Fine 1986, p. 130). But there is something in "progressivism". The significance of science for us lies in part in its progressive character. At a minimal level this progress is a matter of increased empirical success. But given the legitimacy of talking about truth in the context of theoretical discourse the realist of the MOR variety raises the question of progress in some stronger sense, and in so doing parts company with the instrumentalist and with NOA. What is supposed to be wrong with progressivism? There are certainly good reasons for rejecting the faded realist's attempt to explain scientific progress. But rejecting explanatory realism is compatible with recognizing that there is scientific progress in a stronger sense than increasing empirical adequacy. There are also good reasons for rejecting certain formulations of this very idea of progress. The faded realist deploys a notion of approximate truth which would be objectionable if that is based on an underlying metaphysical notion of truth as correspondence. Even if we restrict the underlying notion of truth to what can be captured the Tarski/Davidson semantics, we have to face the fact that any attempt to explicate such a notion (including my own) either gives the wrong answer or is laughably simplistic. But that was true of attempts to explicate the notion of truth before Tarski. In philosophical terms it is early days with regard to the notion of approximate truth. If the notion of truth is legitimate, it may well be that some notion of approximate truth is equally legitimate. One might expect NOA with its respect for the discourse of science to be particularly interested in developing a notion of approximate truth. For scientists do deploy some such notion with respect to individual hypothesis and arguably with regard to theories. They do not think that the relativistic mechanics is true simpliciter but they do think it is better with respect to truth than Newtonian mechanics. For the faded realist, approximate truth or verisimilitude was tied to a doctrine about the constancy of reference of theoretical terms across theory change. A rejection of this doctrine should not lead to a hostility to approximate truth. For one should aspire to a notion of approximate truth which permits the comparison of theories with different referents.

If one is totally pessimistic about the possibilities of explicating a notion of approximate truth, one might seek to recognize the progressive character of science as follows. Our current theories, given the evidence we have available, are better supported than the theories we have rejected. hence the best thing for us to believe is that our theories are more worthy of belief than any others we know of. What we have produced is more worthy of belief than what our predecessors produced and that remains so even if our descendants will develop theories which will be more worthy of belief than our own are now. This progressivism is within the spirit of NOA. It is argued for in scientific and not philosophical terms. And it captures the attitudes of scientists (save those corrupted by positivism). The realist of the MOR variety sees science as having significance for us and thinks that in part that significance derives from our belief in its having progressive character which exceeds progress merely at the observational level. This may be represented as our having more approximately truth theories; or, it may be represented as our having theories which we are more entitled to believe true. NOA without something more, without a thesis about progress, is too economical with the truth.[1]

NOA corrects the faded realist's hermeneutical attempts to give a certain construal to "true" in scientific discourse. But in so militantly advocating a "hands-off" for philosophers of science, it mitigates against raising certain legitimate questions of which two examples are given. The first concerns underdetermination. This issue is not the grand issue raised by the SFW but concerns the identification and evaluation of those marginal elements of underdetermination. Suppose we have a decent theory which involves the assumption, h, that space and time are continuous and not merely dense. Suppose it is possible to develop an evidentially equivalent but incompatible theory using the assumption, h', that space and time are merely dense. There may be no reason at all for scientists to take an interest in this. There are, after all, no Nobel prizes for producing a second, evidentially equivalent theory to an established theory. But there is an issue of interest that a hands-off attitude would leave untouched. How should we think of the choice between h and h'? Should we think of it as a case where truth is at stake but rational belief as to what the truth is, is not possible? Or should we think that truth is not at stake - that we add h or its rival as a little artifice of our theory to make things come out nicely? This difference in attitude is one of the things buried in some of the realist-instrumentalist debates and it looks like a issue which remains regardless once we adopt NOA. Such issues do not affect the praxis of science and so we might as well call them "philosophical". MOR recognizes that there are genuine "philosophical" questions about scientific discourse to which it seeks answers, even if the answer is that there is no answer. Unless one is an unregenerate empiricist one will have to allow that there may be legitimate questions arising about science, the resolution of which has no practical bearing on scientific activities. The question noted above concerning underdetermination is one example. Another example is the issue of ontological commitment. A live scientific theory does not come with a canonical first-order representation. For any actual theory there will be debate (possibly but not necessarily irresolvable) concerning what it actually quantifies over. This issue, which cannot be considered in this paper, may again be an issue of no scientific consequence and hence one which a "hands-off" approach will tend to ignore.

Finally there is the matter of foot-stamping. The faded realist wanted to say that our theories are true or false in virtue of an external reality and she or he used the vocabulary of correspondence in an effort to convey this. Tarski/Davidson semantics combined with a causal theory of reference does explicate a sense in which our theories are about the world and are true or false in virtue of the world. The faded realist wanted more than this. Fine imagines realists stamping their feet, shouting "Really" to invoke an external reality (Fine 1986, p.131). As Fine argues, there is nothing more to say. And in conclusion I offer a somewhat different explanation of the illegitimacy of the impulse to say more.

To this end imagine the existence of a social institution called "Dr Who Clubs". The members of the clubs, who not overly intelligent meet to discuss the escapades of their hero, Dr Who, as revealed in a long series novels, i.e. did the Doctor's robotic dog need its batteries recharging during his first encounter with the Darlicks. Such matters are settled consulting the texts. Within the clubs there is a mode of discourse involving notions of truth and belief. Some members say they believe that the batteries needed recharging, others say they believe not. On resolving this disagreement they come to say that it is true that the batteries needed recharging. When, having left the Club House, someone asks if there are any Darlicks to be feared in Chicago, they answer unhesitatingly say that there are no Darlicks in Chicago. This institution is a form of life and there is nothing we need do as philosophers to give it significance. We do not need to impose any philosophical interpretation on their discourse: it is all right as it stands. But within our philosophical tradition there is a deep-seated impulse to give fully general descriptions of forms of life or modes of discourse. In this case we can safely give into that impulse. We can do something more than provide a Tarski/Davidson semantics for this mode of discourse. We can say that in that discourse things are true or false as the case may be in virtue of what is said in the texts of the novels. The sentences which may be true in this way are not true in virtue of how the world is apart from the texts in question. In saying this we are not invoking any objectionable metaphysical notions of the World and Correspondence. But we are acceding to the philosophical impulse to say something at the most general level about the character and point of a form of life. It is crucial to note that what gives the philosophical description its force is the contrast provided between this discourse and the discourse of common sense realism. In the former unlike the latter, truth claims are responsive only to texts.

To give another but controversial illustration of this desire to "say something more", consider a certain view of mathematics. Someone considering the function of "true" and "believe" in mathematical discourse might be struck by an apparent contrast between mathematical assertion and assertion in everyday discourse. In the one case there are objects which are examined in the evaluation of assertions, assertions which may be false notwithstanding the outcome of the most careful examination. In the other case there are proofs not objects which control the assertions, assertions which if backed by a genuine proof must be true. One struck by this could summarize by saying that in mathematics truth amounts to proof. Again, this general description gains its force through the contrast it draws between ordinary discourse and mathematical discourse.

In a similar fashion, the philosopher looking at the language game of science, looking at that form of life, seeks to give a general description of the function of truth and belief. But given NOA or MOR it is unlikely that any enlightening general description is possible. For NOA and MOR see scientific discourse as being of a piece with ordinary discourse. Unlike the cases of the Dr Who Clubs or mathematics (under a certain construal) there is no possibility of giving content to the general description through a contrast with ordinary discourse. It is interesting to note that if one were an instrumentalist, there one could give a general description with content. For the instrumentalist contrasts the point of theoretical assertion with the point of ordinary assertion. In the one case it is to say true things about the objects, in the other it is simply to say things that give rise to successful predictions. But if one's philosophy of science is NOA or MOR scientific assertion is but an extension of ordinary discourse and consequently there is no possibility of contrast. There is no further step back we can take into order to say something general about the basic mode of assertion. Hence, the frustration and the impulse to floor stamp. The fact that the opposition, the instrument, can use a contrast with ordinary discourse to say something general only adds to the frustration.

Faded realism should be allowed to fade away entirely to be replaced by MOR. MOR begins with CSR and so rejects any philosophies of science such as internal realism which are incompatible with that starting point. There are no impediments to the extension of the stance of CSR to scientific discourse. In view of the fact that ordinary discourse (discourse about the medium-sized dry goods of the world) and scientific discourse constitute a seamless web, some special reason would be needed to partition the web. Attempts by the instrumentalist to provide such a reason fail. And so we arrive at MOR which amounts to NOA with two important additions. First, for the sake of descriptive accuracy some thesis of progress must be added, whether formulated in terms of truth or approximate truth. Secondly, once narrow empiricism is rejected we recognize that scientific discourse gives rise to questions of no practical concern to the praxis of science which are legitimate. These include questions of underdetermination and ontology which animated traditional realists. And they remain to be answered by one who has made the transition to NOA or MOR. So MOR adds to NOA a judicious degree of "hands-on" philosophy.

Notes

[1]On this point I am particularly endebted to Del Ratzsch.

References

Churchland, P.M. and Hooker, C.A. (1985), *Images of Science*. Chicago: Chicago University Press.

Fine, A. (1986a), *The Shaky Game*. Chicago: Chicago University Press.

_ _ _ _ . (1986b), "Unnatural Attitudes: Realist and Instrumentalist Attachments to Science", *Mind* XCV: 149-179.

Jardine, N. (1986), *The Aims of Inquiry*. Oxford: Clarendon Press.

Newton-Smith, W.H. (1980), *The Structure of Time*. London: Routledge.

_ _ _ _ _ _ _ _ _ _ . (1981), *The Rationality of Science*. London: Routledge.

_ _ _ _ _ _ _ _ _ _ . (forthcoming), "The Truth in Realism", *Dialectica*.

Putnam, H. (1982), "Three Kinds of Scientific Realism", Philosophical Quarterly 32: 195-200.

Van Fraassen, B.C. (1980), *The Scientific Image*. Oxford: Clarendon Press.

A Case Study in Realism: Why Econometrics Is Committed to Capacities

Nancy Cartwright

Stanford University

Who is a realist about capacities? Everyone, I maintain, who employs the methods of Galilean idealization typical of modern science. But I am not going to defend this broad thesis today. (For a defense, see N. Cartwright 1989.) Instead I am going to concentrate on one tiny corner of modern science — econometrics — where I hope I can show what it is we do in science that commits us to capacities. Notice that I say, "what we do" and not "what we say." That is because I subscribe to the current movement among historians, philosophers, and sociologists that is turning its attention from scientific theory to science as a whole, and especially to concrete scientific practices. This bears on issues of scientific realism. For surely our knowledge of nature is encoded as much in our methods and applications as in our theoretical structures. We learn about what a particular science is committed to not just by listening to what its laws say, but by looking to see what is required in nature to make its methods work.

That is what I am going to do in econometrics. I am going to write down some laws, or rather some abstract forms for laws, and I will claim that the parameters which appear in these laws represent causal capacities. But I do not expect you to be able to read that from the laws themselves. Instead you must look at the methods by which these laws are tested and at the applications to which they are put. I maintain that these two processes, considered jointly, do not make sense together unless these parameters do represent capacities. Seeing first what the gap is between these complementary processes, and second, what is needed to fill it, will, I hope, help make clear part of what my notion of capacity is. In the meantime, think of a capacity as a causal power that a system possesses by virtue of having a certain property or being in a certain state.

I will turn briefly to modern econometrics near the end, but I begin with traditional ideas developed at the Cowles Commission in Chicago immediately after the second world war, which set the program for a good deal of the econometric work in the U.S.A. for the following twenty years. Indeed I begin even a little earlier than that, with the ideas of two of the founders of econometrics, Ragnar Frisch and his student and colleague, Trygve Haavelmo.

First I want to explain a particular concept, the concept of autonomy, as it originated in the work of Haavelmo and Frisch. Then I want to tell you something about the Cowles Commission's doctrines about autonomous equations — namely, that autonomous equations are structural — and what that means. Last, I want to contrast these Cowles' doctrines with a modern British view about autonomy. The aim is to make clear why I say that Cowles methods are committed to capacities: that is, they presuppose a world in which specific, quantitative, measurable causal powers can be associated with particular properties or sets of properties.

There are two related features of these powers that I want to stress, features that make clear why the language of capacities is appropriate. The first is that the powers are attributed to separately identifiable properties (or sometimes to separately identifiable sets of interacting properties); and secondly that the attribution is stable across change. The properties carry their capacities with them from one kind of situation to another.

I begin with a famous example of Haavelmo's.

Here is where the problem of *autonomy* of an economic relation comes in. The meaning of this notion, and its importance, can, I think, be rather well illustrated by the following mechanical analogy:

If we should make a series of speed tests with an automobile, driving on a flat, dry road, we might be able to establish a very accurate functional relationship between the pressure on the gas throttle (or the distance of the gas pedal from the bottom of the car) and the corresponding maximum speed of the car. And the knowledge of this relationship might be sufficient to operate the car at a prescribed speed. But if a man did not know anything about automobiles, and he wanted to understand how they work, we should not advise him to spend time and effort in measuring a relationship like that. Why? Because (1) such a relation leaves the whole inner mechanism of a car in complete mystery, and (2) such a relation might break down at any time, as soon as there is some disorder or change in any working part of the car. . . . We say that such a relation has very little *autonomy*, because its existence depends upon the simultaneous fulfillment of a great many other relations, some of which are of a transitory nature. (1944, pp. 27-28)

The two reasons that Haavelmo gives show that the two features I am interested in were already combined in his thinking about autonomy: First, the autonomous laws will be structural. Unlike the function that connects speed and accelerator-depression distance, autonomous relations will give the "inner mechanism." Second, autonomous laws will be stable. They will not be liable to "break down at any time, as soon as there is some disorder or change."

The two are similarly joined for Frisch. Consider for example Frisch's characterization of the most autonomous features of a structure as those that "could be maintained unaltered while other features of the structure were changed" (1948, p. 17). Here again we see the emphasis on stability across change. Frisch continues: "The higher this degress of autonomy, the more *fundamental* is the equation, the deeper is the insight which it gives us into the way in which the system functions, in short the nearer it comes to being a *real explanation*" (Frisch 1948, p. 17, underlining original throughout). So for Frisch as well as for Haavelmo, the stability of a relationship is a consequence of the fact that the relationship is fundamental and describes the inner workings of the economy.

This is particularly apparent when we consider the source for unstable or non-autonomous relationships. Typical examples in Frisch and Haavelmo involve transformations of one set of equations into another. The first equations are autonomous: they are fundamental, or structural, and they are stable. Lack of autonomy arises because each of these equations must operate subject to the constraints imposed by each of the others. In the end the actual behavior of the system will be described by a single equation (the so-called "reduced form" equation, which in Frisch's language of 1948 is a *coflux* equation) which is the consequence of the joint operation of the various structural equations.

This equation will clearly not be autonomous with respect to the structural equations. While a change in one structural equation will leave the others unaltered, in general one expects just the reverse with respect to the reduced form.

The derivative nature of the reduced form gave rise not only to problems of autonomy, but to epistemological problems as well, problems well-known for economics. I describe the problem briefly because the shift from the realistic methods of the Cowles Commission to more instrumentalist views, one of which I shall describe in a moment, is usually attributed to a loss of confidence in our abilities to solve these epistemological problems. The Cowles Commission had an ambitious idea: they expected first that they could figure out the correct *form* for the laws that governed economic phenomena (or correct enough to be able to make headway); and secondly, they thought it reasonable to suppose that laws of the prescribed form could be identified from the data. Identifiability is a very strong notion. Using the language of Clark Glymour, identification is a kind of boot-strapping from data to hypothesis. It means that the equation itself can be inferred from the data, once the form of the equation is given. The Cowles Commission hoped that structure would be identifiable. But Frisch thought that would be a happy accident. The reason is that we cannot conduct controlled experiments in economics:

> This is the nature of passive observations, where the investigator is restricted to observing what happens *when all* equations in a large determinate system are actually fulfilled simultaneously. The very fact that these equations are fulfilled prevents the observer from being able to discover them, unless they happen to be coflux equations." (Frisch 1948, p. 15)

To see exactly how autonomy fails in Haavelmo or Frisch's picture, consider the following transformation described in Frisch. He considers two structural equations of the form

$$\Sigma_{i\phi} \, a_{ki\phi} \, x_i \, (t - \phi) = 0,$$

where **k** represents different equations. So they look like this, letting **i** vary over 1 and 2, and ϕ run over two lag periods.

1. $a_{110}x_1(t) + a_{111}x_1(t - \phi_1) + a_{112}x_1(t - \phi_2) + a_{120}x_2(t) + a_{121}x_2(t - \phi_1) + a_{122}x_2(t - \phi_2) = 0$

2. $a_{210}x_1(t) + a_{211}x_1(t - \phi_1) + \ldots + a_{222}x_2(t - \phi_2) = 0$

Frisch imagines that the particular values of the parameters are arranged so that the two equations are linearly independent. Hence they can be non-trivially transformed into a new equation of the same form. In his examples, a new equation results in which a number of the parameters take on the value 0. This equation is called a "coflux" equation, because it is not itself fundamental.

$$x_1(t) = ax_1(t - \phi_1) + bx_2(t) + cx_2(t - \phi_2)$$

where

$$a = a(a_{110}, a_{111}, a_{112}, a_{112}, a_{121}, a_{122}),$$

$$b = b(a_{110}, a_{111}, a_{112}, a_{112}, a_{121}, a_{122})$$

$$c = c(a_{110}, a_{111}, a_{112}, a_{112}, a_{121}, a_{122}).$$

Notice that different parameters in the new equation are functions of the same structural parameters. Imagine then that one wants to implement a policy that would shift the effect which, x_2's occurring at $t - \phi_2$ has on x_1's occurrence at t. With the coflux equation in hand, the obvious strategy is to try to bring about a change in the value of **c**. But if **c** were to change, knowledge of the original coflux equation would no longer be sufficient for predicting $x_1(t)$. That is because the parameters in the coflux equation are not independent of each other. If **c** changes, so may **b** and **a**. What happens to $x_1(t)$ depends, through **a** and **b**, on the way **c** is changed. Since the coflux equation does not give us information about how the parameters relate, it is of no use for predicting the effects on $x_1(t)$ of making "structural" changes in the value of **c**. It is impossible to predict the effects from the new equation alone.

Now that we have some idea of the concept of autonomy inherited by the Cowles Commission from Frisch and Haavelmo, let us turn to a more modern view, that of David Hendry, who is an influential theoretician in Britain. There is also a considerable movement in the United States opposed to various modern descendants of Cowles, and especially prominent are those who defend the use of vector-auto-regression techniques. But their philosophical position seems to me to be a kind of crude pragmatism, and makes a less instructive contrast for us than the Hendry school. The contrast between the ideas of Hendry and his colleagues on the one hand, and the Cowles Commission on the other is very instructive for my concerns about capacities. That is because Hendry's work, while it has much in common with Cowles, is not committed to capacities; whereas, I maintain, the Cowles Commission was.

I begin with worries shared by Frisch and Hendry. Frisch asks of our reduced form equation above:

What does the equation mean? It means that *so long as x_1 and x_2 continue to move with the same time shapes as they had in the past* I can compute the value of x_1 at any point of time t from the knowledge of x_2 at this same point and x_1 and x_2 at certain earlier moments as indicated in the formula. In other words the equation is simply a description of the "routine of change" which x_1 and x_2 follow. The equation determined in this empirical way does *not* state that if a situation occurs where $x_1(t - \phi_1)$, $x_2(t)$, and $x_2(t - \phi_2)$ have some arbitrary values I can again compute $x_1(t)$ by [this equation]. (1948, p. 16)

In order to secure equations — autonomous equations — which can do that job, Frisch says "we are led to constructing a sort of super-structure." Hendry uses the term "super" as well to mark autonomous equations, in the expression "super-exogeneity." Exogeneity is a complex notion. It means intuitively "determined outside the system." But what kind of determination is this? In a well-known paper from 1983, Robert Engle, David Hendry, and Jean-Francois Richard distinguish three conceptually distinct but related notions. The first is straightforwardly causal: exogenous variables in an equation represent causes; endogenous, effects. The second concerns questions of estimation which are at the core of the econometrician's interest, but not of special relevance here. The third is super-exogeneity. A variable is super-exogenous for Engle, Hendry, and Richard just in case it is exogenous in the first two senses — it is a cause of the endogenous variables and the right criteria for estimation are satisfied — and it is stable across change. That means that the laws describing superexogenous variables can be relied on for policy manipulations. These laws do not represent what Frisch called a "mere routine of change" which holds good so long as nothing else critical is altered. Rather they allow us to predict what effects the cause will produce even when other factors are manipulated. As Engle, Hendry, and Richard urge, the concepts they use to characterize super-exogeneity are meant to "guarantee the appropriateness of 'policy simulations' or other control exercises" (1983, p. 284).

With respect to capacities two features of the Engle, Hendry and Richard view are important. The first is a feature they do want; the second, one they deny. What they want, as we have been seeing, is a certain kind of stability of causes, and that is one essential ingredient in accepting capacities. They want to distinguish causal relationships which, though they are perfectly genuine, may hold only so long as no other essential ingredient of the system changes in a basic way, from causal relationships that hold across time. Recall Haavelmo's example of the automobile. Pressing on the accelerator is a genuine cause of the increased speed of the car — if that is not a causal process, what is? Yet the causal relationship is clearly not stable if the parts of the engine are rearranged. It is not a natural capacity of this small lever to be able to make a car go faster; it is rather a fortuitous consequence of the arrangement of circumstances.

What is missing in the Engle, Hendry, and Richard scheme? Why do I say that their scheme has no capacities, whereas the Cowles Commission did? What is missing is the commitment to structure. I will turn in a moment to some specific aspects of the structure that matter for capacities, but first I would like to make a more general point about realism. Engle, Hendry and Richard are concerned with problems of policy and manipulation. Yet they do not pursue this concern in the conventional way. The Cowles Commission believed in the simple Baconian idea: we can control nature, but to do so we must understand the fundamental principles by which it works. Engle, Hendry and Richard also believe in control, but they want to bypass fundamentals and go direct for control. That is what a model is for Hendry and his school. Models do not aim to represent how nature works, but rather to describe immediately a control procedure. From their equations we are supposed to be able to tell what would happen to y if we were to manipulate x. And that's it. This is not meant to be an easy task; yet they take it to be a more possible task to seize control than to win it through understanding.

I call this way of thinking the "new instrumentalism." Science is an instrument for control, not a vehicle for understanding. When I say new, I mean new relative to the post-war thought with which most of us were brought up. From Ernst Nagel and others we learned to contrast realism and instrumentalism. But it was always in the context of *theory*: we talk about the realist versus the instrumentalist interpretation of theories. But the real opponent of an instrumentalism of the kind we see in Engle, Hendry and Richard is not realism. Admittedly they have the kind of epistemological concerns that motivate many anti-realists. The Cowles Commission hoped that the inner structures themselves could be identified from the data; these more modern opponents of Cowles methods despair of that.

But two points must be taken into account. The first concerns the idea of "inner mechanisms." Despite Haavelmo's metaphor of the inner workings of the engine, hidden under the hood, the variables in a Cowles structure were not in any way hidden. They did not represent theoretical entities or theoretical properties in our usual sense of the term, for they were supposed to stand for macroscopic features of economic life which were not just observable but measurable. It was not the fundamental entities that might be hidden on a Cowles account, but rather the fundamental relationships. That is because, as we learned from Frisch, each relationship operates under the constraints of all the others, and only the overall result can be observed passively. So the difference between Cowles' views on the one hand and those like Engle, Hendry and Richard on the other is not like the difference between realists and instrumentalists that we see drawn over theoretical entities.

More important, I think, is the realization that in one sense central for our contemporary philosophical debates, programs like those of Engle, Hendry and Richard are realist, for they are objectivist. They believe after all in autonomy, and they believe that autonomous laws reflect objective processes in nature. (Hendry calls them "data generating processes" in Hendry 1979.)

The point is simply made by contrasting Hendry with some of the Americans I mentioned earlier. Consider for example the famous paper by Edward Leamer, "Let's Take the Con out of Econometrics" (1983). The con that Leamer is worried about is just the attempt by econometricians to establish autonomous and objective laws of the kind I have been discussing. Leamer takes a familiar line, now highly fashionable. All inference makes presupposition. We are never in a position to know which presuppositions are correct. Hence we are never in a position to know which conclusions are right. So, concludes Leamer, econometricians shouldn't make inferences and shouldn't draw conclusions. "The job of the researcher is then to report economically and informatively the mapping from assumptions into inferences" (1983, p. 38). Here we see a real example of the California dictum: "Different strokes for different folks."

I do not want to dwell on this difference, for this kind of dispute is already familiar to us all. I describe it rather by way of contrast with a distinction that is less well rehearsed among us. Views like those of Hendry's are instrumentalist. But they are not, I want to say, anti-realist or anti-objectivist. They are rather anti-theoretical. The Hendry program admits the possibility of knowledge, and it does not in any way try to divorce knowledge from justification. Claims to autonomy must be reasoned about, argued for, and tested. Laws which go wrong must be corrected, and the corrections are not to be made, as Leamer suggests, by whim, but rather by reason. But are the reasons trustworthy? Well, we know how to carry on with that debate.

The more novel contrast I want to point to is the contrast with the work at the Cowles Commission. For the Cowles Commission, claims about autonomous laws are to be defended by the knowledge that is contained in a theoretical structure, a structure which is supposed to represent the fundamental principles which govern the phenomena. But the knowledge which Hendry's program relies on comes in little bits and is highly specialized to subject matter. It is not represented in a unified scheme and in fact, it is sometimes not represented at all. It often reposes in techniques and methods and practices, and it is only from these that it can be read. In fact this too is an increasingly familiar story. I have talked about it before; it is the central theme of Ian Hacking's *Representing and Intervening*; and with people like Peter Galison, Norton Wise, Tim Lenoir, and Steven Shaffer it is at the core of the movement in history of science, that looks to practice rather than theory to find out what we know in science. But the Hendry program helps focus the issues. True, compared to Leamer, Hendry and his coworkers are realists. But compared to the Cowles Commission views like those of Hendry are instrumentalist, and not because they refuse to interpret theories realistically. Rather they eschew theory altogether. It is not scientific theory which is to yield power and control, but rather science itself which is to serve as an instrument, science with its entire tool-kit of tiny refined pieces of knowledge and techniques specialized to specific problems and specific tasks.

This has been a long aside. Having made my plea for a new way of organizing the debate between realism and instrumentalism, let me return finally to my concern with capacities. I said that Hendry's program missed out on capacities because, unlike the Cowles Commission, its models are not structural. Frisch believed in structure, so let us turn back to his equations to see what I find of significance there. I said in the introduction that the parameters in these equations — $a_{110}, a_{111}, \ldots a_{210}, \ldots$ — represent capacities. What are these parameters? Think about what is probably the most well known example of such a law — a demand equation.

$$q = ap + u$$

Here **q** is quantity, **p** is price, and **u** some kind of random shock factor. The parameter **a** is the price elasticity of demand. It tells us how much effect price has on the quantity demanded; or, in my language, it measures the strength of price's capacity to produce (or inhibit) demand.

One central reason for calling it a capacity is the one I have been stressing. It is treated as autonomous. It is assumed that this parameter will not change even if other parameters in other equations do. As a result of manipulations elsewhere in the structure, the total demand may well shift or even the relationship between demand and some of its other causes. But that should not affect the relationship between price and its contribution to demand.

How do we know that this parameter is supposed to be autonomous? Not by looking at the equation itself, but rather by considering how the equation is tested, and then how it is used. As I mentioned, for the Cowles Commission this kind of parameter is to be bootstrapped from the data. It is boot-strapped from the behavior which is observed under one fixed arrangement of the rest of the structure; yet is is used for predictions about what will happen when other pieces of the structure are altered. It is this kind of peculiar stability across change that leads me to use the language of capacities, this stability that Frisch saw as a happy accident. I say that **a** here represents not just the size of **p**'s effect on **q**, but rather its *capacity* because this is an ability that **p** carries around with it, from situation to situation.

We still have no contrast with Hendry though. I turn to that last. The second essential ingredient in the concept of capacity I am exploring is connected with the fact that for Frisch and Haavelmo and the Cowles Commission, an effect is attributed to the joint action of separate individual causes, and the parameters measure the influence of each cause separately i.e. the strength of its capacity. That is not so for Hendry. Look at the difference between what counts as a model for him and what counts as a model for the Cowles Commission. We have seen in Frisch an example of the kind of model developed at the Cowles Commission. It consists of a set of structural equations. According to Hendry,

A crude schematic structure for econometrics is as follows: To a first approximation . . . data generation processes in economics can be written as . . .

$$\mathbf{y}_t / \mathbf{z}_t \sim NI \, (\mathbf{\Pi} \, \mathbf{z}_t, \, \Omega) \qquad (t = 1, ..., T)$$

where $\mathbf{y}_t$ is a vector of endogenous variables, $\mathbf{z}_t$ is a vector of all relevant past and present information so that $E(\mathbf{y}_t / \mathbf{z}_t) = \mathbf{\Pi} \mathbf{z}_t$ and $(\mathbf{\Pi}, \Omega) = \mathbf{p}$ is taken as approximately constant by working in a sufficiently large (but assumed finite) dimensional parameter space . . .

An "economic theory" corresponds to asserting that . . .

$$\mathbf{p} = \mathbf{f}(\mathbf{\Theta}) \qquad \mathbf{\Theta} \, \varepsilon \, \Theta$$

where $\mathbf{\Theta}$ contains fewer parameters than $\mathbf{p}$. (Hendry 1979, pp. 17-18)

So for Hendry giving an economic theory consists in specifying the parameters of a normal distribution as some function of a smaller set of (hopefully measurable) parameters. That is, a theory is just a gigantic probability distribution — the conditional distribution of the endogenous variables given the exogenous. What is important for my point here is that the autonomous parameters characterize the process as a whole. They are in no way associated with a partition of the total influence into separate causes. The process under study may remain stable while changes are made in the basic behavior of the central variables; but there is no suggestion that that stability is due to the operation of separate causes whose influence remains fixed while other factors change. In this respect the Hendry program is holistic, and I think that holism precludes capacities, at least in one natural sense of the idea.

Perhaps I can make this clearer with some cryptic remarks about Galilean idealization, hopefully cryptic only because brief. I think an attempt like Hendry's to divorce autonomy from structure will not work. But that is not for the reasons of the traditional scientific realist who wants to defend the place of fundamental theory. I do not want to defend the view that we need to find some fundamental set of equations for nature in order to secure autonomy. I do not think that instrumentalist programs like Hendry's will fail because it is anti-theoretical, but rather because it is anti-Galilean, and Galilean methods are all we have. In the Galilean method you study features individually, in isolation. You strip away all the impediments, as best you can, in order to see how the feature behaves on its own. But why do you think that this isolated situation is special among all others? The answer must be that you think that in this situation you have learned *about the feature itself*; you have learned what its natural capacities are. That is why you can predict from what it does in this particular situation to what it will contribute when you set it back in far more encumbered situations. What I want to stress is that this is the method we have — this method that presupposes that individually identifiable features carry with them enduring capacities. We have no method for studying nature holistically; so we had better hope that nature has provided the enduring capacities which our methods are competent to find.

References

Cartwright, N. (1989), *Nature's Capacities and their Measurement*. Oxford: Oxford University Press.

Engle, R., Hendry, D. and Richard, J. (1983), "Exogeneity", *Econometrica* 51: 277-304.

Frisch, R. (1948), "Statistical versus Theoretical Relations in Economic Macrodynamics", in *Autonomy of Economic Relations*, R. Frisch, T. Haavelmo, T.C. Koopmans and J. Tinberger (eds.), a collection of photocopied articles issued by the University of Oslo.

Haavelmo, T. (1944), *The Probability in Econometrics*, supplement to *Econometrica* 12.

Hendry, D. (1979), "Econometrics — Alchemy or Science?" *An Inaugural Lecture*, London School of Economics, November 1979 (Typescript).

Leamer, E. (1983), "Let's Take the Con Out of Econometrics", *American Economic Review* 73: 31-64.

Part VII

HISTORY AND METHODOLOGY

Modern Physics and the Philosophy of Science

Dudley Shapere

Wake Forest University

The focus of this paper will be on certain aspects of three concepts, those of existence, explanation, and force, together with some related ideas. Specifically, I want to look at some of the roots of those concepts both in written history and, somewhat speculatively, in more primitive circumstances of the history of the human species; and I want to compare certain aspects of the three concepts with their rational descendants in the theories and expectations of contemporary physics. Although these topics are interesting and important in themselves, and there is much more to say about them, my primary aim in the present essay will be to extract some methodological implications from them regarding the problems and methods of the philosophy of science. Indeed, some of the ways in which the topics are important for philosophers of science will be brought out by the discussion of those methodological implications.

I will begin with the concept of explanation.

1. Explanation, Existence, and Force: Traditional Concepts

In Greek philosophy, a number of approaches are found to the question of how explanation, or at least ultimate explanation, is to be given (or perhaps how it can or must be given): for example, in terms of conflicting opposites; or in terms of the unchanging, the permanent; or in terms of the natures of things; or in terms of the intrinsically active. Although the first two approaches did not die out entirely in later thought, the third and fourth are certainly the most important historically. The view that an explanation, if it is to be truly ultimate, must be in terms of that which is permanent and unchanging is the legacy of Parmenides; it is ubiquitous in later thought, from the theology of an immutable god to the doctrine that atoms must be indestructible and unchangeable, and in a multitude of other ideas. The idea that ultimate explanation must be in terms of the intrinsically active has a more complex history. Its generic origins are, albeit vaguely, in Anaximenes, but there are two distinct lines in the tradition. In one line of thought, passive types of entities are taken to exist; but whatever happens in the universe must be due to another type of existent, ones which are intrinsically capable of action. Activity is then distinguished from existence, which includes the passive as well as the active; and the activity of certain existents (the active ones) becomes the explanation of aggregations and events. This idea, while peripheral to the main trends and motivations of Aristotle's philosophy, is found in his occasional distinction between active and passive elements; the idea is carried out far more thoroughly by the Stoics, in whose hands the doctrine becomes central. In the other version, which finds its fullest expression in Leibniz, the idea of existence itself becomes identified with that of intrinsic activity: to be is to be active, what we take to be passive entities consisting, when properly understood, only in the limitations imposed on active agency due to (but not caused by, says Leibniz) the existence of other active agents.

Thus we see that the concept (or concepts) of existence was tied, early in the history of thought, to certain conceptions of explanation. There has been more to traditional views of existence, however, than comes out of their connections to that of explanation. In particular, two aspects of many traditional ideas about existence will be important in the later stages of this paper. The first is the thesis that there exist individual things which are, in some important sense that varies from doctrine to doctrine, separate and distinct from ourselves and from one another. This idea is an easy inference from everyday experience; perhaps the most fundamental experiential motivation comes from our attributing to certain recurrent aspects of our experience properties which we later find not to be characteristic of those recurrent aspects; conversely, our finding that those recurrent aspects have properties which we did not originally suspect them to have; and, finally, from our attributing to or finding in some recurrent aspects of our experience properties which we do not attribute to or find in other recurrent aspects. The idea of entities existing separately from us and from each other emerges quite naturally from such experience (which is, as Piaget and others have shown) the experience of early childhood), and indeed comes to constitute such everyday human experience. It has a rational descendant in that aspect of the scientific enterprise which I call *the given*: the results of observation or experiment which do or do not agree with our theories, which are independent of the mass of interpretation which selects, describes, and shows how to gain access to that given. This independence of something in our experience from ourselves and our ideas is one of the deepest motives behind (though it is by no means either an explication or a justification of) philosophies which refer to themselves as "realism." As a "common-sense" belief, it is undoubtedly a product of early childhood learning; as an explicit philosophical doctrine, with variations and problems of its own, it emerged gradually in history, from increasing reflection on such characteristics of our most primitive experience. It is also primitive in the sense that the roots of the belief - its having become inexplicitly embedded in everyday activity - lie far back in human history, and indeed must ultimately be a legacy of far earlier evolution.

The second aspect of traditional ideas of existence to which I will want to return later is the supposition that, in order to exist, an entity must have a single precise value or determinate degree of every property it possesses. This idea, which I will call *The Postulate of Complete Determinateness*, also has its roots, as a thesis embedded in everyday life, in the possibility of error or correctness in attributing properties to those recurrent aspects of our experience which we call entities.

As these ideas about what is involved in existing can be traced to more basic and primitive circumstances of human existence (with some speculation, tightly controlled by what we know of paleoanthropology and prehistory), so, also, the two views of explanation I have discussed did not emerge full-blown from the brows of the Presocratic philosophers. Greek scholars have found deeper roots in earlier myths, and others have traced the sources still further into the past. Hans Blumenberg locates the origins of both myth and reason in the fears and uncertainties arising in the primitive circumstances of the human species after its emergence from the close security of the forest into the open savannah. Such hypotheses involve a great deal of speculation. Nevertheless, the speculation is of a sort that is constructed in accordance with what we know of paleoanthropology and prehistory; and speculation that is thus controlled by a background framework of our best available relevant knowledge (in senses of "best" and "relevant" that we have also learned) cannot be condemned as *mere* speculation. But however one may respond to Blumenberg's specific hypothesis, it takes no great imagination to suppose the origins of the desire for permanence to have lain somewhere in circumstances of primitive fear and uncertainty, and to postulate a conversion of that desire into a requisite for understanding. As to the doctrine that explanation must be in terms of the active, the supposition that such ideas as force, energy, power are, or were originally, objectifications of subjective feelings of muscular or volitional exertion has long been a darling of empiricist philosophers, and it, too, is plausible when seen as a source of primitive myths attributing agency to various aspects of nature which were important to human beings.

Our third topic, the concept of force, is thus traceable to the doctrine that explanation must be in terms of active agency. That source is seen clearly in the Platonic doctrine of the soul, and in later doctrines as mind, as being an entity which can initiate its own activity; it becomes one way of making a distinction between the living and the non-living. For present purposes, we can consider only the role of the concept of activity in the origins of modern scientific thought. The concept of soul was transformed into its rough scientific form by Kepler, when he converted the idea of a planetary soul from one of spontaneous self-initiating activity (intelligently seeking its proper path) into that of force, as a cause of law-governed behavior. Subsequently it entered into the new Mechanical Philosophy of Nature as one of the four candidate-concepts (along with space, time, and matter) for the status of being an ingredient in any fundamental explanation of nature. Its interpretation and status relative to those other three occupied a central place in the debates among the advocates of that new philosophy. Were all four of the candidate-concepts for fundamental explanatory role really necessary, or could some be dispensed with in favor of one or more of the others? If matter is wholly passive (a correlate of its geometrical nature, according to the Cartesians), and still more if it is equated with spatial extension, the role of self-initiating activity in the realm of matter is denied, and what we take to be forces and their effects must be accounted for in some other way than by granting them fundamental ontological status. At the other extreme was Leibniz, for whom, as I have noted, it is only through the concept of active force that existence can be understood; what we think of as space, time, and matter are simply well-founded phenomena, manifestations of the limits which channel the expression of active forces. Newton's view, advocating both passive matter and active forces, lay between these two extremes. Those debates would have their descendants - altered, it is true - in the nineteenth century, when the emphasized contrast would be not that between the active and the passive, but rather between the immaterial and the material, and in reactions against materialism, from *Naturphilosophie* and metaphysical idealism to the ruminations of scientists as diverse as Faraday and Helmholtz. And there remain echoes of the debates in the variations found in modern textbooks of mechanics as to what the basic concepts of mechanics are, force being, for example, sometimes said to be eliminable "by definition" (rather than, as earlier, by ontology) in terms of matter (mass) through the second law of motion.

Despite these debates, the concepts of space, time, matter, and force still remained distinct in working scientific theories at the end of the nineteenth century. And it must be confessed that there remains, even today, an aura about the four that seems (at least to those not acquainted with current physics and cosmology) to make them a "natural" set of categories, implicit, almost, in the very concept of a world or universe, of reality. The reason is not hard to find. For as the concept of force can be traced to the notion of activity and its more primitive roots in everyday experience, so also, though in some cases with more complex lineages, can the other three be traced to deeper sources, despite the transformations they have undergone in more sophisticated thought. The roots of the concept of space must lie, at least partly, in those same experiences which produce the unexpressed postulate of objects separate and distinct from ourselves and from each other; and the concept of time arises, equally clearly, from the changes those objects undergo. Matter is more complicated, emerging only partly from the concept of separate objects; very roughly, it also involves the idea of the continued existence, the permanence, of those objects or some deeper stuff of which they are made. The concept of matter, like that of force, is thus a descendant of a primitive notion of what fundamental explanation consists in - only its ancestor is a notion of explanation different from that from which the concept of force descends.

The naturalness of picturing the world in terms of concepts of space, time, matter, and force thus arises in the same way as does the naturalness of so many classical metaphysical and scientific ideas, incompatible though they often are: in all such cases, with all their primitive vaguenesses and potential conflicts, the ideas are (and as I shall argue later, must be) natural outgrowths of everyday human dealings with the world of experience. And even though more sophisticated thought subjects the concepts to great varieties of alterations, the naturalness of their origins appears undiminished: the glosses may be odd, but the idea which they have altered seem not to be so. Yet in the physical theories of the past two decades, the concept of force and its three associates, space, time, and matter, and the views of explanation and existence with which they were originally associated, no longer have unquestioned fundamental status. They are explained, or appear apt to be explained, as products of something more fundamental; and furthermore, the explanations are of a sort utterly foreign to the sorts of explanations, stemming from the primitive, which I have discussed so far, even though they are, in a sense I will explore, descendants of those sorts of explanations.

2. The Three Concepts in Modern Physics

What, then, does become of our three concepts, and associated ones, or rather their rational descendants, in contemporary physics? It will be convenient to begin with the concept of force.

In the three decades before the 1960's, physicists gradually came to the understanding that there are four fundamental forces at work in the universe today: the gravitational, holding large masses or aggregations of masses together; the electromagnetic, responsible for most of the phenomena of the everyday world, such as the cohesion of bodies, chemistry in general, and light; the strong force, governing the properties and behavior of atomic nuclei and much of those of elementary particles; and the weak force, responsible for radioactivity, but far more importantly involved in the conversion of the simple primordial elements (known from cosmology), hydrogen and helium, into the more complex chemical elements of the periodic table which go to make up solid objects. The great achievement of physics since the end of the 1960's has been the taking of large steps toward constructing a unified theory embracing three of those forces, and of the raising of scientifically-reasonable expectations and possibilities for a unification of all four. We now have a highly successful unification - not quite complete, but close - of the electromagnetic and weak forces into a theory of the "electroweak" interaction. Modelled on that theory, a new theory of the strong force has been developed, in which (far more than was the case in earlier theories of that force) calculations are possible in a wide variety of problems. Further, the formal similarities of the new theories of the electroweak and strong interactions lead to their juxtaposition in what has come to be known as the "standard model" of elementary processes. A variety of alternative ways have been proposed for subsuming the group-theoretic structures of the standard model under a more inclusive group, thus producing an even tighter unity. These are the so-called Grand Unified Theories, GUTs for short. Although serious problems afflict such unifying theories, it is widely believed that some specific variation, generalization, or subsumption of the highly-successful standard model will ultimately prove successful.

The standard model, together with well-founded aspects of the idea of a deeper unifying GUTs theory, have been given a physical basis through application in a detailed account of the early history of the universe. Among the results is an account of how the three forces of the standard model came to be distinct. For in this application of physics to the now well-supported Big Bang cosmology, as the temperature of the early universe cooled with its expansion, the strong force, in a dynamically intelligible process of spontaneous symmetry breaking, broke away from its unity with the electroweak. At a still later epoch and still lower temperature (all still well within the first second of the universe's history), the electroweak unity was also broken, giving us three of the four distinct forces we know in our present cold universe.

The fourth fundamental force, gravitation, has not yet been brought into successful unification with the standard model or the GUTs theories. Several theoretical approaches have been developed, most notably supersymmetry (along with supergravity), superstring theories (also incorporating supersymmetry), Kaluza-Klein theories, and the Penrose twistor program. All involve serious difficulties. Nevertheless, many considerations lead to the supposition that that unification is possible. I will mention two such considerations. First: the contemporary theories of each of the three forces brought together in the standard-model unifications is expressed in the same general type of mathematical form, that of a local gauge theory. It was in fact the gauge-theoretic formulation of the electromagnetic and weak forces (coupled with certain other necessary innovative ideas) that made possible the electroweak unification; and it was the gauge-theoretic formulation of a theory of the strong force (again, coupled with other important theoretical developments) that in turn made the standard model possible. But general relativity, our best-by-a-long-shot theory of gravitation, is subsumable under supersymmetry in a supergravity theory. And supersymmetry is not only applicable to the standard model, but is also a gauge theory. Superstring theory also incorporates supersymmetry, and therefore supergravity, and therefore these virtues. Second: many of the mathematical difficulties that plagued earlier quantum field theories - and also general relativity - are removed or at least alleviated in the new theories. Gauge theories, for example, are renormalizable; the infinities of general relativity are removed or reduced in supergravity. Superstring theories eliminate other mathematical problems, in particular, in some cases, anomalies.

Although, as I have said, the alternatives for higher unification all have their problems, many of them still remain promising. For present purposes, the important point is the possibilities they raise, the conceptual innovations they make scientifically plausible, whether they are ultimately accepted or not. The theories have in common the same applicability to the early history of the universe that GUTs theories have had. In particular, the vastly earlier time and correspondingly higher temperature (i.e., energy) at which superunification would have been in effect can be calculated.

Among the central ideas of these new quantum field theories relevant to the present discussion are ones having to do with the reinterpretation of our three fundamental concepts, and the erasure of distinctions which existed in previous scientific and common-sense views. Consider first the concepts of classical mechanics. Already earlier in the century special relativity fused the concepts of space and time into one, space-time, the separation into space *and* time being no more than a product of viewing nature from the perspective of a particular reference-frame. That fusion pales in philosophical significance compared to those brought about already by the field-theoretic unifications that have been achieved in the last two decades. Quantum field theories view matter and force as very much on the same footing: what we have classically thought of as matter consists of particles called fermions (but "particles" in the quantum-theoretic, not the Newtonian-Laplacean sense); while forces are carried by other sorts of particles, bosons. In the well-developed GUTs theories and their various quantum-theoretic components, fermions and bosons are, however, still fundamentally distinct, fermions having, as in earlier theories, half-integer spin or integral multiples thereof and obeying one type of statistics (Fermi-Dirac), while bosons have integer spins and follow a different type of statistics (Bose-Einstein); and that is what remains of the distinction between matter and force in those theories. Supersymmetry, if that turns out to be an ingredient in a superunified theory, would make even that difference superficial. In such a theory, fermions and bosons would be interconvertible, so that there would no longer be the absolute bifurcation of matter and force that the seventeenth century saw as a distinction between the passive and the active, and the nineteenth as one between the material and the immaterial. Finally, a fully-unified theory of the fundamental forces may yet eliminate even the ultimate distinction between force-matter and space-time. A central aim of Kaluza-Klein and superstring theories - an aim which, it must be confessed, has not yet been achieved - is to demonstrate how a higher-dimensional space could be split (by some dynamical process) into two components. One would be four-dimensional space-time, which general relativity makes the source of gravitation. The remaining dimensions would compactify to produce matter-force. Thus the separation of the gravitational from the other forces would also break the primordial unity with the fermions and non-gravitating bosons that we know today, in our cold universe, as the sharp difference between matter and the residual forces. While Penrose's twistor approach remains four dimensional from the outset (but employing a complex twistor space-time), it nevertheless aims at the same kind of ultimate primordial unity from which space-time and the forces and particles of the standard model ultimately emerge, as products rather than primitives. (Some recent superstring theories have also been beginning with four dimensions.)

In discussing the concept of force and its relations to those of space, time, and matter, I have focussed on the unification of those concepts, rather than on the changes those concepts have undergone in the process and in their earlier history. Though time prevents detailed discussion of those conceptual changes, a few can be singled out for brief mention as relevant to the general view which will be advanced in the remainder of this paper. Though analogues and successors remain, the ancient contrast of active and passive has largely dropped out of the modern picture. What is left of Parmenidean permanence is located in symmetry principles, not substances. Force has become interaction, involving such new possibilities as particles interacting with themselves. (Particle decay is thus an interaction.) When I come to discuss the concepts of explanation and existence, however, I will discuss aspects of such conceptual changes.

Despite the occurrence of such conceptual changes, I do not speak of "concepts" in vain, as if there were a single concept of force stretching across history. For as I have argued elsewhere in connection with the concept of a concept, where a succession of more specific concepts (sets of properties associated in the light of prior relevant belief) are related by chains of reasons, then despite the alterations brought about in accordance with those reasons, we can and do speak of the entire heritage of reason-related specific concepts also as being a "concept."

3. Some Lessons for the Philosophy of Science

We saw in Part I that there have been deep historical links between the three concepts on which I have focussed, force, existence, and explanation. Nevertheless, in some later historical contexts those links were broken and, often, forgotten, and each has frequently been assigned to a field of investigation distinct from that of the others - distinct in the sense that the field in question is supposed to be dealt with independently from the others, and by methods different from those employed in the other fields. Existence has traditionally been assigned to the province of metaphysics, or more specifically of ontology, a subject whose inquiries, both traditionally and in their modern logical and linguistic guises, has often been supposed to be conducted independently of science. Force is most immediately associated with physical science, whose methods are supposed to be empirical, as opposed to the *a priori* approach of traditional metaphysics and its descendants. And finally, explanation has been supposed by many modern philosophers of science to be a metascientific concept, which, although applicable to science, is to be analyzed independently of any specific scientific results or methods.

We have seen that, in their roots, the three concepts had deep links, of various sorts, to one another. They become linked again in the new and rationally-projected unifications in contemporary physics and cosmology, though in ways, and for kinds of reasons, that were not present or suspected in earlier thought. We have already seen that those modern unifications include the fusion of the concept of force with those of space, time, and matter. It is a unification in the sense that, at specifiable high energies, which would be found in the very early universe, those four would not be distinct or distinguishable. Rather, like the distinctions of the four forces and of the varieties of elementary particles, they would be products of symmetry breaking, or, in the case of the splitting of space-time from force-matter, perhaps of some other dynamical process. In the final two sections of this paper, I want to bring out some ways in which my other two concepts, explanation and existence, have been incorporated into the developments I have described, and how they have also been altered by the new theories. More generally, I want to show how, by being internalized into the scientific process, even such concepts as explanation and existence can be subjected to alteration in the light of what we learn. And I want to do this by bringing out some lessons, or at least suggestions, about how the problems and methods of the philosophy of science, too, may require reinterpretation and reconception in the light of new scientific developments.

Deeply involved in the new physical theories is a mode of explanation, and a fundamental sort of entity, foreign to pre-twentieth century, and in some respects foreign even to quantum mechanics, even though it descends directly from that theory. It is explanation in terms of Yang-Mills, or more generally, of gauge fields. The paradigm of this type of theory was originally devised (to be sure, with some antecedents) in 1954 as a means of dealing with what was then the intractable strong interaction. The structure of gauge theories is such as to require the existence and character of forces in a natural way, as (roughly speaking) compensations for what would otherwise be local distortions. Supplemented by a mechanism, missing from the original theory of Yang and Mills, for endowing particles with masses, the gauge-theoretic approach was applied successfully in the construction of the electroweak theory in 1968. A number of considerations conspired, in the early 1970's, to convert the gauge-theoretic approach from being merely another type of theory into a *program for the construction of new theories* of fundamental particles and forces. These considerations included the following, among many others: the establishment of the renormalizability of gauge theories by t'Hooft and others; the success of the electroweak theory in a series of experimental tests; a deeper appreciation of the advantages of recasting the theory of the electromagnetic interaction, quantum electrodynamics, as a gauge theory; the construction of a gauge theory of the strong interaction, a theory which, through its asymptotic freedom, had the additional advantage of calculability in a wide range of cases; and the realization that many mathematical difficulties (infinities and anomalies) that had plagued earlier quantum field theory disappeared in the gauge formulations and the sorts of theories that could build further on them, such as supersymmetric string theory. Such considerations have counted as *reasons* giving the gauge-theoretic program a normative status: because of their successes and reason-backed promises, gauge theories are the kind of theories that *should* be sought. That normative principle itself is one that has been learned through the investigative enterprise of science. It is a reason-based change of goals - of cognitive goals - brought about in the light of what we have learned about how to understand.

Gauge fields, as descendants and extensions of quantum mechanics, have the characteristics, and the mysteries, of superposition: they possess all values of their properties, and thus violate what I earlier called the Principle of Complete Determinateness. Particles are excitations of the field. But also, in the light of Bell's Theorem and its empirical successes, they apparently may not have two further basic characteristics of classical objects, distinctness and independence from one another. The explanations offered are incompatible with the classical ideas of explanation which I discussed earlier, even though this new form of explanation emerged in a gradual process of criticism and modification out of scientific theories that did embody those classical patterns. But if what I have said in the opening section of this paper is correct, those classical patterns were forged from the circumstances of practical, everyday dealings with human experience, experience which it has become customary to call "middle-sized," with more than a hint that it is *merely* middle-sized. The universe need not operate in those ways, need not be understandable in any of the senses extracted from, and designed to deal with, everyday experience. Far from being segregated from the evolving content of scientific theories, far from being a subject to be dealt with independently of that content, the issue of what counts as an explanation has been incorporated, internalized, into the scientific process. Whereas the earlier debates about what constitutes a proper fundamental explanation were largely inconclusive, their fruits really being only difficulties for all the theories, we have now become able, through that internalization, to learn, from scientific investigation, something about how to explain. We have gained understanding of what it is, in this universe, to understand, and thereby of the limitations and errors in what we earlier demanded in an explanation in order to count as such - in what we earlier believed understanding to consist in. Whether the present explanatory approaches succeed or not, that fact will remain; for if we learn that those approaches are incorrect or limited, that itself will have been a lesson learned through the investigative enterprise.

Similar cautions should be gleaned from the new physics regarding what philosophers of science call "the problem of realism" of scientific theories. Often in discussions of this problem one cannot avoid the impression that the methodology involved consists in laying down what it is for something to be "real" (or "realism") and then to see whether scientific theories, or certain specific scientific theories, or certain scientific concepts, satisfy those conditions. Yet again, perhaps that is to put the cart before the horse; perhaps the new theories are telling us something that we did not know before about what it is to exist - something that contradicts the concepts of existence which are products of everyday, practical, "middle-sized" experience. We must be deeply sensitive to the possibility that, far from having to *satisfy* some set of antecedently-stipulated conditions in order to count as "real," these highly successful theories may be trying to tell us something, perhaps only cryptically yet, considering their still incomplete state, about what we may have to accept as real. We thus must *learn* what it is for something to be an "individual thing," a "universe," and even what it is for something to be "real."

I do not mean to be passing over the substantial difficulties in interpreting quantum theories in anything like a "realistic" way, whatever may be involved in that. But there are other aspects of the situation that also provide intuitions worth exploring. For many years, philosophers of history have fretted about the status of historical or narrative explanations, as opposed to deductive ones. But the gauge theories, which have at least some features of (or analogies with) deductive explanations, have acquired a historical, narrative dimension of explanation in addition, through their application to cosmology. For what that application does (especially when supplemented by an inflationary episode or some successful analogue of that) is to give a highly successful historical account of what happened in the earliest stages of the universe (dark matter excepted). Indeed, it does far more than that: for by providing an account of the production and proportions of hydrogen and helium, it serves as what I call a *background framework*, a theoretical setting within which theories of galaxy formation, stellar formation and evolution, the evolution of the other chemical elements, and the evolution of planetary systems and of life must be placed in order to be acceptable. It seems implausible to maintain that theoretical entities which make possible so much explanation of that kind can be dismissed as mere fictions, or whose reality, in a sense that those theories themselves may prescribe, can be denied or ignored as an irrelevant issue. And though plausibility arguments are not telling, they must be given due attention. Following up on this intuition (while simultaneously remaining sensitive to the problems in the interpretation of any quantum theory) would involve, among other things, reconsidering the nature and appeal of narrative or historical explanations in the light of the achievements of the new physical/cosmological theories.

4. Methodological Implications

Like the scientist himself, the philosopher of science is concerned with the interpretation and justification of scientific ideas. But with regard to both of these, the philosopher must deal with issues beyond those with which the scientist is occupied, at least ordinarily. Let me take each in turn, beginning with the problem of justification.

A central component of the justification of scientific ideas is the observational evidence relevant to an idea. In other work, I have shown how what counts as observational evidence is shaped by a background of other ideas, theoretical and non-theoretical, and how changes in that background of ideas produce changes in what is counted as observational. It is not there is no such thing as a "given" in observation, independent of theoretical presuppositions; on the contrary, I have argued that there is, in the sense that (a) having been marked out as significant by our best available background ideas, (b) having been appropriately described in terms of those background ideas, and (c) having been made accessible by application of background ideas in an account of how to observe or otherwise get evidence concerning it, the specific character or value we find it to have is independent of - not determined by - those background ideas. But this very conception of a "given" against which theories can be tested and, if need be, altered, brings out the extent to which what counts as observational, as evidential, depends on background ideas; and those background ideas are, in many cases, taken from new and advanced scientific ideas.

Thus, in addition to the problem of ascertaining the justification offered for a theory by a piece of observational evidence, there is a problem of understanding why *that* piece of information (which can be as bizarre from a common-sense viewpoint as the theory it tests) counts as evidence. That problem requires turning to the background ideas which determine the item in question to be a piece of evidence. But here again, more is involved in the interpretation (that is, in the understanding) of those background ideas than is found only in the interrelations of those scientific ideas with others in the theory in which those ideas occur, or even in the contemporaneous family of theories of which that theory is a member; there is more, say, in understanding quantum chromodynamics than *just* seeing the way in which its mathematical structure implies (for instance) asymptotic freedom, and more than is involved in the broader gauge-quantum-field-theoretic and associated apparatus that constitutes the conglomerate background in terms of which that particular gauge theory and its variations are formulated along with the gauge-theoretic version of quantum electrodynamics, gauge theories of the weak interaction, and the gauge-theoretic unifications of these. The scientist can get along with just that, ignoring the history of the ideas involved, because he or she concentrates on manipulating those ideas. But for the philosopher, it is not enough to understand only the theories or the research programs within which those theories are constructed: there is also the problem of the contrast between the scientific concepts and other, prior, and even ancestral, concepts, theories, and contexts of thought; there is, further, the problem of the reasons why the contrasts have had to be introduced and accepted, however tentatively. Given what human beings more naturally tend to believe, given the practicalities of their everyday experience and the long history in which those practicalities have given rise to all sorts of descendant ideas about such matters as how to explain, what to explain in terms of, what it is to exist, and a host of similar questions, there is the problem of how we ever got from those to their weird descendants in modern science. That is the sense in which the philosopher must be concerned, as the scientist need not be, with the backgrounds of the theories and the programs, and the backgrounds of those backgrounds in turn, all the way back even to paleoexplanation and paleometaphysics, for all their speculative aspects.

Of course, such philosophical investigation must not proceed Whiggishly, looking merely for antecedents of present ideas: the question must not be, "What led us to our present ideas?" Rather, it must be, "Given the ideas which were there at some beginning-point, how did we get from there to subsequent stages, and ultimately to our present one?" (Even what we today call "common sense" must be subjected to that question.) This investigation must differ, however, from some other approaches to the history of science. It does not, for example, look at historical cases for the purpose of testing general hypotheses about the nature of science, or for grounds upon which to induce such general hypotheses; for the approach I have described, by recognizing the full depth to which scientific change can occur, makes historical development integral to the very concepts of reasoning and knowledge. To put the point in another way, the very fact that science proceeds in the light of background ideas - in the light of what it has learned or takes itself as having learned - requires that the intellectual products of science be understood historically.

But that remark requires elaboration; for the problems I have described require attention to a historical phenomenon all too often ignored by today's historians of science, as well as by certain sociological approaches to the study of science: namely, the undeniable fact that over its history, science has become increasingly able to develop independently of external influences. It has been able gradually to demarcate external from internal considerations, and to be able to rely more and more fully on the latter in developing its ideas. This is not to deny the existence and influence of external (psychological, sociological, institutional, economic, political, etc.) considerations in science, even in contemporary science; they have their role in a *full* understanding of science. But it is to say that there is something to consider in addition to such factors, namely the role of considerations which are - more exactly, have become - internal to science, and indeed whose historical emergence constitutes the emergence and character of science as a distinct discipline or body of disciplines. Indeed, as I have argued elsewhere, the character of those internal considerations - the character of science as a distinct discipline - must be understood as a prerequisite for making full sense of the role of external considerations. Dealing with the kinds of problems I have described - and in particular, the contrast between scientific and prior conceptions that might appear more natural, and the reasons why the prior conceptions have been rejected in favor of the scientific - requires, in addition to everything else I have mentioned, a concern with the development of the concept of reason(s) itself.

Thus, instead (as in many philosophical approaches) of attempting to understand such concepts as explanation, existence, reality, justification, observation, evidence, and reason, in terms of their logical or conceptual structure, or their role in language, or on the basis of *a priori* or transcendental arguments, or the like, we must trace all these, and human thought in general, to the circumstances in which human beings find themselves; and since later ways of responding and thinking stem from more primitive ones, as descendants from ancestors, our investigation of the nature of human reasoning, knowledge, and the knowledge-seeking enterprise must go ultimately to the earliest confrontations with the world about which we have written or other evidence, or about which we can make *reasoned* speculations in the light of current relevant knowledge. In contrast to philosophical approaches such as those I just described, this is one that is fully consistent with our knowledge that humankind, as a product of evolution, could have had no means of developing patterns of explanation, or ideas about existence, other than those that could be forged in dealing with the uncertainties of his experience. We must examine, for example, the various implicit conceptual ways in which human beings responded to their most primitive circumstances that evolved into ideas of what it is like to understand, to explain, or what it is for something to exist. And we must try to investigate how it is that modern scientific thought has come from there to be what it is, not only in regard to what it says about the world, but also in regard to how it goes about explaining, giving reasons, counting something as observational evidence, and the other such activities that go to make up the knowledge-seeking enterprise.

References

Blumenberg, H. (1985), *Work on Myth*, Cambridge: M.I.T. Press.

Shapere, D. (1987), "External and Internal Factors in the Development of Science," *Science and Technology Studies*, Vol: 1, 1-9.

_____ . (1985), "Objectivity, Rationality, and Scientific Change," in P.Kitcher and P. Asquith (eds.), *PSA 1984*, East Lansing, Philosophy of Science Association, Vol. 2: 637-662.

From Logical Formalism to Control Structure:
The Evolution of Methodological Understanding

C. A. Hooker

University of Newcastle, Australia

1.Introduction

Over the past two centuries the character of science has changed markedly and crucially. When humans had relatively small amounts of information and energy available
they perforce had no option but to attempt to describe objectively a world which behaved
independently of them. The basic practical problem was to discover what the pattern of
the world was so as to adapt themselves as best they may to it, a little in anticipation of
events, but mostly in reaction to them. It is these circumstances which give most sense to
the empiricist paradigm of objective knowledge.

But the more information and energy *is* available to us, the more we are able to ascend
from merely correct description of what is to a grasp of what *is possible*. Newton's laws
don't merely describe the paths of the actual projectiles that have fallen and will fall, they
describe the paths of all possible projectiles. Correspondingly, the more information and
energy available to us the more we are then able, on this epistemic basis, to intervene in
the course of the world so as to bring about a chosen future from among the possible ones.
The switch from fact to possibility as the content of science is matched by a switch from
adaptation to intervention by anticipative design in our practical policies.

Moreover, these two switches are mutually reinforcing: We only grasp possibilities
by intervening to disturb the course of the world, but the more we do this the more practically necessary it is to intervene with anticipative designs. Conversely, the more we are
able to transform the future by anticipative design the more we have a culture oriented
around the exploitation of intervention and change and the more central to such a culture
is the development of science and technology. The result is a doubly self-reinforcing
change engine driven by scientific/technical development. This change engine has been
running, and gaining in power, for three centuries now, resulting in today's planetary science and technology institutions and in science/technology-dominated economies. These
are our 21st century circumstances and they make earlier conceptions of science as passive acquisitor of facts both theoretically and practically obsolete.

This is the setting in which a theory of scientific method has now to be put forward.
Any conception of science which cannot do justice to these profound historical transformations in its nature and role in human affairs will fail to provide a basis of understand-

PSA 1988, Volume 2, pp. 211-221

ing that will take us into the 21st century. The philosophy of scientific method must evolve along with the evolution of the nature and role of science.

Now I believe that over the course of time not only the content of science evolves, and not only the nature and role of science evolves, but the methods of science evolve and the philosophy of science evolves — and all of them co-evolve together in dynamic interaction. But traditional philosophy of science, inherited from the Platonic-Cartesian tradition, is anti-evolutionary. That tradition aims to specify scientific method in terms of eternal logical rules. Thus I argue for the rejection of this conception of scientific method and its replacement by another: that of a complex control system.[1]

2. The Failure of Formal Method

Empiricists wish to avoid the incorporation of error into scientific knowledge by restricting science to the operation of logic on objective observation. This is attractive since logic is held to be a purely formal, hence subject-matter free, theory of truth-tracking. That this approach leads to a powerful, persuasive and formally elegant approach to understanding science, is well-known. Equally well known are its failures, inadequacies and incompleteness. (The literature is vast — see e.g. Hooker 1987 and the extensive bibliography therein.) For present purposes we may summarize a variety of complex issues in this way: There are two primary failure modes for empiricism, the need to appeal to communal judgement and the need to appeal to theory. It turns out that without both of these appeals no reasonable account of either scientific observation and evidence, or of scientific method can be given. Both of these elements (judgement, theory) are substantive and both undermine the empiricist ideals of formal method and objectivity.

With respect to the appeal to community judgement, because observations are finite, perspectively structured and fallible, even severe empiricists were forced to presume communal cooperation in the accumulation, communication and selective filtering of observation reports. (Even the cleverly agnostic empiricist van Fraassen does not escape an irreducible appeal to an 'epistemic community' — van Fraassen 1985, p.284.) Popperian theory is riddled with additional ineliminable appeals to such human decisions: to accept basic observational statements; to choose tests of theories, and theoretical conjectures themselves to test; where to lay the error when theory and experimental result clash; when sufficient testing has been done for the moment and even to cooperate in pursuing the rational life. After Popper, through Kuhn, Feyerabend and all the others, the appeal to community widens rapidly — so rapidly, that all of these latter have been accused of abandoning reason.

With respect to the appeal to theory, it is fundamental to the failure of empiricist assumptions about observation that our current best theory of perception models it as an essentially theory-forming process. (This point is typically illustrated by gestalt switching, the role of expectations in perceptual reports, etc.) And, of more direct interest here, the fundamental reason for the failure of the inductive logic program, is that rational inductive inference is context-dependent, the context being primarily fixed by theory. (The essential point is nicely illustrated by Bertrand Russell's chicken. Suppose a young chick born in January and fed each day until December 22 of that year. This feeding evidence has no exceptions and thus, when combined with Reichenbach's straight rule of induction, yields the conclusion that the fowl will be fed on the morning of December 23. But insert the very same evidence into a wider theory of cultural practices and we may deduce with at least as high a probability that on the morning of December 23 this bird will be slaughtered for the dinner table. Note that the very evidence which one may take in one context to support continued feeding is in itself evidence, in another context, that feeding will not continue: the better the feeding, the fatter the chicken, the more attractive it is as dinner.) Method then is a function not only of logic but of substantive theory.

It does not follow from this conclusion that there is no such thing as scientific method. This inference would require a suppressed premise, The Meta-Method Dichotomy (MMD): either method can be specified in terms of purely formal logical rules or there is no scientific method. But MMD is false. This is conveniently illustrated by Feyerabend's attack on scientific method (e.g. Feyerabend 1978). Feyerabend shows quite effectively that there are no purely formal universal methodological rules which could be applied across the history of science and which would generate all and only the actual decisions made by scientists which we now regard as clearly part of the best examples of science (in particular in respect of the Copernican-Galilean revolution). From this he concludes that there is no scientific method. But suppose instead we switch from a logical to a decision-theoretic model of methodology. In this latter scientists have cognitive utilities which they pursue through the construction of science, accepting those beliefs and methods which are utility-maximizing in the circumstances. In this setting, Feyerabend's various studies amount to various illustrations (many of them insightful) of a nonetheless obvious truth: the methods one uses and the decisions one takes are functions of both the values pursued and the believed circumstances (theory) in which one is deciding. This switch is a profound one, introducing a substantive, and evolving, methodology (see Hooker 1987).

The appropriate paradigm for method is that of a regulatory or control system. In section 3 I illustrate some key features of this conception, in contrast to the formal logic paradigm, through an episode in 20th century physics.

3. Substantive Method

The formalist paradigm had autonomous method operating on facts (evidence) to yield factual generalizations (theory). The new paradigm represents method as operating on utilities (aims, values), evidence and theories to yield, via feedback, transformed utilities, evidence, theories and methods. To illustrate these issues more concretely, I have decided to choose as example my own doctoral work in physics in the 1960s. The system under investigation was a gaseous plasma (i.e. an ionized gas) of moderate temperature (10^4 - 10^6 °C). The gas was confined in a metal cylinder with a strong axial magnetic field and the plasma formed by initiating a high electrical discharge at one end which caused a rotating shock wave to travel down the cylinder. During the small time interval over which the subsequent plasma existed (1 or 2 milliseconds), its characteristics had to be determined and all subsequent experiments on its behavior carried out. It is not easy to obtain information about the interior of a plasma at these temperatures, all normal measuring instruments cease to function (and their disintegration would in any case pollute the plasma!). Like all contemporary physical research a diversity of goals were involved: we were interested in an intrinsic understanding of the plasma state, in understanding the multifarious plasma processes in the upper atmosphere and astronomically and in the useful terrestrial harnessing of thermonuclear fusion to produce energy — but these interests lie beyond the scope of the present paper.

For the moment I wish to concentrate on one relatively low level piece of method, the use of Langmuir probes to measure plasma density and temperature. The central idea behind the Langmuir probe is a simple one: one pokes into the plasma a small piece of wire connected to some electrical circuitry; by measuring voltage and current in the wire one hopes to deduce the state of the plasma surrounding the wire. Even at this stage one requires the use of electromagnetic theory to understand plasma/wire interactions, the use of solid state theory to understand the transmission characteristics of the wire, the use of electromagnetic circuit theory to understand the accurate measurement of the wire responses and so on. The appearance of the probe in the plasma affects the plasma in two relevant ways: it perturbs the distribution of electrons and ions in the plasma and it causes the formation of a ion sheath around the probe which partially screens off the plasma characteristics from the probe. Both of these effects systematically distort the value of the probe responses as a measure of intrinsic plasma characteristics. The really hard constructive work that has gone into developing the Langmuir probe as a useful instrumental methodology in plasma physics research has focused around the development of theory and experimental practice to understand and correct for these two distorting effects. The resulting practices include the use of multiple probes of various sizes and with various electrical biases on them, the extraction of information from transient as well as average probe responses and so on. These practical and theoretical methods for obtaining useful plasma information have in turn required the extensive further application of the theories noted above.[2]

It should be noted that it is these very same theories which are in principle testable through plasma physics experiments, e.g. electromagnetic theory through its application to ion-electron collision characteristics. (The application of electromagnetic theory to ion-electron collisions will specify the conditions under which those collisions are significantly inelastic (involve radiative transfer and emission of significant quantities of energy) and the contribution of inelastic collision processes of various kinds - e.g. single ion-electron, multiple electron-single ion, etc. - to the resulting characteristics of the plasma can be calculated and then measured using devices such as Langmuir probes.) This fundamental interdependence between theory under test and test methodology is a commonplace throughout 20th century physics.

It took many years to develop the theory and practice of Langmuir probes to the point where a consistent and useful experimental methodology was available. During this time theory, theoretical methods and instrumental practices developed in delicate interaction with one another. New specific applications of electromagnetic theory to the probe sheath were developed. New statistical methods were developed for the extraction of data from probe responses. And of course new engineering procedures and experimental probing procedures were developed to meet the conditions laid down by the foregoing pairs of specialized theories. During this period the fundamental electrodynamics was never seriously questioned — but it might have been (e.g. had it predicted seriously wrong collision cross-sections for plasma processes) and if it had been, then a fourth element would have been opened up for interactive adjustment as plasma information evolved. The idea that there is some neat distinction between theory and observation/information/data and that the former evolves by free conjecture and is then simply tested against the latter — this idea provides no useful illumination of the development of Langmuir probes and like episodes at the heart of modern physics. Instead, one must understand theories at various levels (e.g. basic electrodynamics and applied probe sheath theory), theoretical methods (e.g. statistical methods of probe response comparison) and instrumental practices as a mutually interacting evolving system.

The overall goal is to arrive at a set of theories, theoretical methods and instrumental practices which satisfy certain values, for example (i) that the theories are consistent, (ii) that theoretically predicted data is consistent with instrumentally derived data, (iii) that theories have a wide range of applications in diverse situations across which their fundamental parameters are invariant (in this case notions of mass, charge, electromagnetic Hamiltonian, etc.) and (iv) that specific instrumental practices can be cross-checked by a diversity of related instrumental practices (in this case, e.g., by spectroscopic analysis of plasma radiation, laser scattering, microwave transmission probes) and that the several groups of derived data will show cross-situational invariance, (v) that the information extracted is that maximally available and of most interest as specified by theory and ultimate control objectives (which includes here both capacity for further scientific exploration and for practical application in energy production), and so on. These values are all whole-system properties of the science of plasma physics, they cannot be attached simply to this theory or that instrument. The result of this interactive process is a collection of fundamental and applied theories, theoretical methods and instrumental methods which between them achieve the values which the whole of plasma physics pursues, but the final characteristics of each of the components is a function of the characters of the other components. In particular,the specific methods which receive approval are strongly dependent on our theoretical conceptions of the nature of the reality which those methods explore and the material means through which we do the exploring.[3]

The development of Langmuir probes is only one small component in the overall development of plasma physics. A whole range of like instrumental methods are required, furnished e.g. by laser probe and scattering, spectroscopic analysis of plasma radiation, microwave transmission probing and so on. The development of each of these in turn requires a complex, interlocking set of applied theories (e.g. the relations between plasma collision processes and laser scattering, or the relation between plasma temperature, density, confining axial magnetic field and microwave transmission near resonance frequencies, etc.), the development of a set of theoretical methods for extracting relevant information from signals emitted from the plasma and the development of a set of engineering and instrumental practices coherent with these two. At the same time, the development of appropriate applied theory requires in itself a set of theoretical methods concerned with the formation of systematic approximations, the consistent joint application of theories and so on. For example, the full theoretical specification of plasma/microwave interactions leads to an equation which contains several millions of terms; it is impossible in practice to extract any meaningful theoretical understanding from the equation until most of the terms have been eliminated through a judicious systematic approximative simplification. Only then is it possible, e.g., to isolate the dominant resonant structure for microwave transmission from the myriad smaller perturbative effects which complicate it. Correspondingly, the instrumental methodology for microwave transmission probing and the mathematical methods for extracting information from those probes must be such as to cohere with the approximations made, i.e. such as to capture and reveal just the dominant mode transmission characteristics. Thus we should think of the domain of plasma physics as a complex, interacting hierarchy of practices, methods and theories, ranging all the way from detailed experimental techniques such as the use of Langmuir probes through methods of applied theory formation, experimental design and statistical information extraction, to generalized theoretical methods of Hamiltonian mechanics, perturbation expansions and the like, then finally to the most general methods, e.g. the use of logic for tests of consistency, exercise of rational choice in defense of the valuable goals of scientific research sketched above and so on.

The regulatory systems model of science illustrated by plasma physics research applies to all levels and timescales. A nice illustration is provided by the nature and status of observation in science. We see in the historical development of science two great interactive feedback cycles connecting sensory observation and theory/method development. On the one side, the use of the unaided senses leads to the development of both theory and method which in turn leads to restrictions on the use of unaided sensory observation. The development of method leads to the restriction to systematic observational procedures. The development of theory leads to the specification of the conditions for illusion and hallucination in unaided observation and to the still more widespread conditions for the stochastic appearance of observational error and, thereby, to the restriction of unaided observation to circumstances which minimize these epistemically deleterious occurrences. Thus it is that scientists increasingly restrict themselves to the observation of simple pointer readings, and preferably even numbers on computer outputs, since under these conditions they almost never make an error. All the remainder of the theoretically important features of systematic observation are then literally built into the theoretical design and engineered implementation of the observing instruments themselves. On the other side, the use of unaided senses leads, by the development of theory, to the development of observational techniques and instruments which enhance our natural sensory capacities, or even replace them entirely. The light microscope and telescope are examples of the former, radar and travelling electron microscopy are examples of the latter. In this way the senses come to be increasingly deeply embedded in the complex evolving collection of theories and instruments which comprise science and come to play an increasingly restricted and specialized set of roles within that complex.

We can understand this developmental process as one aspect of science as an evolving regulatory or control system, in this case the evolution of our control over the acquisition and disposition of information. It is the overall objective of the system, in this case to achieve maximal quantity of information of the maximum theoretically specified interest and with the maximal epistemic reliability, which is the means through which we can understand the evolution of the overall scientific process. Once again it is more fruitful to think of science as a complex cognitive system as a whole in which the role played by any one component, in particular the senses, can only be understood in the systems context in which it occurs.

4. A Conservative Modification of Empiricist Methodology:
   Bootstraps, and their Systems Methodological Generalization

One component of the empiricist paradigm is the claim that evidential confirmation of a hypothesis is purely a matter of the logical relations between the statement of the hypothesis and the statement describing the evidence. For example a common rule was: E is a logical instantiation of H. Approaches of this kind, pursued vigorously over the previous six decades, have not succeeded. The basic problem, as noted above, is that methodological choice and appraisal are theory dependent. The net result is that the confirmational value of a piece of evidence in relation to a specific hypothesis is a function of the wider theoretical context in which that relation is embedded. Confirmation, just like method, explanation, approximate truth and a host of other cognitive constructions within science, are functions of theory. Again one has the sense of a complex system evolving as a whole, rather than a neat set of initial building blocks from which an edifice is constructed 'upwards' logical step by logical step.

Recently, however, Glymour 1980 has proposed a specific modification of this approach, a conservative relativization of the confirmation relation to theoretical context. The basic idea of the bootstrap is that evidence E will confirm hypothesis H relative to some Theory T if E conjoined with T yields instances of H. Glymour's central example is the case of the inverse square law of gravitation; here descriptions of planetary orbits (the evidence E) together with Newton's general laws of motion (T) entail specific instances of the inverse square law of gravitation (H). Thus the data on each planet provide confirming instances for the view that there is a universal square law of gravitational attraction applying to all material objects. Notice that the same data could lead to very different consequences when combined with theories other than Newton's specific laws, hence the confirmation relation between it and the universal square law of gravitation is theory (Newton's laws) dependent. On the other hand, this is a purely logically specified rule, providing only a highly restricted theory dependence for the confirmation relation — it is thus a very conservative modification of the original empiricist ideal.

The bootstrap idea has been subject to two kinds of criticisms. On the one hand it is argued to be too wide, to include too many relations as confirmatory. Partly this is because one can develop trivial confirmation relations, e.g. when T = H. Partly this is because it is possible to provide examples where there is a bootstrap confirmation relation but no increase in the probability assigned to H. On the other hand, the conception has been criticized as too narrow, and that on four grounds: (i) Glymour has provided a working out of the notion only for those cases where mathematical formulae are involved and the instantiations are numerical; but this seems to exclude the social sciences and qualitative parts of the natural sciences. It may well be possible to overcome this objection. I let it pass here. (ii) The bootstrap confirmation relation fails to provide confirmation of theories as wholes. It applies only to a hypothesis related to some theory in a moderately superficial way. E.g. the universal square law of gravitation is not a fundamental principle of Newton's dynamics, rather it is simply a specific force law which is added to those principles. It is evidently not possible to non-trivially bootstrap confirm Newton's fundamental laws of motion themselves. For this reason, (iii) *a fortiori*, it is impossible to compare confirmation of competing fundamental theories of this kind. (iv) Bootstrap confirmatory support cannot in general be inherited. Thus, if H is bootstrap confirmed by E relative to T and H is then embedded in some more fundamental hypothesis H' which, together with evidence E', can explain H, then it does not follow that E bootstrap confirms H' relative to T. Nor if T is substituted by a more fundamental theory T' which can explain T does it follow that E will still confirm H relative to T'.

The upshot of these last three criticisms is that bootstraps cannot adequately handle confirmation of sufficiently unified theories. But it is precisely unified theories which are desirable in science and central to understanding both confirmation and explanation (cf. Hooker 1987, section 8.3.5.). The bootstrap relation evidently provides a limited, if valuable, insight into the complex of confirmatory relations in science, hobbled by its own empiricist conservatism. However, I believe that the bootstrap relationship generalizes to provide a valuable methodological insight into contemporary science and it is to this that I now turn.

Let us return to my plasma physics research work. In conjunction with my supervisor, I decided to study microwave transmission through plasmas, especially at or near electron cyclotron resonance frequency in plasmas confined by strong magnetic fields. I next designed a modified version of the plasma machines then operating in the laboratory. A considerable time was then spent testing the resulting machine performance and measuring the typical plasma characteristics produced, and adjusting the machine design until a reproducible plasma with suitable parameter values was achieved. Following this a variety of microwave transmission experiments were conducted and the results, delivered by a collection of various instruments, compared with the results of theory (as focused through suitable applied, approximate theories).

Conceptually, what was happening is something like this: *one aimed to develop a reproducible plasma which could be (i) conveniently provided with a theoretical model which clearly (ii) instantiated selected key physical principles such that (iii) these were selectively testable.* These three features are the key to the methodological procedure. One tries to provide controlled circumstances in which one can, as it were, make specific aspects of nature relatively more available to selective testing than the many other aspects which are in fact part and parcel of the complex reality under study.

How is the laboratory situation selected? Which circumstances are conveniently providable with theoretical models instantiating ...? Well, the choice depends on four factors: (1) the particular sorts of principles which have been selected for investigation; (2) the mathematical structure of the primary physical theory T whose key principles are under examination; (3) the corresponding structures of the auxiliary basic theories which are also required to be jointly applied in order to understand the situation; (4) the theories of the measuring instruments which may be used and which specify which features of physical systems can be measured, under what conditions they can be measured, and with what validity they may be measured (see also Hooker 1987, Chapter 4).

Suppose, e.g., I am interested in the accuracy of the theoretical description of microwave behavior at frequencies at or near the cyclotron resonance frequency in a plasma. The basic theory, T, here is classical electromagnetic field theory. (One has already convinced oneself that known quantum effects will be very small and that therefore the correct approximate simplification of quantum electrodynamics is in fact the classical theory.) But electromagnetic behavior cannot simply be studied in isolation, it is always necessary to consider the interaction of electromagnetic fields with matter; thus one requires also the theory of plasma dynamics, itself a joint application of electromagnetic field theory, electrodynamics and statistical mechanics. In addition, to understand the way in which the plasma behaves as it interacts with the walls of the containing vessel it is necessary to add a theory of metal surfaces (and here we need to begin with a simplified version of a quantum mechanical treatment) and some useful macroscopic thermodynamic relations. Thus in this case a primary auxiliary theory T_A will be the conjunction of all these theories, except T. (The thermodynamic relations are suitably approximately consistent with the statistical mechanics despite the fact that statistical mechanics is well known not to adequately reproduce a variety of testable macroscopic thermodynamic phenomena). The theories used in the various instruments will again include electromagnetic theory for the description of the microwave senders and receivers, the collection of theories appropriate to Langmuir probes as described in section 3 and those for a variety of related techniques which we can leave unexplored here. Label the conjunction of all these theories T_I.

Having assembled this collection of theories one can start the delicate dialectical business of deciding whether a plasma produced in a particular way can in fact be perspicuously theoretically characterized and, if it can, whether it is suitable for the study of cyclotron resonance and, if it is, whether instruments are available which could conceivably produce a sharp investigation of the phenomenon. It is essential to add in here a diverse array of initial data D_1 which includes readings of plasma characteristics on machines of related design, constraints on the production of magnetic fields and plasma electrical discharges with the engineering technology available to the laboratory and so on. One then juggles the design of the machine, keeping the foregoing data in mind, until it is likely that a plasma structure emerges which, so T and T_A informs us, can be suitably characterized for study and, according to T_I, is suitably measurable.

The net result is a choice of a particular machine design, producing on each 'firing' a plasma characterized by a particular theoretical model drawn from a class of theoretical models T_M which differ only in some subset of their parameter values, and an array of plasma parameter values (for temperature, density, etc.) given by D_I (initial data). Let us label I_L the set of approximations and parameter value assumptions used to apply T and T_A to their laboratory object so as to form the theoretical models T_M. Similarly, label i_c the corresponding set for T and T_I used to derive D_I. T_M and D_I are then combined to provide a characterization $T_{M,E}$ of the specific plasma produced on a particular occasion of some experiment E. Finally from $T_{M,E}$ and T_I one can deduce the theoretically expected measurement results for the feature of interest, D_P (predicted data). In general this last measurement will be distinguished from those measurements which led to D_I, but may on occasion utilize the same physical processes and devices. (In the microwave research, e.g., it turned out to be convenient to measure the plasma density distribution using microwave transmission, but in a regime well away from cyclotron resonance and where microwave behavior had already been carefully studied. But that methodology was adopted only after the density results were independently confirmed using other instruments, such as Langmuir probes.)

These predictions can then be compared with the actual results of measurements, D_2. Any discrepancy will be fed back to the study of i_c, I_L and, in the last instance, to T itself. But until it has been decided that an irresolvable clash has emerged, there will follow a process of cross-checking and mutual adjustment for consistency among the approximations used in T and T_A, the corrections applied for imperfections in T_I and the constraints D_1 in an effort to arrive at an overall consistent characterization of the laboratory situation in which D_P indeed agrees with D_2. Only when this kind of feedback stability has been achieved does a scientist say that he really understands the laboratory situation with which he is working. Let us call this process dialectical coherency, D_C.

When understanding has been achieved there ensues a methodologically most important phase of the scientific investigation. New diagnostic instruments are brought to bear on the plasma and on the microwave behavior, and new sampling methodologies and other methodological procedures are used with the old instruments themselves (e.g. using amplitude modulation diagnostics to check cyclotron frequency rather than merely phase delay). One then asks whether it is possible to achieve an understanding of the laboratory plasma which not only achieves dialectical coherence for each fixed collection of instruments, but *whose theoretical characterization is cross-situationally invariant across the various individually coherent laboratory arrangements*. If this can be achieved, then we have a strong basis for saying that the theoretical behavior of interest is not only well understood but has been objectively established.

If it is impossible to obtain dialectical coherence with several different instrumental arrangements, or if it is impossible to obtain cross-situational invariance to any significant degree, then methodological attention will fall on the possibility that there is an error in T_A or T. This will initiate a series of comparative investigations of the applications of T and T_A to other kinds of laboratory situations in an effort to isolate the location of the fault by picking up a pattern of recurrent failure to achieve dialectical coherences. But note that an isolated failure to achieve dialectical coherence will instead focus attention on either inappropriate I_L, i_c, or mistaken D_1. And an isolated failure of cross situational invariance will do the same.

The reason for these judgements has itself to do with a certain kind of cross-situational coherence, namely a fault in T_A or T must not only appear in this isolated case but be such as to be disguised or submerged in all the others. This discussion raises a number of important and complex methodological issues which it would be worth pursuing. Why does isolated failure of dialectical coherence focus attention on incorrect initial conditions? Just when does it do this and when ought one to look elsewhere? Cross situational invariance itself, which forms the focus of the comparison of coherent situations, has widespread methodological implications and connects to such key concepts as explanation and objectivity in science. (For the connection to explanation see Hooker 1987, section 8.3.5 and for the discussion of objectivity see ibid. 8.3.8.) Approximations are methodologically central to scientific understanding of a complex world and they contain many surprises when the details are worked out. Just as the search for a purely formal model of method has failed, so too I believe that the search for a purely formal theory of approximate truth must fail — approximateness is theory dependent. (For an elegant summary of one group of approximations which show complex structure see Rohrlich 1988, cf. Hooker 1981. For the discussion of approximate truth see Hooker 1987, sec. 8.4.3.9.) However these and other methodological issues raised cannot be pursued here.

Here I want to focus attention on the move from T_M and D_I to $T_{M,E}$. Since T_M follows from T and T_A, we can think of $T_{M,E}$ as derived by the application of D_I to T and T_A. And in the methodological setting $T_{M,E}$ is an instantiation of some particular physical principle under test (e.g. cyclotron resonance behavior of microwave radiation). Thus the derivation of $T_{M,E}$ has the general structure of a bootstrap. It is best thought of as a quasi- or partial bootstrap because in general only those parameters will have specified fixed values which are specifically relevant to deducing the specification of the measurable data D_P which bears on the testing of the physical principle in question.

This setting reveals a more general methodological significance for the bootstrapping operation. Whereas the usual empiricist hypothetico-deductive structure moves from data via theory to the prediction of further data, a bootstrap moves from data via theory to the development of a specific class of theoretical models. In short, bootstrapping is generally to do with building testable models for laboratory methodology; it is one component of the complex methodological structure sketched above. It is, I suggest, methodologically misleading to lift it out of this complex context to make it play an independent role as a confirmation principle. Doing so can be made plausible for carefully selected cases, as Glymour's example of Newtonian gravitation does, but this is to misdirect attention away from the complex systems structure of contemporary methodology in physics. And note that in its full context bootstrapping is consistent with the hypothetico-deductive structure, this latter structure remains the overall form for the development of the predicted measurements D_P. What is wrong with hypothetico-deductivism again lies in its fragmentation and simplification of complex methodology, in its destroying of the systems structure and the key features of the feedback processes involved (such as cross-situational invariance) in the pursuit of purely formal relations. Conversely, however, in articulating the bootstrap relationship Glymour has brought to philosophical consciousness a ubiquitous and important component of scientific methodology. But it is the overall specification of methodology as a complex control system for the development of physical understanding that is the more appropriate focus for further investigation.

5. Methodological Regulation: Two Examples

In the conference address I drew attention to the complexity of the methodological hierarchy of physics through two further examples: (i) methodological operationalism vis-a-vis methodological realism, and how they are consistent, and can even be mutually supportive, and (ii) the fundamental methodological role of invariance structures and spaces, e.g. Minkowski and Hilbert spaces, and the methodological ideals of intelligibility to which they relate. But length constraints preclude their presentation here.

6. Conclusion

This completes my reflections on methodological features of contemporary physics. They have been designed to show the necessity and usefulness of the notion that methodology is to be thought of as a complex many-leveled regulatory hierarchy of principles, interacting with theory also viewed as a complex many-leveled hierarchy. This in turn requires that we view science as a complex regulatory system for creating content, and creating content which exhibits a variety of valuable regulatory features.

Notes

[1] I have been developing elements of this approach over the past two decades. The major philosophical components are summed up in Hooker 1987. Their insertion into a more explicitly evolutionary context is summarized in Hahlweg/Hooker 1988a.

[2] For a brief review of plasma diagnostic techniques see e.g. Glasstone/Lovberg 1960. A working through of the many theories involved in one well-known scientific instrument, the cloud chamber, is presented in Hooker 1987, chapter 4, Appendix 2. (N.B. Interchange C4 and C5 in the diagram.)

[3] See also Hooker 1987, chapters 4, 5, 8. Of course this is true even of those simple sciences based on elementary unaided observation.

References

Glasstone, S. and Lovberg, R. H. (1960). *Controlled Thermonuclear Reactions*. New York: Van Nostrand.

Glymour, C. (1980). *Theory and Evidence*, Princeton, N.J.: Princeton University Press.

Hahlweg, K. and Hooker C. A. (1989). "Evolutionary Epistemology and Philosophy of Science", in *New Issues in Evolutionary Epistemology*, K. Hahlweg and C. A. Hooker (eds.). New York: State University of New York Press.

Hooker, C. A. (1981). "Towards a General Theory of Reduction, Part I: Historical Framework", Dialog XX: 38-59.

_ _ _ _ _ _ _ _. (1987). *A Realistic Theory of Science*. New York: State University of New York Press.

Rohrlich, F. (1988). "The Logic of Reduction: The Case of Gravitation". Preprint, Department of Physics, Syracuse University, June 1988.

van Fraassen, B. C. (1985). "Reply" in *Images of Science: Essays on Realism and Empiricism*, P. M. Churchland and C. A. Hooker (eds.). Chicago: University of Chicago Press.

Part VIII

DESCARTES

Descartes and Method in 1637

Daniel Garber

University of Chicago

The *Discours de la Méthode* and the three essays that were published with it, the *Dioptrique*, the *Météores*, and the *Géometrie* make up a very curious book. The very title page emphasizes the preliminary discourse, and that discourse, the *Discours de la Méthode*, emphasizes method, the importance that method had for Descartes in making the discoveries he made, the importance that the method Descartes claims to have found will have for the progress of the sciences and for the benefit of humankind as whole. Descartes is not, of course, telling us that we are obligated to follow his method; the *Discours* is, after all, proposed "as a story, or, if you prefer, as a fable." (AT VI 4).[1] But Descartes expects that we will all see the light, the light of reason, of course, and follow his example. It is curious, then, that Descartes gives the reader only brief hints of what that method is, four brief, vague, and unimpressive rules that, taken by themselves, would hardly seem to justify Descartes' enthusiasm, not to mention a whole discourse in their honor. Furthermore, explicit methodological concerns are hardly in evidence in the *Dioptrique*, the *Météores*, and the *Géometrie*, which are, Descartes claims, "essays in this method," as he identifies them on his title page. Indeed, one is hard pressed to find much evidence of the method at all after 1637, either explicit discussions of the method or explicit applications of the method in any of Descartes' writings, published or unpublished. Very curious.

These observations raise quite a number of questions about the development of Descartes' thought and the state of his program as of 1637. In this essay I shall address two of these questions: (1) What precisely was the method Descartes had in mind in 1637, when he sang its praises so enthusiastically? and (2) Why does that method appear so little in the publications of 1637 and appear to drop out altogether after that? Briefly, I shall argue that the method of 1637 was just the method Descartes had put forward more clearly in the earlier *Regulae*, or, at least, the dominant method that shows through the latest stages in its composition. But, I shall argue, perhaps by 1637 and certainly after that the method began to show its limitations, and the method that was one of Descartes' first discoveries, one of his first inspirations proved itself inadequate to the mature program that it led Descartes to undertake. Obviously there is not the time to present the detailed discussions these questions require. But I shall try to present in broad strokes one way of understanding the development of Descartes' methodological thought as he passed from youth to maturity.

1. The Method of the *Regulae*

I have claimed that the method of 1637 is essentially the method of the *Regulae ad Directionem Ingenii*, and to make good on that claim, we must first turn to that work. The *Regulae*, started as early as 1619 and abandoned in 1628, is a very difficult work;

PSA 1988, Volume 2, pp. 225-236
Copyright © 1989 by the Philosophy of Science Association

despite its superficial organization, it is often strikingly unmethodical and disorderly for a work that is supposed to be Descartes' most systematic exposition of his method. It is blatantly a work in progress that never progressed to anything like a finished draft, and the text we have shows obvious signs of having been picked up and put down at different times throughout the period of composition.[2]

To begin unraveling Descartes' complex thought on method in the *Regulae* we must look to the earliest strata of the work, where Descartes sets out the goal of the method in passages likely to have been written in November 1619, shortly after the dreams of November 10.[3] Descartes wrote:

The goal (*finis*) of studies ought to be the direction of one's mind (*ingenium*) toward having solid and true judgments about everything which comes before it. (AT X 359)

But, Descartes thinks, such "solid and true judgments", such "certain and indubitable cognition" (AT X 362) as he calls it in the following rule, can come to us in only two ways, through intuition, or through deduction, "for in no other way is knowledge (*scientia*) acquired." (AT X 366). And so, what we should seek is "the undoubted conception of a pure and attentive mind" (AT X 368), intuition, or through "a certain movement of our mind (*ingenium*)", inferring one thing from another, (AT X 407), a chain of such intuitions, grounded in intuition, what Descartes calls a deduction. To find such knowledge, though, Descartes thinks that we need a method (AT X 371). But what is this method and how is it supposed to work?

From the start Descartes had in mind a two-stage process. Writing in Rule V, again from late 1619[4], Descartes summarized his rule of method as follows:

This rule is observed exactly if we reduce involved and obscure propositions step by step to simpler ones, and thus from an intuition of the simplest we try to ascend by those same steps to a knowledge of all the rest. (AT X 379)

The rule of method thus has two steps: a reductive step in which "involved and obscure propositions" are reduced to simpler ones, and a constructive step, in which we proceed from an intuition of the simplest back to the more complex.[5]

But what in concrete terms does the method come to? How is it to be used in specific cases? It is quite possible that Descartes' vision in 1619 was cloaked in poetic enthusiasm, and that Descartes himself may not have had a clear and distinct conception of precisely how the method was to work in actual practice. But matters are clarified considerably in an example Descartes gave late in the composition of the *Regulae*, where the programmatic bravado of the earlier years is translated into practice. The example I have in mind is the anaclastic line, which Descartes discusses in Rule 8. The example is closely connected with optical investigations Descartes probably undertook between 1626 and 1628, and probably dates from that period.[6] But whenever it dates from, it displays what I take to be the method as Descartes understood it at the time he abandoned the project of the *Regulae*, and represents what he means by method in 1637, I shall argue.

The argument is set out in Table 1. The problem Descartes poses is that of finding the shape of a line (lens) that focuses parallel rays of light to the same point. (AT X 394). Now, Descartes notices — and this seems to be the first step in the reduction — "the determination of this (anaclastic) line depends on the relation between the angle of incidence and the angle of refraction." But this question is still "composite and relative," and we must proceed further in the reduction, to the question of how this refraction is caused by light passing from one medium into another, which in turn raises the question as to "how the ray penetrates into the whole transparent thing, and the knowledge of this penetration presupposes that the nature of illumination is also known." (AT X 394-5). But in order to understand what light is, Descartes claims, we must know what a *potentia naturalis* is. This is where the reduction ends. At this point Descartes seems to think that we can "clearly see through an intuition of the mind" (AT X 395) what a natural power is, something that we understand in terms of local motion.[7] Once we have this intuition, we can then begin the constructive step, and follow back in order through the questions raised until we have answered the original question, Q6 allowing us to deduce an answer to Q5, Q5 allowing us to deduce and answer to Q4, and so on until we reach an answer to Q1, deductively.[8]

Anaclastic Line Example

(Q1) What is the shape of a line (lens) that focuses parallel rays of light to the same point?

(Q2) What is the relation between angle of incidence and angle of refraction (i.e., the law of refraction)?

(Q3) How is refraction caused by light passing from one medium into another?

(Q4) How does a ray of light penetrate a transparent body?

(Q5) What is light?

(Q6) What is a natural power?

Intuition: A natural power is ...

Construction: The construction consists in traversing the series of questions from (Q5) to (Q1), deducing the answer to each question from that of the preceding question.

Table 1

This example suggests the following conception of method. Methodical investigation begins with a question. This question is reduced to simpler questions, questions whose solution is presupposed for the solution of the question originally posed. That is, Q1 is reduced to Q2 if we must answer Q2 before we can answer Q1. Descartes thinks that this process leads us from more specific questions to more general, more basic, more fundamental questions, from the shape of a specific lens, to the law of refraction, to the nature of light and the nature of a natural power. Descartes thinks that when we follow out this reductive series, we will ultimately reach an intuition. Here the reduction ends and construction begins. At this point we can turn the procedure on its head, and begin deducing answers to the questions that we have successively raised, in an order the reverse of the order in which we raised them. When we are finished it is evident that we shall have certain knowledge; the answer arrived at in this way will constitute a conclusion deduced ultimately from an initial intuition.

Descartes' strategy here is extremely ingenious. The stated goal of the method is certain knowledge, a science deduced from intuitively known premises. What the method circa 1628 gives us is a workable procedure for finding an intuition and a deductive chain from which such knowledge can be attained. This workable procedure is the reduction of a question to more and more basic questions, questions we can identify as questions whose answers are presupposed for answering the question originally posed. The efficacy of the reductive step of the method depends upon a substantive assumption about knowledge, the assumption that knowledge, *scientia*, is structured in a very specific way, a doctrine that Descartes seems to have held in one form or another since the crucial night of 10 November 1619. (cf. AT X 204, 215, 255, 361). It is not at all clear how in detail Descartes may have seen this structure in 1619. But Rule 12 of the *Regulae* suggests that by 1628 Descartes saw all knowledge grounded in intuitions about the very most general features of the world, thought, extension, shape, motion, existence, duration, etc. On these intuitions are grounded layers of successively less general propositions. If knowledge is structured in this way, then Descartes thinks we should be able to solve any problem in an orderly and methodical way, tracing step by step through the layers, back toward the intuition, and deducing down from there to the question that interests us.

My account of method in the *Regulae* ignores numerous complexities. I have said relatively little about the stages of composition of the *Regulae*, and nothing about simple natures or the use of experiment in the method (though I will touch on that a bit later). Also, I have said nothing about the *mathesis naturalis* of Rule 4 and Rule 14, which some argue is identical to the method (they are wrong, I think, but it would take me too far from my main theme to argue the case). And finally, I have neglected to mention the numerous other assumptions, largely unwarranted, I think, that Descartes needs to make his method work. But what I have given is an account of the method Descartes held in 1628 or so when he stopped work on the *Regulae* to turn to the construction of his system.

2. The Method in 1637 and Beyond

It is this method, I claim, that Descartes had in mind in 1637 when he published the *Discours de la Méthode*. The method I attributed to Descartes in the *Regulae* agrees well enough with the brief exposition of the method, the four rules that he gives in part IV, particularly the second and third of those rules. The second rule require us "to divide every difficulty ... into as many parts as one can", and the third suggests that I conduct "my thoughts in order, beginning with the simplest objects, those easiest to understand, to rise little by little, as by degrees, up to the most composite knowledge". (AT VI 18). Although I think that commentators have not, in general, grasped the method Descartes recommends in the *Regulae*, the obvious correspondence between the two-stage method Descartes recommends in the *Regulae*, the reduction followed by the construction, and these two rules he recommends in the *Discours* have often been noted.[9]

But our account of Descartes' views on method in 1637 cannot stop with the *Discours*. The *Discours*, Descartes tells his correspondents, does not contain a genuine exposition of the method. Descartes wrote to Mersenne on 20 April 1637, discussing the title he chose for his *Discours*:

I did not call it *Treatise on the Method* but *Discourse on the Method*, which is the same as *Advice on the Method*, to show that I did not intend to teach it, but only to discuss it. Since, as one can see from what I said about it, it consists more in practice than in theory. (AT I 349)

The method, then, "consists more in practice than in theory". But what "practice" should we examine? In writing to P. Vatier about the method on 22 February 1638, Descartes makes a suggestion: "I have given a glimpse (of the method) in describing the rainbow." (AT I 559). The reference here is to the eighth discourse of *Les Météores*, where Descartes gives his celebrated account of the rainbow. Descartes there tells the reader that

I could not chose material more appropriate to show how, by the method I use, one can arrive at knowledge which those whose writings we have didn't possess. (AT VI 325)

A study of the account Descartes gives of the rainbow is, then, supposed to teach us the method by showing us how it works "in practice". But, as Descartes also told Vatier, "the matter is very difficult" (AT I 559), and it is not at all easy to discern the clear outlines of Descartes' method in the mists that surround the rainbow.

Very briefly[10], Descartes uses a combination of reasoning and experiments with spherical flasks of water and with prisms to lead him to an explanation of the two principal features of the rainbow, the colors we see, and the fact that the rainbow is always composed of two separate regions of color that are separated by a dark space. From the experiments with prisms, Descartes concludes that colors arise when light is bent in refraction; he argues that the color is caused by the tendency to rotate that the balls receive during refraction. From observations on the flask of water, and calculations made with the help of his law of refraction, Descartes concludes that most sunlight passing into a droplet of water and following certain paths will leave at one of two angles, about 42° and about 52°. Putting these together we have the rainbow, roughly speaking, two regions of color that arise through refraction, separated by about 10°.

It is not easy to extract a method from this morass of detail. But one can see in Descartes' account, though, the outlines of the two-step method, the reductive step followed by the constructive step that constitutes the core of Descartes' method in the *Regulae*. In Table 2, I have rearranged the argument a bit to show its structure. In the diagram, Q1 - Q5 constitute the reduction of the initial question, the ordered succession of questions Descartes would answer to answer the question originally posed. The reduc-

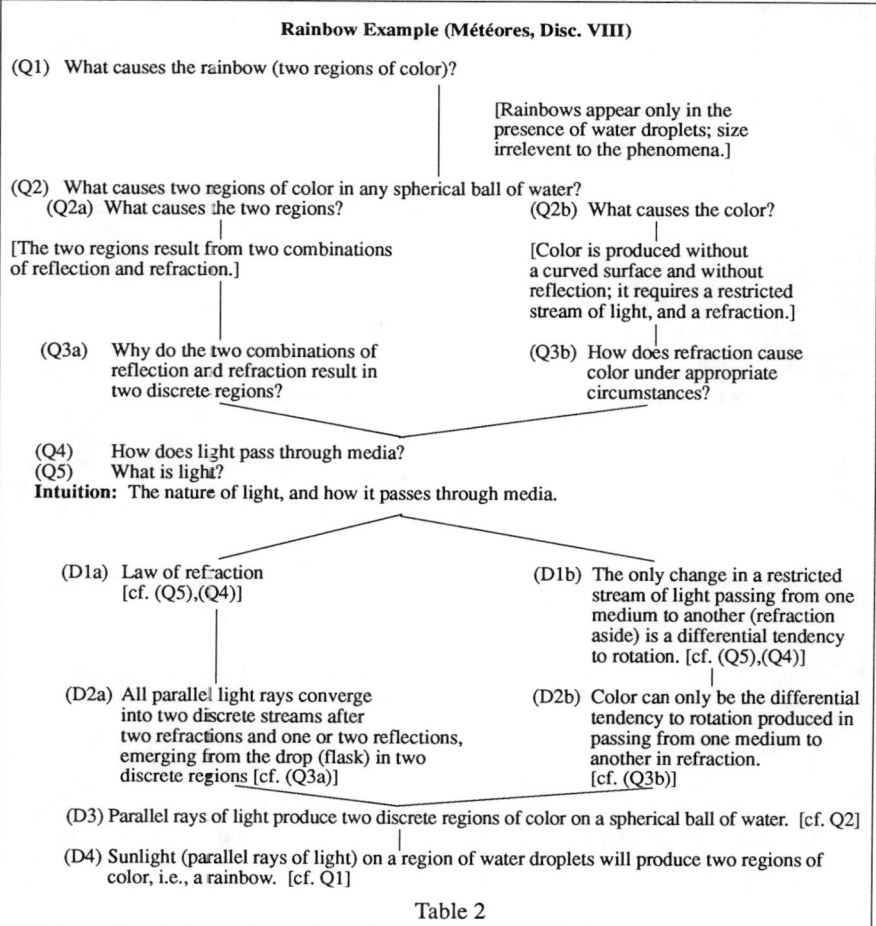

Rainbow Example (Météores, Disc. VIII)

(Q1) What causes the rainbow (two regions of color)?

[Rainbows appear only in the presence of water droplets; size irrelevent to the phenomena.]

(Q2) What causes two regions of color in any spherical ball of water?
 (Q2a) What causes the two regions?

[The two regions result from two combinations of reflection and refraction.]

(Q2b) What causes the color?

[Color is produced without a curved surface and without reflection; it requires a restricted stream of light, and a refraction.]

 (Q3a) Why do the two combinations of reflection and refraction result in two discrete regions?

(Q3b) How does refraction cause color under appropriate circumstances?

(Q4) How does light pass through media?
(Q5) What is light?
Intuition: The nature of light, and how it passes through media.

 (D1a) Law of refraction [cf. (Q5),(Q4)]

(D1b) The only change in a restricted stream of light passing from one medium to another (refraction aside) is a differential tendency to rotation. [cf. (Q5),(Q4)]

 (D2a) All parallel light rays converge into two discrete streams after two refractions and one or two reflections, emerging from the drop (flask) in two discrete regions [cf. (Q3a)]

(D2b) Color can only be the differential tendency to rotation produced in passing from one medium to another in refraction. [cf. (Q3b)]

 (D3) Parallel rays of light produce two discrete regions of color on a spherical ball of water. [cf. Q2]

 (D4) Sunlight (parallel rays of light) on a region of water droplets will produce two regions of color, i.e., a rainbow. [cf. Q1]

Table 2

tion ends with an intuition about the nature of light and how it passes through bodies. (In the *Météores* Descartes actually appeals to the *Dioptrique*, where the nature of light is presented as a hypothesis. See AT VI 331 and 84.). D1 - D4 constitute the constructive stage of the method, where Descartes goes from the intuition to the solution to the problem originally posed. (Again, Descartes actually appeals here to results that are derived in the *Dioptrique*, the law of refraction. See AT VI 337 and 93ff.) Viewed in the way I suggest, the account of the rainbow nicely displays the method of the *Regulae* that we saw in the anaclastic line example.

While it does not pertain to my main theme in this paper, I should point out how nicely the rainbow example shows us the role of experiment in the method. It is worth noticing that experiment enters only in the reductive stage of the method; it helps us to find a path from our complex question to the intuition from which that question will be answered. But the answer itself is purely deductive, and makes no use of experiment. The chain of causes that the Cartesian scientist seeks in reason is exemplified in the causal connections one finds in nature itself. Insofar as these later connections are open to experimental determination, we can use experiment to sketch out the chain of causes and find what causally depends on what, and thus use the connections we find in nature as a guide to the connections we seek in reason. It may not be obvious how we can go deductively from the nature of light to the rainbow, but poking about with water droplets, flasks, and prisms may suggest a path for the deduction to follow. But this does not make the deduction superfluous; while it may be through effects that we are led to causes, it is from knowledge of those causes and the deductions we can make from them that our knowledge actually derives.

But let us return to the main theme. An examination of the rainbow example, Descartes own announced example of the method in 1637, strongly suggests that the method Descartes had in mind in the context of the *Discours* and *Essais* was just the method of the *Regulae*, the two-stage method we saw in the anaclastic line example, the reduction of a question to an intuition, and the construction of an answer to that question from intuition. But it is interesting to note that the account of the rainbow we have been discussing is probably not contemporaneous with the *Discours*; while it is impossible to be certain, it is likely that that portion of the *Météores* dates from late 1629, not long after the *Regulae* were set aside.[11] When the account of the rainbow appears eight years later in the *Météores*, it appears as a kind of ghost from an earlier period. This is significant, for the account of the rainbow is the only place in the *Essais* where Descartes explicitly calls attention to the method of his preliminary discourse and it is the only example of the method to which he calls attention in his letters. Though the method "consists more in practice than in theory" (AT I 349), the practice in question is not exemplified elsewhere in the *Essais*. The *Essais* are, of course, not unconnected in Descartes' mind with the method. Descartes wrote to Mersenne in April 1637:

> I call the treatises that follow *essays in this method* because I claim that the things they contain couldn't have been found without it, and that through [what I have discovered] one can know the value [of the method]. (AT I 349)

But though they show the value of the method, the *Essais* do not themselves use the method. Writing to Vatier on 22 February 1638, Descartes explains this as follows:

> I couldn't show the usage of the method in the three treatises which I gave because [the method] requires an order for investigating things that is very different from that which I thought necessary to use to explain them. (AT I 559)

The mode of exposition Descartes chose for the *Dioptrique* and the *Météores* was, of course, hypothetical. Both works begin with appropriate hypotheses which ground the results which follow, hypotheses that allow Descartes to show some of his results, but in a way that does not force him to divulge the first principles of his system, something for which, he believed, the public was not ready. (AT I 370, 563-4; AT III 39).

But it is interesting to note that even in other contexts, where Descartes is not so shy to divulge the foundations of his system, the method is hardly in evidence. In the earlier *Le Monde*, for example, Descartes divulges more of the foundations of his physics than he will do later in the *Essais*; though certain metaphysical issues that Descartes was concerned with at the time are hidden, he is forthcoming about the nature of matter, the nature of light, the role God plays in maintaining the world and so on. But it is difficult to discern the formal method of the *Regulae* in *Le Monde*. And when a few years later Descartes sets aside his scruples and presents his system in its full and proper form in the *Meditationes* and the *Principia Philosophiae*, there is as little of the method as there was in *Le Monde* and the *Essais*. Descartes does continue to build on first principles, to start with intuition (ultimately the *cogito*), and deduce down from there, from the more general and more metaphysical to the more specific. This, of course, is a feature of the order of reasons that M. Guerolt emphasized, and it looks a great deal like the constructive stage of the method. And his continued interest in experiment and observation show that he is still keenly aware of the problem of finding an appropriate path from intuition to the solution of particular problems in physics. For example, Descartes' interest in embryology and sexual reproduction in the 1640s was, I think, part of an attempt to bridge physics and biology[12]; perhaps an understanding of how purely mechanical processes result in the genesis of a new organism will show how in nature organisms arose from lifeless matter, Descartes thought. But there is little evidence of the earlier method in his later writings, in particular, little evidence of a formal reduction that precedes the constructive deduction of conclusions from intuition, the reduction that earlier had constituted the principle secret of the method. This is so even in the *Meditationes*, a work whose origin, Descartes tells the Doctors of the Sorbonne, was in part a response to a request for him to apply his celebrated method to God and the soul (AT VII 3), a work written in the analytic mode, Descartes tells the second objectors, a work that is intended to follow "the true way through which a thing was ...discovered". (AT IX-A 121) In the *Meditationes*, the intuition that constitutes the starting place of the deduction, the *cogito*, is carefully prepared in the First Meditation. But the preparation does not seem to be a reduction in the precise sense of the term. The First Meditation does many things; it clears away prejudice, establishes a standard for certainty, introduces the problem of knowing our creator as the essential preliminary for any further knowledge. But it does not sketch out the sequence of steps to be followed in resolving a question, the way a proper reduction is supposed to do.

One cannot deny that the *Meditationes* are carefully organized and ordered. But even though there is an order, this order is not evident to the meditator at the start of the *Meditationes*. From the *cogito* of Meditation II to the end of Meditation VI there are numerous places where the meditator tries to lead the argument into a dead-end, where the meditator begins an argument that simply does not pan out. For example, at the beginning of Meditation III, the meditator tries to demonstrate the existence of the external world, before giving the proof for the existence of God. However, at this point in the argument of the *Meditationes*, the meditator doesn't have the means to make his proof work, and he must set the question aside, and turn to another question, to God, leaving aside the question of the external world until Meditations V and VI, where it can finally be settled. These digressions are very important to the structure of the *Meditationes*.[13] The *Meditationes* are addressed, in part, at a very specific audience that Descartes knows quite well, at the unconverted, readers full of prejudice for their senses and for the material world, and these digressions are very important to convince them that the arguments that they are inclined to accept, arguments that take for granted a faith in the senses, arguments that take for granted a priority in belief in the external world—these arguments Descartes wants to show are mistaken. And the way he does this is by letting the meditator try to show that they work, only to show that they don't. This is the function of the failed argument for the existence of body in Meditation III, for the wax example of Meditation II, and for other arguments in the *Meditationes*.[14]

There is method in this procedure, to be sure, but the method is not the strict method of the *Regulae* or the *Discours*. In the method of his youth, the reductive step brings it about that the entire constructive step is sketched out, before the first deduction, and the construction follows directly the order as set out in the reductive step; this, indeed, is the main point of having a reduction, so that one will know how to perform the deduction, and this reductive step is the principal secret of the method, what makes it work. In this method there is no place for the sort of digressions so important to the purpose of the *Meditationes*. Furthermore, it is not clear to me that one can isolate one well-defined question to which Descartes addresses himself in the *Meditationes*—a minimal condition required for the method of the *Regulae* to apply. In this sense one can say that the meditator doesn't follow the method, nor can the reader learn the method by reading the *Meditationes*.

In claiming that Descartes' later works do not display his earlier method I am making a controversial claim, one that would be challenged by other scholars, who have claimed to find the method of the *Regulae* and *Discours* in the *Meditationes*, at very least.[15] But even if they are right (and I don't think they are), it is beyond dispute that Descartes himself hardly mentions his method after the *Discours* and the letters that immediately follow its publication. If method is the key to knowledge and the key to the later Cartesian system (as it seemed to be in 1637), Descartes himself does not call attention to that fact. Indeed, when the earlier method comes up in his later writings, it has a decidedly subordinate role in his thought. In the Letter to Picot that serves as a preface to the French *Principes de la Philosophie* of 1647, Descartes recommends that the student of philosophy "ought to study logic, not that of the schools, but that which teaches one how to conduct his reason to discover truths that one doesn't know." (AT IX-B 13-14). It would be good, Descartes says, for him to "practice the rules concerning easy and simple questions for a long time" until "one acquires a certain habitude for finding truth in these questions." (AT IX-B 14). But in this respect, the method has roughly the status of the provisional morality (which immediately precedes it in the Letter), one of those preliminaries that should be undertaken by the student of nature before undertaking the serious business of philosophy; it is an exercise useful primarily in sharpening the mind and helping us to recognize truth, an exercise that has in 1647 roughly the same role that Descartes earlier gave the scholastic logic he otherwise rejected in the *Regulae*. (cf. AT X 363-4). Whatever it is, it is clearly not nearly as important to Descartes in the 1640s as it appeared to be in 1637.

How can we account for these curious facts? How can we explain the fact that method gets such little play in Descartes' actual scientific writings? How can we explain the fact that the method, the central focus of his theory of knowledge and inquiry in 1637, is barely mentioned in later writings? My claim is this. The method was Descartes' first inspiration, and was crucial for the first results of his system, as he reports in the *Discours*. But, I shall argue, two basic changes in Descartes' thought made the method largely obsolete.

3. The Abandonment of the Method

Descartes' method first dates from mid and late November 1619, it is generally agreed, the days and weeks following the crucial three dreams. It had been a year since Descartes had run into the young Isaac Beeckman in Breda and had had his first sympathetic introduction to the mechanical approach to nature that was later to dominate his thought. Beeckman was not a systematic thinker, it is fair to say, in the sense that he had no large, overarching system. He was interested in the solution of individual problems, and it is with the discussion of individual problems, taken one by one, that his notebooks are filled.[16] It was this way of doing physics and mathematics that he transmitted to the young Renatus Picto, or René du Perron as Descartes styled himself at the time. Beeckman's notebooks show that Descartes worked on a number of such problems, set for him by his older friend, including the behavior of water contained in a vessel, the behavior of a body in free-fall, and numerous problems in music and geometry. (AT X 46-78)[17]. It is not surprising, then, that the method that Descartes first attempted to formulate in November 1619 and developed in the 10 years that followed, that method was a method for the solution of individual problems. To make use of the method, we must first set a specific question for ourselves, what is the shape of a lens with such-and-such properties, or, what causes the rainbow, or whatever. Once we have a specific question, we can then apply the method, reduce the question to simpler questions until we reach an intuition, and deduce back up to an answer to the question originally posed. The method is a method for doing science as, say, Beeckman conceived of it, as a series of discrete questions about the natural world.

But as I noted earlier in discussing the method of the *Regulae*, the method presupposes a certain conception of the structure of knowledge. All knowledge, for Descartes, is interconnected, grounded ultimately in a small number of intuitively knowable propositions from which all else follows deductively. This, as I noted, was one of the things that Descartes probably learned in that night of enthusiasm in November 1619, and this is the key to the method he developed in the years following. It is precisely because all knowledge is interconnected in this way that the method is possible, that it is possible to take a question and reduce it to an intuition from which an answer could be deduced. But this very doctrine that makes the method possible leads to its demise. For if all knowledge is interconnected, then what we should be doing is not solving individual problems, but constructing the complete system of knowledge, the interconnected body of knowledge that starts from intuition and comes to encompass everything capable of being known. Though he may have recognized this implication from the start, in 1619, it will will be 10 years before he begins such a system, in 1629 with the first metaphysics, unfortunately lost, followed immediately by the composition of *Le Monde*. This project is what is striking and distinctive about Descartes' mature system, the system we find sketched in parts IV and V of the *Discours* and developed in the *Meditationes* and *Principia Philosophiae*. Unlike others, Galileo, for example (cf. AT II 380), Descartes' strategy is to start not with individual questions, but to start at the beginning, with the intuitively graspable first principles that ground the rest, and progress step by step from there downward to more particular matters. No longer a mere problem solver, Descartes has become a system-builder.

But as a system-builder, what role can there be for a method whose goal is the solution of individual problems? With this crucial change in Descartes' conception of scientific activity, a change motivated by the same doctrine of the interconnection of knowledge that motivated his method, the method becomes obsolete; or if not obsolete, at very least it is less central than it once had been.

This is one way in which the evolution of the Cartesian program led to the demise of method. But there is another consideration as well. The method is a procedure for answering a question by deducing an answer from intuition; it tells us how to find the appropriate intuition, and how to find a path from intuition to the answer we seek. But this naturally leads us to the question as to why we should trust intuition and deduction at all, and why we should consider them to be the only source of knowledge. The history of Descartes' struggle with this problem is very complex and I can only sketch briefly some of the most important stages. The issue first arises in Rule 8, in what is probably the very last stage of the composition of the *Regulae*, just before Descartes set it aside in 1628 or so.[18] There it appears as the "noblest example" of the method, something useful for preventing ourselves from attempting to solve problems beyond our ability, or preliminary to the actual use of the method in the same way that it is useful for the blacksmith to build sturdy tools before attempting to make horseshoes. (AT X 395-8) It is not altogether clear what status this investigation had in 1628, whether it was a mere preliminary to investigation, or part of the system of knowledge itself, whether it is essential in order for us to have any knowledge, or whether it is simply a practical suggestion about where we might begin. The status of this investigation of the grounds of intuition and deduction, now clear and distinct perception, is also difficult to determine in the *Discours*, where it appears to be something of a digression, part of the answer Descartes gives the reader who objects to the metaphysics presented earlier in part IV of the *Discours*. If you continue to think, contrary to what I have written (Descartes says) that the material world of suns, stars, and tables is better known than God and the soul, reflect on the fact that if we did not know that God exists and is not a deceiver, then we could know nothing at all. (AT VI 37f). But by the 1640s, the epistemological project, the investigation of the trustworthiness of intuition and deduction, clear and distinct perception has become the essential foundation of all knowledge; the tree of knowledge from the 1620s, grounded in the intuition of the most general notions concerning extension and thought, has grown roots, and it is essential for us to understand the foundation of our beliefs in God's veracity for us to have any genuine knowledge at all. (AT IX-B 14).

But with this change, method by itself can no longer lead us to genuine knowledge. The reductive stage of the method starts with a question, and then takes us back to questions presupposed, until we finally reach an intuition. But when the reduction has reached an intuition it goes no farther. Thus the method of the *Regulae* can at best give us imperfect knowledge, the moral certainty we get when we take intuitions for granted, rather than the metaphysical certainty that comes from knowing that our clear and distinct perceptions are the creation of a God who does not deceive. (AT VI 37-8; AT VII 141).

I have argued that two changes in Descartes' thought conspired to make the method of the *Regulae* largely inapplicable to the system of knowledge he hoped to build: (1) the change from a problem-solving conception of scientific activity to a system-building conception; and (2) the adoption of the idea that intuition cannot be taken for granted and must be validated, and that this is the essential preliminary to any system of knowledge. Given these features of Descartes' mature system of the 1640s, it is no wonder that Descartes came to have relatively little use for the method of the *Regulae*, oriented to the solution of individual problems, and incapable of leading us to metaphysical certainty.

But these considerations may also help to explain why the method does not appear very much in the *Essais* either. Descartes suggests that he does not use the method in the *Essais* because he did not want to reveal the foundations of his physics. But this cannot be the whole story. On the one hand, he was quite capable of using his method without revealing any more of his foundations than he wanted to, as he did in the rainbow example. And, on the other hand, even when he was not especially worried about exposing the foundations of his physics, as in the earlier *Le Monde*, method seems to play no substantive role. My own suspicion is that many of the changes in Descartes' thought that make the earlier method obsolete in the 1640s may also be present as early as the first sketch of metaphysics Descartes attempted in 1629-30. Not that Descartes was aware of what was going on. I suspect rather that starting perhaps as early as the winter of 1629-30 method is no longer relevant to his scientific practice, and is simply not used in the project. And so, I suspect, when in the mid 1630s he sat down to gather together some of this material and present it in his *Essais*, the method had as little role to play as it did in its sources. But Descartes was perhaps not aware of the change his thought had undergone. And so, when he sat down to compose the preliminary discourse, out came the *Discours de la Méthode*, a work that expressed a conception of scientific inquiry that belonged to the earlier and somewhat more naive M. du Perron. Though cognizant of the fact that his *Essais* did not make much use of the method, he may not have realized why. This is a conjecture, of course, and a very risky one. But it is indisputable that as his system grew, perhaps from the first metaphysics of 1629-30 onward, method became, first in practice, and then after 1637 in theory, less and less important to Descartes.

If I am right, then, the volume Descartes published in 1637 is a curious work, a beginning and, at the same time, an end. It is, of course, the beginning of Descartes' public career, and it contains a preliminary sketch of the full system he will develop in succeeding years, the interconnnected body of knowledge grounded in first philosophy. But it also marks the end of a period, the last work in which Descartes was to emphasize method as the key to knowledge. Descartes in 1637 is, in a sense, like the butterfly, emerging from his cocoon, spreading his new wings to dry in the sun, not yet fully aware that he is no longer a caterpillar.[19]

Notes

[1]References to the works of Descartes will be given in parentheses in the text, following Adam and Tannery (1965-1975), abbreviated 'AT', with the volume number followed by the page number.

[2]For questions of dating, see Weber (1964). Weber believes that Descartes wrote the text of the *Regulae* in ten discrete "phases". Though the stages of composition are difficult to distinguish with such exactitute, Weber's arguments are often useful for dating particular passages of the *Regulae*. I have also used datings suggested by John Schuster in Schuster (1980).

[3]See Weber (1964), sections 13, 55.

[4]See Ibid. sections 19, 55.

[5]To avoid confusion, I am breaking with most commentators, who refer to these as the analytic and synthetic steps, following the distinction Descartes draws in the *Second Replies*. See, for example, Serrus (1933), chapter I; Rodis-Lewis (1971), pp. 173ff; Beck (1952), chapter XI, etc. This is a distinction that has little direct relevance to the two stages of the method of the *Regulae*. In the *Regulae* we are dealing with a distinction between two parts of a single method; though they are distinct, both are necessary for a true application of the method. But the distinction between analysis and synthesis in the *Second Replies* is completely different. There we are dealing with different ways of setting out a single line of argumentation, and we must choose one or the other. See AT VII 155-56 or AT IX-A 212.

[6]See Schuster (1980), pp. 55, 88 n.68.

[7]In Rule 9 Descartes says that "if one wants to examine" this natural power, one must turn to "local motions of bodies" (AT X 402). According to Schuster, this passage probably dates from the same period as the anaclastic line example; see Schuster (1980), p. 87 n.60.

[8]For a lucid discussion of the anaclastic line example, see Costabel (1982), pp. 53-58.

[9]See, for example, Gilson (1967), p. 205; Beck (1952), pp. 149ff; etc.

[10]This paragraph is a summary of the argument given in *Discours* VIII of *Les Météores*, AT VI 325ff.

[11]In a letter of 8 October 1629 Descartes wrote to Mersenne that he is working on "a small treatise which will contain the explanation of the colors of the rainbow (to which I have given more care than all the rest) and generally the explanation of all sublunar phenomena." (AT I 23). This small treatise will doubtless become the *Météores*, and Descartes' words to Mersenne suggest that Descartes probably solved the problem of the rainbow before October 1629.

[12]See, for example what Descartes says in the *Discours*, AT VII 45-46, and the commentary in Gilson (1967), p. 393ff.

[13]Other important digressions include the celebrated piece of wax in Meditation II and the argument for the existence of the external world drawn from the faculty of imagination at the beginning of Meditation VI.

[14]See Garber (1986) for an elaboration of some of these themes.

[15]See Serrus (1933), chapter III; Beck (1952), chapter XVIII; Schouls (1980), chapters IV-V; etc.

[16]See the summary of the questions on which Beeckman worked between 10 November 1618 and January 1619, when he was in contact with the young Descartes, as given in his journal, AT X 41-45. The very variety of the questions is very impressive. But it is also interesting to note the form of the articles in his journal. Most often the questions are quite specific and deal with specific phenomena. There is little interest in any comprehensive system encompassing all of the sciences.

[17]It is also probable that a part of the *Cogitationes Privatae* containing the *Parnassus* presents problems that Descartes discussed with Beeckman. See AT X 219-48, and Gouhier (1958), pp. 15, 24.

[18]See Schuster (1980), pp. 58-59.

[19]A French version of this paper appeared as "Descartes et la Méthode en 1637" in Grimaldi, N. and Marion, J.-L. (eds.), *Le Discours et sa méthode: Colloque pour le 350e anniversaire du Discours de la Méthode*. Paris: J. Vrin, 1987.

References

Adam, C. and Tannery, P. (eds.) (1965-1975), *Oeuvres de Descartes* (*nouvelle présentation*). Paris: J. Vrin.

Beck, L.J. (1952), *The Method of Descartes*. Oxford: Oxford University Press.

Costabel, P. (1982), *Démarches originales de Descartes savante*. Paris: J. Vrin.

Garber, D. (1986), "*Semel in Vita*: the Scientific Background to Descartes's Meditations", in A. Rorty (ed.), *Essays on Descartes' Meditations*. Los Angeles and Berkeley: University of California Press, 1980.

Gilson, E. (1967), *René Descartes: Discours de la méthode, texte et commentaire, 4me éd*. Paris: J. Vrin.

Gouhier, H. (1958), *Les premiéres pensées de Descartes*. Paris: J. Vrin.

Rodis-Lewis, G. (1971), *L'oeuvre de Descartes*. Paris: J. Vrin.

Schouls, Peter (1980), *The Imposition of Method*. Oxford: Oxford University Press.

Schuster, J.A. (1980) "Descartes' *Mathesis Universalis*, 1619-28," in S. Gaukroger (ed.), *Descartes: Philosophy, Mathematics and Physics*. Sussex: the Harvester Press, 1980.

Serrus, C. (1933), *La méthode de Descartes et son application à la métaphysique*. Paris: Librarie Félix Alcan.

Weber, J.-P. (1964), *La constitution du texte des Regulae*. Paris: Société d'édition d'enseignement supérieur.

Geometry, Time and Force in the Diagrams of Descartes, Galileo, Torricelli and Newton

Emily R. Grosholz

Pennsylvania State University

Mathematics plays a central role in the description, explanation and manipulation of natural phenomena. To what extent, and how and why mathematics applies to nature is a problem that has long occupied philosophers. Descartes, Leibniz, Kant, Mach and Poincaré, to mention some of the most distinguished names, offer global solutions to this problem that are based on deep-lying metaphysical assumptions. In this essay, I would like to suggest an alternative approach, which is piecemeal rather than global, and historical before it is metaphysical.

I want to propose, first, that the question of applied mathematics be recast as a question about how mathematics and physics, a physics "always already" mathematized, are partially unified at various points in history, in such a way that they can share certain items, problems and methods while nonetheless remaining quite distinct. And, second, I suggest that these unifications may be quite heterogeneous and variable over time. If we consider Archimedes' combination of geometry and statics, the Bernoulli's development of the theory of differential equations in the service of mechanics, and the twentieth century marriage of logic and computer technology in all their rich historical detail, we may decide that the factors that distinguish them are philosophically more interesting than those they have in common. In short, perhaps philosophers ought to reason upwards from case studies of the multifarious ways in which mathematical and physical domains can be joined, before they attempt to make global pronouncements about that union.

The present essay is one such case study. Its focus is the project of a geometrical physics presented in Descartes' *Principles of Philosophy*, and which also apparently depends on his *Geometry*. My argument is that Descartes' conception of the "order of reasons" both organizes and impoverishes his mathematics and his physics, and moreover interferes with his own intention to unify them in a novel and more thoroughgoing way. Then I will show how his contemporaries Galileo and Torricelli profit from possibilities that Descartes' strongly reductionist methodology has excluded, and so manage to achieve a deeper unification of mathematics and physics, specifically with respect to the parameters of time and force.

In a famous passage in the *Principles of Philosophy*, Part II, section 11, Descartes announces a kind of identity between the object of geometrical study, space, and the object of physics, *res extensa*, matter. He writes: "If we concentrate on the idea which we have of some body, for example a stone, and remove from that idea everything which we

PSA 1988, Volume 2, pp. 237-248

know is not essential to the nature of body; we shall easily understand that the same extension which constitutes the nature of body also constitutes the nature of space, and that these two things differ only in the way that the nature of genus or species differs from that of the individual." (Miller and Miller 1983/4, p. 34). A physical object is thus precisely and merely an instantiation of a region of three-dimensional Euclidean space. And he reiterates this identification of the subject matter of physics and that of geometry quite strongly in the last section of Part II of the *Principles*: "For I openly acknowledge that I know of no kind of material substance other than that which can be divided, shaped, and moved in every possible way, and which Geometers call quantity and take as the object of their demonstrations." (Miller and Miller 1983/4, p. 77).

Behind this conflation of physical with mathematical objects lies Descartes' desire to purify physics of the anthropomorphic, intentional and psychological qualities and explanations of late Renaissance science, of the iron filings which long for the loadstone and the planets which keep turning themselves to avoid a sunburn on one side. And therein lies the origin of the austere and noble project of modern science, to know nature apart from the accidents of human perception and perhaps even our conceptual categories, to know nature without projecting a human face on it. Matter, according to Descartes, has no attributes besides the quantifiable ones that stem from its extendedness. The essence of matter is therefore also mathematical; matter has an inherent structure articulable as Euclidean geometry. And since Descartes' great mathematical work, the *Geometry*, is designed to reformulate and rationalize Euclidean geometry, his project of mathematizing nature would seem here to find its appropriate grounding.

Philosophical historians of the seventeenth century however have not failed to notice that Descartes' successes in mathematizing physics are few and far between: he enunciates a theorem in optics, a characterization of inertial motion and something like a conservation of momentum principle for impact. The rest of his physics, expounded at length in the *Principles*, the *World* and the *Treatise of Man* (for Descartes, biology was a part of physics), is surprisingly qualitative and inexact; it advances by loose analogy and a quite imaginative array of "mechanisms".

I would like to explain this puzzling incongruity at the heart of Descartes' project as a consequence of the way his method leads him to organize geometry and physics, for his strongly reductionist and therefore homogenizing way of arranging a subject matter impedes the development of his mathematics and generates severe conceptual problems for his physics. And since it also leads him to conflate the domains of geometry and physics, it ultimately tends to block their unification.

Descartes holds that a subject matter can and should be organized according to "the order of reasons," as a linear progression from simples to complexes, such that each item in the chain is known without the aid of succeeding items and all items are known solely on the basis of those that precede them. Thus a subject matter begins with items that are known in themselves, and becomes a progression of successively more complex entities that are simples in some kind of association. The simples for Cartesian geometry, as he announces on the first pages of the *Geometry*, are rectilinear line segments, and their form of association is proportions (Smith and Latham 1954, pp. 2-5). The complexes are then problems (like the trisection of the angle, or the instances of Pappus' problem discussed below) and higher algebraic curves (like the conic sections, and some cubics), which Descartes ranges into hierarchies in Books II and III of the *Geometry*.

Descartes' choice of starting points, straight line segments which alone can stand as terms in relations of proportionality, helps to streamline and reorganize geometry. The closed algebra of line lengths that opens the *Geometry* allows him to use algebra in the solution of geometrical problems, and to define the multiplication of line segments for any number of factors, where classical Greek mathematics limited the number of factors to three. Thus, to choose the example central to the *Geometry*, he is able to solve in a more general way than possible theretofore a locus problem from the canon of classical antiquity called "Pappus' problem".

Nonetheless, Descartes' choice of such "simple" and homogeneous starting points also excludes certain other items from serving as terms in his proportions: areas and volumes (which the Greeks used in their proportions), and curves and infinitesimals (which contemporary mathematicians were using in their proportions). Areas and volumes, aside from a fleeting mention in one paragraph which is left undeveloped, are never treated in the *Geometry*, nor are infinitesimals. Curves are in one sense the subject matter of Book II of the *Geometry*, but Descartes' treatment of them keeps sliding over to the problems, nexuses of straight line segments, in which they figure merely as constructing curves. As is well known, Descartes banishes transcendental curves from geometry, but it is equally true that the tendency of his reductive method to pull the complexes back to the simples, in this case, the investigation of curves back to nexuses of straight line segments, severely restricts his investigation of higher algebraic curves.

Nor does Descartes see in a clearly focussed way the power of his own innovation in the *Geometry*, which allows the investigation of curves as algebraic-geometrical-numerical hybrids, a multivalence that is the key to their investigation and their employment in physics in the latter half of the seventeenth century. Descartes' intuitionism makes him distrust the formal apparatus of algebra, and his insistence on the homogeneity of his subject matter leads him to conflate numbers and line segments, rather than presenting them as two distinct domains linked by a shared structure, that is, the algebra of line lengths given at the beginning of Book I. For Descartes, curves are primarily constructing curves, to be used to construct points in the solution of problems, and hence he never recognizes their rich multivalence and variety.

The mathematics of Descartes' *Geometry*, pruned and homogenized as it is by the demands of method, is then curiously inapt for the representations required by contemporary physics. I can best illustrate my argument thus far by comparing his treatment of Pappus' problem with Galileo's analysis of free fall in the *Two New Sciences*. Pappus' problem is really a (countably infinite) set of problems, each of which asks for the determination of a locus whose points satisfy one of the following conditions illustrated by the following diagram: Let the d_i denote the length of the line segment from point P to L_i which makes an angle of ϕ_i with L_i. Choose α/β to be a given ratio and a to be a given line segment (Bos 1981). (Diagram 1)

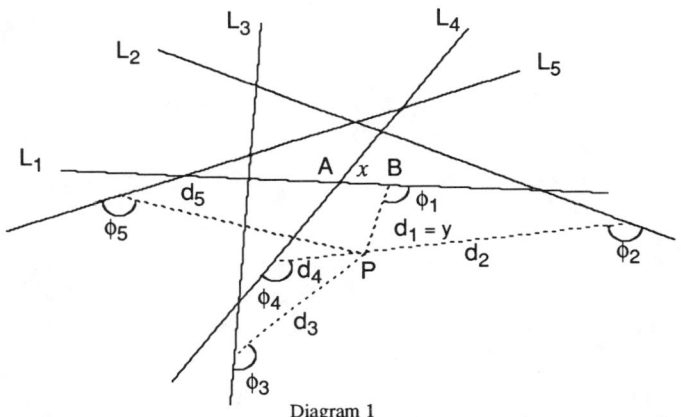

Diagram 1

The problem is to find the points P which satisfy the following conditions. If an even number (2n) of lines L_i are given in position, the ratio of the product of the first n of the d_i to the product of the remaining n d_i should be equal to the given ratio α/β, where α and β are arbitrary line segments. If an odd number (2n - 1) of lines L_i are given in position, the ratio of the product of the first n of the d_i to the product of the remaining (n-1) d_i times a should be equal to the given ratio α/β. (The case of three lines is an exception, since it arises when two lines coincide in the four line problem.) There are in fact points which satisfy each such condition, and they will form a locus on the plane. Since the Greeks interpreted the products of two and three lines respectively as areas and volumes, Pappus, reporting on the work of Apollonius, hesitated to generalize beyond the case of six fixed lines.

In the middle of Book I, Descartes describes his attack on the problem and then proudly announces, "I believe that I have in this way completely accomplished what Pappus tells us the ancients sought to do," (Smith and Latham 1954, pp. 26-7) as if he had solved the problem in a thoroughgoing way for any number of lines. While it is true that his combination of algebraic-arithmetical and geometrical results produces an important advance in the solution of the problem, his treatment of the problem in the *Geometry* is hardly complete, for he adds only one new locus, the solution to a five-line version of the problem, to those already known, that is, the conic sections, which correspond to four-line versions of it.

Descartes' explanation of how he proposes to solve this problem occurs at the end of Book I, accompanied by a diagram of its four line version (Smith and Latham 1954, pp. 26-35). (Diagram 2) He chooses y equal to BC (which is d_1) and x equal to AB, and shows how all the other d_i can be expressed linearly in x and y. Then the proportions defining the conditions for the cases of 2n and (2n - 1) lines given above can be rewritten as equations in x and y. For 2n lines, the equation will be of degree at most n; for (2n-1) lines, it will be of degree at most n, but the highest power of x will be at most (n-1). (For 2n and (2n-1) parallel lines, where y is the sole variable involved, the result is an equation in y of degree at most n.)

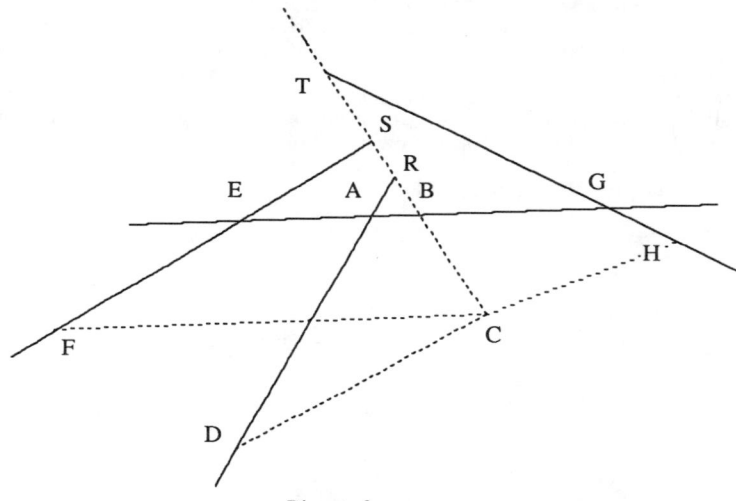

Diagram 2

The point-wise construction of the locus is then undertaken as follows. One chooses a value for y and plugs it into the equation, thus producing an equation in one unknown, x. For the case of 2n lines, the equation is of degree at most n and for (2n-1) lines, it is of degree at most (n-1). The roots of this equation can then be constructed by means of intersecting curves which must be decided upon. This procedure, infinitely iterated, generates the curve point by point (Bos 1981). Thus it seems that Pappus' problem has been reduced to the geometrical construction of roots of equations in one unknown: the construction of line segments on the basis of rational relations among other line segments.

The combination of algebra and geometry is worthy of note. By rewriting the conditions of the problem as an equation, Descartes has converted it from a proportionality involving lines, areas or volumes as terms (as it was in the classical formulation) to an equation about line segments. It has become an algebraic problem to which techniques for simplifying and solving equations can be applied. Yet it has not ceased to be a geometric problem as well, though the algebraic conversion has altered the geometry. The diagram is still centrally present, though it only involves rectilinear line segments; no areas or volumes intervene here or as the focus of auxiliary constructions. The auxiliary constructions will be instead the construction of each x for a given y, using certain chosen constructing curves as well as various geometrical theorems. Because the problem can be viewed simultaneously as algebraic and geometrical, results from both domains can be brought to bear upon it, thus organizing and facilitating its solution. Also, Descartes' abstract statement of the procedure seems to impose no limits on the number of lines initially "given in position," and thus to escape the strictures of the Greek formulation entirely.

And yet Descartes' solution to Pappus' problem does not result directly in the discovery and investigation of a rich collection of new algebraic curves. For one thing, the construction of roots of higher algebraic equations, and therefore points on the relevant loci, is not as easy as Descartes' naive faith in the step-wise advance of reason envisages. And secondly, the strongly reductionist drift of Cartesian method keeps deflecting his interest from curves back to nexuses of line segments. The very first thing that Descartes says about his approach to Pappus' problem may seem odd if we expect him to be primarily interested in the loci which the problem generates (Smith and Latham 1954, pp. 24-5). (I have corrected the translation of the word "*degré*".)

First, I discovered that if the question be proposed for only three, four, or five lines, the required points can be found by elementary geometry, that is, by the use of the ruler and compasses only, and the application of those principles which I have already explained, except in the case of five parallel lines. In this case, and in the cases where there are six, seven, eight, or nine given lines, the required points can always be found by means of the geometry of solid loci, that is, by using some one of the three conic sections. Here, again, there is an exception in the case of nine parallel lines. For this and the cases of ten, eleven, twelve, or thirteen given lines, the required points may be found by means of a curve of level next higher (*degré plus composé*) than that of the conic sections. Again, the case of thirteen parallel lines must be excluded, for which, as well as for the cases of fourteen, fifteen, sixteen, and seventeen lines, a curve of level next higher (*degré plus composé*) than the preceding must be used; and so on indefinitely.

For in this passage, he is classifying cases of the problem not by some feature of the locus generated, but rather by what kind of curve can be chosen in the point-wise construction of the locus, that is, in the construction of the line segment x given the relevant equation in x and y and a definite value for y. He iterates this classification of cases at the very end of Book I in more explicitly algebraic terms. Otherwise stated, this classificatory scheme does not pertain to curves (describable by indeterminate equations in two unknowns) but to problems (describable by determinate equations in one unknown). Curves intervene in this passage only as constructing curves; each higher level of problem will require a constructing curve of higher level (*degré plus composé*).

The diagram just given contains no hint of the locus, only the nexus of line segments with their specified relations to an arbitrary point C of the locus. The implied auxiliary construction would be the determination of the line segment x (for a given value of y) by means of certain constructing curves. The official subject of these diagrams are line segments; constructing curves also intervene, but we have never been told *quid juris* and they are not what the diagram is about. Descartes' commitment to the methodological presupposition that his geometry begins with proportions among rectilinear line segments only, structures and narrows his whole enterprise.

Thus, Descartes' very first announcement concerning the Pappian cases is that he has discovered a way to generalize the classification of problems. And he reiterates this classification at the end of Book I and indeed on the final page of the *Geometry*. However, the announcement leaves vague what it means to say that the constructing curves which are required for each successive level of problems are of *"degré plus composé"*. Thus, Descartes' first way of grouping Pappian cases is supplemented by a second, which explicitly classifies loci. But I want to argue that the fact that this classification of loci is given second is quite significant. Descartes is not interested in classifying curves for their own sake; he undertakes the second classification in order to clarify the first, and regards the curves primarily as constructing curves.

Galileo's treatment of free fall in the Third Day of his *Two New Sciences* requires diagrams that could have no counterpart in Descartes' *Geometry*. Theorem I, Proposition I, states : "The time in which any space is traversed by a body starting from rest and uniformly accelerated is equal to the time in which that same space would be traversed by the same body moving at a uniform speed whose value is the mean of the highest speed and the speed just before acceleration began." (Crew and deSalvio 1954, pp. 173-4). (Diagram 3)

As Koyré points out, the genius of this diagram is that AB represents not the distance traversed (that role is played by the separate line CD) but the time elapsed. Galileo has

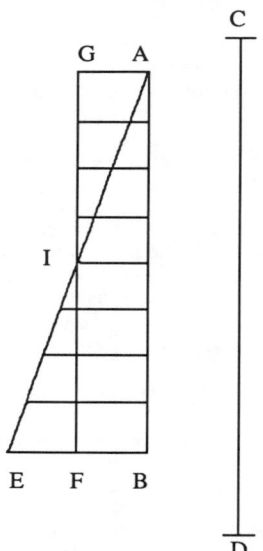

Diagram 3

wrested geometry from the geometer's preoccupation with extension, and put it in the service of the essentially temporal processes of physics (Koyré 1939, pp. 67-73). What I want to stress about the diagram is that it involves areas and a process like integration with respect to time; the parallels of the triangle AEB perpendicular to AB represent velocities, and the area of the triangle as a whole, taken to be a summation of instantaneous velocities, therefore represents distance elapsed. In other words, in this diagram distance is represented in two different ways, as the line segment CD and as the area of triangle AEB; because the second representation is a two-dimensional figure, it can exhibit the way that uniformly increasing velocity and time are related in the determination of a distance. Moreover, the triangle AEB is a summation of infinitesimal *momenti*, and is able to be so in virtue of the painstaking discussion of the possibility of continuously and uniformly accelerated motion that begins the Third Day. Koyré observes that Descartes the physicist rejects Galileo's arguments about the continuity of motion because of his commitment to instants (Koyré 1939, pp. 62-3); I want to point out that Descartes the mathematician has short-circuited the possibility that diagrams such as Galileo's have a place even in geometry.

So too with the diagram to Theorem II, Proposition II, which supplements the one just discussed. The theorem states: "The spaces described by a body falling from rest with uniformly accelerated motion are to each other as the squares of the time-intervals employed in traversing these distances." Once again there are two components to the diagram (Crew and deSalvio 1954, pp. 174-6). (Diagram 4) The line HI stands for the spatial trajectory of the falling body, but it is articulated into a sort of ruler, where the intervals representing distances traversed during equal stretches of time, HL, LM, MN etc., are marked off, forming the sequence of odd numbers, 1, 3, 5, 7... as Corollary I notes. AB represents time (divided into equal intervals AD, DE, EF, etc.) with perpendicular instantaneous velocities raised upon it, generating a series of areas. Distance traversed again has two distinct representations; this time one is geometric and the other numerical. The distinction is as important as the correspondence in the investigations of mechanics and mathematics leading from Galileo to the calculus at the end of the century, particularly in the reasoning that gave Leibniz his central insight into the calculus (Grosholz 1982). But Descartes' conflation of number and geometry stands in the way of such thinking. Nowhere in the *Geometry* does Descartes discuss how to combine the combinatorial patterns of number theory and geometrical results.

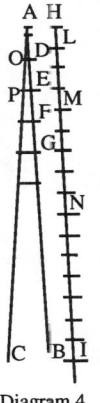

Diagram 4

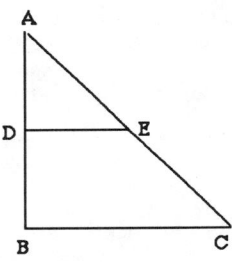

Diagram 5

As it happens, Descartes in his youth (in 1619 and again in 1629) considered the problem of free fall. Beeckman proposed a version of it to him, and in his private journal, Descartes sketched out a solution to it (Adam and Tannery 1964/74, X: 219f). (Diagram 5) Descartes, however, takes the line ADB to represent, not the time elapsed but the distance traversed. Since he considers the lines parallel to AB as representing velocity, triangle ADE stands for the "quantity of motion" expended as the body traverses AD, and the quadrilateral DECB the same as the body traverses DB; since the latter is three times the former, he concludes that the body moves through the second interval of distance three times as quickly. The reasoning thus leads him to suppose that in uniformly accelerated motion like free fall the velocity increases proportionally to the distance traversed, not to time. This was a common mistake in the late sixteenth and early seventeenth century; but Descartes' later attack on the problem in 1629 and his continuing criticisms of Galileo show that he never learned to see the crucial difference between taking time rather than space as the important parameter (Koyré 1939, pp. 40-5.)

And though in these early solutions, the geometrical representation is suggestively close to Galileo's, Descartes' interest in it never revives, especially after what Koyré calls the "revolution of 1630" when Descartes decides that knowledge should be reconstructed according to the "order of reasons." Up to this point, I have tried to indicate how Descartes' reconstruction of mathematics forecloses upon mathematical structures that might have been useful to him in the mathematization of physics. Now I want to argue more in the vein of Koyré that his reordering of physics also interferes with that project of mathematization.

In the *Principles of Philosophy*, the starting points, the simples, for Cartesian physics are bits of matter in uniform (unaccelerated) rectilinear motion. Bits of matter are associated when they share a common (uniform, rectilinear) motion; Descartes says repeatedly in sections 27 - 32 of Part II that the unity of a material body is precisely the common motion, or common rest, of its parts, and can be attributed to no other mode (Miller and Miller 1983/4, pp. 52-5). Material unities are deflected or disrupted when they collide with other bits of matter; collision for Descartes is the only kind of physical interaction. The three laws of motion given in Part II (sections 37 - 52) define inertial motion, which inheres in and signals the unity of material bits, and mathematize impact as seven rules, in each of which particles of various sizes and speeds collide along straight lines (Miller and Miller 1983/4, pp. 59-69). The consequent changes of speed are instantaneous, and conserve the total "momentum," the product of "bulk" and speed, of the system. Finally, since Descartes identifies space and matter, he denies the void; so the jostling of bits of matter packed together as a plenum creates vortices, matter flowing in circuits, which then serve as a further level of associated complexes.

This version of physics afforded the first systematic competitor theory for outmoded Aristotelianism in the seventeenth century; Descartes' *Principles* stood as the primary physics textbook until Newton's deliberately named *Principia* supplanted it. Descartes' articulation of inertial motion and of a conservation law for impact are scientific developments of the highest order. But his physics is also an impoverishment, primarily because it leaves out of account processes which are temporal and dynamical. Moreover, Descartes' field of geometry, so ordered, and his field of physics, so ordered, are strangely ill-suited to each other, an odd couple that doesn't really quite generate the new science.

For example, the shaped volumes that are supposed to serve as the geometrical genera for specific bits of matter have no explicit place in the *Geometry*, which is concerned primarily with straight line segments and secondarily with plane curves. And even if Descartes' geometry did contain such items, deep problems concerning the coherence of associated bits of matter that run all through the *Principles* reveal that geometrical shape, even when taken together with common motion, is insufficient as a principle of unity for physical objects.

By contrast, the straight line segments of the *Geometry* seem to correspond nicely to the inherently uniform, rectilinear motion of the bits of matter which are the simples in the physics. Descartes' quantified model of impact specifies that bodies in uniform rectilinear motion collide along straight lines, and that the change in their motion, to new uniform speeds according to formulae that preserve the total quantity of what we would call momentum, is instantaneous. In this model, however, there isn't really any interesting geometrical context for the line; it is not the side of a triangle, the diagonal of a parallelogram or the ordinate of a curve. Just the straight line itself, along which two material particles bump into each other, does not bring any further mathematics into play, which might illuminate the physical situation. So far, the links between Descartes' geometry and physics seem to be missing or trivial. And yet, having stipulated that all material particles are shaped volumes and that all interaction is collision and thus covered by the seven rules, Descartes appears satisfied that in principle the work of mathematizing physics is complete. (However, in his (forthcoming) Alan Gabbey argues that Descartes indeed envisaged a more complete physics, that would relate the precise microscopic phenomena of the *Principles* to the macroscopic world of machines, free fall, projectile motion, etc., more fully and mathematically.)

What about the curves that figure in the *Geometry*? Descartes discusses curves in the *Principles* only as the trajectories of bits of matter stemming not from the nature of matter or motion, but from external exigencies imposed by the plenum: motion in a plenum can only take place, if at all, in a circuit. And the boundary condition imposed by the existence of the plenum is not strong enough to determine what precisely the curve might be, so that then the peculiar geometrical properties of that curve might be exploited in the service of physics, as Newton exploits the properties of the ellipse in Proposition XI, Book I of the *Principia*, where he derives the inverse square law (Motte and Cajori 1934, pp. 40-2).

Moreover, Descartes' inability to focus on curves as algebraic-geometric-numerical hybrids contributes to his inability to regard curves as representative of the relations among continuously varying parameters. Nothing in Descartes' *Principles* is comparable to Galileo's famous analysis of projectile motion (Crew and deSalvio 1954, pp. 248-50), which takes the parabolic curve of a projectile's trajectory to express relations among time, distance, velocity and the acceleration of gravity, or to Newton's Proposition XI, where the elliptical trajectory of a point mass circling a center of force does much the same. The most significant employment of curves in early modern physics doesn't occur to Descartes. And of course some of the most important such curves were transcendental, curves which he had excluded from mathematics altogether.

Descartes' organization of physics excludes the investigation of accelerated linear and curvilinear motion. The problems concerning continuously varying forces which pose such thorny and fruitful problems for his contemporaries and successors is simply avoided. Significantly, Descartes' most mathematically sophisticated attempt to quantify physics occurs in a context where the temporal and dynamic dimensions of the subject matter are irrelevant; optics is very close to a pure physical geometry, with light rays playing the roles of lines. In a sense then Descartes never makes the transition from kinematics to dynamics, as his contemporaries Galileo and Torricelli succeed in doing. For Descartes, no physical parameter, including of course what he calls "force," varies in any essential or interesting way with time, and strictly speaking bodies never accelerate. (The account of the acceleration due to gravity near the surface of the earth that his vortex theory provides for macroscopic phenomena is too complicated to be quantifiable, as Descartes himself admits.)

Torricelli learned from his master Galileo the importance of the parameter time in the analysis of physical situations, and develops a more sophisticated account of percussion. He argues that the percussion of a falling object exercises an infinite force accumulated in the interval of time required for its fall, because any such finite interval contains an infinity of instants, in each of which the body exerts the simple impulsion of its weight; this infinite accumulation is extinguished as the object struck absorbs the shock in a finite interval of time, containing an infinity of instants as well. Thus equilibrium is reestablished. According to Torricelli's analysis, then, each *momento* requires an instant of time for its generation or its extinction. Though he supposes that *momenti* are finite and thus their summation is infinite, he reasons about force as a continuously varying magnitude, something like an integral taken with respect to time. Moreover, as Galileo also does on other occasions, Torricelli studies the continuous accumulation of impulsions as a limiting case of a finite number of successive small blows. (My exposition here is indebted to

DeGandt (forthcoming).) This is just the kind of reasoning Newton uses in Proposition I, Book I of the *Principia*, which is Newton's version of Kepler's Law of Areas (Motte and Cajori 1934, pp. 40-2). (Diagram 6)

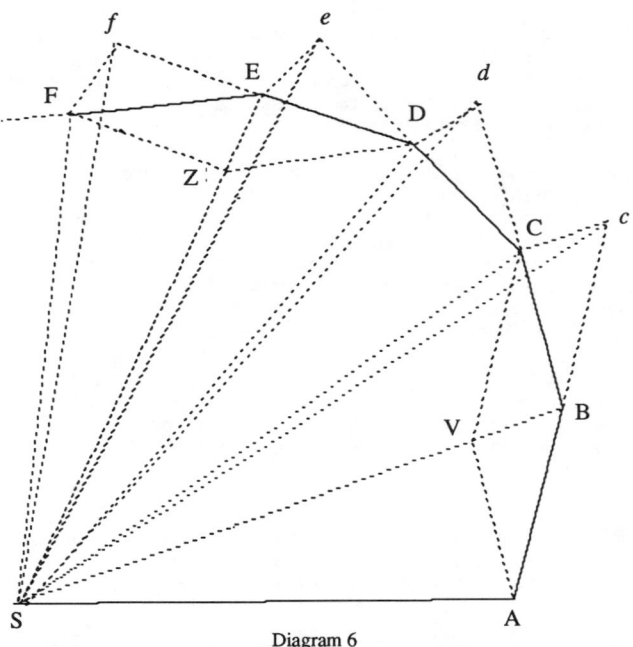

Diagram 6

The claim is: "The areas which revolving bodies describe by radii drawn to an immoveable center of force do lie in the same immovable planes, and are proportional to the times in which they are described." Newton's proof is illustrated by the figure. S is the center of force. A body proceeds on an inertial path from A to B in an interval of time; if not deflected, it would continue on in a second, equal interval of time along the virtual path Bc. However, Newton continues, "when the body is arrived at B, suppose that a centripetal force acts at once with a great impulse," so that the body does not arrive at c, but at C. Then cC (=BV) represents the deflection of the body due to the force; indeed, cC = BV becomes the geometrical representative of the force. The perimeter ABCDEF... represents the trajectory of the body as it is deflected at the beginning of each equal interval of time by discrete and instantaneous impulsions from S. Newton then uses the Euclidean theorem that triangles with equal bases and equal elevations have equal areas, to show that the area of triangle SAB = the area of triangle SBc = the area of triangle SBC; this equality extends to triangles SDC, SED, SFE... by the same reasoning, so that equal areas are described in equal times. We have only, Newton concludes, "to let the number of those triangles be augmented, and their breadth diminished in infinitum" for this result to apply to a continuously acting force and a curved trajectory.

Descartes' diagram of Pappus' problem is ambiguous because his treatment of it allows it to represent not only a geometric locus, but also an algebraic equation in two unknowns. Because it can be read in both ways, both geometric and algebraic results can be brought to bear on the problem, and their combination is the key to Descartes' success. Here I would like to urge an interesting parallel between the partial unification of two mathematical domains, and of a mathematical domain and a physical domain.

Newton's diagram is likewise ambiguous; it must be read both as a collection of finite lines and areas, where the perimeter is composed of rectilinear line segments, and as a collection of infinitesimal as well as finite lines and areas, where the perimeter is a curve. The first reading allows the application of Euclidean theorems to the problem. The second makes the diagram relevant to the kind of continuous motion and force Newton is interested in, as well as Kepler's Law of Areas. And it moreover centrally involves reasonings about proportions involving areas and curves, and infinitesimal lines and areas. Thus in this diagram, Newton can combine the resources of geometry and mechanics in the demonstration of a fundamental result. And when in Propositions VI and XI he builds into his diagrams the geometric structure of the ellipse on the one hand, and Galileo's result about free fall as well as a subtle representation of force on the other, the outcome of the combination is rich indeed.

The only diagrams in Descartes' *Geometry* that have physical significance are the "optical ovals" at the end of Book II (Smith and Latham 1954, pp. 114-149). When he discusses them in order to exhibit their optical properties, he generates them by point-wise construction. Nowhere does he give an equation for one of them, though he does use algebra to discuss their properties. Optics presents Descartes with a series of problems concerning the conditions under which reflected light rays, straight line segments, will converge at a given point. It's the construction of the point of convergence which interests Descartes, not the curve of the curved reflecting surface. Moreover, since for Descartes the propogation of light is instantaneous, the parameter of time doesn't enter into the conceptual situation at all.

And in the *Principles*, none of the diagrams have any geometry that hooks up in any interesting way with the results of the *Geometry*. The spheres and concentric circles that show up in many of them are mathematically inert and isolated; they have none of the suggestive algebraic structure that Descartes might have conferred on them if he had been interested in curves (or surfaces) as geometrical-algebraic hybrids. Furthermore, since the vortex hypothesis doesn't specify the geometrical shape of planetary circuits, the choice of circles as representations of trajectories is arbitrary and so disconnected with the mechanical context. The spherical shape of the Cartesian corpuscles, second element matter, is likewise arbitrary and unmotivated by his physical theory. There is simply no area of overlap, accompanied by appropriately bivalent diagrams, on which Descartes' mathematics and physics can meet and mingle.

The point I want to make about the geometrical-dynamical reasonings carried on by Galileo, Torricelli and Newton is that they take time as the central parameter and try to analyze force as a summation of infinitesimals; their diagrams involve proportions whose terms include areas, curved lines, numbers, infinite magnitudes, and infinitesimals (lines or areas); and they use curves, sometimes transcendental, to represent the continuously varying relations between physical parameters including time and force. All these aspects of their work Descartes excludes from his own, because of the strongly reductionist way in which he has restructured his geometry and physics. Thus he is never able to integrate mathematics and physics in fulfillment of the inspiring promissary note given in Part II of the *Principles*.

References

Adam, C. and Tannery, P. (eds.) (1964/74), *Oeuvres de Descartes*. Paris: Hermann.

Bos, Henk (1981), "On the Representation of Curves in Descartes' *Géométrie*", *Archive for History of Exact Sciences* 24: 295-338.

Crew, H. and deSalvio, A. (eds.) (1954), *Dialogues Concerning Two New Sciences*. New York: Dover.

DeGandt, F. (forthcoming), *"L'analyse de la percussion chez Galilée et Torricelli"*.

Gabbey, A. (forthcoming), "Descartes' Physics and Descartes' Mechanics: Chicken and Egg?".

Grosholz, E. (1982), "Leibniz' Unification of Geometry with Algebra and Dynamics", *Studia Leibnitiana* Sonderheft 13: 198-208.

Koyré, A. (1939), *Etudes Galiléennes: La loi de la chute des corps*. Paris: Hermann.

Miller, V.R. and Miller R.P. (eds.) (1983/4), *Principles of Philosophy*. Dordrecht: Reidel.

Motte, A. and Cajori, F. (eds.) (1934), *Principia*. Berkeley: University of California Press.

Smith, D.E. and Latham, M.L. (eds.) (1954), *The Geometry of René Descartes*. New York: Dover.

Science, Certainty, and Descartes

Gary Hatfield

University of Pennsylvania

It is difficult to determine the place of experience in Descartes' philosophy and science. His *a priori*, rationalist metaphysics would seem, on the face of it, difficult to reconcile with the explicit appeal he makes to sensory evidence in both his scientific practice and methodological remarks. Although earlier descriptions of Descartes as a pure a priorist in natural science (e.g., Koyré 1978, pp. 89-94) have rightly been rejected, it would be a mistake to embrace the other extreme, as does Clarke (1982), with his bold revisionist thesis that Descartes was, in actuality, an empiricist. And yet the possibility of combining the rationalist and empiricist elements in Descartes' thought has also seemed problematic, for it has been assumed that Descartes required that experientially based knowledge meet the standard of absolute certainty set by the the method of doubt in the First Meditation (Garber 1978). This difficulty notwithstanding, any adequate treatment of Descartes' mature philosophy must accommodate both his rationalist metaphysics and his acknowledgment that sense experience plays an essential role in natural philosophy, as the work of scholars such as Buchdahl (1969, chap. 3) and Williams (1978, chap. 9) has shown.

Attempts to develop a satisfactory understanding of the role of experience in Descartes' philosophy have been hindered by two myths. The first of these is the assumption just mentioned: that Descartes applied the standard of hyperbolic doubt from the First Meditation to the sensory evidence used in natural philosophy. Let us call this the myth of absolute sense certainty. It has been pervasive, and has led to a deep misunderstanding of Descartes' conception of the role of experience in the acquisition of knowledge, because it has encouraged the assimilation of Descartes' position to the doctrines of twentieth-century sense-data philosophers (for one of many instances of this tendency, see Rorty 1978, chap. 1, sec. 2). The second myth, which may be called the myth of method, consists in the belief that Descartes subscribed to a single method, announced in the *Discourse on Method* but only fully articulated in the posthumously published *Rules for the Direction of the Mind*, to which he credited his achievements in both metaphysics and natural philosophy (Beck 1952, chap. 16, 18). These two myths reinforce one another. For if it is assumed that Descartes continuously subscribed to a single method, and that this method required that all knowledge meet the standard of hyperbolic certainty, then it will seem that he must have expected absolute certainty from the sensory evidence used to support conclusions in natural science.

Recently, Garber (1987) has rejected the claim that a unified method extends from the *Rules* to the *Meditations*, only to replace it with the claim that a continuous methodologi-

PSA 1988, Volume 2, pp. 249-262

cal doctrine extends from the *Rules* to the *Discourse* and the scientific practice of the *Optics* and *Meteorology*. He supports his belief that the *Rules* describe the method Descartes used in his scientific program of the 1630s by providing a careful analysis of Descartes' account of the primary and secondary bows of the rainbow. According to this analysis, Descartes could reasonably claim that his explanation of the rainbow constituted a successful application of his method; only later did he realize that in other cases his demand for absolute certainty could not be squared with his appeals to sense experience. This realization caused him, in Part Four of the *Principles of Philosophy*, to admit the failure of his scientific program (Garber 1978).[1]

In the following discussion, I will argue that early in his mature period Descartes recognized the need to accept less than absolute certainty in the investigations of the particular sciences, and that this recognition was a secondary consequence of his discovery, in 1629-30, of general metaphysical foundations for his physics. This discovery led Descartes to develop what was for him a radically new conception of nature and of the sciences that describe it. Before 1629, he spoke of various sciences of nature; his examples came from the traditional Aristotelian "mixed" mathematical sciences, from Archimedean sciences such as fluid statics, and from Gilbert's work on the magnet. After 1629, he thought of physics as a single, general science of nature, which could provide framework principles for particular sciences such as optics and meteorology. At the most general level, his new physics shared some features of the traditional mathematical sciences, in that it posited general laws or rules of action, as expressed in Descartes' laws of motion. But, perhaps to his surprise, the mode of explanation that Descartes now found appropriate to wide ranges of particular phenomena was mechanistic, rather than mathematical—where "mechanistic" indicates explanations framed by analogy with the explanation of machines through the interaction of their parts, rather than explanations cast in the form of derivations from a formal mechanics. Although Descartes at first hoped that his physics would allow the details of such explanations to be deduced *a priori*, he soon came to realize that such a deduction would not be forthcoming. Consequently, his conception of the method appropriate to work in the particular sciences changed. He came to recognize the legitimacy and importance of conjectural hypotheses in particular areas of physics, and to acknowledge that, in deciding among competing hypotheses, absolute certainty could not to be expected.

The interpretation of the development of Descartes' conceptions of science and certainty just sketched cannot be fully supported in the space allotted here. In previous work, I have argued that Descartes recognized that a diminished standard of certainty was appropriate to those portions of his physics which could not be derived *a priori* from his metaphysics (see also Clarke 1982), and that as a consequence he abandoned the traditional conception that scientific knowledge consists in necessary demonstrations (Hatfield 1985). Recently, I have urged that in 1629-30 Descartes radically reconceived the relations among physics, mathematics, and metaphysics (Hatfield forthcoming; see also 1986, pp. 65-69). In the present paper, I hope to establish a connection between the beginning of Descartes' mature project in natural philosophy and his acceptance of a lowered standard of certainty. I will contend that after 1629 Descartes abandoned the idea that a single method was suitable for metaphysics and the whole of physics. His attention to method became divided. In metaphysics, he developed the "method of doubt" with its standard of absolute certainty; in physics, reflection on his actual practice in formulating mechanistic hypotheses resulted in a lowered standard of certainty for such hypotheses.

1. Method in Descartes

As do several of the myths about Descartes, the myth about method has a basis in his own words. Indeed, one might say that Descartes himself initiated the myth, as early or earlier than his debate with Chandoux in 1628 (Descartes 1964-74, hereafter AT, vol. I, p. 213), and that in the anonymous *Discourse on the Method* of 1637 he publicly affirmed the rumor that he possessed a special method. Moreover, it can be granted that the *Rules* provided the basis for Descartes' sketch of his method in Part Two of the *Discourse,* so that the former work (abandoned in 1628) provides an interpretive key for the latter. The story that Descartes once claimed to have discovered a special method, indeed, the secret method of the ancients, is, then, far from a myth. The ascription of a special method to Descartes achieves mythic proportions, however, when it is made the primary basis for understanding the *Meditations* and *Principles* (Beck 1952, 1965; Gueroult 1984-5, chap. 1), and when it is read laterally into the entire *Meteorology* or into Part Six of the *Discourse* itself (Garber 1978). The problem is only compounded when the epistemological project of the *Discourse* and *Meditations* is read back into the *Rules* (Schuster 1980, 1986), thereby creating an illusion of programmatic continuity which implicitly supports the thesis of methodological continuity.[2]

Certain resemblances among the *Rules*, the *Discourse*, the *Meditations* (1641), and the *Principles* (1644) have lent *prima facie* plausibility to the thesis of methodological continuity. In all four works Descartes portrays philosophical knowledge as a deductive system in which all claims to knowledge arise through small, certain steps from self-evident beginnings. Following Schuster (1986, p. 41), we may call this picture the latticework conception of knowledge. Descartes remained faithful to this picture as a demand on first principles. The requirements placed on the principles of philosophy in the preface to the *Principles*, that "they must be so clear and so evident that the human mind cannot doubt their truth when it attentively concentrates on them," and that they be such that "the knowledge of other things must depend on them" (AT IX, 2; translations, here and below, are from Descartes 1984-5), might equally well have come from Rule Four, or from the end of Part Two of the *Discourse*, or from the Dedicatory Letter to the *Meditations*. And the wording of Rule Two, that "we should attend only to those objects of which our minds seem capable of having certain and indubitable cognition," would have fit seamlessly into the later works.

Despite the constancy of Descartes' ideal of philosophical knowledge from the *Rules* through the *Principles*, his conception of the relation between certainty and deductive structure in natural philosophy changed dramatically during the 1630s. The change was not the product of isolated reflections on method per se; as we shall see in sections 2-4, it resulted from a shift in the type of explanation he found himself providing for a variety of natural phenomena. But the change was reflected in Descartes' discussions of method, for it altered the role he assigned to method in descriptions of his philosophical achievements. Whereas in the *Rules* and in Part Two of the *Discourse*, he claimed to have discovered a special method, in Part Six of the *Discourse* (AT VI, 63-64), in the *Meditations*, and in the *Principles* he made a quite different claim. There he boasted of having discovered the true principles of philosophy, principles which extend to everything in the created world. He no longer characterized his achievement as the discovery of a special method that applies to all cognition; rather, he claimed to have achieved fundamental cognitions that apply to all existing things, and thereby to have answered certain substantive questions about the natural world once and for all. Prominent among the principles so established were the equation of matter with extension and the three laws of motion.

From a methodological perspective, the important consequence of this change was its effect on the latticework picture. In the period of the *Rules*, Descartes conceived typical explanations as involving appeal to mathematical laws, or to mathematically-expressed rules of action; optics was the paradigm case (Rule Eight). In explanations involving such rules or laws (e.g., of reflection and refraction), once the law had been determined, its application required simply that empirically determined values of variables (e.g., the angle of incidence) be plugged in. Although in Part Six of the *Discourse* and thereafter, Descartes claimed, in language superficially similar to that used in the *Rules*, to "deduce" his physics from first principles, such deductions no longer resembled a mathematical derivation. Beginning in the early 1630s, the explanations in Descartes' writings on natural philosophy involved the conjectural positing of microstructures; the primary explanatory work was performed not by laws, but by the positing of configural mechanisms, like so many gears in a clock, or like so many valves in a hydraulic machine. While such explanations may presuppose the lawfulness of material interactions, they are not couched in terms of lawful relations. In Descartes' new vision of a unified physics of the entire universe, micro-mechanistic explanations came to the fore. The fundamental metaphysical principle of this new physics was the equation of matter with extension, which, in Descartes' conception, implied the denial of the void, and so the physics of a plenum. This conception led to a picture of the universe as a tissue of fluid vortices irregularly studded with large chunks of congealed matter (planets), on which various hydraulic machines were located, along with various groupings of particles having sufficiently similar microstructures to produce regular effects and thus to be denominated as varieties and subvarieties of elemental kinds. In section 2 we shall see that although Descartes had hoped to be able to "deduce" the details of this physics from general principles in a strictly *a priori* manner, he came to realize that a strict deduction was not possible. Yet he continued to use the term "deduce" for the relation between the first principles of his physics and particular mechanistic explanations, and to do so even in close proximity to the passages acknowledging that a strict deduction was not possible (*Discourse*, Part Six, AT VI, 64-65). In such contexts, his use of the term "deduce" is best understood through the etymological sense of the term: his micro-mechanical explanations were "lead out" from his metaphysical principles, in the sense that those principles provided a description of the fundamental concepts, or the building blocks, out of which an explanation was to be be conjecturally constructed.

The method of the *Rules* did not disappear all at once. As mentioned above, Descartes summarized it in Part Two of the *Discourse*. This summary, however, should not be accepted without question as a description of his actual or preferred method in natural philosophy at the time the *Discourse* was composed. Indeed, there are good grounds for reading the methodological remarks in Part Two as an autobiographical account of Descartes' thinking during the time when the *Rules* were composed. As a characterization of a method associated with practice, the early method describes the achievements of Descartes' early work in mathematics and optics. In particular, his account of reflection and refraction in the rainbow is of interest precisely because it illustrates the earlier method (the work dates from 1629; AT I, 23). By the mid 1630s Descartes had abandoned further mathematical work (AT II, 268), and he had expanded his conception of physics in such a way that the methodological discussions developed in conjunction with his work in optics no longer described his primary methodological problems. It is true that the physics of Descartes' new *World* (written from 1629-1633) was not without mathematically expressed laws, in the form of laws of motion. Further, Descartes contended that the subtle aether of light was fine enough to act unhindered by other matter, and thus to exhibit mathematically precise behavior. But explanation by appeal to precise laws of action had become the exception, not the rule. As Descartes' dismissal of Galileo's law of fall made clear (see Koyré 1978, pp. 92-94), in practice the picture of nature described in *The World* could discourage the search for mathematical laws. Descartes began to realize by the mid 1630s that his earlier methodological conceptions no longer applied to his mechanistic explanations.

Although Descartes' new vision of nature was first expressed in *The World*, before beginning that work he had developed a general metaphysical justification for its fundamental doctrines. The attempt to use metaphysical argumentation to justify substantive claims about nature—an attempt made public in the *Meditations* and the *Principles*—marked a radical departure from the philosophical program of the *Rules*. Let us see how.

2. From Method to Principles

It is interesting to note that although he claimed in the *Rules* that his method would reveal "every truth for the knowledge of which human reason is adequate" (AT X, 395), Descartes did not claim that human reason was adequate for every truth (see AT X, 396). He left open the possibility that nature contains powers or agencies which lie beyond human comprehension. The fact that he should leave open this possibility—a possibility he would later foreclose—may be understood by considering the aim of the *Rules*. That work took as its object of investigation the conditions for application of the "knowing power," rather than the natures of the things to be known. It granted primacy to method over metaphysics. The methodological ambition and metaphysical timidity of the *Rules* are most evident in the analysis of perfect and imperfect problems in Rules Twelve and following. Rule Twelve also explicitly disavows the search for a metaphysical guarantee that the method of the *Rules* will yield insight into the essence of corporeal things.

The distinction between "perfectly understood" problems and those "imperfectly understood" occurs near the end of Rule Twelve. Perfectly understood problems are those for which all of the following are known: (1) the criteria for a solution, (2) the premises (or "data") for a solution, and (3) the means by which it is to be proved that premises and solution "are so mutually dependent that the one cannot alter in any respect without there being a corresponding alteration in the other" (AT X, 429). Imperfect problems are defined by contrast with perfect ones; presumably, they are problems for which one or more of these conditions have not been met. We learn further that perfect problems pertain only to the purest of simple natures, and are found primarily in arithmetic and geometry. Imperfect problems pertain to compound natures; the examples given, including magnetism and acoustics, suggest that such problems arise primarily in natural philosophy or physics (AT X, 431).

Unfortunately, the third part of the *Rules*, which was to have been about imperfect problems, was never written. However, the existing Rules contain indications of what Descartes had intended to argue in the unfinished portion. In Rule Thirteen he purports to show how imperfect problems can be transformed into perfect ones. In particular, he says that the problem of discerning the nature of the magnet can be made perfect by limiting the given elements of the problem to "the experiments which Gilbert claims to have performed, be they true or false" (AT X, 431). In the subsequent Rule, he further observes about this problem that "if the magnet contains some kind of entity the like of which our intellect has never before perceived, it is pointless to hope that we shall ever get to know it simply by reasoning; in order to do that, we should need to be endowed with some new sense, or with a divine mind. But if we perceive very distinctly that combination of familiar entities or natures which produces the same effects which appear in the magnet, then we shall credit ourselves with having achieved whatever it is possible for the human mind to attain in this matter" (AT X, 439). These two comments speak, in reverse order, to the first two conditions on perfect problems. The remark from Rule Fourteen indicates how to proceed if the problem of the magnet is imperfect because sure criteria for a solution are lacking, given that the magnet may contain something unfamiliar to the human mind; by requiring proposed solutions to appeal only to known entities, we can achieve "whatever it is possible for the human mind to attain" by way of knowledge of the magnet. Interestingly, it is not claimed that by restricting solutions to known entities, we can be assured of achieving truth about the magnet (as opposed to truth about the solution to the problem, as restricted). Similarly, the remarks from Rule Thirteen suggest that if the problem is imperfect because we are unsure whether all of the data have been observed, we may restrict the data for the problem to that given in a particular source.

These remarks on the problem of the magnet reveal that Descartes was quite willing to trade scope for power in the *Rules*; he was willing to limit the scope of the knowledge attained, in order to be assured that what he did attain possessed certainty. If we look ahead to the correspondence from 1637 and 1638, and to the end of the *Principles*, it would seem that the "imperfection" of the problem of the magnet should in part be ascribed to the difficulty in inferring a true cause from known effects, especially since Descartes demanded of perfect problems that data and solution be tied as necessary and sufficient conditions. But the *Rules* show no concern with this difficulty; either Descartes believed there was no problem in inferring true causes from effects when he wrote that work, or he was satisfied merely "to save the phenomena." Moreover, the fact that he was willing to turn imperfect problems into perfect ones by limiting investigation to the proffered "data," whether they be true or false, reveals a limitation to the scope of the claim that in following the *Rules*, the human mind will attain a knowledge of everything it can know. For now it can be seen that among the things known will be certain conditional truths, such as: if these are all the relevant data for the magnet, then the nature of magnetism is such and such. Under this condition the problem becomes "perfect," but at the expense of assurance that a solution will yield the truth about the nature of the magnet.

The problem of limited data may seem a merely practical one, to be overcome by making the requisite observations. Whether it is merely practical depends upon what may be expected from sensory observation. The *Rules* do not suggest that sensory data is inherently defective; indeed, some sensory experience is included within the domain of intuitive knowledge (AT X, 383, 423), and so is granted certainty equivalent to that of mathematics. But such certainty pertains only to experiential knowledge of things that are "entirely simple and absolute" (AT X, 394). The "simple and absolute" includes the mathematical dimensions of things (size, shape, number; AT X, 419, 439), where the things having such dimensions are as varied as stars, sounds, weight, and speed (AT X, 378, 447-448). But, in any case, no reason is given for expecting that all natural phenomena will present data sufficiently clear to render perfect the problem of assigning a cause for the phenomena. Where such data is wanting, Descartes' message is clear: the truth about the cause of the phenomena lies beyond the scope of human knowledge.

It may seem, however, that there is no need to despair about either the possibility of knowing what is required for a solution, or the possibility of discovering sufficiently clear data. For it may seem that elsewhere in the *Rules* Descartes has provided reason to hope that there is nothing more to be met with in nature than compounds of common and familiar simple natures; and if it could be known that each observed thing is a compound of perfectly knowable natures, that would provide reason to hope that all phenomena could be rendered clear and intelligible. Such hopes would seem to receive encouragement in Rule Twelve, in which Descartes asserts that the simple natures found in bodies include "size, extension, motion, etc." (AT X, 419). This pronouncement would appear to be the familiar Cartesian doctrine that the essence of matter is extension. On the assumption that it is, the *Rules* make a quite powerful claim; for if it is granted that bodies possess geometrical properties only, it follows that the solutions to all imperfect problems regarding compound bodies can be restricted to such properties. Any given data could be truthfully explained through some combination of perfectly intelligible simple natures.

It is at this point that the metaphysical timidity of the *Rules* comes into play. Nowhere in that work does Descartes claim that the essence of matter is extension. Although he recommends that we treat colors, sounds, etc. as if they could be equated with geometrical figures, and contends that if we do so, we will be able clearly to understand their effects upon the sense organs, he doesn't claim, and so doesn't argue, that the qualities of bodies are actually constituted by purely geometrical properties (AT X, 412-414). Indeed, further on in Rule Twelve he explicitly eschews such ontological conclusions, explaining that he is concerned with things "only as they are perceived by the intellect," and not with "how they exist in reality" (AT X, 418; see also AT X, 399). The "simple natures" discerned in the *Rules* are epistemically simple; there is no guarantee that they correspond to something ontically simple. Indeed, it is not clear that Descartes was committed to the central metaphysical tenets of his mature physics when he wrote the *Rules*. In Rule Fourteen, for instance, he treats weight as something real, that is, as something "having a real basis in bodies" (AT X, 448). Although this description contradicts his later denial that weight is a real property of bodies, it accords with his treatment of weight in a document surviving from the early part of the period during which the *Rules* were composed, the physico-mathematical writings undertaken in connection with Beeckman (AT X, 69-74). Because the denial that weight is a real property of bodies is a paradigmatic instance of Descartes' doctrine that the essence of matter is extension, it would seem that the latter doctrine postdates the *Rules*.

The *Rules* describe the requisites for knowledge, without implying that the world is entirely knowable. Perhaps Descartes believed prior to 1629 that the world contains nothing unknowable, but had no argument for his belief, or none that he was willing to divulge. Or perhaps he had no such belief.

Be that as it may, in the spring of 1630 Descartes announced that he had discovered the foundations of physics during the previous year (AT I, 144). I have contended elsewhere that this discovery included the fundamental arguments of the *Meditations* (Hatfield, forthcoming). I will not rehearse that story here. But it is worth noting that by the early 1630s, Descartes was prepared to state explicitly the doctrine that was missing from the *Rules*—that matter is to be equated with extension. But he would state it only as a hypothesis. Thus, in *The World*, he assigned only the properties of size, shape, and motion to the matter of his new, hypothetical, imaginary world (chap. 6). At the outset of the Fifth Part of the *Discourse* he alluded to this hypothesis, and he presented it again in the First Discourse of the *Meteorology*. As Descartes remarked in *The World*, according to this hypothesis matter contains "nothing that you do not know so perfectly that you could not even pretend to be ignorant of it" (AT XI, 35). In correspondence dating from 1638 (February 22, to Vatier) he claimed he could prove this hypothesis in his metaphysics (AT I, 563), and he attempted to do so in the *Meditations* and *Principles*.

It has been maintained, on reasonable grounds, that the arguments presented in the *Meditations* and redescribed in the *Principles*—both those attempting to prove that the essence of matter is extension, as well as others—are generic instances of the regressive method described in the *Rules* (Beck 1952, chap. 18; 1965, chap. 13). Indeed, Descartes himself characterized the method of the *Meditations* in this manner (though without, of course, explicit reference to the *Rules*; AT VII, 3; compare AT VII, 156 and X, 376). There is, then, some methodological continuity between the later works and the *Rules*. To acknowledge this continuity is not, however, to equate the method of the *Meditations* with that of the earlier period. There are significant differences between the two. The hyperbolic doubt of the *Meditations* was new; it was not explicit in *Discourse*, and it was opposed to the teaching of the *Rules* (according to which the intuitively evident should be accepted straightaway; Rule Three). Correspondingly, the certainty that was methodologically required prior to the *Meditations* was not hyperbolical, but ordinary certainty; it follows that the certainty accorded to sense experience prior to the *Meditations* also was ordinary certainty. And so even though it is true that a descendant of the early method appears in the *Meditations*, and that the *Meditations* provide the foundations for the physics, it should not be assumed that the early method therefore extends to the physics so grounded. In fact, it is in connection with the new physics that the greatest methodological discontinuity occurs.

3. Deductive Structure and *a priori* Physics

In order for the thesis of methodological continuity to be true, it must be the case that Descartes understood the explanations of natural phenomena in his mature physics to be strict deductions from basic principles—deductions that meet the standards of the latticework model in the *Rules*, a model in which each link in a chain of reasoning is connected to the succeeding one as its necessary and sufficient condition. In the present section I shall argue that as Descartes began to construct his *World*, he actually did envision the possibility of a thoroughly *a priori* science, one which would allow a suitably tight *a priori* deduction of the structure of various kinds of earthly body.

Two letters to Mersenne in 1632, which report progress on *The World*, reveal Descartes' vision of a completely *a priori* science. In the first, written in April, Descartes described an addition to his original project:

> In the treatise which I now have in hand, after the general description of the stars, the heavens and the earth, I did not originally intend to give an account of particu-

lar bodies on the earth but only to treat of their various qualities. In fact, I am now discussing in addition some of their substantial forms, and trying to show the way to discover them all in time by a combination of experiment and reasoning. This is what has occupied me these last days; for I have been making various experiments to discover the essential differences between oils, ardent spirits, common and strong waters, salts etc. (Descartes 1970, p. 22)

In fact, the addition here described has come down to us not as part of *The World*, but as part of the *Meteorology*. In any event, the brief allusion to experiment and reason in this letter is not by itself methodologically distinctive. One month later, however, Descartes wrote that he hoped to discover a means of knowing the forms and essences of earthly bodies in an *a priori* manner:

For the last two or three months I have been rapt in the heavens. I have discovered their nature and the nature of the stars we see there and many other things which a few years ago I would not even have dared to hope; and have now become so rash as to seek the cause of the position of each fixed star. For although they seem very irregularly distributed in various places in the heavens, I do not doubt that there is a natural order among them which is regular and determinate. The discovery of this order is the key and foundation of the highest and most perfect science of material things which men can ever attain. For if we possessed it we could discover *a priori* all the different forms and essences of terrestrial bodies, whereas without it we have to content ourselves with guessing them *a posteriori* from their effects. (10 May; 1970, pp. 23-24)

In this puzzling passage Descartes held out the hope of an *a priori* physics, one that would enable him to discover even the microstructures of the various types of terrestrial bodies. What could he have had in mind?

Consider the account of the development of the universe given in *The World* and later in the *Principles*; in each work, the world as we see it now arises from a hypothetical beginning in a chaos of matter in motion. A completely *a priori* science, as Descartes envisioned it, would begin with a general description of this chaos and would show how our present world could arise. Without relying on any observations of the present world, this science would succeed in "deducing" the features of that world, including the particular "forms" and "essences" of its minerals. Perhaps the passage just quoted expressed Descartes' belief that an understanding of the principles of star location could provide insight into vortical formation, and that this insight would be sufficient to allow a completely *a priori* deduction of the particular species of matter which, by hypothesis, were churned out by the vortex and which then congealed to form the earth (a deduction of the sort found in a more limited form in the *Principles*).

While such speculations have the makings of an interesting fable, sadly, as Descartes went on to admit in the letter just quoted, there was little hope that the principles of star placement could be discovered. This pessimistic presentiment seems to have been borne out, for in the *Principles* he confided that the vortices exhibit an "inexplicable variety" (1983, Part III, art. 68). In that work he provides what is presumably an *a priori* account of the formation of the matter surrounding the earth; he explains the chief qualities of such matter, which were vortical motion, weight, light, and heat (IV, 15-31). But as the account descends to the explanation of more particular qualities of matter, it becomes less clear whether it should to be construed as *a priori* or *a posteriori*. (See, for example, the account of the formation of air and water; IV, 32-48.) Descartes refers back to the "particular hypotheses" introduced in the First Discourse of the *Meteorology* (to be distinguished from the more general hypothesis of the corpuscular constitution of matter) in order to support his conclusions regarding the microstructures of various minerals, including those structures that explain the difference between water and salt water (IV,

48). In attempting to explain the properties of various minerals and metals, he is forced to appeal to experience (IV, 16), to rely on the hypotheses of the *Meteorology*, or to curtail his presentation for want of experiments (IV, 63). Thus, he fails to advance beyond the position he had envisioned in April, 1632; he can do no better than to guess causes from effects, *a posteriori*. His initial suspicion that it would be necessary to employ the method of hypothesis in describing his new world, a suspicion that arose not long after he turned his attention to the explanation of terrestrial particularities, was proven sound.

4. *A posteriori* Physics, Experience, and Certainty

As numerous commentators have observed (Buchdahl 1969, chap. 3; Clarke 1982; Garber 1978; Laudan 1966), Descartes himself explained that it was the explanatory fertility of his own first principles which drove him to the method of hypothesis. The properties of salts, oils, ardent spirits, common and strong waters, and so forth could be explained in many ways; Descartes' problem was to see which among the potential explanations were true. On occasion, he characterized each of the variety of potential explanations as a "deduction" from his first principles (e.g., AT VI, 65). As already noted, the notion of a deduction expressed in such remarks was not very strict; it was less strict than that found in the *Rules,* and bore little resemblance to formal deduction (see also Buchdahl 1969; Clarke 1982). Descartes' claim to have provided multiple "deductions" of the phenomena from his principles may be read as the claim that starting from size, shape, and motion, together with the laws of motion and the assumption of an initial chaos, a variety of micro-mechanisms, any one of which would be adequate to explain the known phenomena, can reasonably be imagined to arise. Given this imagined variety, the problem becomes that of discovering tests based upon sense experience that could serve to determine which of the micro-mechanisms is actual. Although hypothetical posits tested in this manner might in some cases be established with a high degree of confidence, in the letters of 1637-38 and in the final remarks to the *Principles* Descartes refrained from granting them certainty comparable merely to the mathematical variety—let alone hyperbolic certainty (Hatfield 1985).

My claim that Descartes refused to grant mathematical or metaphysical certainty to the postulation of micro-mechanisms must be distinguished from the claim that he believed all sense-based knowledge of the external world lacks certainty. Contemporary appreciations of the Cartesian spirit notwithstanding, Descartes was willing to grant ordinary certainty—again, as opposed to the hyperbolical variety—to claims about material objects that go beyond the "immediately given" in sensory experience. This willingness is foreshadowed in the First Meditation. Descartes offers an initial sceptical challenge to the senses based on their occasional deceptiveness, only to rebut this challenge by suggesting that under good conditions the senses are perfectly reliable. The dream argument and the hypothesis of an evil deceiver are successively stronger skeptical responses to this rebuttal; it seems fair to assume, therefore, that once these grounds for doubt have been removed, the senses may be treated as reliable sources of knowledge of things which are close at hand, which can be touched as well as seen, etc. And indeed, the senses are vindicated in the Sixth Meditation (AT VII, 80; see Hatfield 1986), in which it is suggested that even factual details about the material world, such as the size of the sun, can be known by means of the senses (in conjunction, of course, with reasoning), as can the properties of objects which are relevant to the preservation of one's bodily well-being (AT VII, 80-1). Further, in Parts One and Four of the *Principles* (I, 69-70; IV, 200), Descartes affirms that size, shape, and motion can be "clearly perceived" in bodies by means of the senses (in conjunction with the understanding).

Descartes was prepared to allow that, when sufficient care is taken, sensory observation can yield certainty. But the possibility of such certainty does little to solve the problem of how to achieve certain knowledge about those micro-mechanisms that must, as remarked in 1632, be "guessed" from their effects, *a posteriori*. The task of attaining certainty about hypothetical posits could be completed only by means of extensive experimentation, if at all; Descartes was aware of the underdetermination of micro-mechanical hypotheses by sensory evidence. Of course, he thought he had a head start toward solving the problem of underdetermination in framing his conjectural hypotheses, for he believed that he had established, once and for all, what the theoretical vocabulary of any physical hypothesis must be: theoretical posits must be couched in the language of size, shape, and motion. But the fact that he claimed to be metaphysically certain that his general approach to physics was correct does not alter the fact that he granted, from near the inception of his fully mechanistic conception of natural phenomena, that he was not entitled to claim such certainty about the particular mechanisms he posited.

The use of mechanistic hypotheses became central to Descartes' natural philosophy in the early thirties. Its absence from the primary examples of method in the *Rules* helps to explain how Descartes could there envision the attainment of certainty with respect to particular problems in physics. Let us put to one side the certainty that arises from the transformation of imperfect problems into perfect ones, for such certainty is purchased at the price of forfeiting knowledge that the solutions of the problems are true. There remain other examples from the *Rules*, or from the period of its composition, which offer the possibility of certainty for natural scientific conclusions that rely on experience. The determination of the anaclastic as envisioned in Rule Eight is a case in point. To it may be added the explanation of the angle at which the rainbow is seen. As Descartes observed in Rule Eight, in the *Optics* (Second Discourse), and in the *Meteorology* (Eighth Discourse), experience must play a role in each case.

Despite this fact, Descartes claimed in such cases to have achieved the sort of "deductive" certainty required by the *Rules*. The notion of "deduction" in these cases is regressive (Buchdahl 1969, chap. 3, sec. 2d; Garber 1987); it requires working back to the explanatory basis from which the given problem can be progressively derived. Descartes could claim that certainty is achievable in the instances cited because the relevant experientially based claims provide evidence not for postulating micro-mechanisms, but for determining a mathematical law governing observable relations among phenomena. And although the law presumably cannot simply be "read off" from experience, experiential elements can and must play an intrinsic role in the process of deduction. Experience can play a role, because Descartes allows that sensory experience can count as an instance of "intuition" if the objects of the experience are pure and simple natures (AT X, 383, 394), as the dimensions of things listed in Rule Fourteen presumably are (AT X, 447-8). The angular relations between the paths of a light ray before and after refraction might also count as an instance of an intuited dimension (within the limits of accurate measurement; see the table of measurements in the Eighth Discourse of the *Meteorology*). In this last example, experience must play a role in both the regressive and progressive phases of the "deduction," for the refractive indices of various media such as water and air can only be determined through experience, as Descartes himself made clear (AT VI, 102, 337). (This point about refractive indices does not contradict Descartes' claim to have derived the sine law of refraction on *a priori* grounds, on which see Smith 1987.)

If it is correct that the *Rules* permit sense certainty only in the case of the sensory intuition of medium-sized objects, it becomes apparent that once Descartes accorded primacy to his vision of a universe of little machines, this route to certainty would become less central to his philosophy of science. Once this change in emphasis had occurred, most cases of reasoning in Cartesian physics ceased to be like the case of determining the position of the rainbow; typical problems no longer involved the use of direct measurement to determine a law or relation. Because the majority of explanations involved an appeal to micro-mechanisms, the mode of inference depended on analogical reasoning from the known to the unknown, from observable mechanisms to posited microstructures (Galison 1984). Admittedly, in his discussion of the position of the rainbow, Descartes did reason from the big to the little, inasmuch as he used glass globes to model raindrops. But in this case, the "analogy" involved merely shrinking the scale of a model which exhibited a "macro" version of the phenomenon in question. By contrast, when inferring hidden mechanisms, one is free to posit any of a great variety of mechanisms as a means of explaining a phenomenon; the methodological constraints enjoyed when investigating the operation of a single mathematical law in larger and smaller spheres are no longer present. The possible posited mechanisms are virtually endless, under the limitation that any mechanism posited be truly mechanical, that is, that such mechanisms be ascribed only the properties of size, shape, and motion (AT I, 420; *Principles* IV, 200, 201). Given these circumstances, even if a particular posited mechanism saves all of the phenomena, one may assert that it is the true mechanism with merely "moral" certainty; that degree of certainty may suffice for the rational fixation of belief, but it falls short of the notorious absolute certainty of Cartesian metaphysics.

5. Science, Hypothesis, and Descartes

Descartes recognized the need to proceed by the method of hypothesis from near the inception of the mechanistic program described in the *World*, the *Meteorology*, and the *Principles*. As Laudan (1966) and Buchdahl (1969) have observed, he adopted a "hypo-thetical-deductive" mode of supporting his claims about particular mechanisms. The description of Descartes' strategy as "hypothetical-deductive" cannot be left unqualified, for Descartes did not leave the domain of allowable hypotheses unrestricted (see Buchdahl, pp. 144-7); it was restricted by metaphysically determined principles, including the equation of matter with extension, and the laws of motion. Assuming this qualification, it may be recognized as a point in the history of methodology that Descartes explicitly described the limitations on certainty that result from a hypothetical mode of reasoning.

Despite the presence of the method of hypothesis in Descartes' mature science, it would be misleading to portray Cartesian science as directly continuous with post-Newtonian science. Indeed, the method of hypothesis which Descartes applied to mechanisms contrasts both with the method he used in explaining the rainbow's position—and so with the method of the *Rules*—and with the type of scientific explanation inspired by Newton's *Principia* and *Optics*. From a methodological point of view, the central examples of Newtonian explanation, such as the derivation of elliptical orbits from the inverse-square law, seem closer to explanations contained in the *Rules* than to those found in Descartes' mature work.

The difference in spirit between the mature Descartes' primary model of scientific explanation and that of Newton may be brought into relief by briefly considering an aspect of Descartes' mature scientific program that has been cited as a common feature of Cartesian and Newtonian science: the "mathematization" or "geometrization" of nature. In the attempt to understand Descartes' claim that his science was "geometrical" or "mathematical," two interpretations might naturally come to mind. The claim might be taken to refer either to the alleged deductive structure of his science, or to its use of mathematical laws, such as the sine law of refraction or the laws of impact. However, when Descartes wrote in the *Principles* that the "only principles" of his physics are those of "geometry and pure mathematics" (II, 64), he had yet a third point in mind: he was simply expressing the doctrine that the essence of matter is "pure extension." For this doctrine amounts to the view that matter is wholly describable in terms of the geometrical properties of size, shape, position, and motion. Inasmuch as size or volume may be measured, these properties may be described as "quantitative"; but the ascription of such properties to matter does not lead necessarily to a "quantitative" approach to nature of the sort that we are likely to associate with Newton, or with Galileo's discovery of the law of fall. It does not lead inevitably to a notion of nature as governed by quantitative laws, or to the notion that scientific reasoning is principally concerned with mathematical derivation. As we have seen, in Descartes' mature natural philosophy the ultimate emphasis was on mechanisms, not laws; "mathematization" amounted to "mechanization" (see also Hatfield, in press).

Of course, a mechanistic account of nature such as Descartes' is unthinkable unless it is supposed that matter in motion moves with lawful regularity, and Descartes did invoke the required regularity in his three laws of motion and seven rules of impact. These laws *are* cast in quantitative terms—as relations between the fundamental magnitudes, "quantity of matter" and speed. But the possibility of such quantitative laws is not established by the doctrine of extension, even if this doctrine is taken to include the idea that motion is a geometrical (kinematic) property and hence is a "mathematical" property; a kinematic (purely descriptive) treatment of motion does not entail that the patterns of motion in the world are describable by simple laws. As is well known, Descartes went outside the notion of matter as extension and introduced God as a "dynamic" element to fix these laws—the laws are understood as a manifestation of God's immutability, specifically, of his conservation of the same quantity of motion in the material world as at the creation (see Hatfield 1979 for further discussion).

The introduction of universal laws of motion governing the interaction of all particles in the universe was, indeed, a step of some significance in the march of the world spirit toward Newtonian science: it provided a prominent example of the conception of nature as a single, law-governed system, to replace the conception of a universe bifurcated into two regions governed by different principles; methodologically, it encouraged replacement of the conception of physics as a series of particular sciences of nature with the picture of one basic science of physics, possessing various branches. But these valid points of comparison between Descartes' program and the development of post-Newtonian science should not lead us to be misled by other seemingly valid comparisons. In particular, we should not be led to believe that the laws of motion played the same explanatory role in each instance. A mistake on this point would explain why the role of hypothesis in Descartes' mature science was at one time routinely overlooked. To an interpreter imbued with a post-Newtonian conception of science, the character and the extent of the role that Descartes accorded to hypothesis in his physics would seem insignificant in comparison to the things Descartes claimed could be known independently of experience, through metaphysics. In particular, Descartes' claim to know his laws of motion *a priori* would seem of especial importance; for to a reader with post-Newtonian sensibilities, it would seem that once the laws governing motion (including the law of gravitation, which has no counterpart in Cartesian physics) had been established, the explanatory basis of physics was largely complete. For if Newton's laws could be known *a priori*, it would seem of small consequence that, say, the masses of individual bodies could not be. But in Descartes' mechanistic account of natural phenomena, the chief explanatory work is performed by the configuration of the particles in various types of body. There are no cases in which the laws governing motion serve to explain an important class of phenomena, such as the planetary orbits; moreover, given his explanatory ambition, the postulation of micro-mechanisms—which might seem to a Newtonian like so many speculations about "initial conditions," which should be avoided if the conditions cannot be determined through measurement—becomes the central activity of Cartesian science. The postulation of such configurations proceeds by a method of hypothesis, based upon analogy. Hence Descartes' mathematical conception of matter, far from permitting him to adopt a geometrical style of proof in natural philosophy, and far from leading him to retain a mathematical or metaphysical standard of certainty, led him to acknowledge that absolute certainty may be beyond reach throughout much of the science of nature, because that science must proceed conjecturally, by means of hypothesis.

Notes

[1]Garber continues to hold the view that there was a fundamental and persistent tension between Descartes' demand for certainty and his appeal to experience in science, and that the methodological points made at the end of the *Principles* constituted an admission of defeat for his program, as is evident from his "Descartes and Experiment in the *Discourse* and *Essays*," presented at San Jose State University during April, 1988 (p. 10 of the distributed draft).

[2]Schuster (1986) chastises others for seeking to find methodological continuity in Descartes' work; yet he is guilty himself of adopting the position that a single "problem of justification"—the problem of justifying the applicability of mathematics to matter—may be found in both the *Rules* and the *Meditations*, being met in the first by the "optics-psychology-physiology nexus" (1980, pp. 59-73), the failure of which led to the metaphysical arguments of the second (1980, pp. 75-79).

References

Beck, J. (1952). *The Method of Descartes*. Oxford: Clarendon Press.

_ _ _ _ . (1965). *The Metaphysics of Descartes*. Oxford: Clarendon Press.

Buchdahl, G. (1969). *Metaphysics and the Philosophy of Science. The Classical Origins: Descartes to Kant*. Cambridge: MIT Press.

Clarke, D. (1982). *Descartes' Philosophy of Science*. Manchester: Manchester University Press.

Descartes, R. (1964-74). *Ouevres de Descartes*. Edited by C. Adam and P. Tannery. New edition. Paris: Vrin.

_ _ _ _ _ _ _. (1965). *Discourse on Method, Optics, Geometry, and Meteorology*. Translated by P. Olscamp. Indianapolis: Bobbs-Merrill.

_ _ _ _ _ _ _. (1970). *Philosophical Letters*. Translated by Anthony Kenny. Oxford: Clarendon Press.

_ _ _ _ _ _ _. (1983). *Principles of Philosophy*. Translated by V. Miller and R. Miller. Dordrecht: Reidel.

_ _ _ _ _ _ _. (1984-5). *Philosophical Writings*. Translated by J. Cottingham, R. Stoothoff, and D. Murdoch. Cambridge: Cambridge University Press.

Galison, P. (1984). "Descartes's Comparisons." *Isis* 75: 311- 326.

Garber, D. (1978). "Science and Certainty in Descartes." In *Descartes: Critical and Interpretive Essays*. Edited by M. Hooker. Baltimore: Johns Hopkins University Press, pp. 114-151.

_ _ _ _ _ _. (1987). "Descartes et la Methode en 1637." In *Le Discours et sa Methode*. Edited by N. Grimaldi and J. Marion. Paris: Presses Universitaires de France, pp. 65-87.

Gueroult, M. (1984-5). *Descartes' Philosophy Interpreted According to the Order of Reasons*. Translated by R. Ariew. 2 vols. Minneapolis: University of Minnesota Press.

Hatfield, G. (1979). "Force (God) in Descartes' Physics." *Studies in History and Philosophy of Science*. 10: 113-140.

_ _ _ _ _ _. (1985). "First Philosophy and Natural Philosophy in Descartes." In *Philosophy, Its History and Historiography*. Edited by A. J. Holland. Dordrecht: Reidel, pp. 149-164.

_ _ _ _ _ _. (1986). "The Senses and the Fleshless Eye: The *Meditations* as Cognitive Exercises." In *Articles on Descartes' Meditations*. Edited by Amèlie Rorty. Berkeley: University of California Press, pp. 45-79.

_ _ _ _ _ _. (in press). "Metaphysics and the New Science." In *Reappraisals of the Scientific Revolution*. Edited by D. Lindberg and R. Westman. Cambridge: Cambridge University Press.

_ _ _ _ _ _. (forthcoming). "Reason, Nature, and God in Descartes." *Science in Context*.

Koyré, A. (1978). *Galileo Studies*. Translated by J. Mepham. New Jersey: Humanities Press.

Laudan, L. (1966). "The Clock Metaphor and Probabilism." *Annals of Science*. 22: 73-104.

Schuster, J. (1980). "Descartes, *Mathesis Universalis*, 1619-28." In *Descartes: Philosophy, Mathematics and Physics*. Edited by S. Gaukroger. New Jersey: Barnes and Noble, pp. 41-96.

_ _ _ _ _ _. (1986). "Cartesian Method as Mythic Speech: A Diachronic and Structural Analysis." In *The Politics and Rhetoric of Scientific Method*. Edited by J. Schuster and R. Yeo. Dordrecht: Reidel, pp. 33-95.

Smith, A. (1987). "Descartes's Theory of Light and Refraction: A Discourse on Method." *Transactions of the American Philosophical Society*. Vol. 77, Part 3.

Williams, B. (1978). *Descartes: The Project of Pure Inquiry*. New York: Penguin.

Part IX

BIOLOGY

Theoretical Models, Biological Complexity and the Semantic View of Theories[1]

Barbara L. Horan

University of Maryland

In this paper I discuss how, given the complexity of biological systems, reliance on theoretical models in the development and testing of biological theories leads to anti-realism. This is the result of the uniqueness and hence diversity of biological phenomena, in contrast with the uniformity of items in the domain of physics. I have argued elsewhere (Horan, 1986, 1989a, 1989b) that the use of theoretical models creates an unresolvable tension between the explanatory strength and predictive power of hypotheses, and I review this argument here. My discussion is in part motivated by the claims of Nancy Cartwright (1983), who has argued that the use of *ceteris paribus* laws in physics creates an antagonism between truth and explanation that requires theoretical models to figure centrally in scientific explanation, thereby precluding realism. I argue instead that in biology it is the use of theoretical models that creates this conflict, and conclude that adequate biological explanation cannot rely on the modelling approach alone. Finally, I claim that if we accept the semantic view of theories, according to which models are an integral part of our conception of theories, we must accept anti-realism as well.

1. The Semantic View of Theories and Theoretical Models

The semantic view of theories has gained popularity recently because of its demonstrated superiority to what is called the "received view" or "syntactic view" of scientific theories. Because this has been discussed in great detail by others (Beatty 1980, 1981; Lloyd 1984, 1986, 1987, 1988; Thompson 1983, 1986, 1988; Suppe 1977, 1989), I will not pursue it here. I want to concentrate instead on explicating the claim that the structure of scientific theories is best understood in terms of classes of models.

The semantic view of theories is a view about what kinds of things scientific theories are. Introduced by Fred Suppe (1977, 1989), Patrick Suppes (1960, 1962, 1967), and Bas Van Fraassen (1967, 1970, 1972, 1980) to supplant the traditional syntactic account of theories, the semantic view emphasizes "the essential role of models in science" (Van Fraassen, 1970, p. 337). According to this view, theories describe abstract systems called models that can be used to represent real systems, but which do not make any direct claims about those real systems themselves.[2]

Suppes makes it clear that when he talks about models of a theory, e.g., "One of the simplest ways of providing...an extrinsic characterization [of a scientific theory] is simply to define the intended class of models of the theory," (1967, p. 60), he is referring to the

PSA 1988, Volume 2, pp. 265-277

theory's *logical* models. Following Tarski, we may regard a logical model of a theory as a possible realization in which all valid sentences of the theory (i.e., all of its axioms together with their logical consequences) are satisfied (Tarski 1953). A possible realization is a nonlinguistic structure: for Van Fraassen it is a state space together with a set of trajectories; for Suppes it is a set-theoretical entity, "a certain kind of ordered tuple consisting of a set of objects and relations and operations on these objects" (Suppes 1967, p. 290).

As Suppes points out, it is possible to define the models of a theory by first determining the valid sentences of the theory, and then defining a class of structures (models) that satisfy them. But because the class of models can be defined directly, without recourse to the syntax of the theory, and because "the explicit consideration of models can lead to more subtle discussion of the nature of a scientific theory," a direct or "extrinsic" characterization of theories is to be preferred (*ibid.*, p. 62).

Subsequent defenders of the semantic view have cited its ability to capture important features of scientific theories as a reason for preferring it over the "received" view. John Beatty, for example, wrote that "given that optimality models in their usual adaptive-landscape format simply specify a kind of system, it seems reasonable that we should characterize those models in terms of the semantic view of theories" and that "the semantic view also satisfactorily accommodates the explanatory use of optimality models" (Beatty 1980, p. 554).

Elisabeth Lloyd used the semantic view of theories to illustrate the way in which empirical claims about models are confirmed. In describing the fit between models and data she wrote

> The most obvious way to support a claim of the form "this natural system is described by the model," is to demonstrate the simple matching of some part of the model with some part of the natural system being described. For instance, in a population genetics model, the solution of an equation might yield a single genotype frequency value. The genotype frequency is a variable in the model. Given a certain set of input variables (e.g., the initial genotype frequency value, in this case), the output values of variables can be calculated using the rules or laws of the model. The output set of variables (i.e. the solution of the model equation given the input values of the variables) is the outcome of the model (1987, p. 280).

Her discussions of how assumptions of a model can receive independent support and of the role of variety of evidence in the confirmation of ecological and evolutionary models are intended to demonstrate that "the greater complexity and variety [of the semantic view], as compared to, for example, a Popperian approach, can facilitate detailed analysis and comparison of empirical claims" (*ibid.* p. 291).

According to Paul Thompson, "the most significant advantage of the semantic account [of evolutionary theory] is that it quite naturally corresponds to the ways in which biologists expound, employ, and explore the theory" (1983, p. 227). Biologists' views about the "as-if" character of theories, he claimed, "are more faithfully represented by the semantic account." More importantly, the structure of the evolutionary theoretical framework, which he characterized as "a family of interacting models," is distorted by syntactic accounts, but can be accurately represented by a semantic view of theories:

> ...the semantic conception of theories provides a framework within which a formalization of 'evolutionary theory' understood as a family of interacting theories can be given, whereas a syntactic conception does not. The semantic conception is, consequently, a richer account of theory structure and the more appropriate and promising account within which to formalize evolutionary theory. The semantic

conception, unlike the syntactic conception, accommodates complex explanatory frameworks because theories — which are extralinguistic entities that define a class of models — can interact (1988).

According to Beatty, Lloyd, and Thompson, we should prefer the semantic view of theories, which emphasizes the models of a theory, over a syntactic account, which emphasizes its axiomatic structure, if we want to capture the prominent role of theoretical models in the development and testing of scientific theories and in scientific explanation. Indeed, the semantic view claims to be a successful account of theory structure precisely because it recognizes the important place models have in the actual practice of science. It is this emphasis on theoretical models, I argue, that leads to anti-realism. I return to this problem in section 5.

2. An Example of a Theoretical Model in Behavioral Biology

In this section I want to present a theoretical model in behavioral biology to illustrate the anti-realist consequences of biologists' reliance on model building.

There has recently been in behavioral ecology and sociobiology a debate about the best explanation for polygyny, a mating system in which males establish a pair bond and mate with more than one female (Simmons 1988). The majority of mammalian species, but that less than 10% of avian species, are polygynous.[3] Why females choose to mate with males who already have mates is somewhat puzzling. It would appear that female reproductive success, and hence female fitness, would decrease when the attention of the male is diverted by needs of other females and their offspring. Theoretical models have been proposed that attempt to explain the existence of polygyny in terms of fitness compensation females might receive from mating with males who already have mates.

One of the earliest models hypothesized that polygyny evolved as a result of natural selection favoring females choosing to be the second mate of superior males instead of the first mate of inferior males. In particular, the model proposed that females would cross a "polygyny threshold" if the difference in quality of male territories was so great that females would have greater reproductive success in the better territory, even if this meant less (or no) assistance from the male, than if they were to mate with a monogamous male in a territory of poorer quality (Verner 1964; Verner and Willson 1966; Orians 1969). Although this model made several important successful predictions (e.g., females mated to polygynous males reared, on average, as many offspring as females mated to monogamous males), other predictions were not successful. For instance, it was found that in one polygynous species (red-winged blackbirds, *Agelaius phoeniceus*), harem size, the average number of females mated to a single male, does not increase with territory quality as predicted (Weatherhead and Robertson 1977). The original model was therefore modified: females must be choosing mates not simply on the basis of territory quality, but with an eye to the attractiveness of the male himself. By choosing an attractive mate females can produce "sexy sons," male offspring who inherit some of their father's attractive qualities. If, in turn, these sons mate polygynously, females might recoup an initial loss in fitness through an increased number of grandoffspring (Weatherhead and Robertson 1979).[4]

According to the sexy-son model, after producing two generations of offspring, females mated to polygynous males will have greater fitness than females mated to monogamous males if the polygyny threshold, x, is greater than 0.77. That is,

$$x = \frac{(\text{\# offspring fledged/polygynously mated female})}{(\text{\# offspring fledged/ monogamously mated female})}$$

must be such that for every offspring successfully reared by a female mated to a monogamous male, at least 0.77 offspring must be fledged by the female mated to a polygynous male, or polygyny will be maladaptive.[5] Whereas the original model predicted that polygyny should evolve only when $x \geq 1$, (that is, when polygynously mated females have at least as many offspring as monogamously mated females), the sexy-son model shows that the polygyny threshold can be lowered by differences in male attractiveness. This is especially true if there is a strong tendency for sons to inherit their father's attractive traits. If this tendency does not exist, the value of the polygyny threshold, x, again becomes close to one.

In a reply to criticisms of the model its proponents claimed that evidence ruled out the alternative explanations of the data (e.g., that the lower breeding success of secondary females is due to inexperience). Weatherhead and Robertson insisted that the model's successful prediction of the value of the polygyny threshold in several species confirmed the sexy-son hypothesis:

> In summary, the best data to asses whether male quality influences female choice are those from lark buntings...and from bobolinks..., and both studies support the hypothesis. As well, while not without ambiguity, all the other studies cited are at least suggestive that male quality is a factor in female mate choice (Weatherhead and Robertson 1981, p. 352).

Granted, the authors of the sexy-son model claimed that they are only "postulating" the model as the explanation of polygyny, and are only "suggesting" that male quality is a factor in female mate choice. But should the model continue to make accurate predictions of polygyny thresholds, they would surely conclude that the model did, in fact, explain polygynous behavior, and that male quality was, in fact, a factor in female mate choice, just as they claimed that their model did, in fact, explain the evidence that refuted earlier models.

The conclusion they draw about the role of male attractiveness in female mate choice is quite general — it is not even restricted to avian species, much less to territorial passerines. Most importantly, it is a conclusion about the success of the hypothesis as an *explanation,* and this conclusion is drawn solely on the basis of its *predictive* success. It is precisely this assumed connection between predictive success and explanatory power that I wish to challenge.

We want to know what the mechanism is by which polygyny evolved, and what function polygyny serves. The fact that the model can accurately predict the polygyny threshold for some species does not mean that polygyny evolved because it increases female fitness through the production of sexy sons, nor that the function of polygyny is to increase female fitness through the production of sexy sons. The predictive success of the model does not mean that the sexy-son hypothesis is true. Indeed, there is some reason for thinking that it means that the hypothesis is false. It is to this problem that I now turn.

3. Realism vs. Robustness in Theoretical Models

The modelling approach is central to some biological theories, e.g., optimal foraging theory, and to research programs like E.O. Wilson's "new synthesis," sociobiology. In population biology, the theoretical framework on which sociobiology is built, empirical data are used to demonstrate that some real system (e.g., a particular population) is similar or "homomorphic" to a kind of abstract system defined by the theory (e.g., a Mendelian population in Castle-Hardy-Weinberg equilibrium). In behavioral biology, it has become fashionable, and even a mark of methodological sophistication, to put forward a theoretical model that describes some hypothetical system with empirical data collected to show that the behavior of some species of interest fits the predictions made by the model. This was clearly illustrated by the "sexy son" model of polygyny.

The semantic view of theories accommodates this practice: it accurately characterizes biological research in which theoretical modelling has a central role. I thus grant the claims of semantic view theorists that the account is descriptively accurate. My objection to it is that by virtue of this descriptive accuracy, it licenses a rather problematic inference, viz., the inference from the successful prediction by a theoretical model in one case to the conclusion that the model affords a general explanation that covers a wide variety of cases. This is a problem that arises whenever a theoretical model is used to justify theoretical generalizations on the basis of the model's predictive success in individual cases, for it is a consequence of the fact that the predictive success of a theoretical model competes with its explanatory power.

The sexy-son model stated that in polygynous species, females choose a mate not only on the basis of the quality of the territory that males defend, but also on the basis of the male's "attractiveness." The justification for this generalization was the model's *robustness,* its success in predicting values of the polygyny threshold in several species of polygynous, territorial passerines. The proponents of the model took its predictive accomplishments as an indication of the model's *realism,* its explanatory power, and they concluded that in *all* polygynous species females choose polygyny because their initial loss of fitness can be made up by the production of "sexy sons." The robustness of the model — its predictive power — is thus made the basis for concluding that the model is also *realistic,* that is, that it can provide an explanation that not only covers what is to be explained, but does so using hypotheses that are *true.*

However, far from the predictive success of the model being grounds for thinking it is genuinely explanatory, it is a reason for thinking that the hypotheses of the model are false, and hence cannot be explanatory. The reason is this. It may be possible to use a theoretical model developed in one case as a successful predictor in other, different, and perhaps even more general cases; however, it is unlikely that hypotheses produced in this way will be true. Those hypotheses therefore cannot be explanatory. Robust models are usually not realistic.

This tension between the realism and robustness of a theoretical model is best understood in terms of the strategy of theoretical model building itself (see Richerson and Boyd 1987 for an informative and thoughtful discussion of the model building approach). No one, except possibly Wilson and the more fanatical of his adherents, believes that just because a theoretical model about the incidence of homosexuality or aggressive behavior in laboratory rats can make some successful predictions about the appearance of these behaviors in human beings that these behaviors are to be *explained* in the same way. Two natural, and methodologically sound, responses to such claims are (1) to want more information about the apparent *similarities* of these cases — is aggression in Norway rats *really* the same as human aggression? and (2) to want more information about the *differences* between these cases, is it plausible to attribute what *appear* to be similar effects (homosexuality or aggression) to similar causes (stress induced by overcrowding)? But because the model building approach to theory construction has been employed, such information is not forthcoming: abstraction, simplification, and neglect of the details of individual cases is what the model building approach requires. When theoretical models are constructed, one eliminates as much of the detail about the case one wants to explain as possible — complex models are costly and cumbersome, and in biology seldom pay back the effort in terms of robustness. Instead, the model builder selects a few variables and uses them to make the model a successful predictor. Thus all "unnecessary" or "nonessential" details about the cases being studied have been omitted in order to yield a theoretical model that is predictively successful, i.e., that is robust.

But ignoring the details of other cases about which one wishes to generalize will allow one to construe different effects as similar, and to attribute a (what may be only an apparent) concordance of effects to an actual identity of causes. The method of model building will permit us to conclude that apparently similar effects (homosexuality or aggression) have similar causes (stress), and thus to postulate that the hypothesis proposed as the explanation in the one case (rats) is the explanation in the other (human beings). But one would have to be extraordinarily insightful — or very lucky — to have picked out from one case exactly those factors that were explanatory of the observed effects in very different cases — unless one knew a good deal about each of the cases. I argue below that this is especially true of biological cases, since one cannot assume, as is done in physics, that one case will be very much like another in "important" respects.

While it might be tempting to think that the predictive success of a theoretical model entailed the explanatory power of its hypotheses in those cases for which successful predictions were made, it is hardly credible that a conclusion about the *general* explanatory ability of the model is warranted.[6] After all, showing that the predictive scope of a hypothesis can be increased is not to have provided a guarantee that the hypothesis has captured the relevant causal factors in a diversity of cases, factors that would provide a genuine explanation. And it is precisely *because* of their diversity that apparently similar effects are not reliable indicators of the identity of causes. It is diversity that makes predictive success unable to guarantee the success of an explanatory hypothesis. So in the absence of details about the individual cases to be covered by the explanatory hypothesis, its predictive success cannot be taken as sufficient grounds for concluding that it will be the correct explanation.

Usually hypotheses must jockey for predictive and explanatory success by undergoing numerous and extensive experimental tests, in which the experimenter tries to falsify them outright or to point to modifications that must be made in order to accommodate new cases. Picking out from one case precisely those factors that supply the explanation of what is observed in all cases is an accomplishment that requires either a deep understanding of the phenomenon to be explained, or a great deal of luck. Model builders extol theoretical models for their ability to make relationships between important properties of a real system "more apparent, predictions more precise, and hypotheses more subject to well-defined experimental tests," (Elseth and Baumgardner 1981, p. 3) without being sufficiently cautious about the method by which those "important properties" have been chosen. The perils of inferences from the specific to the general, from what might arguably be a uniform set of cases to what is surely not so, are blithely ignored. As one critique of theoretical models noted, model builders in ecology frequently assume that a model that is successful on a small scale can be transposed, without modification, to phenomena occurring on a much larger scale:

> The tremendous mismatch in size between, for example, a laboratory microcosm in which the rotifer *Asplanchna* preys on *Paramecium* and an ecosystem of continental extent in which lynxes prey on snowshoe hares has not deterred theorists from using a single model to "explain" both systems. While admitting that models designed for small, self-contained systems may be inapplicable to unbounded systems several orders of magnitude larger, theoreticians still do apply them, even if only to provide "crude caricatures" (in the words of May *et. al.*) (Pielou 1981, pp. 18-19).

My point is that if we hope for more than crude caricatures as scientific explanations we will have to relinquish a reliance on theoretical models as a means for obtaining them.

4. Realism, Robustness, and *Ceteris Paribus* Laws

I argued in the previous section that the predictive success of a theoretical model should lead us to suspect the truth, and thus the explanatory power, of its hypotheses. This is because when a model is a good predictor, its hypotheses are likely to be false. They therefore *cannot* be explanatory. I regard this as a strong defect of the use of the model building approach to scientific theories, and, as I shall argue below, as a reason for rejecting the semantic view of theories in biology.

In this section I consider an objection to my claim that we can avoid the conflict between prediction and explanation by eschewing theoretical models in the development and testing of theories. Nancy Cartwright has argued that where so-called "*ceteris paribus* laws" are concerned, successful explanatory hypotheses *must* be false. On her view this is a necessary consequence of the way laws explain, not an avoidable error. Moreover, this antagonism between truth and explanation *requires* theoretical models to figure centrally in explanation.

According to Cartwright, "covering laws" are scarce. (These laws are required by a logical empiricist covering-law account of explanation; they are so called because the explanatory laws are required to "cover," or deductively subsume, empirical phenomena). They are scarce because most empirical phenomena are complex, resulting from the interaction of several causes. A theoretical law that tells us what one cause contributes cannot be a covering law, for in order to perform its explanatory function, it must describe the effect of a single cause while denying the action of others. But then the law will have falsely described the actual, complex phenomenon. Hence, the law will not cover the phenomenon to be explained.

Most physical laws, Cartwright concludes, are "*ceteris paribus* laws." They state that a given cause will produce a given effect only if all other things are equal. These laws cannot be both true and explanatory. If, on the one hand, they are read with the *ceteris paribus* qualifier, they clearly cover no actual cases, and so do not explain; if, on the other hand, they are read without the qualifier, they are clearly false, and so cannot explain. Her conclusion is that we should abandon the covering-law model of explanation. If we want to maintain that scientific explanations are possible, we will have to give up the requirement that our explanatory laws must be true.

On Cartwright's view, physics allows two kinds of explanations: causal explanations, in which causes of a phenomenon are cited, and what might be called "theoretical" explanation, in which a phenomenon is "fit...into a general theoretical framework" (Cartwright 1983, p. 16). On the covering law account of explanation, phenomena are fit into a theory when empirical generalizations about them (Cartwright calls these "phenomenological laws") are derived from the basic laws of the theory. By contrast, the "simulacrum" account that she defends construes explanation as fitting phenomena into a theoretical framework by means of models. The fitting is accomplished by means of "bridge principles" that tell us which theoretical or mathematical principles of the theory to use, given the description we have of the phenomenon to be explained. Importantly, the bridge principles require that the description supply a particular kind of information, with a particular structure: the descriptions must be "prepared" in the right way, or the theory cannot be applied at all. But the "right" kind of description is rarely, if ever, a *true* description of the phenomenon: "generally the prepared and unprepared descriptions cannot be made to match" (Cartwright 1983, p. 17). The only way to fit the phenomena into our theoretical framework is to distort the true picture of them. Only when we create theoretical models of the phenomena, on whose objects theories can act, do we have any hope that those phenomena will treated by the theory:

> The fundamental laws do not govern reality. What they govern has only the appearance of reality and the appearance is far tidier and more readily regimented than reality itself (Cartwright 1983, p. 162).

On Cartwright's account, therefore, theoretical models are necessary stand-ins for descriptions of actual phenomena. They are distortions of the true descriptions, falsehoods tailored to fit the equations and principles the theory provides as an explanatory network. Only when there is a sufficiently good fit between phenomenon and theory via a theoretical model can the theory supply an explanation of the phenomenon. Yet strictly speaking it is the behavior of objects in the model that the theory explains. The model bears a relation to the actual phenomenon that is like the one a simulacrum bears to a real thing, "having merely the form or appearance of a certain thing, without possessing its substance or proper qualities" (Cartwright 1983, p. 17).

It would require much more space than I have here to give an adequate reply to this set of considerations. I will therefore merely state two points that a longer account would treat in detail. First, in my opinion, Cartwright reverses the problem and its solution. She claims that there is an unresolvable tension between truth and explanation that demands a critical role for theoretical models in explanation. I would maintain, rather, that it is our having given theoretical (especially mathematical) models such an important role that created the conflict between truth and explanation, as well as the tension between prediction and explanation discussed above. Once we recognize this, it is open to us to relinquish theoretical models as the *sine qua non* of scientific explanation, and to develop theories that are both realistic (that provide true accounts of the important factors underlying real systems whose behavior we want to understand) and robust (that are predictively successful).

Second, Cartwright's reasons for thinking that the conflict between truth and explanation is inescapable are circular, albeit subtly so. Ever since Galileo introduced idealized systems as *explananda*, the use of mathematical models has had tremendous advantages in the search for scientific explanations. As Cartwright points out, real systems are seldom (save possibly in the case of a very well-controlled experiment) as tidy as the models we create of them. In my view, the reason that Cartwright regards theoretical models as necessary in the end is that scientific explanation via theoretical models has been tacitly presupposed at the outset. She accepts as a basic fact about scientific explanation that fundamental laws only cite isolated causes, and do not cover complex situations with multiple causes. But only begging the question licenses the inference from "Fundamental laws do not fit real, complex cases" to "Fundamental laws cannot fit them." I want to claim that if we reexamine our commitment to the role of mathematical models in scientific practice, we will see that their ubiquity is not so much a necessity as a legacy. Their initial advantages, so obvious to the founders of modern science, have disappeared in the face of our need to fine-tune our explanations to the complexities of real cases.

5. Biological Complexity, Realism, and the Semantic View of Theories

I have argued that inference from the predictive success of a theoretical model to a general theoretical explanation is inadvisable. Such an inference depends on similar effects in diverse cases having identical causes — an unlikely state of affairs. Furthermore, there is by and large among scientific researchers a much greater emphasis on predictive success than on explanatory power. Many research scientists, when asked, for example, whether they regard their hypotheses as *true* or merely as *useful* predictive instruments, voice the latter, instrumentalist, and anti-realist view. Explanatory power — the *realistic* character of scientific theories — is usually forsaken for the sake of the theory's *robustness*, or predictive scope. It is because of this importance of robustness over realism in the minds of practicing scientists faced with a choice between them that I claim that reliance on theoretical models in the development and testing of scientific theories leads to anti-realism.

The argument given in section 3 was a probabilistic one. I claimed that the technique of model building, because it requires abstracting from the details of a case in order to pick out the "important" or "essential" features of a system, forces theorists to choose between predictive success and explanatory power. Predictive power is generally regarded as the more important immediate goal. As a consequence, it is unlikely that using the model-building approach will enable one to "guess" the true explanatory hypothesis. I now want to point out that one is far *less* likely to do this in biology than in the physical sciences. The reason for this is the complexity of biological systems that is due to the uniqueness, and hence diversity, of biological phenomena, in contrast with the uniformity of the subject matter of physics. Ernst Mayr's account of biological complexity locates its source in the uniqueness of biological systems. He wrote, for example, that

> The uniqueness of biological entities and phenomena is one of the major differences between biology and the physical sciences....If a physicist says "Ice floats on water," his statement is true for any piece of ice and any body of water. The members of a class usually lack the individuality that is so characteristic of the organic world where all individuals are unique, all stages in the life cycle are unique, all populations are unique, all species and higher categories are unique, all inter-individual contacts are unique, all natural associations of species are unique, and all evolutionary events are unique....Uniqueness, of course, does not entirely preclude prediction. We can make many valid statements about human attributes and human behavior and likewise about other organisms. But most of these statements (except taxonomic ones) have purely statistical validity....It is quite impossible to have, for unique phenomena, general laws like those existing in classical mechanics (Mayr 1961, p. 1502).[7]

There are at least three respects in which biological systems exceed physical systems in complexity. First, experiments in physics often attempt to study simplified or "stripped-down" systems in which all but one or a few influences have been removed. Here experimental systems can be made almost as simple as one pleases — indeed, one of the hallmarks of an ingenious experimenter is insight about which factors can be ignored in the design of the experiment. In order to study the effects of gravity on falling bodies, Galileo simplified his experimental system by using inclined planes to slow the motion of the balls and reduce the air resistance acting on them. In laboratory studies, to say nothing of hypothesis testing in field biology, one can only simplify the experimental set-up to a certain point. An evolutionary biologist studying enzyme production might choose to study *E. coli* instead of white mice, and a field ecologist might elect to study symbiosis in lichens instead of in orchids, but even these simplest of biological systems are more complex than the simple systems physicists study.

Second, physicists can *isolate* the systems they study, whereas in biology, biotic and abiotic environments cannot be ignored. It makes sense for physicists to study the interaction of lasers with atomic structure using an apparatus designed to shield the interaction from magnetic fields, but it would be absurd for biologists to ignore the influence of *E. coli* on each other or the exchanges between lichens and their algal and fungal symbionts. These interactions are essential parts of the system being investigated.

Third, biological things are complex and diverse, while physical things are complex and uniform. Therefore, the physicist can profit by developing complicated theoretical models that require much time and money to analyze, because his or her results are very likely to be *robust*. What is discovered about one population of atoms will hold, all other things being equal, for every other like population. In general this will not be the case for the biologist who attempts to use complex theoretical models for predictions about different populations of organisms: because biological systems are complex and diverse, a model developed with respect to one will more than likely not successfully predict what will be observed about another.

The solution proposed by two of the strongest advocates of theoretical models, Peter Richerson and Robert Boyd, is to begin with *simple* models — models that are simple enough to be completely understood, yet that are to some extent both *realistic*, i.e., that "reflect how the world does work," and *robust*, i.e., that yield predictions "that are at least qualitatively correct, at least for some range of situations."

However, simple models will not be both realistic and robust any more than complex models will be, for in building theoretical models one is abstracting from the details of the biological case, in order to obtain a result that matches observations of other cases. The more robust a model becomes, that is, the more it can successfully predict what will be observed in different, more general cases, the less realistic it will be, for the trade-off of generality for detail is greater. That is, the greater the coincidence of results (the greater the predictive success of the model), the less likely it is that the model puts forward what is the actual explanation in each case. Similarly, supposing that explanatory power were the primary goal, the more realistic the model becomes (the more attention it pays to the details of a particular case in order to ascertain the actual causes of the result to be explained), the less robust the model will be. To the extent that the model is sufficiently complex to incorporate enough of the details of a particular case to make probable the identification of the actual causes of the result to be explained, the less able it will be to make predictions about other, different cases, simply because their details will differ.

Diversity complicates biological systems by introducing complexity at the level of description. This is what makes it unwise to think, *a priori*, that a principle found to be explanatory in one case will be explanatory in others. This point has been noted by at least one other author:

> ...robust explanations in evolutionary biology have proved more difficult to obtain than was hoped. In my view, a good part [of] the reason difficulties have arisen is that the complexity of evolutionary (and ecological) systems has not always been confronted with sufficient respect. In attempting to achieve explanatory closure in terms of simple, one-dimensional mechanical systems, orthodox neo-Darwinians have cut off access to a full understanding of the phenomena they study. This self-limitation was indeed for a long time necessary and productive, but as the program has tried to extend itself to issues in molecular biology and sociobiology, the limitations have begun to look particularly serious (Dyke 1985, p. 108).

I do not intend my remarks on the uniformity of physical systems to suggest that I believe that *explanations* in physics will be simple. Nor do I want to draw the same skeptical conclusion about the possibility of biological laws to which Mayr was led by the recognition of biological diversity. My conclusion is that in order to discover reliable explanatory generalizations in biology, we must use methods other than that of theoretical modelling — in particular, we must return to the practice of natural history and the use of inductive methods such as the method of comparison. If we accept the semantic view of theories, a view that makes theoretical models an integral part of our conception of scientific theories and of our methods for developing and testing theories, we must accept anti-realism as well. Realists must therefore forgo the semantic view of theories in biology.

Notes

[1] I wish to thank Peter Achinstein, John Dupré, Robert Hilborn, Robert Rynasiewicz, Dan Rothbart, and especially Fred Suppe for helpful comments on earlier versions of this paper.

[2] On Suppe's (1989) view, in the end scientific theories make claims only about real systems as they *would be* were their behavior determined by a more limited set of parameters than are actually present. His account, by his own admission, is therefore only "quasi-realistic." Whether it is an account that is sufficiently realistic to satisfy other contemporary realists is an important question requiring further discussion. Limitations of space preclude a more extended discussion of his "semantic conception" of theories in this essay; however, it should be noted that his account does appear to escape the anti-realist consequences to which I call attention here.

[3]Among North American passerines, only 5% are polygynous. *Passerines,* "perching birds," are birds belonging to the order *Passeriformes* (familiar examples include sparrows, swallows, larks, thrushes, starlings, and wrens) (Robbins, Bruun, and Zim 1966).

[4]Because of several "inconsistencies" in the model proposed by Weatherhead and Robertson (1979), "misleading predictions" were made. These inconsistencies are corrected by Heisler (1981). Both Weatherhead and Robertson's model and Heisler's model give the same qualitative results. My discussion focuses on the Weatherhead and Robertson model, primarily because its simplicity serves the purpose of expositional clarity; conclusions about methodology are unaffected by the modifications Heisler made. I cannot give a more detailed account of the debate about this hypothesis and the evidence for it here. Wittenberger (1981) offers an especially thoughtful discussion; the same issue of the *American Naturalist* contains a number of other insightful articles, as well as a reply by Weatherhead and Robertson.

[5]Heisler shows that under the assumptions that polygynous and nonpolygynous males exist in equal proportions, harem size is 2, and that all males mate exactly like their fathers, taking the disadvantage to nonpolygynous males into account can lower the polygyny threshold to 0.33.

[6]In my view, and for similar reasons, not even the more limited conclusion is justified by predictive success. See my 1986 and 1989a,b.

[7]On biological diversity see also Kemp (1985), who notes that a basic assumption of evolutionary biology is "that evolution is the cause of diversity" (p. 144).

References

Beatty, J. (1980), "Optimal-Design Models and the Strategy of Model Building in Evolutionary Biology", *Philosophy of Science* 47: 532-561.

_ _ _ _ _ . (1981), "What's Wrong with the Received View of Evolutionary Theory?", in *PSA 1980*, Vol. 2, P. Asquith and R. Giere (eds.). East Lansing, MI: Philosophy of Science Association: 397-426.

Cartwright, N. (1983), *How the Laws of Physics Lie.* New York: Oxford University Press.

Dupré, J. (ed.) (1987), *The Latest on the Best: Essays on Evolution and Optimality.* Cambridge, MA: MIT Press.

Dyke, C. (1985), "Complexity and Closure", in *Evolution at a Crossroads*, D.J. Depew and B.H. Weber (eds.) Cambridge, MA: MIT Press, pp. 97-131.

Elseth, G.D. and Baumgardner, K.D. (1981), *Population Biology.* New York: Van Nostrand.

Heisler, I.L. (1981), "Offspring Quality and the Polygyny Threshold: A New Model for the "Sexy Son" Hypothesis", *American Naturalist* 117: 317-328.

Horan, B.L. (1986), "Sociobiology and the Semantic View of Theories", in *PSA 1986*, Vol. 1, P. Asquith and P. Machemer (eds.). East Lansing: Philosophy of Science Association, pp. 322-330.

_ _ _ _ _ _ . (1989a), "Functional Explanations in Sociobiology", *Biology and Philosophy*, Vol. 4, forthcoming.

_____ . (1989b), "Functional Explanations in Sociobiology: A Reply to Critics", *Biology and Philosophy*, Vol. 4, forthcoming.

Kemp, T.S. (1985), "Models of Diversity and Phylogenetic Reconstruction", in *Oxford Surveys in Evolutionary Biology*, Vol. 2, R. Dawkins and M. Ridley (eds.). New York: Oxford University Press, pp. 135-158.

Lloyd, E. (1984), "A Semantic Approach to the Structure of Population Genetics", *Philosophy of Science* 51: 242-264.

_____. (1986), "Thinking About Models in Evolutionary Theory", *Philosophica* 37: 87-100.

_____. (1987), "Confirmation of Ecological and Evolutionary Models", *Biology and Philosophy* 2: 277-293.

_____. (1988), *The Structure and Confirmation of Evolutionary Theory*. Westport, CT: Greenwood Press.

Mayr, E. (1961), "Cause and Effect in Biology", *Science* 134: 1501-1506.

Orians, G.H. (1969), "On the Evolution of Mating Systems in Birds and Mammals", *American Naturalist* 103: 589-603.

Pielou, E.C. (1981), "The Usefulness of Ecological Models: A Stocktaking", *Quarterly Review of Biology* 56: 17-31.

Richerson, P.J. and Boyd, R. (1987), "Simple Models of Complex Phenomena: The Case of Cultural Evolution", in *The Latest On the Best: Essays on Evolution and Optimality*, J. Dupré (ed.). Cambridge MA: MIT Press.

Robbins, C.S., Bruun, B. and Zim, H.S. (1966), *A Guide to Field Identification of Birds of North America*. New York: Golden Press.

Simmons, R.E. (1988), "Food and the Deceptive Acquisition of Mates by Polygynous Male Harriers", *Behavioral Ecology and Sociobiology* 23: 83-92.

Suppe, F. (1977), *The Structure of Scientific Theories*, 2nd ed. Champagne-Urbana: University of Illinois Press.

_____. (1989), *The Semantic Conception of Theories and Scientific Realism*. Champagne-Urbana: University of Illinois Press.

Suppes, P. (1960), "A Comparison of the Meaning and Uses of Models in Mathematics and the Empirical Sciences", *Synthese* 12: 287-301.

_____. (1962), "Models of Data", in *Logic, Methodology, and Philosophy of Science: Proceedings of the 1960 International Congress*, E. Nagel, P. Suppes and A. Tarski (eds.). Stanford: Stanford University Press, pp. 252-261.

_____. (1967), "What is a Scientific Theory?", in S. Morgenbesser (ed.) *Philosophy of Science Today*. New York: Basic Books.

Tarski, A. (1953), "A General Method in Proofs of Undecidability", in *Undecidable Theories*, A. Tarski, A. Mostowski, and R.M. Robinson (eds.). Amsterdam: North-Holland Publishing Co.

Thompson, P. (1983), "The Structure of Evolutionary Theory: A Semantic Approach", *Studies in History and Philosophy of Science* 14: 215-229.

_ _ _ _ _ _ _ . (1986), "The Interaction of Theories and the Semantic Conception of Evolutionary Theory", *Philosophica* 37: 73-86.

_ _ _ _ _ _ _ . (1988), *The Structure of Biological Theories*. Albany, NY: State University of New York Press.

Van Fraassen, B. (1967), "Meaning Relations among Predicates", *Nous* 1: 161-179.

_ _ _ _ _ _ _ _ _. (1970), "On the Extension of Beth's Semantics of Physical Theories", *Philosophy of Science* 37: 325-338.

_ _ _ _ _ _ _ _ _. (1972), "A Formal Approach to the Philosophy of Science", in R. Colodny (ed.) *Paradigms and Paradoxes*. Pittsburgh: University of Pittsburgh Press.

_ _ _ _ _ _ _ _ _. (1980), *The Scientific Image*. Oxford: Oxford University Press.

Verner, J. (1964), "The Evolution of Polygamy in the Long-billed Marsh Wren", *Evolution* 18: 252-261.

_ _ _ _ _ and Willson, F. (1966), "The Influence of Habitats on Mating Systems of North American Passerine Birds", *Ecology* 47: 143-147;

Weatherhead, P.J. and Robertson, R.J. (1977), "Harem size, territory quality and reproductive success in the redwinged blackbird (*Agelaius phoeniceus*)", *Canadian Journal of Zoology* 55: 1261-1267.

_ _ _ _ _ _ _ _ _ _ _ _ _ _ _ _ _ _ _ . (1979), "Offspring Quality and the Polygyny Threshold: the Sexy-Son Hypothesis", *American Naturalist* 113: 201-208.

_ . (1981), "In Defense of the 'Sexy Son' Hypothesis", *American Naturalist* 117: 349-356.

Wittenberger, J.F. (1981), "Male Quality and Polygyny: The 'Sexy Son' Hypothesis Revisited", *American Naturalist* 117: 329-342.

The Semantic Approach and its application to Evolutionary Theory[1]

Elisabeth A. Lloyd

University of California–Berkeley

1. Elements of the Semantic View

The semantic view of theory structure, as developed by Suppes (1957,1967), Suppe (1974, 1976, 1977, 1988), and van Fraassen (1970, 1972, 1980), represents theories as classes of models or structures. These models are, on the version of the semantic approach used here, defined by specifying their laws, parameters, and variables. The semantic approach to theory structure is simply a method of formalizing the content of scientific theories.

2. Application of the Semantic View to Evolutionary Theory

In a series of articles and a book, I have analyzed the structure of modern evolutionary theory using the semantic view as a framework (Lloyd 1984, 1986a, 1986b, 1987a, 1987b, 1988, forthcoming; cf. Thompson 1983, 1985,1988). I shall briefly recap the analyses I have done, in order to demonstrate the range of problems accessible to the semantic view.

First, I have used the semantic view to analyze the structure of population genetics models, including kin and group selection models. There is a heated debate in the genetics literature about whether kin selection models should be interpreted as group or as organismic selection models. I have offered an analysis of this problem utilizing the semantic view (1988, Ch. 5).

The semantic view is not applicable exclusively to mathematical models, however. I have also analyzed the basic structure of natural selection models, and the interrelations among the components of these models; these models are characterized non-mathematically (see esp. 1988, Ch. 6). I use this characterization of the basic structure of selection models to offer a new definition of a unit of selection in terms of its actual role in models. This definition, in turn, allows a precise formulation of several controversial problems involving units of selection.

For instance, I use my structural definition of a unit of selection to compare species selection models with other hierarchical selection models (Eldredge and Gould 1972; Gould and Eldredge 1977; Eldredge and Cracraft 1980; Vrba and Eldredge 1984; Vrba 1984; Vrba and Gould 1986). I find that there are certain discrepancies between the structure of species selection models and other selection models, and I suggest a new formulation of species selection models which is consistent with the general structure of selection models (1988, Ch. 6; Lloyd and Gould ms.).

PSA 1988, Volume 2, pp. 278-285

My analysis of the structure of population genetics models is also useful in understanding what is wrong with genic selectionism (1988, Ch. 7). The semantic view allows a precise formulation of exactly how and why genic selection models are bound to be empirically inadequate, under one interpretation, or trivially different from other models, under an alternate interpretation. In particular, I demonstrate why the attempted resurrection of genic selectionism by Sterelny and Kitcher (1988) misses the point of the debate.

Use of the semantic approach to theory structure allows precise formulation of various different questions about units of selection. Questions about which entities are functioning as replicators, and which as interactors, need to be kept completely distinct, and the semantic view allows the precise translation of this distinction in terms of laws and state spaces.

Finally, I also use the semantic view of theory structure to develop a schema of theory confirmation that is more subtle and sensitive to genuine scientific concerns than traditional approaches in philosophy of science (1988, Ch. 8).

3. Is the Semantic View doing any Real Work?

Anyone who wants to argue that the semantic view has done nothing for analyzing biological theories needs to show either that the use of the semantic view is doing no real work in the analyses reviewed above, or that the analyses themselves are ineffective. I would certainly not claim that using the semantic view formally is the only way to make progress on issues in philosophy of biology. I do find it suggestive, however, that the fine analytic work of Brandon, for instance, is completely compatible with, and in many ways, suggestive of, an informal use of the semantic view of theory structure (e.g., Brandon 1981, 1982). At any rate, I have no attachment to the "manifest destiny" of the semantic view. Workers using the semantic view have created careful reconstructions of some important parts of evolutionary theory using this approach, and the use of this analytic framework has, I would argue, contributed to the clarification of a number of issues in the philosophy of biology.

I know of a number of people who have dismissed the utility and value of the semantic view on account of some basic misconceptions. I would like to take some time today to review what I see as the three most common misconceptions about the semantic view.

4. Three Common Misconceptions about the Semantic View

By far the most common complaint is the following: "The semantic view necessarily involves anti-realism. I am a realist, therefore I must reject the semantic view".

As a matter of fact, the semantic view, as a view about theory structure, is *neutral* on the issue of scientific realism. There is nothing in the claim that scientific theories are usefully and clearly reconstructed as classes of model types which entails either realism or anti-realism. The epistemic attitude towards these models and the entities within them is distinct from the description of the structure of the models themselves.

It is especially surprising that people cling to the belief that the semantic view of theories is necessarily anti-realist, in the face of Ronald Giere's visible and long standing advocacy of both the semantic view of theory structure and realism (see Giere 1988). Fred Suppe's quasi-realist approach is yet another possible epistemic approach associated with the semantic view (Suppe 1988).

I do take it that Barbara Horan believes that the semantic view and anti-realism must go together. She, at least, has arguments for this view, rather than just assuming guilt by association. I shall address her argument later.

The second most common myth about the semantic view is that "the semantic view can only work with mathematical theories". This was never true, as evidenced by Suppe's analysis (1974) of biological taxonomy early on, and also by my work on Darwin (1983) and on species selection, and most recently, by James Griesemer's work on laboratory and museum models (Griesemer and Wade 1988, Griesemer ms.).

The supposed restriction to mathematical theories is no more true for the semantic view of theories than it is for model theory in general. It is easier to see how the semantic view would represent mathematical models, but this is in no way exclusive.

The third most popular misunderstanding of the semantic view, according to my informal survey is: "there is no substantive difference between the semantic view and the standard, received view of theory structure."

First, one logical point. If the received view is taken to require the use of first order logic only—which is the way it is often conceived—then any theory involving the real numbers is not representable within it, but would be within the semantic view. This formal point does not, however, strike me as getting to the heart of the matter. The sensible response is simply to lift the restriction to first order languages. Then the two approaches would seem to be equivalent.

But this is not quite right, because one of the advantages of the semantic view is that it, unlike syntactic approaches, is not restricted or committed to a particular *linguistic* formulation. This point has consequences for some rather important issues in theory identification and theory change. On the syntactic view, a change in linguistic formulation means that there is now a new theory. One can imagine many cases in which such a change would be trivial, yet the syntactic approach would still call the entity a new theory. On the semantic view, in contrast, two different axiomatic systems that are semantically equivalent—that is, they share the same set of models—constitute *one* theory, not two. In other words, the fact is that you can change the names without changing the meaning relations, and under the semantic view it is the relations which are taken to be *essential* in the description of the theoretical systems. Hence, there is an important difference between the semantic view and various syntactic approaches, namely that the semantic view avoids the worries about theory change and definition which arise from having particular linguistic formulations of a theory.

But let us step aside from these formal issues for a moment. Suppose someone were to say that the problems with committing to a particular language did not bother them. Let us imagine that the two systems of description of scientific theories are formally equivalent, that they contain precisely the same information, and are completely intertranslatable. Are there still reasons for preferring the semantic view? That depends on what you want to do. If you want a view of theory structure that can be used to analyze the empirical content of theories, discuss their interrelations with other theories, and examine how they are used by scientists, the semantic view has a clear advantage.

First, the semantic view is closer to the practices of scientists, and it provides a natural and convenient way of reconstructing theories and claims about those theories.

Second, the semantic view does not require laws of nature—a problematic concept, especially in evolutionary biology—though it also does not preclude the formulation and use of laws of nature.

Third, the semantic view has a better chance of representing scientists' problems in terms accessible to them, because it is closer to the form of scientists' own reasoning.

Fourth, the semantic view allows either a realist or anti-realist (or quasi-realist) interpretation of theories.

Fifth, the semantic view provides the framework for a much more subtle, faithful, and powerful analysis of how data can support or confirm an empirical claim.

In summary so far, I claim that rejections of the semantic view of theory structure often rely on mischaracterizations of this view. Furthermore, I think that Barbara Horan's objections to the semantic view lie in this camp, and I am especially sorry about that, because I also think Horan raises some important issues about theoretical explanations, testing, and the construction of scientific theories.

5. Horan's Criticisms of the Semantic View

I see two main problems with Horan's argument. The first is that she does not distinguish between philosophers' reconstructions of scientific theories and the scientific theories themselves. A *philosophical* preference for reconstructing and representing theories a particular way, as classes of models (semantic entities) is identified by Horan with the value of the *scientific* preference for certain abstract, general theories. But normative claims made by scientists about theory construction are *not* the same as normative claims made by philosophers about theory reconstruction. Without this identification, furthermore, Horan has no case against the semantic view.

The basic problem is that Horan identifies high-level theoretical models in population biology with the semantic models of the semantic view. Hence, she seems to think that (1) the semantic view demands these very high level models, and (2) that the semantic view can fairly be saddled with any problems associated with these high level models.

Let us consider the first claim, in which Horan identifies the semantic view with the demand for very high-level, abstract models. Horan writes, "when theoretical models are constructed one eliminates as much of the detail about the case one wants to explain as possible...instead, the model builder selects a few variables and uses them to make the model a successful predictor...ignoring the details of other cases about which one wants to generalize..." (1989, p. 269.)

Horan claims that these models are unrealistic, since they are so abstract and removed from the biological details and complexity of each case. Therefore, according to Horan, the use of models is supposed to result in anti-realism. *This* is the basis of her argument that the semantic view is necessarily anti-realist. In her words, if we accept the semantic view, "which makes theoretical models an integral part of our conception of theories, we must accept anti-realism as well". (1989, p. 274.)

But the claim that the semantic view of theories demands high-level theoretical models is simply false. What a semantic view theorist actually says is this: give me a scientific description of a system—any system—and I prefer to represent that system in terms of its meaning structure. The semantic approach can be used at *any* level of scientific theory, including extremely detailed, low-level theories.

In fact, I take this flexibility of the semantic view to be one of its *virtues*. Furthermore, in my discussion of genic selection, I criticize the reductionist approach by showing that the genic level models do not ordinarily contain enough information to describe accurately the systems they are intended to explain (1988, Ch. 7). In other words, the fact that more complexity and detail are required in the model is clearly delineated and *defended* using the semantic view of theories.

I conclude that the complexity of biological phenomena emphasized by Horan is no problem at all for the semantic view. In fact, it is rather good for business. With many, many mid- to low-level descriptions of natural systems running around, we are unlikely to run out of work anytime soon.

Let me also mention a second basic problem I see with Horan's account of the semantic view, which has to do with theory testing. I find that Horan uses the terms 'realism', 'robustness', and 'explanatory power' differently than I do—but here is my interpretation of her argument. She claims that the semantic view "licenses a rather problematic inference", namely, "an inference from the successful prediction by a theoretical model in one case to the conclusion that this model affords a general explanation that covers a wide variety of cases". (1989, p. 269.)

Frankly, I do not know anyone who would want to buy that type of inference without evidence, but let us consider whether it really does arise out of the semantic view. The problem highlighted by Horan is that predictive success is not sufficient evidence for truth. In her words, "in the absence of details about the individual cases to be covered by the explanatory hypothesis, its predictive success cannot be taken as sufficient grounds for concluding that it will be the correct explanation"; and for Horan, the correct explanation is the true one.

I believe that Horan is unjustified in blaming models themselves for the problems with realism. This point should be clear from the fact that the tension between explanation and truth is completely divorced from any issues about models. It is a logical point, emphasized by van Fraassen: as explanatory power goes up, the probability of truth goes down (see esp. van Fraassen 1985). That is, the logically stronger claim is less likely to be true. This is a problem for any realist position; it has nothing to do with whether the theory is presented as a class of models.

Horan then presents and ridicules the overreaching claims of sociobiologists (such as E.O. Wilson on human evolution) as examples of this problematic inference. Basically, the problem is that the details of causal mechanisms of these models cannot be assumed to be real, simply because the predictions turn out well.

I think that Horan is dead right about the problems with these cases. But I think it is ironic that she is trying to lay the sins of the sociobiologists at the doorstep of the semantic view, given that both Thompson and I have used the semantic view to analyze, in great detail, precisely what the evidential and theoretical problems are with sociobiological claims (Thompson 1985, 1988; Lloyd 1988, Ch. 8).

The claim at stake is that the high-level model can describe a range of natural systems. I made a point of looking at just this sort of case in my paper on confirming evolutionary models (1987a). There, I argued that providing a range of cases, i.e., a variety of evidence, is one form of confirming empirical claims made about models. But there is also the problem of supporting the assumptions of the model *independently*. In that paper, this type of independent testing was explicitly presented as a corrective to the overemphasis on prediction practiced by most philosophers, and by some biologists as well, particularly the ones Horan is talking about.

The usual hypothetico-deductive views of confirmation concentrate on the accuracy of the model outcome, that is, the prediction. This exclusive focus on the outcome is extremely misleading, I have argued, in many cases in evolutionary biology. As the semantic approach to theory structure makes very clear, assumptions made in constructing any model play a major role in theory; I have claimed they should likewise play a major role in evaluation of evidence.

In sum, I want to agree with Horan on two major issues. First, an obsession with developing extremely high-level, general models in biology can lead to sloppy and badly confirmed scientific claims that miss essential elements of natural systems. Second, an obsession with the predictive power of models, to the exclusion of concerns about the descriptive accuracy of details of the models, is a bad idea.

But I take it that, since I have argued for these points *using* the semantic view of theories, they provide no evidence against the semantic view. I have argued that Horan's rejection of the semantic view arises from her identifying that view with a particular (and problematic) scientific approach. I can see perfectly well how this might have come about. We semantic view theorists are always saying that the semantic view is good because it is closer to the practice of science. But I would like to set the record straight, and say, "not *that* close...".

I will conclude by mentioning a few areas in which I think the work that Thompson, Suppe, Griesemer and I have done on biological theories can be extended. For instance, what are the relations between the models and explanations in molecular biology and the models of population genetics and macroevolution? The way is paved to explore the question of reductionism with with sophistication and precision. The received view of theories has had a notoriously hard time explaining or describing the apparent reduction of Mendelian genetics to molecular genetics (cf. Hull 1974). Nevertheless, I think Nancy Maull and Lindley Darden defined the interrelations of the fields well, though in general terms (Darden and Maull 1977). Our work on the semantic approach to evolutionary theory provides, I would claim, the analytic tools and background research necessary to analyze the interrelations of these models with precision and sensitivity.

Alternatively, the topic of the interrelation between population genetics and molecular biology could be approached without worrying about reduction per se. The primary task might be, instead, to give a detailed description of how the results of one theory feed into and take from the results of another theory. This sort of analysis is very important, especially for the sorts of work on confirmation advocated by both Thompson and me.

In conclusion, there are many advantages to using the semantic approach to theory structure in biology. I want to urge that this approach can be especially valuable for describing the content of theories, for analyzing the interrelations of complicated theories with many parts (evolutionary theory, for example), and for helping to formulate and clarify the central scientific arguments occurring in evolutionary biology today.

Note

[1]I would like to thank Bas van Fraassen, Michael Dietrich, and James Griesemer for their helpful comments and discussion.

References

Brandon, R. (1981), "A Structural Description of Evolutionary Theory", *PSA 1980*, volume 2, 427-439. East Lansing, Mich.: Philosophy of Science Association.

_____ . (1982), "The Levels of Selection", *PSA 1982*, volume 1, 315-323. East Lansing, Mich.: Philosophy of Science Association.

Darden, L. and N. Maull (1977), "Interfield Theories", *Philosophy of Science* 1: 43-64.

Eldredge, N. and J. Cracraft (1980), *Phylogenetic Patterns and the Evolutionary Process*. new York: Columbia University Press.

Eldredge, N. and S.J. Gould (1972), "Punctuated equilibria: An alternative to phyletic gradualism", in *Models in Paleobiology*, T.J.M. Schopf (ed.). San Francisco: W.H. Freeman, pp. 82-115.

284

Giere, R. (1988), *Explaining Science*. Chicago: University of Chicago Press

Gould, S.J. and N. Eldredge (1977) "Punctuated equilibria: Tempo and mode of evolution reconsidered", *Paleobiology* 3: 115-151.

Griesemer, J.R. (ms), "Ecology and Abstraction: Theoretical Modeling in the Museum of Vertebrate Zoology".

_____ . and M. Wade (1988), "Laboratory Models, causal explanation, and group selection", *Biology and Philosophy* 3: 67-96.

Horan, B. (1989), "Theoretical models, biological complexity, and the semantic view of theories", *PSA 1988*, volume 2, 265-277. East Lansing, Mich.: Phillosophy of Science.

Hull, D.H. (1974), *Philosophy of Biological Science*. Englewood Cliffs, N.J.: Prentice-Hall.

Lloyd, E.A. (1983), "The Structure of Darwin's support for the theory of natural selection", *Philosophy of Science* 50: 112-129.

_____ . (1984), "A semantic approach to the structure of population genetics", *Philosophy of Science* 51: 242-264.

_____ . (1986a), "Thinking about models in evolutionary theory", *Philosophica* 37: 87-100.

_____ . (1986b), "Evaluation of evidence in group selection debates", *PSA 1986*, volume 1, 483-493. East Lansing, Mich.: Philosophy of Science Association.

_____ . (1987a), "Confirmation of evolutionary and ecological models", *Biology and Philosophy* 2: 277-293.

_____ . (1987b), "Response to Sloep and Van der Steen", *Biology and Philosophy* 2: 23-26.

_____ . (1988), *The Structure and Confirmation of Evolutionary Theory*. Westport, Conn.: Greenwood Press.

_____ . (forthcoming), "A structural approach to defining units of selection", *Philosophy of Science* 56.

_____ and S.J. Gould (ms), "Species selection on variability".

Sterelny, K. and P. Kitcher (1988), "The return of the gene", *Journal of Philosophy* 85: 339-361.

Suppe, F. (1974), "Some philosophical problems in biological speciation and taxonomy", in *Conceptual Basis of the Classification of Knowledge*, J.A. Wojcieckowske (ed.). Munich: Verlag Dokumentation, pp. 190-243.

_____ . (1976), "Theoretical Laws", in *Formal Methods of the Methodology of Science*, M. Prezelecke, K. Szaniawski, and R. Wojcicki (eds.). Wroclow: Ossolineum, pp. 247-267.

_____ . (1977), *The Structure of Scientific Theories* (2nd ed.) Urbana, Ill.: University of Illinois Press.

_ _ _ _ _ . (1988), *The Semantic Conception of Theories and Scientific Realism*. Urbana, Ill.: University of Illinois Press.

Suppes, P. (1957), *Introduction to Logic*. Princeton, N.J: Princeton University Press.

_ _ _ _ _ _. (1967), "What is a scientific theory?", in *Philosophy of Science Today*, S. Morgenbesser (ed.). New York: Meridian, pp. 55-67.

Thompson, P. (1983), "The structure of evolutionary theory: A semantic perspective", *Studies in History and Philosophy of Science* 14: 215-229.

_ _ _ _ _ _ _ . (1985), "Sociobiological explanation and the testability of sociobiological theory", in *Sociobiology and Epistemology*, J.H. Fetzer (ed.). Dordrecht: Reidel, pp. 201-215.

_ _ _ _ _ _ _ . (1988) *The Structure of Biological Theories*. Albany, NY: State University of New York Press.

van Fraassen, B.C. (1970), "On the extension of Beth's semantics of physical theories", *Philosophy of Science* 37: 325-339.

_ _ _ _ _ _ _ _ _ _. (1972), "A formal approach to the philosophy of science", in *Paradigms and Paradoxes*, R. Colodny (ed.). Pittsburgh: University of Pittsburgh Press.

_ _ _ _ _ _ _ _ _ _. (1980) *The Scientific Image*. Oxford: Clarendon

_ _ _ _ _ _ _ _ _ _. (1985) "Empiricism in the Philosophy of Science", in *Images of Science*, P.M. Churchland and C.A. Hooker (eds.). Chicago: University of Chicago Press.

Vrba, E.S. (1984), "What is species selection?", *Systematic Zoology* 33: 318-328.

Vrba, E.S. and N. Eldredge (1984), "Individuals, hierarchies and processes: Towards a more complete evolutionary theory", *Paleobiology* 10: 146-171.

Vrba, E.S. and S.J. Gould (1986), "The hierarchical expansion of sorting and selection: Sorting and selection cannot be equated", *Paleobiology* 12: 217-228.

Explanation in the Semantic Conception of Theory Structure

Paul Thompson

University of Toronto

1. Historical Background

The semantic conception of theories has a relatively short history the beginnings of which Frederick Suppe has traced to von Neumann (Suppe 1988). Two other early initiators and advocates were Evert Beth in 1948-49 (Beth 1948, 1949; see also 1961) and Patrick Suppes in 1957 in his *Introduction to Logic* (Suppes, 1957). Beth advanced what has become known as a state space approach while Suppes advanced a set-theoretical predicate approach.

Suppes suggested that scientific theories are more appropriately formalized as set-theoretical predicates. Shortly thereafter in 1961 Robert Stoll in his *Set Theory and Logic* (Stoll, 1961) made a similar claim about the formalization of informal theories of which scientific theories are instances. Suppes wrote a number of papers during the 1960's in which he indicated the features of theories that were better represented on a set-theoretical predicate conception of theories (see, for example Suppes, 1962). And, in 1967, he wrote a brief non-technical account of his view (Suppes, 1967). In this paper he quite clearly sets out his reasons for rejecting what he calls the "standard sketch of scientific theories" and for adopting a semantic conception. His central thesis is that scientific theories are not appropriately or usefully formalized as axiomatizations in mathematical logic but rather in set-theory.

The main thrust of his argument was that correspondence rules (he calls them co-ordinating definitions) "do not in the sense of modern logic provide an adequate semantics for the formal calculus" (Suppes, 1967, p.57). One should instead talk about models of the theory. These models are non-linguistic entities that are highly abstract and are far removed from the empirical phenomena to which they will be applied. As I have argued in numerous papers and most recently in my forthcoming book and will argue again below, this feature is one of the major strengths of this conception since it makes intelligible, indeed necessary, the actual use of a multiple number of theories in the application of the theory to phenomena.

Suppes suggests two reasons why the syntactic view is so widely held, despite what he argues are logical and practical weaknesses with it. First, philosopher's examples of scientific theories are usually fairly simple and, therefore, easily able to be given a linguistic formulation. Not surprisingly most examples used to explicate and defend the syntactic

PSA 1988, Volume 2, pp. 286-296

view are drawn from Newtonian Mechanics - and from a reasonably simple and sketchy account of it. Also not surprisingly, the advocates of the semantic view discuss complex theories such as quantum mechanics (see van Fraassen, 1972), learning theory (see Suppes, 1962), and evolutionary theory (see Beatty, 1980a, 1980b; Lloyd, 1983, 1984, 1986, 1987; Thompson, 1983b, 1985, 1986, 1987, 1988a). Second, there is a much more sophisticated mathematical character to discussions of models of a theory than to discussions of axiomatic-deductive structures partially interpreted by correspondence rules.

During the late 1960's and the 1970's the semantic conception was consolidated and extended by a number of philosophers from a variety of perspectives (see, for example: Suppe, 1967, 1972a, 1972b, 1974, 1976; van Fraassen, 1970, 1972, 1980; Sneed, 1971; Stegmuller, 1976). Despite this coalescing of the conception, there were and continue to be important differences of motivation and structure between the views of these advocates. During the 1980's John Beatty, Elisabeth Lloyd and I have been extending and applying the semantic conception in the context of biology and, in particular in context of evolutionary theory and genetics (see Beatty, 1980a, 1980b; Lloyd, 1983, 1984, 1986, 1987; Thompson, 1983b, 1985, 1986, 1987, 1988a, 1988b, 1989). We, of course, were not the first to apply this view to Biology - Frederick Suppe for example applied it to aspects of taxonomy in 197? - and, I hope, will not be the last. The formulation of the conception that I have employed is mainly due to Frederick Suppe and Bas van Fraassen.

2. A Sketch of the Semantic Conception

The semantic conception is so called because scientific theories are formalized in terms of models (i.e., semantic structures) and, hence, an adequate formal approach to the structure of scientific theories consists in the direct specification of the models (i.e., the semantics) and not in the specification of a linguistic axiomatic-deductive system (i.e., a syntax). The significant differences, therefore, between syntactic and semantic accounts are the nature of an adequate semantics of a scientific theory and the nature of an adequate (logically and heuristically) formalization of a scientific theory. On the syntactic conception, the semantics of a theory are provided by correspondence rules. On a semantic conception the semantics of a theory are provided directly by defining a class of models. For Patrick Suppes, the class of models is directly defined by defining a set-theoretical predicate. For Bas van Fraassen and Frederick Suppe, the class of models is defined in terms of a phase space or state space (i.e., a topological structure). One point of difference between van Fraassen and Suppe is that van Fraassen identifies theories with state spaces whereas Suppe understands state spaces as "canonical iconic models of theories" (Suppe 1972a, p.161, note 18) or "canonical mathematical replicas of theories" (Suppe 1977, pp.227-228, note 565).

One of the major consequences of the differences in the semantics of the syntactic and semantic conceptions is that the class of models, directly specified in terms of set theory or a state space, is an extralinguistic, highly abstract entity which is most often quite removed from the phenomena to which it is intended to apply. The relationship of a model to phenomena is one of isomorphism and the establishment of the isomorphism is a complex task not specified by the theory.

In the syntactic conception, on the other hand the semantics is provided by correspondence rules which *are* part of the theory and directly link the formal system to the phenomenal world. In effect the correspondence rules define an empirical model of the formal system. That empirical model is understood as logically equivalent to the phenomenal system to which the theory applies. It is for this reason that actual phenomena can be deduced from the statements of the theory. That is, the statements of the theory are laws that describe the *actual* behavior of objects in the world. Hence, any behavior deduced from the statements of the theory is either a prediction about what *actually* will happen under the specified circumstance or an explanation of what *actually* did happen under the specified circumstances. The interpreted formal system directly describes the behavior of entities in the world.

In a semantic conception, a theory is defined directly by specifying in mathematic English the behavior of a system. Most importantly, laws do not describe the behavior of objects in the world, they specify the nature and behavior of an abstract system. This abstract system is, independently of its specification, claimed to be isomorphic to a particular empirical system. Establishing this isomorphism, as I shall argue below, requires the employment of a range of other scientific theories and the adoption of theories of methodology (e.g., theories of experimental design, goodness of fit, etc.).

In essence, the substance of the difference between the two conceptions is that the semantic conception calls into question the possibility of providing an adequate semantics for a scientific theory by means of correspondence rules. And, calls into question the need for any reference to a formal system since the semantics can be provided directly by defining a mathematical model. Advocates of the semantic view have seen the separation of the theory and the method of its application as a major logical, heuristic and methodological advantage of the conception.

Before leaving this sketch of the semantic conception, I want to emphasize one important similarity between the syntactic and semantic conceptions, namely, they are both conceptions of the formal structure of theories. There is no comfort to be found in the semantic conception for those philosophers who dispute the appropriateness, and usefulness of formalization in one or all branches of science. Those of us who espouse the semantic conception are, like those who espouse the syntactic conception, committed to the value of formalization in science and philosophy of science. Patrick Suppes provides the clearest statement of this point:

> The sense of formalization I shall use in the subsequent discussion is just that of a standard set-theoretical formulation. I do not want to mean by formalization the stricter conception of a first-order theory that assumes only elementary logic. Such stricter formalization is appropriate for the intensive study of many elementary domains of mathematics, but in almost all areas of science a rich mathematical apparatus is needed. We can properly appeal to that apparatus within a set-theoretical framework (Suppes, 1968, p.653).

The paper from which this quotation is drawn extols the virtues of formalization in the sense set out in the quotation.

3. Problems with Explanation in the Syntactic Conception

A major criticism of the syntactic conception is that specifying the procedures for applying the theory to phenomena by means of correspondence rules which are part of the theory does not provide an accurate representation of the kinds of ways in which theories are applied to phenomena. And, since explanation involves applying theories to phenomena, any inadequacy in the account of the relationship between a theory and phenomena in a syntactic conception is *ipso facto* an inadequacy in its account of explanation. In this paper, I shall set out three ways in which the syntactic conception fails to represent accurately the relationship between a theory and phenomena. The first, introduced by Kenneth Schaffner, is that the role of laws from other theories is ignored. The second, introduced by Patrick Suppes, is that the need for reference to numerous other theories is ignored. The third, introduced by Bas van Fraassen and myself, is that the need, in many cases, for an interaction of several different theories is ignored. I first outline these deficiencies of the syntactic conception, then I offer three examples of the significance of these deficiencies to biological explanation and of how they are eliminated in a semantic conception.

Kenneth Schaffner (Schaffner 1969) has convincingly argued that specifying the ways in which a theory relates to phenomena in terms of correspondence rules ignores the ways in which laws from other independent theories are employed in 'causal sequences' which causally relate theories to phenomena. These causal sequences describe the causal mechanisms underlying the measurement and observation procedures which are used to apply the theory. That is, they explain why the measuring device or observation apparatus behaves the way it does and, hence, why it is acceptable to use a particular procedure to obtain observations relevant to the theory. Since the correspondence rules of the syntactic conception provide no role for laws of other theories in the relating of a particular theory to phenomena, two distinct theories cannot be interactively employed (see, Nagel *et al*, 1962).

For example, the theoretical term 'chromosome' in a theory about chromosomal segregation during meiosis is connected to observation terms by correspondence rules which will link the term to entities which have certain physical characteristics, staining properties, locations in a cell, behaviors, etc. as seen using a light microscope or an electron microscope. In order to explain, however, just why what is seen using the microscope should have anything to do with chromosomes, one needs to provide a causal account of how the microscope works. That is, in order to assert the connection asserted by the correspondence rules one needs to employ laws of optics (in the case of a light microscope) or laws of sub-atomic physics (in the case of an electron microscope). Therefore, relating theories to the world involves employing the laws and correspondence rules of other theories in causal sequences which render comprehensible the rationale for providing the link.

In the syntactic conception these other laws and correspondence rules play no role in relating the theory to the world. Hence, the syntactic conception does not provide an accurate account of the ways in which several theories can be used to provide a causal sequence of laws in which laws of several theories link a designated theory with observation reports.

The second deficiency of the syntactic conception is the distorted picture it provides of the experimental procedures employed in relating theories to phenomena. As Suppes (Suppes 1962) has argued, there is a hierarchy of theories which mediate between a designated theory and an experimental situation.

Suppes had identified three theories which, along with numerous unstated *ceteris paribus* conditions, mediate between a physical theory (which stands at the top of the hierarchy of theories) and phenomena: the theory of the experiment, the theory of data and the theory of experimental design (which is where the ceteris paribus conditions are relevant). Possible realizations of the theories are models of the theories.

A third deficiency of the syntactic view is that two or more theories cannot easily or naturally be employed conjointly. In a syntactic conception the conjoint employment of several theories requires a simultaneous axiomatization in a single theory of the component theories because correspondence rules provide a global meaning structure for an axiomatized formal system. Since different theories will have *different global* meaning structures, they cannot, as separate theories, interact. Some terms of one theory will not occur in the other theory or theories. Hence, those terms will be meaningless in the other theory or whatever meaning they are given will be different from the meaning given in the first theory because they will be given meaning with a different global meaning structure. This problem, as I argue below, is at the heart of the difficulties involved in providing an axiomatization in first order predicate logic of evolutionary theory.

In these three ways correspondence rules fail to accurately capture the complexity and richness of the relationship between a theory and phenomena (see also van Fraassen 1981). I now turn to three biological examples of deficiencies of the syntactic conception and of the richness of the semantic conception.

4. Genetics/Cytology Example

In a semantic conception, unlike in a syntactic conception, the relationship of a theory to phenomena is not specified by the theory and a multitude of laws from other theories can be employed in a causal sequence in relating the principal theory to phenomena. One important consequence of the employment of laws from other theories is that one can justify the assertion of a theory/phenomena isomorphism even though, considered in isolation, the isomorphism seems not to obtain.

Consider a simple example of this kind of role for imported theories. Population genetical theory describes a breeding system, the principal laws of which are Mendel's law of independent assortment and his law of segregation. A wide array of populations of organisms including humans are held to be systems of this kind. However, because of crossing over, linkage, inversion, translocation, meiotic drive and a whole host of other factors, almost no population behaves exactly the way a Mendelian system, as specified by the theory, behaves. What is needed here is a theory about linkage, inversions, etc. that specifies the effects these phenomena will have on the structure of actual populations. In this case the theoretical grounding comes from cytological theory and molecular genetics.

The relationship between population genetics and cytological theory is best understood as follows. Cytological theory, neither the terms of which nor the laws of which are part of population genetics, is employed in applying (relating) population genetics to phenomenal systems (populations of organisms). By employing cytological theory, population genetics can be related in a mediated way to populations. What cytological theory does, in effect, is to provide a basis for asserting that actual populations are systems of the kind described by population genetics even though there are clear differences in the behavior of the two systems. It does this by providing accounts of linkage, translocation, etc. which make clear that the differences between the systems are not a result of the inapplicability of the theory to the particular phenomena but of other causal factors described by other theories.

The upshot of this is that, in the syntactic conception, failure of the phenomena to be deducible from the theory has but two remedies: rejection of the theory - a drastic and seldom employed remedy if no alternative theory is available - or the development of *ad hoc* hypotheses which will permit the deduction of the phenomena.

In a semantic conception, the relationship between the theory and phenomena is one of asserted structural and behavioral identity. Establishing the truth of this assertion is an extra-theoretical task in which it is impossible to employ other theories in a manner similar to the above employment of cytological theory.

5. An Example of Genetic Explanation

Consider another case in which a multiple number of theories are employed in biological explanation. Some persons manifest impaired growth and development, increased susceptibility to infection, severe abdominal pain with normal bowel sounds and no rebound tenderness, paleness and physical weakness. These symptoms are explained in certain cases by reference to hemolytic anemia resulting from the sickling of red blood cells. This explanation appeals to a theory of physiology which describes, among other things, the structure, internal behavior and function of red blood cells in the body as well as the physiological effects of an abnormal red blood cell structure. The sickling of the red blood cells is explained by reference to the fact that, in the case of those with sickle-cell anemia, the nucleotide sequence, which determines the structure of the protein 'hemoglobin', codes for the amino acid 'valine' in the sixth position on the , -chain instead of 'glutamic acid' as is normally the case. This substitution, under reduced oxygen tension, results in the valines as positions 6 and 1 forming a hydrophobic association which leads to a conformation that stacks in such a way as to distort the erythrocyte and thus cause the sickle shape. This explanation appeals to molecular biological theory which describes, among numerous other things, how the nucleotide sequences of DNA determine the structure of proteins and to biochemical theory which describes the behavior of biochemical structures. Hence, by appeal to a number of different characteristics cites and a sequence of nucleotides is made and the physical characteristics are explained. Laws from these theories are used in a causal sequence linking molecular genetical theory (the putative explanatory theory) to the physical characteristic to be explained.

Given the character of the physical condition and the high mortality rate associated with it, an evolutionary explanation of the persistence of this physical condition has been sought. The current explanation of this persistence refers to the increased fitness, in malaria infested environments, of individuals who are heterozygous for the sickle-cell allele (i.e., have one allele for sickle-cell hemoglobin and one for normal hemoglobin at the same locus). According to this explanation, the heterozygote has slightly increased protection against *Plasmodium falciparum* and hence, in malaria infested areas, gains a selective advantage over either homozygote - one homozygote will suffer from sickle-cell anemia and the other will have no protection against malaria. This explanation involves an application of the population genetical component of evolutionary theory to the phenomenon of sickle-cell anemia.

The application of evolutionary theory to sickle-cell anemia in this explanation is mediated by the same causal sequence of theories as outlined in the above discussion of the explanation of the physical condition. That is, the physical condition described above is related to the reduced oxygen transport in the blood stream by a theory of physiology which describes the effects of reduced oxygen transport on the organism. The reduced oxygen transport to the cells is related to the sickling of the red blood cells by biochemical theory which describes the effect of the hydrophobic association of the valines at positions 1 and 6 on the -chain of the protein hemoglobin. The presence of valine at the sixth position on -chain is related to the nucleotide sequence of the protein hemoglobin (the allelic combination) is related to evolutionary selective pressure by population genetics. An adequate explanation, therefore, involves a large number of theories employed in a causal sequential manner.

I hope these two examples illustrate how adequate biological explanations involve the employment of a number of quite difference theories. The pattern that emerges is that one theory is the putative explanatory theory because it is at the beginning of the chain theories employed. This theory is linked (related) to the phenomenon to be explained by means of the other theories. Unlike in the syntactic conception, in the semantic conception the ways in which a theory relates to the world is not specified by the theory through correspondence rules. The task of relating the putative explanatory theory is an extra-theoretical task. Hence, other theories can be employed in relating a putative explanatory theory to phenomena. Such is not the case in the syntactic conception. I turn now to the third example.

6. Evolutionary Explanation

A major reason why I have argued for the semantic conception of theory structure is that I consider it to be the conception which most accurately captures the structure of evolutionary theory. I have argued in a number of papers and most recently in my forthcoming book that the syntactic accounts of evolutionary theory given by Ruse (Ruse 1973, 1977), and by Williams and Rosenberg (Williams, 1970, 1973a, 1973b, 1982; Rosenberg 1981, 1985; Rosenberg and Williams 1986) are inadequate and that at the root of their inadequacy is the fact that they distort evolutionary theory in order to have a unified theory to axiomatize. Ruse distorts it by identifying it with population genetic theory and Williams and Rosenberg distort it by identifying it with selection theory.

I have argued that fundamental to any formalization of an adequate evolutionary theoretical framework is the need to be able to represent it as a family of interacting theories. In particular, the interaction of population genetics, selection theory, ecology, embryology and many others. None of these separately constitute evolutionary theory nor is evolutionary explanation possible using any one in isolation from the others.

Consider, for example, Ruse's account (Ruse 1973, pp. 52-59) of the evolutionary explanation of Darwin's finches - a subfamily of the finch family which is found on the Galapagos Archipelago. Ruse, drawing on work done by Lack (Lack 1947), points out that there is a cluster of phenomena to be explained in the case of Darwin's finches. First, there is the fact that the finches on the islands are often similar to the finches on the mainland but differ in important ways from them. Second, on many of the islands there are endemic forms of Darwin's finches. That is, finches that are similar to those on each other islands but differ in important ways. Third, this similarity is found primarily in birds, reptiles, and insects. In general, islands which are some distance from the mainland do not have mammals and in the rare cases where they do the mammals are very different from those on the mainland. These phenomena all have to do with the geographic distribution of organisms.

The principal explanatory concept used by Lack for explaining these phenomena is 'geographical isolation' which entails 'reproductive isolation'. The more isolated islands are from each other and from the mainland by bodies of water which are difficult to traverse, the greater will be the evolutionary divergence of finches on each island from those on other islands and from those on the mainland. Geographic isolation, therefore, explains the differences: populations of finches reproductively isolated from each other will evolve in different ways. It also explains the absence of mammals since the greater the difficulty in traversing the distance between the islands and the mainland, the less likely mammals are to attempt it or to succeed. It also explains presence of the finches (and other organisms such as reptiles and insects) since finches have some chance of traversing the distance between the mainland and between islands or of being blown off course and ending up on an island different from their home island or their mainland home. This in turn explains the similarity of island finches to each other and to those on the mainland: although infrequent because of the natural timidity of finches to try the crossing, finches do traverse the distance. Hence, founding populations on an island will have come from the mainland or from another island and once a founding population is established there will be an extremely low level of immigration from the mainland and other islands. Hence, there will be sufficient reproductive isolation to allow evolutionary divergence to occur.

Ruse concedes that this explanation is most likely correct. He argues, however, that it is incomplete. What is missing is an explanation of why isolated populations of organisms should evolve into new species. To answer this question Ruse contends that one must turn to population genetic theory. Population genetic theory can explain why isolated gene pools undergo changes that result in the evolution of new species. Indeed, using population genetics one can explain why a very small founding population (perhaps even one gravid female) can very rapidly evolve into a new species. The genetics of this phenomena of rapid speciation is referred to as the founder effect (or principle). For Ruse the essence of the evolutionary explanation of the various features of this phenomenon is population genetical.

Ruse, of course, is correct in pointing out that explaining why isolated populations evolve into new species requires appeal to population genetic theory. However, his charge of incompleteness against Lack, can be leveled at his own account. While the biogeographic account given by Lack explains the various phenomena, it does so by assuming that isolated populations will evolve into new species. Also, however, while the population genetical account given by Ruse explains the phenomena, it assumes, among other things, reproductive isolation and selective pressure. The biogeographic account may need to be supplemented by population genetic theory and selection theory, but the genetic account needs to be supplemented by biogeography and selection theory. How, other than by an appeal to biogeographical theorizing, could one explain adequately the phenomena of isolation which is required for the population genetical explanation to be justified? How, other than an appeal to theorizing about the principles under which selection occurs (which will involve ecological theorizing), could one explain adequately the occurrence of selective pressure (different on each population) on the island populations which is required for the population genetical explanation to be justified? Without input from these other domains of theorizing, the population genetical explanation is as incomplete and speculative as the biogeographic account alone or an ecological account alone. In short all three (and no doubt more) theoretical domains need to be brought to bear in order for an adequate *evolutionary* explanation to be given.

In this case the relevant principal theory were it sufficiently well developed could be biogeography with inputs from population genetic theory and selection theory. It is not, however, in my opinion well enough developed. Hence, population genetic theory can serve as the principal theory with inputs from selection theory and biogeography. There is no primacy of population genetic theory here, only a practical decision based on the developed state of the theory. All the theories are equally required for an adequate explanation to be given. Which one of the theories is chosen as primary in a given case is based on numerous considerations including the domain of theorizing within which the explanation is being developed as well as the sophistication of the various theories relative to the need for sophistication in the particular case.

This example illustrates that a Darwinian evolutionary explanation involves the conjoint employment of several different theories which conjointly comprise what is commonly called Darwinian evolutionary theory. Therefore, any formal account of the logic of Darwinian evolutionary theory must capture this interactive character. The semantic conception of theory structure is well suited to this task. The syntactic conception is not.

In a semantic conception, two theories can be conjointly employed in explanation. The semantic conception, unlike the syntactic conception, accommodates complex explanatory frameworks because theories - which are extralinguistic entities that define a class of models - can interact. Hence, a phenomenon can be explained by choosing from the theories that make up the framework, the theory most relevant to the phenomenon and then using other theories in the framework to determine the values of the required inputs to the principal theory. In this way, other theories can be used in applying a given theory to empirical phenomena even though they are independent of the principal theory.

7. Conclusion

My conclusion on the basis of the features, set out in this paper, of biological explanation and of the semantic and syntactic conceptions of theories is that the semantic conception of theories more faithfully and richly describes explanation and theory structure in biology.

References

Beatty, J. (1980a), "Optimal-Design Models and the Strategy of Model Building in Evolutionary Biology" *Philosophy of Science* 47:532-561.

_____ . (1980b), "What's Wrong with the Received View of Evolutionary Theory?" In P.D. Asquith and R.N. Giere (eds.) *PSA 1980*, vol. 2, East Lansing: Philosophy of Science Association.

_____ . (1987), "On Behalf of the Semantic View," *Biology and Philosophy* 2: 17-23.

Beth, E. (1948), *Natuurphilosophie* Gorinchem: Noorduyn.

_ _ _ _ . (1949), "Towards an Up-to-Date Philosophy of the Natural Sciences," *Methodos* 1: 178-185.

_ _ _ _ . (1961), "Semantics of Physical Theories," in H. Freudenthal (ed.) *The Concept and the Role of the Model in Mathematics and Natural and Social Sciences* Dordrecht: Reidel, pp. 48-51.

Lack, D. (1947), *Darwin's Finches: An Essay on the General Biological Theory of Evolution*. Cambridge: Cambridge University Press.

Lloyd, E. (1983), "The Nature of Darwin's Support for the Theory of Natural Selection," *Philosophy of Science*: 50:112-129.

_ _ _ _ _. (1984), "A Semantic Approach to the Structure of Population Genetics," *Philosophy of Science* 51:242-264.

_ _ _ _ _. (1986), "Thinking about Models in Evolutionary Theory," *Philosophica* 37:87-100.

_ _ _ _ _. (1987), "Confirmation of Ecological and Evolutionary Models," *Biology and Philosophy* 2: 277-293.

Rosenberg, A. (1981), "The Interaction of Evolutionary and Genetic Theory," in L.W. Sumner, J.G. Slater and F.F. Wilson (eds.) *Pragmatism and Purpose: Essays Presented to Thomas Goudge* Toronto: University of Toronto Press.

_ _ _ _ _ _ _. (1985), *The Structure of Biological Science Cambridge*: Cambridge University Press.

_ _ _ _ _ _ _. and Williams, M. (1986), "Discussion of Fitness as Primitive and Propensity," *Philosophy of Science* 53: 412-418.

Ruse, M. (1973), *The Philosophy of Biology*, London: Hutchinson & Co. Ltd.

_ _ _ _ _. (1977), "Is Biology Different from Physics?" in R. Colodny (ed.) *Logic, Laws, and Life* Pittsburgh: Pittsburgh University Press.

Schaffner, K.F. (1969), "Correspondence Rules" *Philosophy of Science* 36: 280-290.

Sneed, J. (1971), The Logical Structure of Mathematical Physics. Dordrecht: D. Reidel.

Stegmuller, W. (1976), *The Structure and Dynamics of Theories*. New York: Spring-Verlag.

Stoll, R.R. (1963), *Set Theory and Logic*. San Francisco: W.H. Freeman and Co.

Suppe, F. (1967), *On the Meaning and Use of Models in Mathematics and the Exact Sciences*. Ann Arbor: University Microfilms International (Ph.D. Dissertation).

_ _ _ _ _. (1972a), "Theories, their Formulations, and the Operational Imperative," *Synthese* 25: 129-164.

_ _ _ _ _. (1972b), "What's Wrong with the Received View on the Structure of Scientific Theories?" *Philosophy of Science* 39:1-19.

_____. (1974), "Theories and Phenomena," In W. Leinfellner, and E. Kohler, (eds.), *Development in the Methodology of Social Science*. Dordrecht: Reidel, pp. 45-91.

_____. (1976), "Theoretical Laws," in M. Prezlecki, K. Szaniawski and R. Wojcicki, Formal Method in the Methodology of Empirical Science, Wroclaw: Ossolineum.

_____. (1979a), *The Structure of Scientific Theories* 2nd ed. Urbana: The University of Illinois Press.

_____. (1979b), "Theory Structure" in P.D. Asquith and H.E. Kyburg, Jr. (eds.) *Current Research in the Philosophy of Science* East Lansing: Philosophy of Science Association.

_____. (1988), *The Semantic Conception of Theories and Scientific Realism*. Urbana: University of Illinois Press.

Suppes, P. (1957), *Introduction to Logic*. Princeton: Van Nostrand.

_____. (1962), "Models of Data," in E. Nagel, P. Suppes and A. Tarski (eds.) *Logic, Methodology and Philosophy of Science: Proceedings of the 1960 International Congress* Stanford: Stanford University Press, pp. 252-261.

_____. (1968), "The Desirability of Formalization in Science," *Journal of Philosophy* 65: 651-664.

Thompson, P. (1983), "The Structure of Evolutionary Theory: A Semantic Approach," *Studies in History and Philosophy of Science* 14: 215-229.

_____. (1985), "Sociobiological Explanation and the Testability of Sociobiological Theory," in J.H. Fetzer (ed.) *Sociobiology and Epistemology* Dordrecht: D. Reidel, pp. 201-215.

_____. (1986), "The Interaction of Theories and the Semantic Conception of Evolutionary Theory," *Philosophica* 37: 73-86.

_____. (1987), "A Defense of the Semantic Conception of Evolutionary Theory," *Biology and Philosophy* 2: 26-32.

_____. (1988a). "The Conceptual Role of Intelligence in Human Sociobiology," in H.J. Jerison and I.L. Jerison (eds.) *Intelligence and Evolutionary Biology* New York: Springer-Verlag.

_____. (1988b), "Logical and Epistemological Aspects of the 'New' Evolutionary Epistemology," *Canadian Journal of Philosophy* 14 (supplementary):235-253.

_____. (1989), *The Structure of Biological Theories*. Albany: State University of New York Press.

van Fraassen, B.C. (1970), "On the Extension of Beth's Semantics of Physical Theories," *Philosophy of Science*, 37: 325.

_____. (1972), "A Formal Approach to Philosophy of Science," in R.E. Colodny (ed.) *Paradigms and Paradoxes*, Pittsburgh: The University of Pittsburgh Press.

_____. (1980), *The Scientific Image*. New York: Oxford University Press.

_____. (1981), "Theory Construction and Experiment: An Empiricist View," in P.D. Asquith and R.N. Giere (eds.) *PSA 1980* vol. 2, East Lansing: Philosophy of Science Association, pp. 663-677.

Williams, M.B. (1970), "Deducing the Consequences of Evolution," *Journal of Theoretical Biology*, 29: 343-385.

_____. (1973a), "Falsifiable Predictions of Evolutionary Theory," *Philosophy of Science* 40: 518-537.

_____. (1973b), "The Logical Status of Natural Selection and Other Evolutionary Controversies," in M. Bunge (ed.) *The Methodological Unity of Science*, Dordrecht: D. Reidel.

_____. (1982), "The Importance of Prediction Testing in Evolutionary Biology," *Erkenntnis* 17: 291-306.

Part X

LAWS IN THE SOCIAL SCIENCES

Confirmation, Complexity and Social Laws[1]

Harold Kincaid

University of Alabama at Birmingham

Two attitudes largely dominate current philosophic thinking about the social sciences. One argues that the social science are not and cannot be real sciences—because the social sciences have not and probably cannot produce anything like the laws of the natural sciences. The other attitude—coming from the hermeneutic tradition—defends the social sciences, but only by arguing that laws are irrelevant to social explanation. Disparate as they are, both traditions share the assumption that laws in the social sciences are beyond reach. In this paper I argue the opposite thesis—viz. that social laws are both a possibility and actuality.

1. How Not to Think About Social Laws

Philosophers have typically approached laws in the social sciences by an old-fashioned route—by means of relatively *a priori* conceptual analysis. One or several necessary conditions for laws are identified, usually by appeal to intuitions. Then it is argued that no alleged law in the social sciences could have the necessary features. Laws are thus shown impossible or improbable. Since writer and reader usually antecedently believe that the social sciences are drastically inferior, the conclusion gains added warrant because it in turn explains the sorry state of the social sciences.

Let me cite two examples of just this approach.[2] John Searle (1984) has recently argued that the social sciences cannot produce laws. Why? The basic predicates of social science do not divide up the world in the right way. In particular, the basic kinds or predicates of social science are "multiply realizable"—the same kind of social event or state can be brought about by diverse physical states. Social kinds have no systematic connection to the physical. Thus social laws are impossible.

A second example comes from Davidson (1970). Real laws must involve a closed system. The social sciences, however, cannot gain closure: outside, non-social factors inevitably interfere. Thus social laws are impossible.

Both these arguments follow the pattern I just described: A necessary condition for lawfulness is invoked (on grounds that are far from clear), the social sciences are shown unable to meet that condition, and laws are ruled out. This way of thinking about social laws is fundamentally wrong headed. We can begin to see why by pressing Searle and Davidson's arguments a bit. Neither argument is plausible—though both have convinced many.

PSA 1988, Volume 2, pp. 299-307

Searle demands that real social laws have a "systematic connection" to the physical. But where does that demand come from? Social kinds probably are *supervenient* on the physical, but it is clear that Searle wants a more systematic connection than that. Should we make some sort of necessary co-extension with the physical a requirement for laws? I hope not, for if we do, many scientific laws are inadequate. Biology uses predicates with multiple physical realizations at every turn. Fitness is notoriously open ended; basic predicates of molecular biology fare no differently.[3] Biology, even at its best, could not produce laws according to Searle's standard. Similarly, computational states of computers have open-ended physical realizations—lawful claims about them would be ruled out as well.

The Davidsonian argument runs into a similar problem. The "closed system" require-ment is itself ambiguous. What is supposed to be "closed," the system being studied or the theory itself? If the object being studied is supposed to be closed, then Davidson's argument rules out laws in much of physics, not to mention biology. Kepler's Laws, for example, could be no such thing—since the solar system is not free of outside influence.

Is it any more reasonable to think that a real law must belong to a closed *theory*—a theory where outside influences can be handled entirely in its own terms? Not that I can see. Biological and psychological factors influence physical events. Unless we can reduce our biological and psychological predicates to physical ones, physics will be stuck with citing outside factors in non-physical terms. Physics itself will thus not strictly be closed either—and thus will have no laws if we follow Davidson. Biology, of course, will be in equally bad shape.

These two arguments make the same basic mistakes. They try to legislate on largely con-ceptual grounds what good science must be like. Davidson and Searle do not argue against social laws by first providing a careful analysis of laws in our best science and then showing in detail that the social sciences are dissimilar. Instead, some general, simple restriction is announced separating laws from other statements. Then very general conceptual considera-tions are adduced to show the social sciences cannot met the necessary condition.

This procedure, it seems to me, ignores the real progress made in the philosophy of sci-ence since Quine's attack on the analytic/synthetic distinction. Philosophy, especially philos-ophy of science, cannot decide on conceptual, *a priori* grounds what science must be like. Any adequate account of science must be closely tied to real scientific practice. These insights have gained general acceptance in the philosophy of science. However, they have not been consistently applied when it comes to laws. Any adequate assessment of laws in the social sciences must rely on a realistic picture of how other sciences actually use laws—as well as a concrete look at social science research. No simple conceptual argument will do.

2. Dealing Successfully with Complexity

That said, let me turn now to much more serious difficulty—the qualified nature of social laws. Any reasonable candidate for a social law is qualified *ceteris paribus*. That fact by itself raises problems. A qualified law seems insufficiently general to be a law. Furthermore, *ceteris paribus* clauses are frequently catchall terms—we cannot cite all the interfering factors. And even for those we can cite, we may know that they seldom hold—as in the rationality assumptions of microeconomics.

Is the *ceteris paribus* problem insurmountable for the social sciences? Clearly there is no simple answer. For *ceteris paribus* clauses to be a fatal flaw, we need to show that apparent laws in physics and biology do not employ them—or at least, that physics and biology can handle *ceteris paribus* clauses in a way the social sciences cannot. I want now to argue in some detail that no such difference can be established: for some parts of the social sciences, the *ceteris paribus* problem can be adequately handled.

Ceteris paribus assumptions are ubiquitous in science. Cartwright (1983) has argued convincing that most physical laws are strictly speaking false if not qualified *ceteris paribus*. Furthermore, explaining real events usually calls for combining multiple laws. Typically, physics provides no automatic and precise way of combining multiple laws. Rather, there are numerous piecemeal and *ad hoc* rules of thumb for specific circumstances. As I will argue later, precisely the same situation exists in biology. *Ceteris paribus* difficulties thus do not seem unique to the social sciences.

These facts have several implications. For one, they argue for a more catholic notion of a scientific law: the old paradigm of a universally quantified, exceptionless statement will not do, not if we want there to be any laws in science. Furthermore, the fact that *ceteris paribus* clauses are common in the natural sciences means that the mere presence of such clauses in the social sciences is not by itself an inherent problem. If *ceteris paribus* clauses undercut social laws, it will have to be because the social sciences cannot adequately handle *ceteris paribus* laws. How could we show that? No simple conceptual argument will do—only a detailed comparison between social research and work in other sciences can decide the issue.

Ceteris paribus laws do, of course, raise numerous conceptual difficulties about how they should be construed. The really important puzzles for my argument are two: (1) how can we confirm laws that presuppose conditions that seldom or never hold? and (2) how can we know that *ceteris paribus* laws, once confirmed, explain— since they, strictly speaking, may not apply to real events, events where other things are not equal? In other words, how do we tell the genuinely explanatory laws from the irrelevant ones?

Good scientific practice does have ways to answer these questions. Let me mention a few.[4] First there are more or less *direct* methods:

(1) Most obviously, sometimes the *ceteris paribus* clauses actually hold—when they do so, we can confirm directly.

(2) We can try to show the law holds—even though other things are not equal. In short, we can find empirical evidence for the alleged relation despite the fact that all things are not equal.

(3) We can show that our law is what Leamer (1983) calls "sturdy". Even if we cannot fully specify the *ceteris paribus* clause, we can plug in the different possible complicating factors and show the law would be unaffected.

(4) We can provide inductive evidence for a *ceteris paribus* law by showing that as conditions approach those of the *ceteris paribus* clause, the law itself becomes more predictively accurate.

(5) We can explain away failures—by citing counteracting factors. Doing so can of course become *ad hoc* rationalization, but there is no reason it must—if the exceptions were to be expected on theoretical grounds.

In addition to these methods, indirect evidence can help as well:

(6) As with theories generally, a *ceteris paribus* law that unifies diverse phenomena, that provides for striking predictions, and that has close ties with other well confirmed explanations gains plausibility.

(7) We may also have domain-specific methodological generalizations that can be brought to bare—generalizations that tell us "*ceteris paribus* qualifications of this sort generally do not cause problems". For Darwin, for example, the

methodological norm "*ceteris paribus* clauses citing non-selective forces can be ignored" perhaps had some good empirical warrant.

These methods are of course loosely formulated and are likewise not foolproof. Nonetheless, they all can help insure that a *ceteris paribus* law is confirmed and/or is relevant to explaining the phenomena under investigation. These techniques are common fare, I would argue, in the natural sciences. Are they available for the social sciences? And can the social sciences employ them "well enough" to confirm laws? No *a priori* answer is possible. Instead, we need to see just how specific social science research uses these methods. We also need to see how well they are employed in the natural sciences as well—if we are to make a comparative judgement.

Obviously such an analysis is a huge task—we can see why philosophers have found simple conceptual arguments so appealing! Since I have abjured such quick arguments, I am obligated nonetheless to make a rough stab at showing some social science research adequately deals with the *ceteris paribus* problem. Let me thus discuss in some detail a concrete piece of social research and compare it to some paradigm work in evolutionary biology.

Jeffrey Paige (1975), in *Agrarian Revolutions*, has produced an exemplary piece of social science research. Paige investigates the causes of agrarian political behavior. He develops a detailed theory explaining how agrarian classes behave and what determines that behavior. Analyzing agrarian societies in class terms, Paige is able to generate and confirm numerous predictions about the political behavior of such classes.

Agrarian classes can be classified according to three distinct criteria: (1) are they composed of cultivators or noncultivators (owners)? (2) if they are cultivators, do they receive their income in wages or by rights to the land? and (3) if they are non-cultivators, do they depend solely upon land ownership for income or are returns from commercial capital dominant? Using this classification along with important auxiliary assumptions, Paige can generate numerous substantial predictions about the political behavior of owners and laborers.

Let's consider owners first. Owners whose income comes predominantly from land, rather than from capital invested in land, face multiple dilemmas. Large landed estates compete poorly with commercial agriculture. Without investing extensively in capital improvements, owners find few prospects for increasing productivity and increasing total output. These facts insure that landlords face constant pressure: they will be economically pressed by larger market forces and will have little room for concessions to laborers. Thus owners dependent on land are economically weak and naturally seek *political* protection of their economic position. They are likewise involved in a zero-sum game with their employees and have little taste for compromises.

Owners whose income results from agricultural capital live in a different world. Capital investments mean increasing productivity and thus room to bargain. Using the most productive techniques, capitalist landowners can successfully compete with the larger market. They have room to bargain and strike compromises; they do not depend strongly on political force to maintain their positions.

Cultivators are likewise influenced by their economic position. Some cultivators draw their income from wages—others from use of the land. Not surprisingly, the latter group avoids conflict while the former finds it necessary. Cultivators drawing their income from land use or ownership tend to be conservative. Their entire subsistence depends on access to the land. Since the risks of rebellion are high, they avoid confrontation. Since they till their plots individually, they have reduced incentives for collective behavior.

On the other hand, cultivators who depend on wages behave differently. They do not have subsistence plots to lose. Furthermore, payment in wages usually means industrial agriculture and thus collective work groups sharing common conditions and interests. Income from wages thus promotes collective political action, income from land inhibits it.

What does all this tell us about agrarian political behavior? Obvious predictions emerge once we combine the various owner and worker income sources. When owners depend upon capital and cultivators depend on wages (as in large plantations), Paige predicts that we will see cultivator resistance ending in compromise. Because of their laboring conditions, cultivators will make demands. Owners, however, are economically strong and generally have an increasing pie to divide. Compromise will thus be the name of the game.

What happens when we combine owners deriving all income from land with cultivators depending on wages? Paige follows the obvious logic. The upper class will be economically weak, reliant on political intervention and unable to make compromises. Cultivators will have little to lose by rebelling and a collective interest for doing so. The end result is "revolutionary" political behavior. Thus a land-dependent upper class and wage-dependent lower class should mean movements for radical economic and political change.

Such are Paige's predictions. They are relatively straight forward. Are they true? Paige does a meticulous job of arguing that they are. His evidence goes considerably beyond the usual anecdotal evidence found, for example, in Marx. Paige divides nearly most of the world's non-industrial countries into sectors. He then classifies those sectors according to their predominate class relation and indexes political activity in them over a 30 year period. From this data, he can thus test his main claims. The results are quite positive. Paige then provides three detailed cases studies—of Peru, Angola and Vietnam—to further verify his hypotheses. Each case appears to nicely confirm his analysis.

Does Paige successfully avoid using *ceteris paribus* clauses? Of course not. His predictions are indeed qualified—in ways both specified and unspecified. So the real question is whether this work succeeds according to the criteria laid down above. Can he adequately handle the exceptions? Is there is reason to think Paige's claims are relatively well confirmed?

Numerous factors complicate Paige's analysis. Increasing population, rapid loss of land, frequent ecological disasters, the presence of urban parties and other events weakening the political hold of the landed classes may turn conservative peasants into ones demanding change. Whether wage income produces revolutionary movements may be influenced by the nature of production (centralized vs. decentralized) and by differences between sharecropping and migratory wage labor. Causal influences may be exaggerated by contagion effects—any protest may up the odds for future events, regardless of the original cause. *Ceteris paribus* clauses abound in Paige's account.

Nonetheless, the apparent complications are not beyond control. Paige's analysis can be refined: he shows that, at least for some complicating factors, more accurate *ceteris paribus* clauses mean more accurate predictions. For example, political behavior of wage earning cultivators depends on whether production is centralized or decentralized. Paige subdivides his units of analysis and shows that controlling for centralization gives more accurate results. Similar results come from separating sharecropping from migratory wage labor. Migratory laborers face conservative influences of distant villages. When Paige controls for these differences, his correlations between income sources and political behavior become even stronger So Paige's work in part meets one crucial criterion: refinability.

We also know that—whatever the counteracting forces—Paige's predictions still hold for the most part. Even when population growth, ecological disasters, rapid loss of land, etc. are not controlled for, the basic predictions are borne out by the data. Paige's laws hold for the most part.

The theory of agrarian political behavior also gains support from its ties to other, firmly entrenched parts of the social sciences. Peasants tend to be conservative because they estimate risks and make a more or less rational decision. Similarly, cultivators paid in wages have little incentive to improve productivity: the profits will not be theirs. So Paige's laws are tied to rational decision theory. We can also predict how landlords behave because of what we know about economic growth and the nature of zero sum conflicts. Furthermore, identifying classes relies on well-established data about differential distribution of income. Thus Paige's *ceteris paribus* laws have real connections to other relatively well-confirmed generalizations.

Finally, a case can be made that Paige's account makes novel predictions, unifies a diverse set of phenomena and holds promise for further research. The theory's ability to unify is obvious. Agrarian movements across time, cultures, religions and regions are explained by one unitary analysis. This unity itself was a novel prediction, previous writers failing to see such organization behind the many world-wide agrarian movements. Finally, Paige's account poses numerous unanswered questions, questions that others have fruitfully pursued (see Bunker 1983).

Thus the theory of agrarian movements has the trappings of respectable science: its laws hold despite the *ceteris paribus* clauses, they can on occasion by refined, counter-examples can be explained away, there exist ties to other confirmed work, and the theory appears to unify, make striking predictions and offer future promise.

No doubt Paige's account can be criticized. But that misses the point. Most scientific theories can be criticized on one ground or another. The real question is whether Paige handles his difficulties roughly with the same success as good science does in general. I think it is obvious that he does. However, to make this claim more compelling, let's look briefly at some good work from evolutionary biology. Doing so will make it clear that Paige's work is "scientific" by any reasonable standard.

The basic components of evolutionary theory are of course well worked out. The Hardy-Weinberg law describes gene frequencies in a population where no net selective forces are present. Natural selection acting on variation causes changes from equilibrium. Theoretically, selective forces can be quantified and resulting trajectories of populations determined for particular situations. For simple selective forces acting on individuals of known fitness, the results are known with mathematical precision.

However, such simple selective forces are not the only ones operative. Once we move to more realistic models, we must add in a host of other factors: genetic drift, kin selection, sexual selection, group selection (potentially at multiple levels), the effects of gene linkage, the possible role of internal or developmental constraints, the effects of gene-culture evolution, and the effects of patchy environments, just to name a few. We know how to combine some of these factors, but some we do not. Models of group selection and sexual selection exist, but adding these factors with more standard selection is a messy affair. For some elements—such as internal constraints and patchy environments—we are just beginning to describe how such forces work *in isolation*, not to mention in tandem with other factors.

Given this situation, we should not be surprised to see that concrete empirical work in evolutionary biology faces all the standard problems of *ceteris paribus* clauses and the like. Consider, for example, a paradigm piece of evolutionary research—Peter Grant's (1986) *Ecology and Evolution of Darwin's Finches*. Grant's book is the definitive research to date on finch evolution in the Galapagos islands. From more than 10 years of study in the islands plus a mass of previous work, Grant develops a detailed evolutionary picture. Finch beaks are adapted to different seed sizes. Seed availability is a strong constraint on population size. Selection is currently occurring among some populations, favoring large body and beak size. Current species apparently originated under gradual directional selection. Speciation was driven by both allopatry (separation with subsequent adaptation to different food sources) and competitive exclusion.

Grant recognizes the many possible complicating factors and attempts to rule them out. Drift seems not to be important, since finches flock and populations are usually relatively large. Internal constraints appear minimal, since beak and body size involve small, accumulative genetic changes.

Nonetheless, Grant's work runs into numerous complications and qualifications. Most of his predictions are qualitative—he can argue for the direction of selection, but not its strength. Mutation rates and gene flow are complicating factors, but their relative importance is unknown. Sexual selection has played some role (p.103), but it is not clear how much nor how exactly. Rain, large seed production, timing of seed production and other food sources may influence to what extent seed actually serves as a population limit and selective force; at best, Grant gives reasons to think these factors are relatively unimportant. Predation and disease are additional selective forces not included in his model; again, ancedotal field experience suggests they are probably unimportant. The relative role of competition and habitat diversity in speciation is unclear. While selection clearly picks out beak size, it is not obvious how it does so—the mechanism is unclear. Finally, most of Grant's major findings have exceptions in the Galapagos themselves that must be explained away by appeal to exceptional circumstances.

What conclusion should we draw from these qualifications? My point is not to challenge the quality of Grant's work—far from it. The correct inference is that good work in the social sciences faces and handles problems very much like those faced by the best work in biology. *Ceteris paribus* clauses, restricted generalizations and the like do not separate the social sciences from the natural sciences. Social research like that of Paige compares favorably with the best work in evolutionary theory and ecology.

3   Universality and Nomic Force

I want now in this last section to consider one last objection to social laws. The objection fails, I think, but it does point out some important restrictions on laws in the social sciences.

Social laws like those described earlier, a critic might argue, are not real laws—merely restricted generalizations. They are simply summaries of correlations in specific populations and as such lack "nomic force".

These objections are still surprisingly popular given the obvious compelling responses to them. References to particulars can be transformed into universally quantified statements, so syntactic universality carries little weight. Furthermore, some prime examples of laws—Kepler's or Darwin's—on the face of it do refer to particulars; numerous apparent laws also have restricted domains of application. Universality per se is not the issue.

Are the results of regression studies and the like merely generalizations and not real laws? Traditionally, accidental generalizations are distinguished by their inability to support counterfactuals. But which counterfactuals are warranted seems itself to depend upon prior knowledge of laws: "All coins in my pocket are copper" does support counterfactuals if I know that the magnetic snap on my pocket always excludes non-copper coins. So the real issue seems to be whether an alleged law plays the "right" role in the overarching theory. Goodman's (1965) account of projectability and Skyrm's (1980) notion of resilience are in part attempts to specify just that relation. Whatever one's favored *analysis* of laws, it seems that something like projectability or resilience constitutes the *evidence* for separating laws from other kinds of statements (cf. Wilson 1986).

Projectability and resilience, however, are relative notions. Thus it seems to me we cannot sharply separate "real" laws from "mere" generalizations—the difference, at least epistemologically, is one of degree, not kind. So Paige's work or similar social research cannot be rejected out of hand because its statements have limited domains or are based on generalizations from statistical data. Much more concrete analysis is needed.

Clearly some or much work in empirical social science carries little nomic force. Generalizations result from statistical analyses of a small number of samples from a restricted population—say post-war American voters. Explanatory variables may have no other motivation than finding a regression equation that is statistically significant. But not all generalizations must be like this— and not all are.

No detailed account exists to tell us just how to separate mere curve-fitting generalizations from generalizations with nomic force. Nonetheless, we have some rough ideas about what is needed. We would like ties to other laws or explanations—both horizontally and vertically. Horizontally, generalizations would ideally be deducible from more fundamental assertions. However, since the social sciences—like all the special sciences—may be irreducible, derivations of laws may be too much to ask. Weaker horizontal ties are still possible, however. We can proceed case by case, showing how any specific instances of our law can be explained in lower-level terms. In short, we can cite mechanisms. Vertical connections minimally involve logical consistency with other well-confirmed explanations. Stronger connections might include evidential relations—as evolutionary biology and parts of geology mutually support one another.

Aside from such ties, deep generalizations will have evidence not found in mere curve fitting. They should hold over repeated tests and over a variety of different kinds of evidence. Restrictions in a generalization's domain should be motivated—and not by the need to fit a curve but by prior independent reasons. Political science generalizations holding only for post-war US make an unmotivated restriction; assertions restricted to feudal societies are prima facie less troublesome, for we have good reason to divide feudal societies from other social structures.

Though I will not argue it in detail here, Paige's work fares relatively well in this respect. Of course his claims hold only for certain kinds of societies—but those societies are not determined in an *ad hoc* way; they are grouped by a kind or predicate that relies on basic structural features and can be determined by means independent of his theory. Paige's basic kinds are intuitively quite general. His alleged laws have ties to other parts of social science like decision theory and cost-benefit analysis.

Although the generality objection fails, it still nonetheless points to a useful insight. The social sciences can, I believe, produce laws, at least on some occasions. Yet, *ceteris paribus* clauses and domain-specific generalizations predominate. Perhaps their prevalence suggests not that laws do not exist but that laws are only part of social science explanation. Social systems may involve unique constellation of variables or unique variables altogether—hence the frequent need for *ceteris paribus* restrictions. Social laws with any generality will involve significant abstraction; they provide a general framework citing significant variables and their rough interrelation. Explanation often requires seeing how things work out in detail. In short, much of the explanatory weight relies on specifics—it is in the application that the real work is done. Some such situation characterizes ecology and evolutionary theory as well—where *ceteris paribus* clauses and other restrictions are also rife. Thus perhaps in the end the best defense for social laws comes from realizing they are not all there is to explanation.

Notes

[1]Work on this paper was begun under a fellowship from the American Council of Learned Societies.

[2]These arguments and others are examined in more detail in my (forthcoming,a).

[3]See my "Molecular Biology and the Unity of Science" (forthcoming, b).

[4]I am borrowing here in part from Hausman's (1981) useful discussion.

References

Bunker, S. (1983), "Center-Local Struggles for Bureaucratic Control in Bugisu, Uganda," *American Ethnologist* 10: 749-769.

Cartwright, N. (1983), *How The Laws of Physics Lie*. Oxford: Clarendon Press.

Davidson, D. (1970), "Mental Events," in Foster and Swanson (1970).

Foster, L. and Swanson, J. (1970), *Experience and Theory*. Amherst: University of Massachusetts Press.

Goodman, N. (1965), *Fact, Fiction and Forecast*. Indianapolis: Bobbs Merrill.

Grant, P. (1986), *Ecology and Evolution of Darwin's Finches*. Princeton: Princeton: Princeton University Press.

Hausman, D. (1981), *Capital, Profits and Prices*. New York: Columbia University Press.

Kincaid, H. (forthcoming,a), "Defending Laws in the Social Sciences," *Philosophy of the Social Sciences*.

_ _ _ _ _ _ . (forthcoming,b), "Molecular Biology and the Unity of Science," *Philosophy of Science*.

Leamer, E. (1983), "Let's Take the Con Out of Econometrics," *American Economic Review* 73:31-43.

Paige, J. (1975), *Agrarian Revolutions*. New York: The Free Press.

Searle, J. (1984), *Minds, Brains, and Behavior*. Cambridge: Harvard University Press.

Skyrms, B. (1980), *Causal Necessity*. New Haven: Yale University Press.

Wilson, F. (1986), *Laws and Other Worlds*. Dordrecht: D. Reidel.

Ceteris Paribus Clauses and Causality in Economics[1]

Daniel M. Hausman

University of Wisconsin–Madison

Explicit or implicit *ceteris paribus* clauses are pervasive in economics. People do not always buy more of x when the price of x decreases. The generalization holds only "other things being equal" or *ceteris paribus*. Not everybody wants more wealth, but economists have held that the generalization holds, *ceteris paribus*. When government imposes price controls, shortages do not always arise, but, *ceteris paribus*, they do. *Ceteris paribus* clauses are common in other sciences, but I shall confine my remarks to economics.

Many have found such *ceteris paribus* clauses problematic. For they are vague, and they seem to insulate theories from empirical criticism and correction. When the observed phenomena are not as the theory predicted, one can (or so it has been alleged) always claim that other things were not "equal", that there was some "disturbance" or "interference" and thus that the *ceteris paribus* condition was not met. If *ceteris paribus* clauses truly provided blanket excuses, then the claims that contained them would be empirically empty. Unless we are to condemn virtually all theory, we must find some other way of construing *ceteris paribus* clauses.

In this essay I shall be concerned with two different sorts of qualified generalizations that are important in social theories. In both cases I shall suppose first that the generalizations involved might (apart from their *ceteris paribus* clauses) reasonably be regarded as laws, and second that the words, "*ceteris paribus*" are not mere abbreviations for some perfectly determinate list of relevant factors. Thus the generalizations with which I shall be concerned are vague but nevertheless genuinely lawlike, unlike rough claims such as "Birds fly", "Scots are thrifty," or "Democrats get the U. S. into wars."

The two kinds of qualified generalizations (and the two kinds of *ceteris paribus* qualifications) that I shall discuss might be called "inexact fundamental laws" and "partial generalizations." Economic examples of the first are claims such as "*Ceteris paribus* people's preferences are transitive," or "*Ceteris paribus* people prefer more commodities to fewer." These are fundamental, but inexact, "behavioral postulates" of economics. The "interferences" ruled out by the *ceteris paribus* clauses are vague and extra-economic. In partial equilibrium work, on the other hand, one often considers the effects of various "economic" causal factors separately from one another. In this case the "interferences" are precisely those factors that economic theory *itself* identifies as other causal influences. For example, contemporary economics implies that demand for widgets depends on the price of widgets, other prices, income and tastes. Yet economists may want to

PSA 1988, Volume 2, pp. 308-316

consider demand for widgets as a function (*ceteris paribus*) of the price of widgets only. The content of the *ceteris paribus* clause in this case is almost completely definite: other prices, incomes and tastes are the "other things" which must be "equal". Although the *ceteris paribus* clauses in derivative laws thus do not, for the most part, pose the same semantic puzzles as do the *ceteris paribus* clauses in inexact fundamental laws, they raise philosophical questions concerning the causal nature of such partial generalizations. As I shall argue in section 3 below, both kinds of *ceteris paribus* clauses are intimately tied up with philosophical questions concerning causality.

1. Inexact Fundamental Laws

Economists, in John Stuart Mill's view, know only the laws of the "greater causes" of the phenomena, and they are unable to infer invariably and precisely what actually occurs. This inability is a consequence of inexactness *within* the genuine causal laws of their theory, not merely of faulty data or mathematical limitations. A claim such as "People's preferences are transitive" seems to be inexact. How should one analyze this inexactness?

I suggest that one should regard the "laws" of inexact sciences as carrying implicit *ceteris paribus* clauses. To assert that people's preferences are transitive is to make a qualified claim. A change in tastes, for example, does not falsify it, since changes in tastes are ruled out by the implicit *ceteris paribus* clause. When Mill enunciates the "psychological law" "that a greater gain is preferred to a smaller", he is not claiming that people always prefer greater gains, but that this is one motivational "force" that often predominates. Economists tell us how agents behave in the absence of various complications.

The *ceteris paribus* clauses that render fundamental economic laws inexact are imprecise and ineliminable. Even if it is sometimes legitimate to employ them (as Hutchison questions 1938, pp. 40f), vaguely qualified claims are surely sometimes false or absurd. It is not the case that, *ceteris paribus*, humans are all immortal or that all dogs have six legs. So one who regards the laws of inexact sciences as vaguely qualified claims must distinguish legitimate (correct) from illegitimate (incorrect) uses of ineliminable *ceteris paribus* clauses. This demands that two problems be solved: (1) What do sentences with *ceteris paribus* clauses mean, and when, if ever, can they be true? (2) When is one justified in regarding them as laws?

1.1 The Meaning or Truth Conditions of Inexact Laws

The basic claims of economics are true (if at all) only under various not-fully-specified conditions. What precisely is such a vague *ceteris paribus* clause? What does it mean to say that "*Ceteris paribus* people's preferences are transitive?" What must the world be like if such a claim is true?

I shall provide only a cursory account here (for more detail see my 1981, pp. 124-30). The same sentence can say different things in different contexts. Following Stalnaker (1972, pp. 390-97), let us distinguish the *meaning* of a sentence—the context-invariant interpretation of the sentence— from its *content*—the proposition expressed by the sentence—, which may vary in different contexts. "I'm confused by this paper" has a single *meaning*, but its *content* depends on who utters it and when it is uttered. The meaning of a sentence is a function from contexts to contents or propositions. It determines a content in a given context.

I suggest that *ceteris paribus* clauses have one *meaning*—"other things being equal," which in different contexts picks out different *propositions* or *predicates,* depending on the grammatical structure of the qualified claim. The scientist's background understanding determines (vaguely) what the "other things" are and what it is for them to be "equal." Although the phrase, "*ceteris paribus*", has an invariant meaning, its content, the *predicate* or *proposition* it picks out, varies from context to context.

What proposition does a vaguely qualified law, such as "*Ceteris paribus* people's preferences are transitive", express? Suppose that the logical form of such a law is "*Ceteris paribus* everything that is an F is a G" where "F" and "G" are predicates with definite extensions.[2] In the case of qualified generalizations with this form, some things that belong to the extension of F do not belong to the extension of G—otherwise the qualification would be unnecessary.[3] In my view, "*Ceteris paribus* everything that is an F is a G" is a true universal statement only if in the given context the *ceteris paribus* clause picks out a predicate—call it "C"—and everything that is both C and F is G.[4] In offering a qualified generalization, one is asserting that, once the qualifications are met, the extension of F lies within the extension of G. In committing oneself to a law qualified with a *ceteris paribus* clause, one envisions that the imprecision in the extension of the predicate one is picking out will diminish as one's scientific knowledge increases.[5] Thus to believe that, *ceteris paribus* everybody's preferences are transitive is to believe that anything that satisfies the *ceteris paribus* condition and is a human being has transitive preferences. Sentences qualified with *ceteris paribus* clauses may be genuine laws.

One should, however, recognize that *ceteris paribus* claims may guide research and help economists to interpret data even if they are not regarded as lawlike assertions, let alone laws. If a theorist believes that *in a certain domain* the interferences inadequately denoted by the implicit *ceteris paribus* clause are absent, he or she can regard the generalization *in that domain* as a restricted law. (See Morgenbesser 1956, chs. 1,2 on "virtual laws".) But the theorist need not regard the unrestricted generalization as true or false at all.

My interpretation of *ceteris paribus* clauses does not rule out this possibility of regarding inexact generalizations as qualified laws only within particular domains, for one need not judge all sentences that contain *ceteris paribus* clauses to be true or false. One may find that the justification conditions (to be discussed briefly below) are satisfied in some domains, while in others they are not. Depending on the circumstances, it may be best to regard the unrestricted "law" as merely a general assumption and not as an assertion to be judged true or false. Since economic theorists use basic economic "laws" to try to explain economic phenomena, they cannot, of course, *always* regard them as mere assumptions. They must rather take them as expressing some truth, albeit possibly quite rough. At some point with respect to some domains, economists must construe some of the assumptions of their models as qualified lawlike assertions.

Countenancing qualified laws forthrightly, one is no longer forced to make invidious comparisons between the natural sciences, which possess laws and provide explanations, and the social sciences, which possess at best restricted laws and whose explanations are unsatisfactory. One has instead gradations of inexactness. Scientists strive for exactness, but possessing, as they typically do, only qualified generalizations, they may nevertheless be able to explain and predict phenomena in significant domains.

1.2 When Should One Accept a Qualified Law?

It is not enough to provide an interpretation of qualified generalizations that enables one to understand what they say and how they can be true laws. One also needs to consider when, if ever, *one has reason to believe* that a qualified generalization is indeed a true law. I have elsewhere (1981, pp. 131-33) defended four conditions that qualified generalizations must satisfy before one can be justified in regarding them as laws:

1. The generalization must be *lawlike* rather than accidental.

2. The generalization must be *reliable*. Without its *ceteris paribus* clause or with only specific qualifications in its place, the generalization must (within an appropriately specified domain) usually have correct implications.

3. The generalization must be *refinable*. Adding additional specific qualifications (in a non *ad hoc* way), the reliability of the generalization in the given domain must increase.

4. The generalization must be *excusable*. When it has a false implication, one should (with only rare exceptions that call for further research) be able to explain why.

Obviously a great deal more can and should be said about these conditions. The basic rationale behind them is that if any is not satisfied, then one's conviction is undermined that the *ceteris paribus* clause really is a place holder for some predicate that would make the qualified generalization into an exact law. These are necessary, not sufficient conditions and ought ideally to be integrated into a general theory of confirmation. Although vague and not fully satisfactory, they rationalize our intuitive judgment that claims such as "*Ceteris paribus* dogs have six legs" or "*Ceteris paribus* Scots are thrifty" are not laws (see also Kincaid 1989).

2. Partial (Causal) Generalizations or Derivative Laws

Economic phenomena depend on many causal factors, only a few of which—the specified antecedents of the generalizations and the listed arguments of the functions they determine—are taken into account. Thus, according to fundamental economic theory, economic phenomena are to be explained in terms of the maximizing efforts of individuals given tastes, endowments with resources and abilities, and the set of production or technological possibilities. Less explicitly in the background are institutional and epistemic constraints. When tastes, endowments and technology satisfy certain conditions, then there exists a general equilibrium in which there are no excess demands on any market. Most theorists concede that there are other significant causal factors and that some of the "givens" upon which properties of the general equilibrium causally depend, such as technology, themselves depend on dependent variables such as factor prices (Hicks 1979, p. 58). As a first approximation, however, tastes, initial endowments, production possibilities, and institutional and epistemic circumstances are treated as primitives that have no economic explanation and do not depend on any economic variables. These primitives may, of course, change, but only the consequences, not the causes of these changes, are open to economic investigation. General equilibrium primitives are supposed to be the ultimate or underlying causes of economic phenomena.

In principle, fundamental neoclassical theory leads to the view that there is just one big explanatory claim to be made: The determination of general equilibrium answers all questions. Indeed some general equilibrium theorists have rejected, at least by implication, all less global explanations (Bliss 1975, pp. 29, 34, 120). But such theoretical purism will not do. For it is possible, at least approximately, to causally order (in Simon's sense (1953)) the basic general equilibrium model. For example, suppose that one adds Smith with Smith's tastes and her tiny endowment to an on-going economy. The economy's price vector will not be much changed. In this case Smith's endowments and the virtually fixed prices jointly determine her income. Her tastes, her income and the prices then determine her consumption. Her income can thus be causally between the general equilibrium primitives and her consumption. In this way the standard theory of consumption is "approximately consistent"[6] with a full general equilibrium account and represents a partial and approximate causal ordering of a general equilibrium model. Supply and demand explanations, the bread and butter of applied economics, involve such partial and approximate causal orderings.

To use economic theory as an "engine for the discovery of concrete truth" (Marshall 1885, p. 159), economists often need to be able to treat single markets or small groups of markets in isolation. But markets are not isolated. Supply and demand for boots, for example, depend on many variables besides the price of boots. To treat them as (approximately) functions of the price of boots only, is reasonable only if the other causal factors remain constant or if one takes into account the shift in supply or demand caused by a change in these other variables.[7] Economists conceptualize this issue, *which essentially involves separating different causal factors*, in terms of defining the *ceteris paribus* conditions for supply and demand functions. Boot supply and demand functions state how many boots would be demanded and supplied at various prices, other things being equal. These other things are other *causes*.

Let us focus on demand functions with price as their only explicty argument. One would like the *ceteris paribus* clause for such a demand function to include (a) all those factors besides the price that (within the given time period) significantly affect the amount of the commodity or service demanded and (b) none which are themselves within the given time period significantly affected by the price.[8] "Significantly" is a vague word, but to define demand as a function separately of price and of other factors impounded in a *ceteris paribus* clause, one must simplify the causal structure.

These two conditions may be defended as follows: If one leaves out a significant causal factor, then one will sometimes be mistaken about the quantity demanded, even though, supposedly, all other things are equal. If, on the other hand, one includes a factor that is not exogenous, but is signficantly affected by the price of the commodity or service, then one's *ceteris paribus* clause can never be satisfied. As price varies, one should move along a single unshifting demand curve. Demand and supply curves may shift because of a change in the values of variables in their *ceteris paribus* clauses, but the *curves* should not *shift* whenever the price of boots changes. In short-run supply and demand explanations, one may also include in the *ceteris paribus* clause of a demand function factors (such as the size of steer herds) that are sensitive to the price of boots, but that take a long time to adjust.

Unfortunately, demand for boots also depends on variables, such as the price of shoes, which cannot be regarded as unchanging even in the short run as the price of boots varies and which thus do not qualify for membership in the *ceteris paribus* clause. There are two options here: either one *provisionally* includes such factors in the *ceteris paribus* clause anyway and then later corrects one's solution for "indirect effects," or one may be able to deal with a small number of markets simultaneously.

Finally, since one wants to be able to use a demand curve to explain the change in prices which results from a shift in supply (and vice versa) the factors included in the *ceteris paribus* clauses for supply and demand curves should differ to some extent. Otherwise every shift in supply would be equally a shift in demand, and there would be little point to categorizing factors into influences on supply and influences on demand (Friedman 1953b, p. 8).[9]

Given the general criteria for membership in a *ceteris paribus* clause— that it should include all significant causal factors that do not themselves significantly depend upon the focal causal factor— how is one to decide which factors to include? Merely possessing the criterion represents little progress on the notion that *ceteris paribus* clauses should specify significant "disturbing causes" or "interferences." But in the case of the theory of consumer choice, for example, the committed neoclassical theorist has the luxury of taking seriously the specification of disturbances or interferences that neoclassical theory *itself* provides. On such a view (which supposes the correctness of fundamental neoclassical theory), the other things which must be equal consist of tastes, technology, endowments, and institutional structure or their relevant implications, such as the values of the other prices. The *ceteris paribus* clauses in specific applications of neoclassical economics reflect the theoretical commitment of neoclassical economists to a particular structure of causal explanation. As better defined and delimited, such *ceteris paribus* clauses seem more scientifically respectable. Although they may still function as excuses, they are not excuses for all occasions. In particular, they would not serve as excuses for apparent disconfirmations of the fundamental laws. This observation should alert us to the fact that the *ceteris paribus* clauses attached to derivative laws cannot be as precise as they seem. Since the fundamental laws remain inexact themselves, derivative laws, such as the law of demand, will fail excusably not only when there are changes in incomes or other prices, but also sometimes when the fundamental inexact laws from which such generalizations are derived themselves fail. Their precision is only hypothetical, supposing that the fundamental theory were not itself inexact. Partial generalizations or derivative laws will thus at least implicitly carry *both* sorts of *ceteris paribus* qualifications. Only within the theoretical framework of a particular fundamental theory do the *ceteris paribus* clauses attached to derivative generalizations such as the law of demand appear to be completely precise.

3. Causality and *Ceteris Paribus*

Various causal notions have been mentioned over and over again in the above discussion. Why? What are the connections between *ceteris paribus* clauses in the two senses discussed above and causal claims?

As John Mackie (1974) has pointed out most clearly, but as many others have also recognized, to say that c is a cause of e is at most to imply that c is only necessary or sufficient for e *in the given circumstances*. Struck matches fail to light and matches that are not struck sometimes light. Striking a match may, however, be necessary and sufficient for its lighting in the circumstances—that is, *ceteris paribus*. Similarly, if we accept counterfactuals such as "If the match had not been struck, it wouldn't have lit" we interpret them with respect to "close" possible worlds in which the other relevant causal circumstances are unchanged—that is, subject to the same *ceteris paribus* condition (Lewis 1973).

Causal thinking involves a focus on only a few of the causally relevant factors in a given case. It thus requires one to separate out the nomologically relevant factors and to relegate many to the background circumstances or to the *ceteris paribus* clause. But what, specifically, is the role *causality* plays in determining what to include in the *ceteris paribus* clause? Causality is crucial because *ceteris paribus* factors may be causes of "the cause", but they may not be effects of it. Otherwise one could not coherently consider the varying effect of the given cause against the background of some fixed circumstances. The asymmetry is essential in order for the focal factor to possess the sort of exogenity or independence relative to the attendant circumstances that is central to the notion of causal priority (Hausman 1984, 1986). These last two claims are, I believe, central and of the utmost importance.

If one attends only to the mathematical form of laws and functions, one will miss the causal asymmetry. *Causal asymmetry is not "in" the laws or functions, but in the details of the actual mechanisms.* Consider two flagpoles, an ordinary one and a telescoping one rigged with a device to adjust the height to keep the length of the shadow constant over some time period. The law that light travels in straight lines is true regardless, and the implication that the height of the flagpole divided by the length of the shadow equals the tangent of the angle of elevation of the sun holds of both. But the causal relationships are quite different. Similarly the demand function "f" in "*Ceteris paribus* $q^d(x) = f(p_x)$" may be invertible, but in the context of consumer choice theory, demand depends on price not (baring some Rube Goldberg set-up) vice versa.

These causal concerns are not philosophers' inventions, and they cannot be safely ignored by economists. Consider, for example, the version of the neoclassical theory of a perfectly competitive firm that attributes to the firm a production function of the form $q_z = f(q_a, q_b, ...)$ where z is the single output, and a, b, etc. are the inputs. f is twice differentiable, and in the economically relevant range of the variables its first partial derivatives are positive and its second partial derivatives are negative. Suppose (as is common) that the first partial derivative functions have inverses. One can then consider the marginal increment in q_a, for example, needed to produce a marginal increment of output (other outputs held fixed). Suppose one multiplies this function by p_a, the price of input a. Can one regard this function now as the firm's marginal cost function? This is a causal question, and the mathematical form of the function provides no answer to it. If the other factors cannot easily be varied and may thus be "held fixed" or impounded in a *ceteris paribus* clause (which is implicit in partial differentiation), then it can, and one can explain or predict the firm's output decisions by noting when the marginal cost equals p_z, the given price of the output, z. But if the other inputs are easily variable, it would be nonsense to regard this function as the firm's marginal cost function. Even though, as a matter of physics, the firm might be able to hold the other inputs constant, as a matter of economics, it will not choose to do so.

It can sometimes be informative and useful to maintain that, *ceteris paribus*, z = f(x,y), even though x, y, and z depend on many other variables. This is so both when this relationship is a fundamental inexact law, and one does not know the other variables or their exact functional relations to x, y, and z, and also when this relationship is a derivative law, and one is only suppressing relatively precise information in order to highlight the particular relations. In either case, if one treats z as the dependent variable, then the constituents of the *ceteris paribus* clause should influence z and should not causally depend significantly on x or y. Otherwise the calculated value of the function as x and y vary will not be the correct value of the variable z. (Similar remarks apply if x or y is treated as the dependent variable.) In any event, generalizations that are qualified with *ceteris paribus* clauses—and most of the generalizations in science are so qualified—presuppose such causal judgments. As I have argued in this paper, these generalizations may be fully meaningful nevertheless, and one may sometimes have good reason to believe that they are correct.

Notes

[1] I would like to thank Harold Kincaid, James Woodward and members of the audience at the PSA meetings for some searching and helpful criticisms and suggestions.

[2] It has been argued that the form of the "neoclassical maximization" hypothesis is more complicated, "There is something that everyone maximizes." See Boland (1981), Caldwell (1983) and Mongin (1986). The account offered in this section can easily be extended to laws with a logical form involving "mixed quantification" such as this.

[3] It might be the case, although I can think of no example, that the unqualified generalization was accidentally true, but only the qualified generalization was a true law.

[4] As I shall argue in sections 2 and 3 below, in most cases the "interferences" excluded by C must be causal conditions of G and must not causally depend upon F.

[5] Insofar as one is able to explain what are now fundamental laws in terms of others, one converts their *ceteris paribus* clauses into the *ceteris paribus* clause of a derivative law, which is an amalgam of a constrained *ceteris paribus* clause and the vague *ceteris paribus* qualifications of the new fundamental laws.

[6] A continuum of traders (Aumann 1964) permits exact consistency.

[7]Simplified aggregative general equilibrium theories which include only a small number of commodities or agents raise the same problems in a different guise. For aggregation presupposes that the relations among the units that are treated as one are not themselves significantly affected by the variables one is attempting to explain.

[8]James Woodward offered the interesting suggestion that one might possibly want to impose as a third condition a requirement that the effects of the various factors that affect demand be additive, so changes in *ceteris paribus* conditions may cause "shifts" in functions such as demand curves, but may not cause changes in their shape. But the variable, "tastes," does not satisfy this condition, and as simple examples such as Boyle's Law (PV = k) show, derivative qualified "laws" in physics do not always meet this condition.

[9]This analysis of *ceteris paribus* clauses is compatible with the two main positions in the literature concerning demand curves (see Yeager 1960, Friedman 1953a).

References

Aumann, R. (1964), "Markets with a Continuum of Traders", *Econometrica* 32: 39-50.

Bliss, C. (1975), *Capital Theory and the Distribution of Income*. Amsterdam: North-Holland.

Boland, L. (1981), "On the Futility of Criticizing the Neoclassical Maximization Hypothesis", *American Economic Review* 73: 1031-36.

Caldwell, B. (1983), "The Neoclassical Maximization Hypothesis: Comment", *American Economic Review* 75: 824-27.

Friedman, M. (1953a), "The Marshallian Demand Curve", in *Essays in Positive Economics*. Chicago: University of Chicago Press, pp. 47-99.

_ _ _ _ _ _ _. (1953b), "The Methodology of Positive Economics", in Essays in Positive Economics pp. 3-43.

Hausman, D. (1981), *Capital, Profits, and Prices: An Essay in the Philosophy of Economics*. New York: Columbia University Press.

_ _ _ _ _ _ _. (1984), "Causal Priority", *Nous* 18: 261-79.

_ _ _ _ _ _ _. (1986), "Causation and Experimentation", *American Philosophical Quarterly* 23: 143-54.

Hicks, J. (1979), *Causality in Economics*. New York: Basic Books.

Hutchison, T. (1938), *The Significance and Basic Postulates of Economic Theory*. rpt. New York: A. M. Kelley, 1960.

Lewis, D. (1973), "Causation", *Journal of Philosophy* 70: 556-67.

Mackie, J. (1974), *The Cement of the Universe*. Oxford: Oxford University Press.

Marshall, A. (1930), *Principles of Economics*. 8th ed. London: Macmillan.

316

Mill, J.S. (1836), "On the Definition of Political Economy and the Method of Investigation Proper to It", rpt. in vol. 4 of *Collected Works of John Stuart Mill*. Toronto: University of Toronto Press, 1967.

_ _ _ _ _. (1843), *A System of Logic*. London: Longmans, Green and Co., 1949.

Mongin, P. (1986), "Are "All-and-Some" Statements Falsifiable After All? The Example of Utility Theory", *Economics and Philosophy* 2: 185-95.

Morgenbesser, S. (1956), "Theories and Schemata in the Social Sciences", Dissertation, University of Pennsylvania.

Simon, H. (1953), "Causal Ordering and Identifiability", rpt. in *Models of Discovery and other Topics in the Methods of Science*. Dordrecht: Reidel, 1977, pp. 53-80.

Stalnaker, R. (1972), "Pragmatics", in D. Davidson and G. Harman, eds. *Semantics of Natural Language*, Dordrecht: Reidel, pp. 380-97.

Yeager, L. (1969), "*Methodenstreit* over Demand Curves", *Journal of Political Economy* 68: 53-64.

Ceteris Paribus Conditions as Prior Knowledge: A View from Economics[1]

Neil de Marchi and Jinbang Kim

Duke University

We take it that what is intended by "laws" in the title of this session—"Confirming Ceteris Paribus Laws in the Social Sciences"—is reliable rather than fortuitous relations, that is, Millian "tendencies," with "forcing" properties. These refer to causes which do their work whether or not interfered with, or even counteracted by, other causes. Early econometricians had this in mind when they referred to "autonomous" relations (or behavioral relations). We entirely beg the question of how we know when we are dealing with genuine causes and with the related question how we come by autonomous relations. Instead, we concentrate upon some implications of our supposing that we can identify the experimental conditions necessary to confirm the operation of whatever tendency is in question.

Another expression for these experimental conditions is the steps necessary in order to transform a theoretical model into an estimable one. We suggest that these steps do two things: on the one hand, they complicate the deductive link between hypothesis and data. On the other hand, without them there is no link. When we make a ceteris paribus statement we are in effect asserting some such link; stressing the steps, then, is merely making that link plain for all to see. The assertion involves claims to prior knowledge; setting out the steps and looking into them enables an assessment of the claim. Below we give an illustration of how these steps infuse testing in modern economics. In the process we are also able to illustrate how ceteris paribus, in the form of the error term—so often thought of negatively, as another way of saying "we don't know" - may be turned to positive use in modern economics. To start with, however, we offer a broad classification of the ways in which ceteris paribus has been invoked by economists down the years.

1. How Economists Use Ceteris Paribus

As we read the history of economics there are basically two ways in which the problem has been regarded, although each category really contains a range of positions.

A. The true causes approach.

In the first approach it is supposed that we know the true causes relevant to a particular effect, but not all the interfering causes have been identified nor their effects calibrated. Thus economic laws are true but incomplete.

Here testing is not much of a problem, because we also know or can obtain a reasonable empirical basis for claiming—that the causes we have identified are also the *main* ones. The role of statistics here, then, is simply to measure the extent of the gap between our predictions and the actual outcome.

This is the view of John Stuart Mill and others among the British classical school of economics (Ricardo, Senior, Cairnes, for example).

The function of the ceteris paribus clause here is to simplify. One may of course simplify because what one excludes is reckoned empirically to be not dominant (the Ricardo-Mill- Cairnes view); or one may simplify to render analysis itself tractable (and communicable). The latter is Alfred Marshall's approach. We lump the two variants under one head because, unless one regards economic theory as a game the view that one knows the true and major causes tends to infuse even the analytical simplifications that one adopts. And in any case, we follow Dan Hausman (in his contribution to this volume) in placing the use of ceteris paribus for *purely* theoretical purposes outside our field of concern here.

B. The design of experiments approach.

In the alternative approach, which dates from the efforts of early econometricians (1930s) to estimate the parameters of autonomous relations, ceteris paribus is understood as the set of experimental conditions that would have to hold for a true test to be conducted. This is the approach that we single out for further examination below.

Here too there is a range of positions. Narrowly, this approach involves spelling out the rules which enable us to go from theoretical entities to measurable ones. The question is, how would we go about getting measurements which corresponded to the "true" (theoretical) entities? This was a main concern of Trygve Haavelmo, father of the probabilistic approach to modern econometrics. The design of experiments notion was borrowed directly from Sir Ronald Fisher.

More broadly, what we are talking about here is the auxiliary hypotheses and approximations introduced to give us some sort of measurable implications in our theoretical models.

In this second hold-all category there is much more attention explicitly given to test conditions; and the stages to be passed through in moving from theory to estimable model involve all manner of choices (e.g., choosing the form of function to embody a theoretical hypothesis) and technical transformations (e.g., using a linear approximation for the chosen functional form, which is very common; or incorporating assumptions such as, that agents are alike in all key respects, so as to render irrelevant strictly unobservable elements, such as the *distribution* of wage offers facing an individual, as is done in formulating the testable implications of job search theory[2]).

2. Ceteris Paribus (category B) and the Deductive Link

It is clear that these choices and transformations under B complicate the deductive link between theory and the proposition(s) actually tested. If the link is to be maintained the choices made and the technical shifts adopted have to be believed in in some sense (e.g., that they are "reasonable" or that they facilitate the testing of but do not destroy the original hypothesis). Otherwise we end up quite agnostic. As it turns out, then, both categories of ceteris paribus formulations necessarily invoke a good deal of "prior knowledge". In the first category we suppose that we know the true and major causes. In the second, we suppose two quite distinct sorts of knowledge. One is that we are actually explicating and not fundamentally destroying the theory-test deductive linkage by introducing all our choices and transformations. Secondly, we are implicitly affirming that the decisions we take about specification (which variables are related) and form of function (how exactly are they related?), about which approximations and estimation technique to use, and so on, represent tried and received wisdom about what is "best" in the circumstances.

Granting all this, it seems that there is a case for interpreting the immediate task ahead more narrowly than is implied in a phrase like "the status of ceteris paribus laws." Instead of focussing on the status of laws, we are led to inquire into the role of our "prior knowledge" in facilitating testing. And this means some quite specific things; for example, finding out about diagnostic statistical tests that help us judge how important or unimportant our choices and transformations are in terms of their effect on the results we measure. Our main message is that prior knowledge should become a focus of attention to a much greater extent than has been the case heretofore.[3]

What follows is simply an attempt to make this suggestion more plausible by giving an example showing just how the move is made from theory to estimable model. The example is taken from the work of Robert Lucas, of the University of Chicago, and one of the founders of the New Classical Economics. It concerns his efforts to test the natural rate hypothesis.

3. Lucas' Test of the Natural Rate Hypothesis

3.1 The Problem.

In observations, we always capture something that looks as if there is a tradeoff between unemployment and inflation (less of the one seems to involve more of the other). If the observed tradeoff is part of the stable structure of the economy, then demand policy may be used to increase output (thereby decreasing unemployment, albeit at the cost of some rise in inflation). What if, however, the tradeoff is not stable/structural as supposed? In particular, what if attempts to stimulate the economy above "normal" (roughly "full capacity") output generate only expectations of inflation? These expectations, provided they are "rational" (to be defined presently) will translate into actual inflation but not into producer decisions to increase output. Then the "true" relationship is this, that there is no structural tradeoff; and any observed tradeoff only reflects producers' misinterpretations of general price increases as relative price increases favoring them as individuals. The consequences of an incorrect belief about which of the relationships is the structural one on the part of policy-makers are obvious and may be serious.

3.2 An Equilibrium Model.

For Lucas to test the natural rate, or no tradeoff, hypothesis, he needs a model which can encompass both tradeoff and no tradeoff possibilities. This encompassing can be effected in a model of aggregate (i.e., economy-wide) supply and demand in which markets are supposed to clear, to which is *added* a hypothesis to the effect that forecasts (expectations) are formed rationally. We shall explain these two components separately, pitching the account at a level that may be followed by anyone who cares to dip into a modern macroeconomics textbook at the intermediate level. A perusal of chapter one of Klamer (1984) would also help. We do need to assume that the reader is comfortable with the notion of a demand curve (relating quantities purchased negatively to prices) and a supply curve (relating offers of work or of goods positively to wage rates or prices).

An aggregate demand and supply model to determine nominal income for the economy as a whole —nominal income is output times an index of prices—is something quite generic. To give it a more specific character, which is a step in the direction of getting testable hypotheses, assume the following.

(i) In Lucas' model economy aggregate output and the general price level are assumed to be determined so that the money and goods markets underlying the *aggregate demand schedule* are in "equilibrium" (without excess demand or supply). The aggregate demand schedule represents such equilibrium output and price level combinations. This demand schedule, it is argued, shifts as certain economic variables alter. Among these variables are monetary and fiscal policy, through which the government can exert control over the aggregate demand schedule. The control, however, is not perfect. The government can set policy, but it does not control all the random influences that may also affect demand. Policy therefore only determines the "center" about which the demand schedule moves in response to unpredictable shocks.

In terms of the theory on which it rests this representation of aggregate demand is fairly standard. It embodies a good deal of underlying theory (consumption/saving theory, investment theory, money demand and supply theory), all of which may be said to be uncontroversial in any sense relevant here, hence prior knowledge. The depiction of policy as "centering" the demand schedule, with further perturbations reflecting random shocks is, on the other hand, anything but traditional. It marks virtually the first clear introduction of stochastic *theory* into macroeconomics. Here an error term is added to an equation, not by way of providing a catch-all for what we don't know or cannot measure—ceteris paribus in a purely formal and theory-saving role—but by way of distinguishing between the (knowable) process underlying price-output observations, and an unpredictable random element. The distinction is critical when it comes to incorporating rational expectations; for the rational forecaster is said to be one who incorporates fully in his or her forecasts all that is known about the *process* thought to underlie the variable(s) of interest. The unpredictable element, although it inhibits accurate forecasting, takes on a central theoretical role in Lucas' explanation of the slope of the aggregate supply schedule, which is the complementary second half of his model. Thus in this case the error term facilitates theorising in an essential way and, as shown later, also contributes to testing.

(ii) The *aggregate supply schedule* represents output and price level combinations under which the labor market is cleared. In other words, the supply schedule shows how workers/producers change their effort/output in response to general price changes. What is it that allows us to think of this schedule as sloping upwards and to the right? Speaking generally, such a slope implies that both prices and output will increase somewhat following an increase in demand (i.e., there is a tradeoff between unemployment and inflation—lower unemployment resulting in higher output). But in Lucas' model, this mixed response is wholly due to incomplete information on the part of suppliers.

Each supplier is assumed to forecast the price level rationally, using a knowledge of the process underlying aggregate demand and supply. In these rational forecasts, the unpredictable random element is assumed to take the mean value of such errors, which is zero (by assumption). But the only information on prices in the current period available to a supplier is his or her own price, not the level of prices as a whole. So each supplier has to place an interpretation on any difference between the forecast price level and the observed own price. Is the difference due to an unpredicted (and unpredictable) change in general prices; or is it that the forecast of general prices is correct and the change reflects an alteration in own price relative to general prices? If the latter, and the belief is that own price has risen relative to prices in general, the supplier will perceive an advantage to increasing output. But if the supplier perceives the difference as due to an unexpected increase in general prices, this will translate immediately into a forecast of wage and cost inflation; the supplier therefore perceives no gain from expanding output and the supply schedule will be vertical. This last is the no tradeoff situation. (Or, if one prefers, the upward-sloping supply schedule itself shifts upwards by the full amount of inflation, reflecting the changed prior information used in forecasting. Either way, there will be an increase in prices, but no change in output.)

(iii) Thus the equilibrium point of output and prices, which is the crossing point of demand and supply schedules, will move in the same direction as a positive shift in the demand schedule, to the extent that this shift is not offset by a change in the supply schedule. The random variations mentioned above may produce such shifts, and hence the observed tradeoff between output (or unemployment) and prices, although only to the extent that they are viewed as changes in relative prices. On the other hand, government policy does not produce such shifts: policies are regarded as part of the prior information of workers/producers and if policy changes then the supply schedule also shifts.

Notice how the analysis has been focussed on the random element in positioning the demand schedule. Stochastic theory is center stage precisely because whether there is or is not a tradeoff depends on how observed own price changes are viewed—as due to stochastic shifts in all prices, or as reflecting relative price movements. By introducing a stochastic term from the beginning of theory construction the econometricians' view of economic data (which is to take data as generated by some process together with a random component) does not have to be justified separately. The economic variables of the theory itself are regarded as embodying stochastic elements.

We might also note that, as with aggregate demand, so too Lucas' treatment of worker/producer responses on the supply side is not controversial (at least within neo-classical economic theory). What is new is his focus on the *perceived* significance of observed own price relative to forecast price levels. This, and his way of formalizing the problem of information/interpretation. The latter is usually called the rational expectations hypothesis, on which more in a moment.

(iv) We noted above that the market-clearing supply/demand model just described is a way of specifying an otherwise quite generic approach so as to have it yield refutable hypotheses. But we still do not have an estimable model. Certain auxiliary assumptions are added to ease us in that direction. An incomplete list includes the following: (a) the aggregate demand curve is of unit elasticity (so that nominal income is uniquely related to given policy); (b) the secular component of output increases exponentially over time (steady compounded growth); (c) the elasticity of the cyclical component of output to the perceived, relative price is constant. These may be viewed as simplifying assumptions.

3.3 The Rational Expectations Hypothesis.

We must still show how the rational expectations hypothesis is linked to the natural rate hypothesis so as to enable a test of the latter to be carried out. First a word about the notion of rationality employed here.

In general, rationally formed expectations are expectations that incorporate fully all available information, including a knowledge of the processes generating observations. Rational forecasts need not be correct, because there is always a random - hence unpredictable—component too. But they will be better on average than any alternative that ignores the data generating processes, not only because these processes yield crucial information about patterns but also because any change that causes a forecast to go awry is fully taken into account in the next round. Again, this does not guarantee complete accuracy, since a new unexpected element may enter; but there is at least no correlation between the successive errors. By contrast, an adaptive expectations mechanism supposes that persistent surprises in one direction will leave the forecaster catching up with the change indefinitely. Adaptive expectations thus cannot encompass a no tradeoff situation, hence is an unsuitable hypothesis for Lucas' test purposes, quite independently of the fact that rational expectations by their nature are more consistent with traditional maximising assumptions.

There are two major arguments in the rational expectations hypothesis:

(i) The first concerns the statistical inference on the observed price made by workers/producers—i.e., how they interpret observed price—given their prior information about the general price level and relative price behavior. As we have seen, the supply schedule slopes upward if a price level increase is viewed as a relative price increase.

This interpretation is not irrational. It is inevitable that errors will be made, since producers do not know the current general price level, and can at best make an inference as to how much of the observed own price change reflects a relative price change which is of interest to them.

(ii) This inference in turn is based on prior information. Workers/producers are assumed to know the distribution of the general price level. It is also assumed that relative prices (own prices relative to general prices) are normally distributed over all suppliers at each moment, and this too is known to workers/producers. Both pieces of information are assumed to be fully utilized in making an "optimal" (rational) inference on the observed own price. In other words, each supplier is said to calculate the mathematical expectation of the relative price of his or her product conditional on an observation on the current own price plus prior knowledge of the distribution of the general price level. This is often referred to as signal extraction.

Prior information here means full and correct knowledge of the relevant history and mechanism of the economy. In particular, suppliers know all past nominal and real outputs. Furthermore, they hold correct conjectures as to the functional relationship between the current demand level and the level of prices. If the conjecture were not correct, the economy by definition would not be in equilibrium, yet in fact we see only supply/demand equilibrium points in market data. The knowledge assumptions thus put our model in line with available data.

In using the rational expectations hypothesis too there are auxiliary assumptions introduced, to facilitate the link between testable implications of theory and actual data. These are: (a) the prior distribution of the current general price level is normal with constant mean and variance; (b) the current relative prices over the whole economy at each moment constitute, and are known to constitute, a normal distribution with constant mean and variance; (c) the nominal demand level is drawn randomly in each period from a normal distribution with a constant mean and variance.

4. The Test

The model has several implications. The first is that observations on the inflation rate and output will form a positively-sloped curve on an inflation-output graph, but this is nothing more than an empirical regularity, without theoretical foundation. Secondly, there is no real tradeoff between these two variables. Any observed relationship is only a result of unexpected changes in nominal demand which happen to be misinterpreted by suppliers. Any intentional or systematic manipulation of nominal demand by the government will affect inflation rates but not real output. Finally, the larger the prior variance of the general price level compared with that of relative prices, the smaller will be the portion of an observed price change that will be viewed as a relative price change, hence as giving the individual supplier an advantage that will induce an increase in output. But since the prior variance of general prices depends positively on the variance of nominal output, a larger variance of nominal output will result in a steeper curve—less of an apparent tradeoff—on the inflation-output graph. This last is a theoretically based and testable prediction, and is the one actually tested by Lucas.

Lucas' test makes use of observations across several countries, including Argentina and the United States; hence, some with histories of volatile demand and others with relatively stable demand histories. Those with a history of volatile demand shifts are observed to have large variances in the level of prices. It is found that the apparent tradeoff between inflation and output is less for the countries with stable demand histories, and this is taken as evidence for his model and against a tradeoff. Suppliers, it is presumed, have inferred that relative price variance is a small portion of general price variance.

5. What Exactly is Tested Here?

What Lucas has tested is—just as Duhem-Quine would have us believe—a package comprising (i) a group of macroeconomic theories which are summarised in the demand schedule; (ii) a neoclassical theory about work and leisure decisions across time; and (iii) the rational expectations hypothesis, which cooperates with the neoclassical theory to yield a supply schedule.

The importance of the rational expectations hypothesis in this context cannot be stressed too much. Before Lucas there had been no test attempted which directly involved the natural rate hypothesis. Nor had any model been built which successfully incorporated both the natural rate hypothesis and the observed tradeoff between inflation and output (or unemployment). Lucas succeeded in building such a model, and his success is due essentially to the rational expectations hypothesis. Moreover, he was the first to formulate it in a form compatible with general equilibrium macroeconomic models. His work, not surprisingly, is properly regarded as revolutionary. But the test we have described is not of the natural rate hypothesis in isolation; it is of that hypothesis expressed with the aid of rational expectations.

Nor is it just the case that what is tested here is actually some mix of conventional theories and the rational expectations hypothesis. In addition, a host of specifications and adjustments are brought into the procedure outlined above. It may be useful to list them:

• A particular form was given the demand schedule, even although there is no a priori or theoretical justification offered.
• A similar kind of specification was imposed on the behavioral equation attributed to workers/producers.
• An assumption was made about the nature of relative prices and demand: each is assumed to be a random variable. Relative prices across the whole economy are said to constitute a normal distribution; and the aggregate demand level for each period is drawn randomly from a normal distribution.
• Agents are assumed to have a correct model of the economy at their disposal: that is, they somehow (how we are not told) know the relationship between demand changes and the price level; they know just how other suppliers will respond to own price observations; and so on.

These technical adjustments are typical of the steps taken in transforming a generic framework of analysis (for example, supply and demand) into a specific theoretical model and thence into an estimable model. Noticing them is nothing new[4]; but their unavoidable presence needs to be stressed, and their number and character noted for two separate reasons. Firstly, and precisely because they do complicate the deductive nature of testing, it is as well to be aware of the specifics of the complexities thus introduced. Secondly, because they are in fact elements in the ceteris paribus pound implicitly or explicitly attached to the theoretical hypothesis to be tested. There being no way to avoid ceteris paribus, the second-best strategy is to detail what is in the pound and lay it bare for examination.

It used to be stressed (e.g., Hutchison 1938) that unless we undertake this laying bare process our hypotheses will have no empirical content. We want to place the emphasis differently: unless we do it we are ignoring the very steps taken to provide some sort of connection between theory and data. For these elements are part of the prior knowledge that we postulate in drawing any conclusion from a test such as the one summarised here. Moreover, because they are always very particular, the case study method seems the only reliable one to pursue if we are to say anything useful about the status of our hypotheses.

6. Summary

What we have in the Lucas procedure, then, is this:

a. A hypothesis to be tested (the natural rate hypothesis);

b. A theoretical model (i) of such sort—equilibrium in character—that we can relate it to available data (supply/demand cross points); (ii) which encompasses both the natural rate possibility and the observed tradeoff relation.

c. A complementary hypothesis (rational expectations) which stresses process plus a random element in forecasting, which supposes a correct knowledge of processes and of the distributions of past and present observable variables, and which supposes that best inference will be made on current observed price by all individuals. These suppositions enable us to focus wholly upon the random shift element in aggregate demand as the problem of interpretation. When the possible interpretations are cast in terms of a history of relative price versus total price variance, a test can be conducted on real world data.

d. A host of auxiliary assumptions, some made with an eye to simplification and others to facilitate a narrowing of the focus, complete the picture.

Our initial claim was that there is an assertion about prior knowledge in each of the steps listed above. (If many detailed steps make up what we have called one category, then correspondingly many specific claims are made.) When the test result is in, debate can begin on its significance and acceptability for further work. Our contention is that debate must necessarily center upon the succession of steps we have listed.

Notes

[1]We wish to thank participants in the PSA session at which a draft of this paper was presented for stimulating comments; and Dan Hausman and Nancy Cartwright who, perhaps unwittingly, drew us in this direction through comments they have made at widely separated times and places on things ranging from the absence of concern by the logical falsificationist Popper for what constitutes supporting evidence to the role of ceteris paribus and prior knowledge in the causal thinking of early econometricians. They bear no responsibility for the use to which we have put their ideas.

[2]This emerges from a detailed study which Jinbang Kim is pursuing into job search theory.

[3]Prior knowledge is a key notion in the writings of Cowles Commission econometricians of the 1940s. See, for example, Marschak (in Koopmans, 1950, p.2), for whom economic theory yields a knowledge of "structure," and Haavelmo (1944, p. 81), who speaks of "a certain a prior knowledge" that we must impose in order to reach nontrivial conclusions. A more general recognition of the importance of prior knowledge in relation to the tastes that shape our judgements of "worth" (of a play, a poem, a sculpture) is to be found in Herrnstein Smith (1988, p. 105). She writes:

[T]he process of testing the adequacy of a scientific model or theory is never only ... a measuring of its fit with what we call 'the data,' 'the evidence,' or 'the facts,' all of which are, themselves, the products of comparable conceptual and evaluative activities already appropriated to one degree or another by the relevant community; it is also a testing, sampling, and, in effect, *tasting* in advance of the ways in which the product will taste to other members of that community...

[4]Nonetheless, the work of spelling out what is involved has *not* been begun until quite recently.

References

Haavelmo, Trygve (1944; mimeo 1941), *The Probability Approach in Econometrics*. Supplement to *Econometrica* 12, Chicago: University of Chicago Press.

Hutchison, T.W. (1938), *The Significance and Basic Postulates of Economic Theory*: Reprint ed., New York: A.M. Kelley.

Klamer, Arjo (1984), *Conversations with Economists*, Totowa, New Jersey: Rowman and Allanheld.

Koopmans, Tjalling C. (ed.) (1950), *Statistical Inference in Dynamic Economic Models*, Cowles Commission Monographs no. 10, New York: Wiley.

Marschak, J. (1950), "Statistical Inference in Economics: An Introduction," in Koopmans (ed.), *Statistical Inference in Dynamic Economic Models*, 1-50.

Lucas, Robert E., Jr (1973), "Some International Evidence on Output-Inflation Tradeoffs," *American Economic Review* 63, 326-34.

Smith, Barbara Herrnstein (1988), *Contingencies of Value. Alternative Perspectives for Critical Theory*, Cambridge, Mass.: Harvard University Press.

Part XI

DECISION AND GAME THEORY

Backward Induction without Common Knowledge

Cristina Bicchieri

Carnegie-Mellon University

1. Information and meta-information

Game theory studies the behavior of rational players in interactive situations and its possible outcomes. For such an investigation, the notion of players' rationality is crucial. While notions of rationality have been extensively discussed in game theory, the epistemic conditions under which a game is played — though implicitly presumed — have seldom been explicitly analyzed and formalized. These conditions involve the players' reasoning processes and capabilities, as well as their knowledge of the game situation.[1] Game theory treats some aspects of information about chance moves and other players' moves by means of information partitions in extensive form games. But a player's knowledge of the structure, for example, of information partitions themselves is different from his information about chance moves and other players' moves. The informational aspects captured by the extensive form games have nothing to do with a player's knowledge of the structure of the game.

Game theorists implicitly assume that the structure of the game is *common knowledge* among the players. By 'common knowledge of p' is meant that p is not just known by all the players in a game, but is also known to be known, known to be known to be known, ... *ad infinitum*.[2] The very idea of a Nash equilibrium is grounded on the assumptions that players have common knowledge of the structure of the game and of their respective priors. These assumptions, however, are always made *outside* the theory of the game, in that the formal description of the game does not include them.[3]

The assumptions about players' rationality, the specification of the structure of the game, and the players' knowledge of all of them should be part of the theory of the game. Recent attempts to formalize players' knowledge as part of a theory of the game include Bacharach (1985, 1987), Gilboa (1986), Mertens and Zamir (1985), Brandenburger and Dekel (1985), Kaneko (1987) and Samet (1987). In these works a common knowledge axiom is explicitly introduced, stating that the axioms of logic, the axioms of game theory, the behavioral axioms and the structure of the game are all common knowledge among the players.

Is it always necessary for the players to have common knowledge of the theory of the game for a solution to be derived? Different solution concepts may need different amounts of knowledge on the part of the players to have predictive validity at all. For example, while common knowledge is necessary to attain an equilibrium in a large class of normal form games, it may lead to inconsistencies in finite, extensive form games of perfect information (Reny 1987, Bicchieri 1989).[4] More generally, if players' epistemic states and their degree of information about other players' epistemic states are included in a theory of the game, which solutions to non-cooperative games can be derived? I believe the consequences of explicitly modeling players' knowledge as part of the theory of the game are far-reaching.

In this paper I examine finite, extensive form games of perfect and complete information. These games are solved working backwards from the end, and this procedure yields a unique solution. It is commonly assumed that backward induction can only be supported by common knowledge of rationality (and of the structure of the game). In section 2 it is proved instead that the levels of knowledge of the theory of the game (hence, of players' rationality) needed to infer the backward induction solution are finite.

That limited knowledge is sufficient to infer a solution for this class of games does not mean it is also a necessary condition. In section 3, I introduce the concepts of *knowledge-dependent games* and *knowledge-consistent play*, and prove that knowledge has to be limited for a solution to obtain. More specifically, it is proved that for the class of games considered here backward induction equilibria are knowledge-consistent plays of knowledge-dependent games. Conversely, every knowledge-consistent play of a knowledge-dependent game is a backward induction equilibrium.

For the class of games considered, there exist knowledge-dependent games that have no knowledge-consistent play. For example, a player might be unable — given what she knows — to 'explain away' a deviation from equilibrium on the part of another player, in that reaching her information set is inconsistent with what she knows.

If the theory of the game were to include the assumption that *every* information set has a small probability of being reached (because a player can always make a mistake), then no inconsistency would arise. In this case, the solution concept is that of *perfect equilibrium* (Selten 1975), which requires an equilibrium to be stable with respect to 'small' deviations. The idea of perfect equilibrium (like other 'refinements' of Nash equilibrium) has the defect of being *ad hoc*, as well as of assuming — as Selten himself has recognized — less than perfect rationality.[5]

The present paper has a different goal. What I want to explore here is under which epistemic conditions a rationality axiom can be used to derive a unique prediction about the outcome of the game. As it will be made clear in the example of section 2, a small variation in the amount of knowledge possessed by the players can make a big difference, in that higher levels of knowledge of the theory of the game may make the players unable to 'explain away' deviations from the equilibrium path. The idea is that of finding the minimal set of axioms from which a solution to the game can be inferred.

Since the players (as well as the game theorist) have to reason to an equilibrium, the theory must contain a number of meta-axioms stating that the axioms of the theory are known to the players. In particular, the theory of the game T can contain a meta-axiom A_n stating that the set of game-theoretic ('special') axioms A_1-A_{n-1} is k-level group-knowledge among the players, but not a meta-axiom A_{n+1} saying that A_n is group-knowledge among the players. If A_{n+1} is added to T, it becomes group-knowledge that the theory is inconsistent at some information set. In this case, the backward induction solution cannot be inferred.

2. Backward induction equilibrium

In this section non-cooperative, extensive form games of perfect information are defined and it is proved that the levels of knowledge needed to infer the backward induction equilibrium are *finite*, contrary to the common assumption that only an infinite itera-tion of levels of knowledge (i.e., common knowledge) can support the solution.

Definition 2.1. A non-cooperative game is a game in which no precommitments or binding agreements are possible.

Definition 2.2. A finite n-person game Γ of perfect information in extensive form consists of the following elements:

(i) A set $N = \{1, 2, ..., n\}$ of players.

(ii) A finite tree (a connected graph with no cycles) T, called the game tree.

(iii) A node of the tree (the root) called the first move. A node of degree one and dif-ferent from the root is called a terminal node. Ω denotes the set of all terminal nodes.

(iv) A partition $P^1, ..., P^n$ of the set of non-terminal nodes of the tree, called the player partition. The nodes in P^i are the moves of player i. The union of $P^1, ..., P^n$ is the set of moves for the game.

(v) For each $i \in N$, a partition $I^{i1}, ... , I^{ik}$ of P^i (I^{ij} denotes the j-th information set ($j \geq 1$) of player i) such that for each $j \in \{1, ..., k\}$:
   (a) each path from the root to a terminal node can cross I^{ij} at most once, and
   (b) since there is perfect information, I^{ij} is a singleton set for every i and j.

(vi) For each terminal node t, an n-dimensional vector of real numbers, $f^1(t), ... , f^n(t)$ called the payoff vector for t.

Every player in Γ knows (i)-(vi).

Definition 2.3. A *pure strategy* s^i for player i is a k-tuple that specifies, for each information set of player i, a choice at that information set. The set of i's pure strategies is denoted by $S^i = \{s^i\}$. Let $S = S^1 \times .... \times S^n$. A *mixed strategy* x^i for player i is a prob-ability distribution over player i's pure strategies.

Definition 2.4. The function $\pi^i : S^1 \times .... \times S^n \to \Re$ is called the *payoff function* of player i. For an n-tuple of pure strategies, $s = (s^1, ... , s^n) \in S$, the expected payoff to player i, $\pi^i(s)$, is defined by

$$\pi^i(s) = \sum_{t \in \Omega} p_s(t) \, \pi^i(t)$$

where $p_s(t)$ is the probability that a play of the game ends at the terminal node t, when the players use strategies $s^1, ... , s^n$.

Definition 2.5. A pure strategy n-tuple $s = (s^1, ... , s^n) \in S$ is an *equilibrium point* for Γ if

$$\pi^i(s|y^i) \leq \pi^i(s) \text{ for all } y^i \in S^i$$

where $s|y^i = (s^1, ..., s^{i-1}, y^i, s^{i+1}, ..., s^n)$.
We also say that $s^i \in S^i$ is a *best reply* of player i against s if $\pi^i(s|s^i) = \max_{y^i \in S^i} \pi^i(s|y^i)$.

Definition 2.6. A subgame $\Gamma_j \in \Gamma$ is a collection of branches of the game that start from the same node and the branches and node together form a game tree by itself.

Theorem 2.1. (Kuhn 1953) A game Γ of perfect information has an equilibrium point in pure strategies.

Proof. By induction on the number of moves in the game. Suppose the game has only one move. Then the player who has to move, in order to play an equilibrium strategy, should choose the branch which leads to a terminal node with the maximum payoff to him. Therefore the theorem is true when Γ has one move. Suppose the theorem is true for games with at least K moves ($K \geq 1$). Let Γ be a game with at most K+1 moves, where T is the game tree for Γ, r the root of T, and k the number of branches going out of r (these branches are numbered from 1 to k). The node at the end of of the j-th branch from r is the root of a subtree T_j of T, where T_j is the tree for a subgame Γ_j of Γ (since the game is one of perfect information).

For each $s^i \in S^i$, let s^{ij} be the actions recommended by s^i at the information sets in Γ_j. Let $S^{ij} = \{s^{ij}\}$ and let $\pi^{ij}(s^j)$ be the expected payoff of player i in the subgame Γ_j, when the players play the combination of strategies $s^j = (s^{1j}, ... , s^{nj})$ ($j = 1, ..., k$). Each subgame Γ_j is a game of perfect information with K moves or less and by assumption it has an equilibrium $\bar{s}^j = (\bar{s}^{1j}, ... , \bar{s}^{nj})$, so that

(*) $\pi^{ij}(\bar{s}^{j}| t^{ij}) \leq \pi^{ij}(\bar{s}^j)$ for all $t^{ij} \in S^{ij}$

Consider now the root r of T. For some player $n \in N$, $r \in P^n$. Let $l \in \{1, ... , k\}$ be a branch departing from r such that

$$\pi^{nl}(\bar{s}^l) = \max_{1 \leq j \leq k} \pi^{nj}(\bar{s}^j)$$

$\bar{s}^i \in S^i$ is defined as follows:

(a) $\bar{s}^i = \prod_{j=1}^{k} \bar{s}^{ij}$ for $i \neq n$ (b) $\{l\} \times \prod_{j=1}^{k} \bar{s}^{nj}$ for $i = n$

Thus at information set I^i, the equilibrium strategy $\bar{s}^i$ of player i tells him to choose branch l if $I^i = \{r\}$, and to play $\bar{s}^{ij}$ (I^i) if I^i is an information set in Γ_j. We have to prove that $\bar{s} = (\bar{s}^1, ... , \bar{s}^n)$ is an equilibrium point for Γ.

For $i \neq n$, and $t^i = \prod_{j=1}^{k} t^{ij} \in S^i$, $\pi^i(\bar{s}| t^i) \leq \pi^i(\bar{s})$ by (*) and since by assumption player n chooses branch l at node r under $\bar{s}^n$. For $i = n$, and $t^n = \{m\} \times \prod_{j=1}^{k} t^{ij} \in S^n$, where m is one of the branches going out of r, $\pi^n(\bar{s}| t^n) \leq \pi^n(\bar{s})$ by (*) and since, by assumption, $\pi^{nl}(\bar{s}^l) = \max_{1 \leq j \leq k} \pi^{nj}(\bar{s}^j)$. Hence $\bar{s}$ is a pure strategy equilibrium for Γ.

The equilibrium can be found by working backwards from the terminal nodes to the root. At each information set, a player chooses the branch which leads to the subtree yielding him the highest equilibrium payoff. To illustrate this method, consider the following two-person extensive form game of perfect information with finite termination.

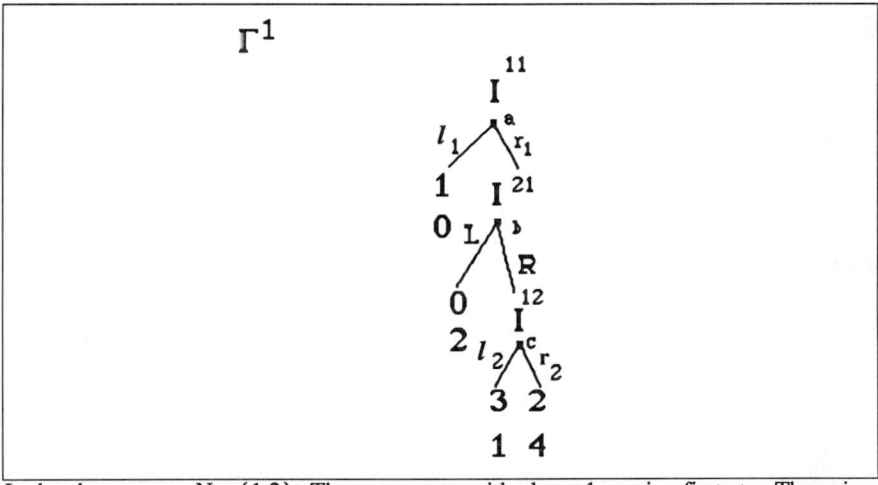

Γ^1

In the above game, $N = \{1,2\}$. The game starts with player 1 moving first at a. The union $P^1 \cup P^2$ is the set of moves $\{a, b, c\}$. $P^1 = \{a, c\}$, $P^2 = \{b\}$. $I^{11} = \{a\}$, $I^{21} = \{b\}$, $I^{12} = \{c\}$. Each player has two pure strategies: either to play left, thus ending the game, or to play right, in which case it is the other player's turn to choose. $S^1 = \{l_1 l_2, l_1 r_2, r_1 r_2, r_1 l_2\}$. $S^2 = \{L, R\}$. The payoffs to the players are represented at the endpoints of the tree, the upper number being the payoff of player 1, and each player is assumed to be rational (i.e., to wish to maximize his expected payoff).

The equilibrium described above for such games is obtained by backward induction as follows: at node I^{12} player 1, if rational, will play l_2, which grants him a maximum payoff of 3. Note that player 1 does not need to assume 2's rationality in order to make his choice, since what happened before the last node is irrelevant to his decision. Thus node I^{12} can be substituted by the payoff pair $(3, 1)$. At I^{21} player 2, if rational, will only need to know that 1 is rational in order to choose L. That is, player 2 need consider only what she expects to happen at subsequent nodes (i.e., the last node) as, again, that part of the tree coming before is now strategically irrelevant. The penultimate node can thus be substituted by the payoff pair $(0, 2)$. At node I^{11}, rational player 1, in order to choose l_1, will have to know that 2 is rational *and* that 2 knows that 1 is rational (otherwise, he would not be sure that at I^{21} player 2 will play L). From right to left, nonoptimal actions are successively deleted, and the conclusion is that player 1 should play l_1 at his first node. Thus $\bar{s}^1(I^{11}) = l_1$, $\bar{s}^2(I^{21}) = L$, $\bar{s}^1(I^{12}) = l_2$, and $(\pi^1(\bar{s}), \pi^2(\bar{s})) = (1, 0)$.

In the classical account of this game, $(l_1 L\, l_2)$ represents the only possible pattern of play by rational players because the game is one of *complete information*, i.e., the players know each other's rationality, strategies and payoffs. Player 1, at his first node, has two possible choices: l_1 or r_1. What he chooses depends on what he expects player 2 to do afterwards. If he expects player 2 to play L at the second node, then it is rational for him to play l_1 at the first node; otherwise he may play r_1. His conjecture about player 2's choice at the second node is based on what he thinks player 2 believes would happen if she played R. Player 2, in turn, has to conjecture what player 1 would do at the third node, given that she played R. Indeed, both players have to conjecture each other's conjectures and choices at each possible node, until the end of the game.

In our example, complete information translates into the conjectures $p(l_1) = 1$, $p(R) = 0$ and $p(r_2) = 0$. The notion of complete information does not specify any particular level of knowledge that the players may possess, but it is customarily assumed by game theorists that the structure of the game and players' rationality are *common knowledge* among them.

Note, again, that *specification of the solution requires a description of what both agents expect to happen at each node, were it to be reached, even though in equilibrium play no node after the first is ever reached.* The central idea is that if a player's strategy is to be part of a rational solution, then it must prescribe a rational choice of action in all conceivable circumstances, even those which are ruled out by some putative equilibrium. An equilibrium is thus endogenously determined by considering the implications of deviating from the specified behavior. The backward induction requirement calls for considering equilibrium points which are in equilibrium in each of the subgames and in the game considered as a whole. This means that it only matters where you are, not how you arrived there, as history of past play has no influence on what individuals do.

Since a strategy specifies what a player should choose in every possible contingency (i.e., at all information sets at which he may find himself), and a player's contingency plan ought to be rational in the contingency for which it was designed, it is necessary to give meaning to the idea of a choice conditional upon a given information set having being reached. Does it make sense to talk of a choice contingent upon other choices that may never occur? What counts as 'rational' behavior at information sets not reached by the equilibrium path depends on how a player explains the fact that a given information set is reached, since different explanations elicit different choices. For example, it has been argued that at I^{21} it is not evident that player 2 will only consider what comes next in the game (Binmore 1987; Reny 1987). Reaching I^{21} may not be compatible with backward induction, since I^{21} *can only be reached if 1 deviates from his equilibrium strategy, and this deviation stands in need of explanation.* When player 1 considers what player 2 would choose at I^{21}, he has to have an opinion as to what sort of explanation 2 is likely to give for being called to play, since 2's subsequent action depends on it. Binmore's criticism rightly points out that *a solution must be stable also with respect to forward induction.* In other words, if equilibrium behavior is determined by behavior off the equilibrium path, a solution concept must allow the players to 'explain away' deviations.

Selten's 'trembling hand' model (Selten 1975) provides the canonical answer. According to Selten, we must suppose that whenever a player wants to make some move a, he will have a small positive probability ε of making a different and unintended move $b \neq a$ instead by 'mistake'. If any move can be made with a positive probability, all information sets have a positive probability of being reached.

What relates Selten's theory of mistakes to backward induction? Since the backward induction argument relies on the notion of players' rationality, one has to show that rationality and mistakes are compatible. Admitting that mistakes can occur means drawing a distinction between deciding and acting, but a theory that wants to maintain a rationality assumption is bound to make mistakes entirely random and uncorrelated. Systematic mistakes would be at odds with rationality, since one would expect a rational player to learn from past actions and modify his behavior. If a deviation tells that a player made a mistake (i.e., his hand 'trembled'), but not that he is irrational, a mistake must not be the product of a systematic bias in favor of a particular type of action, as would be the case with a defective reasoning process.

In our example, when player 2 finds she has to move, she will interpret 1's deviation as the result of an unintended, random mistake. So if 1 plays (but did not choose to play) r_1, 2 knows that the probability of r_2 being successively played remains vanishingly small, viz. $p(r_2) = p(r_2|r_1) = \varepsilon$. This makes 2 choose strategy L, which is a best reply to player 1's strategy after allowing for the possibility of trembles. Player 1 knows that, were he to play r_1, player 2 would want to respond with L, and that there is only a vanishingly small probability that R is played instead. For $p(R) = \varepsilon$, player 1's best reply is l_1.

Thus $(l_1 L \ l_2)$ remains an equilibrium in the new 'perturbed' game that differs from the original game in that any move has a small positive probability of being made.

According to Binmore (1987, 1988), this characterization of mistakes is necessary for the backward induction argument to work, in that it makes out of equilibrium behavior compatible with players' rationality. Otherwise, Binmore argues, a deviation would have to be interpreted as proof of a player's 'irrationality'. Is this conclusion warranted? If common knowledge of rationality is assumed, then one must also offer some argument to explain how a player, facing a deviation, can still be able to maintain without contradiction that the deviator is rational. Selten's 'trembling hand' hypothesis is not the only plausible one, but certainly it is an answer.[6] But is common knowledge of rationality at all needed to get the backward induction solution?

Binmore and Reny have been skeptical of the classical solution precisely because they did not question the common knowledge assumption. In what follows, I show that the backward induction solution can be inferred from a set of assumptions that include a specification of players' knowledge. *The levels of knowledge needed for the solution to obtain are finite, and their number depends on the length of the game.*

A play of the game we have just described makes a number of assumptions about players' rationality and knowledge, from which the backward induction solution necessarily follows. Let us consider them in turn. First of all, the players know their respective strategies and payoffs. Second, the players are rational, in the sense of being expected utility maximizers. Third, the players have group-knowledge of rationality and of the structure of the game. This means that each player knows that the other player is rational, and knows the other player's strategies and payoffs. Is this information sufficient to infer a solution to the game?

It is easy to verify that in the above game *different levels of knowledge* are needed at different stages of the game for backward induction to work. For example, if R_1 stands for 'player 1 is rational', R_2 for 'player 2 is rational', and $K_2 R_1$ for 'player 2 knows that player 1 is rational', R_1 alone will be sufficient to predict 1's choice at the last node, but in order to predict 2's choice at the penultimate node, one must know that rational player 2 knows that 1 is rational, i.e. $K_2 R_1$. $K_2 R_1$, in turn, is not sufficient to predict 1's choice at the first node, since 1 will also have to know that 2 knows that he is rational. That is, $K_1 K_2 R_1$ needs to obtain. Moreover, while R_2 only (in combination with $K_2 R_1$) is needed to predict L at the penultimate node, $K_1 R_2$ must be the case at I^{11}.

Theorem 2.2. In finite extensive form games of perfect and complete information, the backward induction solution holds if the following conditions are satisfied for any player i at any information set I^{ik} : (α) player i is rational and knows it, and knows his available choices and payoffs, and (β) for every information set I^{jk+1} that immediately follows I^{ik}, player i knows at I^{ik} what player j knows at information set I^{jk+1}.

Proof. The proof is by induction on the number of moves in the game. If the game has only one move, the theorem is vacuously true since at information set I^{i1}, if player i is rational and knows it, and knows his available choices and payoffs, he will choose that branch which leads to the terminal node associated with the maximum payoff to him and this is the backward induction solution. Suppose the theorem is true for games involving at most K moves (some $K \geq 1$). Let Γ be a game of perfect and complete information with K+1 moves and suppose that conditions α and β are satisfied at every node of game Γ. Let r be the root of the game tree T for Γ. At information set I^{ir}, player i knows that conditions α and β are satisfied at each of the subgames starting at the information sets that immediately follow I^{ir}. Then at I^{ir} player i knows that the outcome of play at any of those subgames would correspond to the backward induction solution for that subgame. Hence at I^{ir} if player i is rational, he will choose the branch going out of r which leads to the subgame whose backward induction solution is best for him, and this is the backward induction solution for game Γ.

3. Knowledge-dependent games

Theorem 2.2 tells that, for the backward induction solution to hold, we do not need to assume common knowledge but only *limited knowledge* of rationality and of the structure of the game. All that is needed is that a player, at any of her information sets, knows what the next player to move knows. Thus the player who moves first will know more things than the players who move immediately after, and these in turn will know more than the players who follow them in the game. However, if the same player has to move at different points in the game, we want that player's knowledge to be the same at all of his information sets. This requirement has a natural interpretation in the normal form representation of such games.

Consider the normal form equivalent of game Γ^1

$$2$$

		L	R
	l_1	1,0	1,0
1	$r_1 l_2$	0,2	3,1
	$r_1 r_2$	0,2	2,4

In this game, strategy $r_1 l_2$ weakly dominates $r_1 r_2$, so if 2 knows that 1 is rational 2 will expect 1 to eliminate $r_1 r_2$. In the extensive form representation, this corresponds to player 2 knowing that rational player 1, at the last node, will choose l_2. *In order to eliminate his weakly dominated strategy,* player 1 need not know whether 2 is rational. This corresponds to the last node of the extensive form representation, where 1 does not need to consider what happened before, since it is now strategically irrelevant. Player 1 needs to know that 2 is rational only when, having eliminated $r_1 r_2$, he has made L weakly dominant over R. Note that player 1, in order to be sure that 2 will choose L, has to know that 2 is rational *and* that 2 knows that 1 is rational, otherwise there would be no weakly dominated strategy for player 2 to delete. Having thus deleted R, 1's best reply to L is l_1. And this corresponds to the first node, where player 1 has to know that 2 is rational and that 2 knows that 1 is rational. Evidently player 1 needs to know *more* than player 2, even in the normal form, since the order of iterated elimination of dominated strategies starts with player 1's strategy $r_1 r_2$. In the extensive form the backward induction argument makes player 1's previous knowledge irrelevant at his subsequent node, but this does not mean that player 1 knows less. This point becomes even clearer if we remember that we are dealing with static games: a player can plan a strategy in advance and then let a machine play on his behalf.

Given that the solution for this class of games depends upon the information possessed by the players, we may want to know whether variations in the level of knowledge would make a difference. Since only limited knowledge is sufficient to infer the backward induction solution, is it also a necessary condition? We know that assuming common knowledge leads to an inconsistency (Reny 1987; Bicchieri 1989), but is an inconsistency produced by simply assuming levels of knowledge higher than those which are sufficient to infer the solution? In particular, it is worth exploring what would happen were the players to know what the players preceding them know, i.e., what would happen were knowledge to go in both directions.

In order to address this issue, we have to explicitly model players' knowledge of the game, as well as the reasoning process that leads them to choose a particular sequence of actions. The theory of the game will have to include a set of assumptions specifying what the players know about the structure of the game and the other players. The main result of this section is that, for any finite extensive form game of perfect and complete information, *the levels of knowledge that are sufficient to infer the backward induction solution are also those which are necessary to infer it.* Higher levels of knowledge make the theory of the game inconsistent at some information set.

More formally, if we have n players, and some propositions $p_1, ... , p_m$, we can construct a knowledge language L by closing under the standard truth-functional connectives and the rule that says that if p is a formula of L, then so is $K_i p$, (i = 1, ... , n), where $K_i p$ stands for 'i knows that p'. Since we are interested in modeling collective knowledge, we add the group-knowledge operator E_G, where $E_G p$ stands for 'everyone in group G knows that p'. If G = { 1, 2, ... , n} $E_G p$ is defined as the conjunction $K_1 p \wedge K_2 p \wedge ... \wedge K_n p$. K-level group-knowledge of p can be expressed as $E^k_G p \equiv \bigwedge_{i_j \in G, 1 \le j \le k} K_{i_1} K_{i_2} ...... K_{1_k} p$.

If p is E^k_G-knowledge for all k ≥ 1, then we say that p is common knowledge in G, i.e., $C_G p \equiv p \wedge E_G p \wedge E^2_G p \wedge ..... \wedge E^m_G p \wedge ....$ $C_G p$ implies all formulas of the form $K_{i_1} K_{i_2} ...... K_{i_n} p$, where the i_j are members of G, for any finite n, and is equivalent to the infinite conjunction of all such formulas.

In order to reason about knowledge, we must provide a semantics for this language. Following Hintikka (1962), we use a possible-worlds semantics. The main idea is that there is a number of possible worlds at each of which the propositions p_i are stipulated to be true or false, and all the truth functions are computed at each world in the usual way. For example, if w is a possible world, then $p \wedge q$ is true at w iff both p and q are true at w. An individual's state of knowledge corresponds to the extent to which he can tell what world he is in, so that a world is possible relative to an individual i. In a given world one can associate with each individual a set of worlds that, given what she knows, could possibly be the real world. Two worlds w and w' are equivalent to individual i iff they create the same evidence for i. Then we can say that an individual i knows a fact p iff p is true at all worlds that i considers possible, i.e., $K_i p$ is true at w iff p is true at every world w' which is equivalent to w for individual i. An individual i does not know p iff there is at least one world that i considers possible where p does not hold.

The following set of axioms and inference rules provides a complete axiomatization for the notion of knowledge we use

A1 : All instances of tautologies
A2 : $K_i p \Rightarrow p$
A3 : $(K_i p \wedge K_i (p \Rightarrow q)) \Rightarrow K_i q$
A4 : $K_i p \Rightarrow K_i K_i p$
A5 : $\sim K_i p \Rightarrow K_i \sim K_i p$
MP : If p and $p \Rightarrow q$, then q
KG : If $\vdash p$, then $\vdash K_i p$

Some remarks are in order. A_2 tells that if i knows p, then p is true. A_3 says that i knows all the logical consequences of his knowledge. This assumption is defensible considering that we are dealing with a very elementary (decidable) logical system. A_4 says that knowing p implies that one knows that one knows p. Intuitively, we can imagine providing an individual i with a database. Then i can look at her database and see what is in it, so that if she knows p, then she knows that she knows it. A_5 is more controversial, since it says that not knowing implies that one knows that one does not know. This axiom can be interpreted as follows: individual i can look at her database to see what she does not know, so if she doesn't know p, she knows that she does not know it. Rule KG says that if a formula p is provable in the axiom system A_1 -A_5, then it is provable that $K_i p$. A formula is provable in an axiom system if it is an instance of one of the axiom schemas, or if it follows from one of the axioms by one of the inference rules MP or KG. Also, a formula p is consistent if $\sim p$ is not provable.

It is easy to verify that the rule KG makes all provable formulas in the axiom system A_1-A_5 *common knowledge* among the players. Suppose q is a theorem, then by KG it is a theorem that K_iq (i = 1, ... , n). If K_iq is a theorem, then it is a theorem that K_jK_iq (for all $j \neq i$), and it is also a theorem that $K_iK_jK_iq$, and so on. In the system A_1-A_5, if $\vdash p$ then $\vdash Cp$. We call the class of axioms A_1-A_5 *general axioms*.

Beside logical axioms, a theory of the game will include game-theoretic solution axioms, behavioral axioms, and axioms describing the information possessed by the players. This second class of axioms we call *special axioms*.

Let us consider as an example game Γ^1:

A6 : The players are rational (i.e., $R_1 \wedge R_2$)
A7 : At node I^{11}, $(r_1 \vee l_1) \wedge \sim(r_1 \wedge l_1)$
A8 : At node I^{21}, $(L \vee R) \wedge \sim (L \wedge R)$
A9 : At node I^{12}, $(r_2 \vee l_2) \wedge \sim(r_2 \wedge l_2)$
A10 : $\pi^1(l_1) = 1, \pi^2(l_1) = 0$
A11 : $\pi^1(L) = 0, \pi^2(L) = 2$
A12 : $\pi^1(r_2) = 2, \pi^2(r_2) = 4$
A13 : $\pi^1(l_2) = 3, \pi^2(l_2) = 1$
A14 : At node I^{12}, $R_1 \Rightarrow l_2$
A15 : At node I^{21}, $[R_2 \wedge K_2R_1] \Rightarrow L$
A16 : At node I^{11}, $[R_1 \wedge K_1 (R_2 \wedge K_2R_1)] \Rightarrow l_1$
A17: $E^2_G (A_6$ -$A_{16})$

A_6 is a behavioral axiom: it tells that the players are rational in the sense of being expected utility maximizers. A_7 -A_9 specify the choices available to each player at each of his information sets, and say that a player can choose only one action. A_{10} -A_{13} specify players' payoffs. A_{14} -A_{16} are solution axioms, and specify what the players should do at any of their information sets if they are rational and know a) that the next player to move is rational and b) what the next player to move knows. A_{17} says that each player knows that each player knows A_6-A_{16}. We call these axioms 'special' since, even if every player knows that every player knows the axioms A_6 -A_{16}, no common knowledge is assumed.

From A_1- A_{17}, the players are able to infer the equilibrium solution l_1. To verify that this level of knowledge is compatible with a deviation from equilibrium, consider in turn the reasoning of both players. In order to decide which strategy to play, player 1 must predict how player 2 would respond to his playing r_1. The main stages of 1's reasoning can be thus described:

r_1	1
By assumption	
$K_1 K_2 ([R_1 \wedge K_1 (R_2 \wedge K_2R_1)] \Rightarrow l_1)$	2
By axioms A_{16}, A_{17}	
$K_1 K_2 (\sim l_1 \Rightarrow \sim [R_1 \wedge K_1 (R_2 \wedge K_2R_1)])$	3
By 1, 2, A_1, KG	
$K_1 K_2 (r_1 \vee l_1) \wedge \sim(r_1 \wedge l_1)$	4
By A_{17}, A_7	
$K_1 K_2 (\pi^1(l_1) = 1, \pi^2(l_1) = 0)$	5
By A_{17}, A_{10}	
$K_1 (R_2 \wedge K_2R_1)$	6
By A_6, A_{17}	
$K_1 \sim K_1 K_2 (K_1(R_2 \wedge K_2R_1))$	7
By A_5, 3, A_{17}	

For all that player 1 knows, his playing r_1 can be 'explained away' by player 2 as due to $\sim K_1(R_2 \wedge K_2R_1)$. In other words, what player 1 knows of player 2 does not conflict with his knowledge that $K_2R_1 1$. Since

$$K_1 [R_2 \wedge K_2R_1] \Rightarrow L \qquad\qquad 8$$
By A_{17}, A_{15}

player 1 knows that 2 will respond with L to r_1, hence he plays l_1.

What would player 2 think facing a deviation on the part of player 1?

$$r_1 \qquad\qquad 1$$
By assumption
$$K_2 (r_1 \vee l_1) \wedge \sim(r_1 \wedge l_1) \qquad\qquad 2$$
By A_{17}, A_7
$$K_2 (\pi^1(l_1) = 1, \pi^2(l_1) = 0) \qquad\qquad 3$$
By A_{17}, A_{10}
$$K_2 (\sim l_1 \Rightarrow \sim [R_1 \wedge K_1 (R_2 \wedge K_2R_1)]) \qquad\qquad 4$$
By A_{17}, A_{16}, A_1
$$K_2 (R_1 \wedge K_1R_2) \qquad\qquad 5$$
By A_{17}, A_6
$$K_2 (\sim l_1 \Rightarrow \sim K_1 K_2R_1) \qquad\qquad 6$$
By 4, 5

player 2 can 'explain' why r_1 was played, and since this explanation does not conflict with K_2R_1, she will choose strategy L.

What would happen if further levels of knowledge were added? Suppose the following axiom is added to the theory

$$A_{18} : E^2{}_G (A_6 - A_{17})$$

Since there is one more level of knowledge, now both players know that $K_1K_2R_1$ and $K_2K_1R_2$ obtain. This level of information implies that — were r_1 to be played — player 2 would face an inconsistency. As before,

$$K_2 (\sim l_1 \Rightarrow \sim [R_1 \wedge K_1 (R_2 \wedge K_2R_1)]) \qquad\qquad 1$$
By A_1, A_{16}, A_{18}
$$K_2 [R_1 \wedge K_1 (R_2 \wedge K_2R_1)] \qquad\qquad 2$$
By A_6, A_{18}
$$K_2 l_1 \qquad\qquad 3$$
By 2, A_{16}
$$r_1 \qquad\qquad 4$$
By assumption
$$K_2 (r_1 \vee l_1) \wedge \sim(r_1 \wedge l_1) \qquad\qquad 5$$
By A_7, A_{18}
$$K_2 \sim [R_1 \wedge K_1 (R_2 \wedge K_2R_1)] \qquad\qquad 6$$
By 1, 4
$$[R_1 \wedge K_1 (R_2 \wedge K_2R_1)] \qquad\qquad 7$$
By A_2
$$\sim [R_1 \wedge K_1 (R_2 \wedge K_2R_1)] \qquad\qquad 8$$
By A_2

Since the conjunction of the formulas 7 and 8 is false, and in classical logic one can deduce anything from a false statement, player 2 can use this conjunction to construct a proof that "r_1". Adding axiom A_{18} makes the theory of the game *inconsistent for player* 2, therefore 2 is unable to use it to predict how player 1 would respond if she were to play R. Which leaves 2 uncertain as to how to play herself.

Is the theory of the game also inconsistent for player 1? It is easy to verify that the state of information of player 1 does not let him realize that — were he to play r_1— player 2 would face an inconsistency. By A_{18}, player 1 knows $K_2K_1R_2$. But the levels of knowledge assumed in A_{18} do not let 1 know that $K_2 (K_1K_2R_1)$. Therefore player 1 can believe that 2 will explain a deviation by assuming $\sim (K_1K_2R_1)$. If so, he can predict that 2's response will be L, which makes him play l_1. Hence a theory of the game that includes axiom A_{18} supports the backward induction solution.

The backward induction equilibrium cannot be inferred only in the case in which $K_1(K_2K_1K_2R_1)$ obtains. This level of knowledge is brought forth by the additional axiom

$A_{19} : E^2_G (A_6 - A_{18})$

In this case player 1 would know that playing r_1 makes the theory of the game inconsistent for player 2 at I^{21}. If so, player 2 would be unable to predict what would happen were she to play R and 1, knowing that, would be unable to predict what would happen were he to play r_1.

Since a solution concept for the class of games we are examining depends upon the levels of knowledge possessed by the players, we have to introduce a few new definitions:

Definition 3.1. A *knowledge-dependent game* is a quadruple $\Gamma = (N, S^i, K^i, \pi^i)$ where $N = \{1, \ldots, n\}$ is the number of players; S^i is the set of strategies of player i; K^i is the knowledge possessed by player i and is defined as the union of what i knows at each of his information sets, i.e., $K^i = \cup K^{ij}_I$; π^i is player i's payoff.
$$1 \leq j \leq k$$

Definition 3. 2. An n-tuple of strategies $(s^1, \ldots, s^n)$ is a *knowledge-consistent* play of a knowledge-dependent game if, for each player i, every choice s^{ij} that strategy s^i recommends at each information set $I^{ij} \in P^i$ satisfies the following conditions: (i) reaching I^{ij} is compatible with K^i and (ii) it can be proven from K^i that s^{ij} is a best reply for player i at I^{ij}.

Theorem 3.1. For every finite, extensive form game of perfect and complete information, the backward induction equilibrium is a knowledge-consistent play of some knowledge-dependent game and, conversely, every knowledge-consistent play of a knowledge-dependent game is a backward induction equilibrium.

Proof. The first part of the proof is trivial, since Theorem 2. 2 illustrates a specification of the knowledge of each player that makes the backward induction equilibrium a knowledge-consistent play. The second part of the theorem can be proven by induction on the number of moves in the game. Suppose the game has only one move. In order to make a choice, the player who has to move must know his available strategies and payoffs. A rational player knows that he should choose that branch which leads to a terminal node with the maximum payoff to him. Then if the player knows his strategies and payoffs, he can infer his payoff-maximizing solution, which is the backward induction solution. Assume the theorem is true for all games involving at most K moves (some $K \geq 1$). Then it follows that the knowledge-consistent play $(s^1, \ldots, s^n)$, restricted to any of the subgames of Γ having no more than K moves, corresponds to the backward induction solution for that subgame. Let Γ be a knowledge-dependent game with K+1 moves and

let r be the root of the game tree T for Γ. At information set I^{ir} there is a recommendation of play s^{ir} for player i that can be inferred from K^i. Let $K = \cup\ Kj_t{}^{jr+m}$ be the union of the
$$1 \le m \le k$$
knowledge possessed by each player j which has to play at an information set that immediately follows I^{ir}. Then player i's knowledge of K implies the choice of the move that is the backward induction solution at I^{ir}. Therefore the union of K^i and K allows one to derive both the backward induction solution for I^{ir} and the strategy s^{ir}. The two must coincide since the union of K^i and K cannot lead to an inconsistent system.

Notes

[1] Recent attempts to analyze and model the players' reasoning process that leads to the selection of an equilibrium include Harsanyi's 'tracing procedure' (Harsanyi 1977), Skyrms' 'deliberational dynamics' (Skyrms 1986), Harper's application of the notion of 'ratifiable choice' to games (Harper 1988) and models of counterfactual reasoning in games (Shin 1987; Bicchieri 1988). Other studies of players' reasoning that focus on internal consistency of beliefs have led to the notion of 'rationalizability' (Bernheim 1984; Pearce 1984).

[2] The iterative notion of common knowledge was introduced by Lewis (1969), and a different definition, based on the notion of knowledge partition, was applied to game theory by Aumann (1976). Tan and Werlang (1986) have shown the equivalence of the two notions.

[3] Bayesian game theory has the same problem: the players' incomplete information about the structure of the game is simply *described* in the form of an extensive form game with chance moves (Harsanyi 1967, 1968). In this case, too, some basic assumptions of the theory are not treated as part of the theory.

[4] More recently, Gilboa and Schmeidler (1988) proved that in information-dependent games a common knowledge axiom is inconsistent with a rationality axiom.

[5] I have shown elsewhere (Bicchieri 1988) that the various refinements of Nash equilibrium can be uniformly treated as different rules for belief change, and that such rules can be inferred from a richer theory of the game that includes epistemic criteria that allow an ordering of the rules in terms of epistemic importance. In the class of games I am considering, a theory of the game that contains a model of belief change would *always* let the players 'explain away' any deviation from equilibrium (Bicchieri 1988a).

[6] If the players were endowed with a model of belief-change (Bicchieri 1988, 1988a), there would be other hypotheses beside Selten's that make common knowledge of rationality compatible with out of equilibrium behavior.

References

Aumann, R. J. (1976), "Agreeing to disagree", *The Annals of Statistics* 4: 1236-1239.

Bacharach, M. (1987), "A theory of rational decision in games", *Erkenntnis* 27: 17-55.

_____. (1985), "Some extensions of a claim of Aumann in an axiomatic model of knowledge", Journal of Economic Theory 37: 167-55.

D. Bernheim (1984), "Rationalizable strategic behavior", *Econometrica* 52: 1007-1028.

Bicchieri, C. (1988), "Strategic behavior and counterfactuals", *Synthese* 76: 135-169.

‗ ‗ ‗ ‗ ‗ ‗. (1988a), "Common knowledge and backward induction: a solution to the paradox", in M. Vardi (ed.) *Theoretical Aspects of Reasoning about Knowledge.* Morgan Kaufmann Publishers, Los Altos.

‗ ‗ ‗ ‗ ‗ ‗. (1989), "Self-refuting theories of strategic interaction: a paradox of common knowledge", *Erkenntnis* 30: 69-85.

Binmore, K. (1987), "Modeling rational players I", *Economics and Philosophy* 3: 179-214.

‗ ‗ ‗ ‗ ‗ ‗. (1988), "Modeling rational players II", *Economics and Philosophy* 4: 9-55.

‗ ‗ ‗ ‗ ‗ and A. Brandenburger (forthcoming), "Common knowledge and game theory", *Journal of Economic Perspectives.*

Bonanno, G. (1987), "The logic of rational play in extensive games", Disc. Paper no. 16, Nuffield College, Oxford.

A. Brandenburger (forthcoming), "The role of common knowledge assumptions in game theory", in F. Hahn (ed.) *The Economics of Information, Games, and Missing Markets*, Cambridge University Press.

‗ ‗ ‗ ‗ ‗ ‗ ‗ ‗. and Dekel, E. (1985a), "Common knowledge with probability", Research Paper no. 796R, Graduate School of Business, Stanford University.

‗ ‗ ‗ ‗ ‗ ‗ ‗ ‗ ‗ ‗ ‗ ‗. (1985b), "Hierarchies of beliefs and common knowledge", Research Paper no. 841, Graduate School of Business, Stanford University.

Gilboa, I. (1986), "Information and meta-information", Working paper no. 30-86, Tel-Aviv University.

‗ ‗ ‗ ‗ and D. Schmeidler (1988), "Information dependent games", *Economics Letters* 27: 215-221.

Halpern, J. and Fagin, R. (1988), *Modelling knowledge and action in distributed systems.* Technical Report, IBM.

‗ ‗ ‗ ‗ ‗. and Moses, Y. (1987), "Knowledge and common knowledge in a distributed environment", IBM Research Report RJ 4421.

Harper, W. (1988), "Causal decision theory and game theory", in Harper and Skyrms (eds.), *Causation in Decision, Belief Change and Statistics*, Reidel.

Harsanyi, J. (1967-68), "Games with incomplete information played by 'Bayesian' players", Parts I, II, and III. *Management Science* 14: 159-182, 320-332, 468-502.

‗ ‗ ‗ ‗ ‗. (1975), "The tracing procedure: a Bayesian approach to defining a solution for n-person non-cooperative games", *International Journal of Game Theory* 4: 61-94.

‗ ‗ ‗ ‗ ‗. and R. Selten (1988), *A General Theory of Equilibrium Selection in Games.* The MIT Press, Cambridge.

Hintikka, J. (1962), *Knowledge and Belief.* Cornell University Press, Cornell.

Kaneko, M. (1987), "Structural common knowledge and factual common knowledge", RUEE Working Paper no. 87-27, Hitotsubashi University.

Kuhn, H.W. (1953), "Extensive games and the problem of information", in H.W. Kuhn and A.W. Tucker (eds.) *Contributions to the Theory of Games*. Princeton University Press, Princeton.

Lenzen, W. (1978), "Recent work in epistemic logic", *Acta Philosophica Fennica* 30: 1-219.

Lewis, D. (1969), *Convention*. Harvard University Press, Cambridge.

Luce, R. and Raiffa, H. (1957), *Games and Decisions*. Wiley, New York.

Mertens, J.-F. and Zamir, S. (1985), "Formulation of Bayesian analysis for games with incomplete information", *International Journal of Game Theory* 14: 1-29.

Parikh, R. and Ramanujam, R. (1985), "Distributed processes and the logic of knowledge", *Proceedings of the Workshop on Logics of Programs*: 256-268.

Pearce, D. (1984), "Rationalizable strategic behavior and the problem of perfection", *Econometrica* 52: 1029-1050.

Reny, P. (1987), "Rationality, common knowledge, and the theory of games", Working paper, Department of Economics, University of Western Ontario.

Samet, D. (1987), "Ignoring ignorance and agreeing to disagree", mimeo, Northwestern University.

Selten, R. (1975), " Re-examination of the perfectness concept for equilibrium points in extensive games", *International Journal of Game Theory* 4: 22-55.

Shin, H.S. (1987), "Counterfactuals, common knowledge and equilibrium", mimeo, Nuffield College, Oxford.

Skyrms, B. (1989), "Deliberational dynamics and the foundations of Bayesian game theory", in J. E. Tomberlin (ed.) *Epistemology*. Ridgeview, Northridge.

_ _ _ _ _ _. (1986), "Deliberational equilibria", *Topoi* 1.

Tan, T. and Werlang, S. (1986), "On Aumann's notion of common knowledge—an alternative approach", Working paper no. 85-26, University of Chicago.

Van Damme, E.E.C. (1983), *Refinements of the Nash Equilibrium Concept*. Springer-Verlag, Berlin.

Decisions, Games and Equilibrium Solutions

William Harper

University of Western Ontario

1. Decision Theories: a Survey

1.1 Utility

Von Neumann and Morgenstern based their theory of games on the representation of individual preferences for outcomes by utilities generated by preferences for gambles over these outcomes. A utility function (an assignment of numbers to outcomes) represents an agent's preferences just in case the agent's preference relation between any two gambles agrees with the numerical relation between their expected utilities (where these expectations are calculated using the objective probabilities specified in the gambles). This representation constrains the utility assignments up to scale transformations (multiplying each value by the same positive number) and adjustments of the zero point (adding the same positive or negative number to each value). This fixes the ratios of differences between utilities of outcomes.

Von Neumann and Morgenstern proposed qualitative constraints on the agent's preferences among these gambles which are equivalent to the assertion that such a utility representation exists. Here is an equivalent formulation of such rationality postulates.

(1) *Ordering* (weak preference is *connected* and *transitive*).

(2) *An archimedian condition* (this requires that preference differences cannot be finer than can be specified by real numbers).

(3) *A reduction principle* (this treats the various probabilities in a compound lottery as independent so that the compound lottery is regarded as equivalent to the simple lottery corresponding to the result of multiplying out its probabilities).

(4) *Stochastic dominance* (If outcome x is strictly preferred to outcome y and two lotteries differ only in their assignments of probability to x and y then the agent will prefer the one assigning higher probability to x.)

(5) *Substitution* (Preference between two lotteries that differ only in that one has prize (which may be an outcome or a lottery) f substituted for g should agree with the preference between f and g.)

PSA 1988, Volume 2, pp. 344-362

These postulates or axioms are taken to help motivate the idea that an ideally rational agent who understands all the gambles over the outcomes and treats them as value neutral ought to have preferences that can be represented by a utility function.

The substitution principle came under attack early on in work of Allais (1953) and has recently been challenged anew by Allais (1979), Machina (1982), Chew (1983), McClennen (1983) and others who want to provide more sensitive representations of attitudes towards risk than it will allow for. The ordering axioms (especially connectedness) have been attacked by those, e.g. Levi (1980) and Seidenfeld (1988) who argue that even ideally rational agents might have preferences more indeterminate than it allows for. There is a lively dispute between adherents of these two alternative weakenings of the Bayesian ideal for individual decision making (see Seidenfeld 1988 and McClennen 1988). Neither group has dealt explicitly with game theory. I shall, however, discuss a natural application of indeterminate subjective probability allowed for by those who weaken the ordering condition. There is also an application of recent challenges to the archimedian axiom. This application also applies more to subjective probability than to preference directly.

1.2 Subjective Probability

Von Neumann and Morgenstern took the objective probabilities used in their gambles over outcomes as primitive and never discussed explicitly the idea that an agent's expectations about the other players' choices should be guided by constraints on rational subjective probability. Frank Ramsey (1926) had provided a qualitative preference representation sufficient to generate both utilities and subjective probabilities. Savage (1954) provided another such representation based upon what were regarded as especially compelling sure-thing principles constraining relations between preference and conditional preference. Savage's book was taken up by many statisticians, philosophers and others concerned with rational belief. Together with Ramsey's work, the von Neumann-Morgenstern utility theory, and De Finetti's (1936) representation of subjective probability with convergent learning by Bayes' theorem, it has become a central text in what has become the dominate Bayesian approach to individual rationality.

Many Bayesians have followed Levi's (1980) lead in allowing that even an ideally rational agent might have indeterminate degrees of belief. This may happen if connectedness fails as the preference ordering is extended to include event lotteries used to scale degrees of belief, even if the preferences over lotteries on outcomes fix the utilities of outcomes up to scale and zero.

In classical game theory the utilities of the players are assumed to be common knowledge, but no specification is made of any probabilities that might represent their beliefs about the choices of the other players. One natural way to represent this is to have the agent's beliefs about the other players' choices start out indeterminate and then use reasoning about the strategic situation to form them. On this view game theoretic reasoning constrains what are to count as reasonable beliefs about what the players in the game will do.

Harsanyi (e.g. 1967-68 and 1977) has long argued for such a view of the relation between game theory and Bayesian agents. He and Selten (1988) have recently developed this approach into the most comprehensive, even if controversial, solution concept in the entire game theory literature. Kreps and Wilson (1982), Pearce (1984) and Bernheim (1984) all appeal to Savage's account of subjective probability to use constraints on what could count as rational expectations about the other players to inform game theoretic reasoning.

The Krep's-Wilson approach appeals explicitly to an agent's conditional beliefs relative to assumptions about off-equilibrium paths that the agent assumes to be false. These conditional beliefs require an extension of Savage's framework to allow for such conditional probabilities. Karl Popper (1959), A. Renyi (1955) and De Finetti (1970) provided treatments of such probabilities. Harper (1978) and Gardenfors (1988) offered an account of such probabilities in a framework of qualitative constraints on conditional assumption contexts. Giles (1979) showed that this qualitative structure as well as the needed conditional probabilities can be secured in a Bayesian Framework if the archimedian axiom is dropped. Harper, Giles and Hyack (1989) are developing such non-archimedian belief models. Brandonberger and Dekel (1985) have, independently, been developing non-archimedian subjective probabilities, explicitly designed to apply to reasoning of the sort Kreps and Wilson appeal to.

1.3 Savage's Decision Theory

Savage's theory is framed in such a way that neither evidential nor causal relations between acts and the uncertain events their outcomes depend on can be represented. On this theory expected utilities are calculated using the agent's subjective probabilities of the events. If $B_1,...,B_n$ is the relevant partition of outcome determining events for choosing among acts $a_1,...a_k$ and if for each act a_j and each event B_i, $u(A_j,B_i)$ is the utility for the agent of doing a_j if B_i is the case, then we have

$$\text{Savage } U(a_j) = \sum_i P(Bi) \cdot u(a_j,B_i).$$

The same probability multipliers are used to evaluate the utilities of each of the acts; therefore, weak dominance holds so long as some event on which the weakly dominate act is strictly preferred has non-zero probability.

Savage actually takes a kind of weak dominance principle as one of the primitive rationality axioms of his system. Here is a condition equivalent to one of his sure-thing principles (Savage 1954, pp. 21-26).

If acts f and g agree (have exactly the same consequences) in all states in not-A then unconditional preference between f and g ought to agree with their conditional preference given A.

This principle is compelling if the events A and not-A are appropriately independent of the agent's choice between f and g. We shall see, however, that it may lead to unreasonable recommendations if the agent knows his choice will influence the events.

One must be very careful when attempting to apply Savage's theory where the uncertain events are acts of another agent one is interacting with. Consider the following two stage game.

Game #1

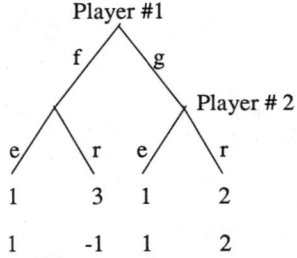

Player #1's payoffs are on top. We assume that the outcome of f given e agrees exactly with the outcome of g given e. If we let event not-A be that player #2 ends up doing e and A be the proposition that he ends up doing r, then the principle would tell player #1 to opt for f.

This, of course, would not be reasonable. Choosing f would influence player #2 to opt for e, while choosing g would influence player #2 to opt for r. The purported sure-thing argument for f is undercut by the causal dependence of these states on player #1's choice.

The partition A (player #2 ends up doing r) and not-A (player #2 ends up doing e) is not appropriate for applying Savage's principle. We can find an appropriate partition for this game by going to the normal form. Player #2 has four strategies, while player #1 has two.

Player #2

	b1	b2	b3	b4
	e, if f	r, if f	e, if f	r, if f
	e, if g	r, if g	r, if g	e, if g
f	(1,1)	(3,-1)	(1,1)	(3,-1)

Player #1

| g | (1,1) | (2,2) | (2,2) | (1,1) |

If we assume the outcomes having the same utility assignments agree then player #2 can legitimately apply the sure-thing principle to rule out the weakly dominated strategies b1, and b2. b1 agrees with b3 given f and player #1's choice between f and g cannot be influenced by player #2's selection of b3 over b1. A corresponding sure-thing argument rules out b2 in favor of b3. b4 is strongly dominated by b3, so it is ruled out as well. Player #1 can see that player #2 has this sure-thing argument for choosing strategy 3 so he can reasonably commit himself to strategy g which is his best reply to what he knows player #2 is going to do. In this game Savage's sure-thing principle can be correctly applied to support the game theoretic solution, the unique equilibrium pair (g, b3).

There can be games with causal dependencies between strategies that show up in the extensive form but are masked in the normal form. Moreover, researchers such as Kreps and Wilson (1982), Pearce (1984) and Bernheim (1984) all want to apply subjective expectations in such a way as to take advantage of the information provided as one reaches various stages in the extensive form of a game. This suggests that one may want to formulate choices relative to partitions of events that take more of this structure into account than the normal form provides for.

4. Evidential Decision Theory

Evidential decision theory formulated by Richard Jeffrey (1965, 1983), would purport to allow any partition whatsoever to be used to formulate a decision problem. This theory has the agent's acts in the algebra of events. On it the incorrect sure-thing argument on the first partition is undercut because the appropriate probabilities for computing expected value are the conditional probabilities of the events on the act. Using v to represent evidential expected utility we would have in general

$$\text{Evidential } v(a_j) = \sum_i P(B_i|a_j) \cdot u(a_j, B_i).$$

In the example we have:

$$v(f) = P(e|f) \cdot 1 + P(r|f) \cdot 3$$

$$v(g) = P(e|g) \cdot 1 + P(r|g) \cdot 2$$

In this case the conditional probabilities are constrained by the agent's knowledge of the causal dependencies so that

$$P(e|f) = 1 = P(r|g).$$

Therefore,

$$v(f) = 1 < v(g) = 2.$$

So, the correct decision can be supported on the original problematic partition. On this theory the appropriate constraint required to support sure-thing reasoning is that the uncertain events be evidentially independent of the acts as represented by independence (e.g. $P(A|f) = P(A)$) in the agent's subjective probabilities.

Causal decision theorists have argued that evidential decision theory is not sensitive enough in its representation of an agent's beliefs about what events are independent of his or her acts. Newcomb's Problem (Gibbard and Harper 1978) and prisoners dilemma with a twin or clone whom you expect to end up making the same choice you do (David Lewis 1979) are examples where an agent might regard his or her own acts as evidentially relevant to events that the agent knows cannot be causally influenced by them.

We can use the following normal form game to illustrate such a situation:

Game #2

	b1	b2
a1	(3,3)	(1,1)
a2	(4,1)	(2,2)

You are Row Chooser. Suppose you assign conditional probabilities such that

$$P(b1|a1) > 3/5 \text{ and } P(b2|a2) > 3/5$$

You may be fairly confident that, whichever of these acts you choose, column chooser will have figured out your choice and will have made a best reply to it. Your background knowledge may be such as to make such beliefs plausible, even in a context like this normal form game where you know that column chooser's choices are causally independent of yours.

Given such subjective conditional probabilities evidential decision theory will recommend a1; however, such beliefs, even if reasonable, would not make it reasonable to choose a1, if you know that your choice cannot actually influence that of the other player.

According to causal decision theorists the independence condition for making sure-thing reasoning valid is that the agent believe the events are causally independent of the acts, not that they are evidentally independent in the agent's subjective conditional probability. In this example, your knowledge that the other player's choices are causally independent of yours underwrites the sure-thing argument for a2.

1.5 Causal Decision Theory

The agent's epistemic conditional probabilities do not allow for a sensitive enough representation of the agent's beliefs about which events are causally independent of his or her choices. Causal decision theorists have introduced additional apparatus to provide what is needed. In the Gibbard-Harper (1978) formulation the appropriate probabilities for computing expected utilities are the agent's unconditional epistemic probabilities for subjunctive conditionals of the form

$$a \square \rightarrow b$$

If I were to do act a then event b would be the case.

Where the relevant partition of outcome determining events is $b_1 \ldots b_n$ and each $u(a,b_i)$ is the utility to the agent of doing a if b_i is the case, the causal utility for a is

$$\text{Causal } U(a) = \sum_i P(a \square \rightarrow b_i) \cdot U(a,b_i).$$

In our example, we have

$$U(a1) = P(a1 \square \rightarrow b1) \cdot 3 + P(a1 \square \rightarrow b2) \cdot 1$$

$$U(a2) = P(a2 \square \rightarrow b1) \cdot 4 + P(a2 \square \rightarrow b2) \cdot 2$$

If the agent believes that the b's are causally independent of the a's then

$$P(a1 \square \rightarrow b1) = P(b1) = P(a2 \square \rightarrow b1)$$

and

$$P(a1 \square \rightarrow b2) = P(b2) = P(a2 \square \rightarrow b2)$$

Thus, the causal utilities will support sure-thing reasoning if the events are believed to be causally independent of the acts. In the example at hand causal utility will deliver the correct prescription of a2, since a2 dominates a1 and the b's are causally independent.

Causal utility agrees with evidential utility where that theory is correct. The following theorem in the logic of our subjunctive conditional

$$a \supset [(a \square \rightarrow b) \equiv b]$$

ensures that

$$P(a \square \rightarrow b|a) = P(b|a)$$

so that

$$P(a \square \rightarrow b) = P(b|a), \text{ if}$$

$$P(a \square \rightarrow b) = (a \square \rightarrow b|a).$$

The probability of our conditional equals the corresponding conditional probability if the conditional is evidentially independent of its antecedent. In game 1 we have

$$P(f\square\!\!\!\rightarrow e) = 1 = P(f\,\square\!\!\!\rightarrow e|f).$$

In general, this result allows causal decision theory to correct the recommendations where evidental decision theory goes wrong without giving up the recommendations where it is correct.

Subjective probabilities of subjunctive conditionals may seem to be introducing obscurities that are too far removed from more familiar treatments of probability. In the context of an extensive form game, however, constraints needed to make these probabilities illuminating can be read off from the causal and informational structure built into the game tree. This can be tied with more familiar treatments of probability by the observation that the subjunctive conditional appropriate for decision making ought to be evaluated by the subjective expectation of the objective conditional chance of its consequent on its antecedent. At a given information set each of the nodes about which the agent is uncertain counts as an alternative objective chance set up determining conditional probabilities of outcome determining events on the agent's acts. We can use the agent's subjective probabilities conditional on the nodes and his acts to represent what the agent takes these objective conditional chances to be. Thus, at an information set with nodes $N1 \ldots Nn$

$$P(a\square\!\!\!\rightarrow b) = \sum_i P(Ni) \cdot P(b|Ni\ \&a)$$

This makes the formulation of causal decision theory using subjunctive conditionals equivalent to an alternative formulations developed by Brian Skyrms (1989, 1984, 1982), which takes a special partition (our $N1 \ldots Nn$) into events that count as alternative specifications of the causally relevant factors that are not in the agent's control as primitive. As long as alternative choices of the other player are causally independent of your choice between two or more strategies then you can use those alternative choices of the other agent as alternative chance hypotheses.[1]

2. Classical Game Theory: a Review

2.1 Von Neumann and Morgenstern

The central idea in classical game theory is that solutions should be equilibria.

A profile $s_1 \ldots s_n$ of strategies (one for each player i) is an equilibrium just in case each s_i is a best reply to the combination of the remaining strategies.

Here is von Neumann and Morgenstern's indirect argument to motivate the idea that in zero-sum games solutions ought to be equilibria.

We are trying to find a satisfactory theory, — at this stage for the zero-sum two person games. Consequently, we are not arguing deductively from the firm basis of an existing theory — which has already stood all reasonable tests — but we are searching for such a theory. Now in doing this, it is perfectly legitimate for us to use the conventional tools of logics, and in particular that of the *indirect proof*. This consists in imagining that we have a satisfactory theory of a certain desired type, trying to picture the consequences of this imaginary intellectual situation, and then in drawing conclusions from this as to what the hypothetical theory must be like in detail. If this process is applied successfully, it may narrow the possibilities for the hypothetical theory of the type in question to such an extent that only one possibility is left, — i.e. that the theory is determined, discovered by this device. Of course, it can happen that the application is even more "successful", and that it narrows the possibilities down to nothing — i.e. that it demonstrates that a consistent theory of the desired kind is inconceivable.

Let us now imagine that there exists a complete theory of the two-person zero sum game which tells a player what to do, and which is absolutely convincing. If the players knew such a theory then each player would have to assume that his theory has been "found out" by his opponent. The opponent knows the theory, and he knows that a player would be unwise not to follow it. Thus the hypothesis of a satisfactory theory legitimizes our investigation of the situation when a player's strategy is "found out" by his opponent. And a satisfactory theory can exist only if we are able to harmonize the two extremes Γ_1 and Γ_2 — strategies of player 1 "found out" or of player #2 "found out".

We can see, fairly clearly, how it is that harmonizing these two extremes requires an equilibrium solution if the solution is unique. Only if player #1's strategy is a best reply to that of player #2 will his choice be reasonable given that player #1 has "found out" player #2's strategy. Similarly, player #2's strategy must be a best reply to that of player #1.

Their most successful results were with two person zero-sum games. A two person game is zero-sum just in case the agent's preferences over gambles over the outcomes are strictly opposed in that if player #1 strictly prefers f to g then player #2 strictly prefers g to f, and if player #1 is indifferent between f and g so is player #2. When this condition is met then there are utility assignments u_1 representing player #1's preference and u_2 representing player #2's preferences such that for any outcome x, $u_2(x) = -u_1(x)$. It is convenient to represent only one player's utilities (player #1's since player #2's utilities can be recovered by just negating those numbers).

Here is a normal form representation of a zero-sum game where the indirect procedure yields a unique answer.

Game #3

	b1	b2
a1	2	3
a2	1	2

Player #1 (Row Chooser) is trying for outcomes with large numbers, while player #2 (Column Chooser) is trying for outcomes with small numbers. The unique saddle point (maximum in its column, minimum in its row) is the outcome corresponding to the unique equilibrium pair (a1,b1). This pair is the only combination of strategies that can pass the test where each player is able to "find out" the strategy of the other. Any other strategy for either player would give the opponent reason to change his or her choice if the opponent were to "find it out".

This application of the "finding out" test of the von Neumann-Morgenstern indirect argument corresponds to what Luce and Raiffa (1957, pg. 63) call an *a priori* demand on the theory of the solution to the game.

It seems plausible that if a theory offers ai and b_j as suitable strategies, the mere knowledge of the theory should not cause either of the players to change his choice: just because the theory suggests b_j to player 2 should not be grounds for player one to choose a strategy different from a_i; similarly, the theoretical prescription of ai should not lead player 2 to select a strategy different from b_j (pg. 63).

Knowledge of the theory should not give either player grounds to change their choice.

Zero sum games may have multiple saddle points. For example,

Game #4

	b1	b2	b3	b4
a1	6	2	2	3
a2	5	2	0	5
a3	3	2	2	4

Here each of the four equilibrium pairs (a1,b2), (a1,b3), (a3,b2), (a3,b3), and no others, satisfy the indirect argument. The set of all four equilibrium points also counts as a solution that satisfies the test of the indirect argument. On this solution each player would be restricted to his or her equilibrium strategies — Row Chooser to {a1,a3} and Column Chooser to {b2,b3}. The result of having each of these strategy sets found out by the opponent would lead neither player to change to a different choice. These equilibria are *interchangeable* in that either player's strategy in any one of them is best against the other player's strategy in any of the others. Von Neumann and Morgenstern proved that the indirect argument would apply to all two person zero sum games with multiple saddle points by showing that in any such game the set of all equilibria are interchangeable in this sense.

The crowning achievement in the theory of two person zero sum games was the use of mixed strategies to generate solutions for games which have no pure strategy equilibria. Consider

Game #5

	b1	b2
a1	1	3
a2	4	2

Here there is a single pair of mixed strategies that satisfies the indirect argument. This is the unique equilibrium pair $x^* = (1/2a_1, 1/2a_2)$ and

$$y^* = (1/4b_1, 3/4b_2).$$

The minimax theorem showed that the indirect argument can be successfully applied to any two person zero sum game. There is always at least one equilibrium point and if there are multiple equilibria the set of all of them will satisfy the interchangeability condition.

Von Neumann and Morgenstern suggest that a solution successfully discovered by this indirect argument must then be independently justified by a direct argument.

After the theory is found we must justify it independently by a *direct argument*. This was done for the strictly determined case in 14.5, and we shall do it for the present complete theory in 17.8.

If we look at what they counted as the direct argument in the strictly determined case we see in section 14.5 (1957, pp. 106-108) the result that where pure strategy equilibria points exist they not only satisfy the interchangeability condition, but they also all have the same payoff to each player and, in addition, the equilibrium strategies are exactly the maximin strategies (strategies which maximize their player's security level — the minimum that the player could expect if the strategy is played). Section 17.8 (pp. 158-160) contains the minimax theorem which establishes not just that equilibria always exist and satisfy the interchangeability condition, but that, once again, they all yield the same pay-offs and the equilibrium strategies are exactly the maximin strategies.

There is, therefore, some ambiguity about which of these results are to count as the direct argument. This may have helped promote the idea that a policy to seek maximin strategies should count as an independent rationality principle, rather than as a mere arti-fact of the special circumstances of perfect competition in zero sum games. It is also not clear what ought to count as the natural extension of von Neumann and Morgenstern's solution concept to non zero sum games.

2.2 Nash

John Nash (1951) provided the most seminal work on extending the idea of equilibri-um solutions to non-zero sum games. He used an ingenious fixed point proof to show that every finite non-cooperative game has at least one equilibrium profile. This is perhaps the most celebrated result in game theory. Von Neumann's minimax theorem follows as a special corollary from Nash's more general theorem.

Nash (1951, pg. 290) also proposed *interchangeability* of equilibria as a solution con-cept that could be applied to non-zero sum games. He defined it for n-person games. Let us introduce some of Nash's convenient notation. Where $s = (s1,...,sn)$ is a profile of strategies (one for each of the n-players) and ri is any strategy for player i, let

$$(s_j, r_i) = (s_1, ..., s_{i-1}, r_i, s_{i+}, ..., s_n)$$

be the result of substituting strategy ri for the strategy assigned to player i in profile s. We shall say that a set G of profiles *satisfies the Nash Solution Concept* just in case

(1) all profiles in G are equilibria
(2) $s \epsilon G$ and $t \epsilon G \rightarrow (t, s_i) \epsilon G$ for all i.

A set of equilibria satisfies (2) just in case the result (t, s_i) of substituting the strategy si assigned to player i in any profile s in G into any other profile t in G is also an equilibri-um profile in G. According to Nash (1951, pg. 290), a game is *solvable* if the set of all its equilibrium profiles satisfies what I have just called the Nash solution concept and the *solution* of a solvable game is the set of all its equilibrium strategies.

Let us say that a strategy is an equilibrium strategy just in case it is a constituent of some equilibrium profile. The result of applying von Neumann and Morgenstern's indi-rect argument to profiles considered as candidates for a possible unique solution will be that each player is left with only his or her equilibrium strategies to choose from. If the game is solvable in Nash's sense then all its equilibria will be interchangeable so that its Nash solution — the set of its equilibria profiles — will also satisfy the von Neumann Morgenstern test. No player will be able to *expect to gain* by defecting to a non-solution strategy, if each is sure the others will play one of their equilibrium strategies.

Many games have no Nash solutions. Any game with multiple non-interchangeable equilibria will provide an example. Here is one pointed out by Nash (pg. 291).

Game #6

	b1	b2
a1	(1,1)	(-10,-10)
a2	(-10,-10)	(1,1)

There are three equilibrium pairs, the two pure strategy equilibrium pairs (a1,b1), (a2,b2), and the mixed strategy equilibrium pair $x^* = (1/2a1, 1/2a2)$ and $y^* = (1/2b1, 1/2b2)$. A solution theory which simply restricted each player to his or her set of equilibrium strategies will not guarantee that they will end up at one of the equilibrium pairs. For each strategy there will be a conjecture about the choice of the other player (consistent with common knowledge of the theory) relative to which that strategy would be non-optimal.

Common knowledge of such a multiple equilibrium theory of the solution to this game would not actually give either player grounds for shifting to a non-solution strategy. This may, however, be an artifact of the uniformativeness of the putative solution. All the pure strategies (as well as x^* and y^*) are equilibrium strategies. Harsanyi and Selten (1988, pg. 13) point out the undesirability of having something so uninformative count as a solution.[2]

Even if this theory did not actually fail the test of the indirect argument, it would seem to undercut the best reply reasoning that the argument appeals to. In games that have a Nash solution agents have grounds for supposing that common knowledge of the theory would lead one to expect the other player's to make best replies to one's own choice of strategy. This fails when the game has multiple non-interchangeable equilibria.

Nash also defined what he called *strong solutions*. Let $u^i(s)$ be the utility to player i of the outcome corresponding to the profile s of strategies. A game is *strongly solvable* if its solution G is such that

$$s \; \varepsilon \; G \text{ and } u^i(s,r_i) = u(s) \rightarrow (s,r_i) \; \varepsilon \; G$$

for any profile s, player i, and strategy ri for player i. If a game is strongly solvable any player who expected the other players to play equilibrium strategies would *expect to lose* if he defected to a non-equilibrium strategy.

Games #2 and #3 are strongly solvable. Game #4 is not strongly solvable, because a2 is as good against b2 as the strategy a1 in the equilibrium pair (a1,b2), but (a2,b2) is not one of the equilibrium pairs.[3] Similarly, Games #1 and #5 are also not strongly solvable. Game #6, of course, is not strongly solvable, since it is not Nash solvable at all.

Perhaps influenced by Nash's terminology Harsanyi (1977) introduced the term "weak" for equilibria such as (a1,b2) and (a3,b2) in Game # 4, where player a could defect to the non-equilibrium strategy a2 without expecting to suffer any loss. In zero-sum games (like #5) where the only solutions are in mixed strategies the equilibria are always weak, because the payoff to one player is the same no matter what strategy he or she chooses so long as the other player plays his or her equilibrium (mixed) strategy. No such game is strongly solvable in Nash's sense.

3. Must Solutions be Equilibria?

3.1 Rationalizability: A Challenge

We have seen that the indirect argument for restricting solutions to equilibria may break down if there are multiple non-interchangeable equilibria. Games with such multiple equilibria have seemed problematic to many researchers. A number of writers, however, have gone beyond this to suggest that one need not restrict solutions to equilibria, even in games that are Nash solvable. At one point Harsanyi (1977, pp. 124-127) suggested that solutions, even if unique, need not be equilbria if some maximin strategy would give an expected payoff as good as that of any equilibrium strategy.[4] E. F. McClennen (1978) suggested that restricting a solution to weak equilibria when defection to a non-equilibrium strategy would lead to no expected loss makes game theory inconsistent with its decision theoretic foundation. These doubts were, in both cases, motivated by the problem of weak equilibria. They have recently been dramatically reinforced by the work of B. D. Bernheim (1984) and D. Pearce (1984). Bernheim and Pearce provided a serious challenge to the Nash point of view by developing a detailed alternative.

Bernheim (1984) appealed to the idea that a rationally supportable strategy ought to be a best reply to a rational conjecture (subjective probability assignment) about the strategies selected by the other players. He required that a rational conjecture be such that each strategy to which it assigns non-zero probability be, in its turn, a best reply to a rational conjecture by the corresponding agent about the strategies selected by the others. These conjectures, in their turn, are to have their non-zero probability strategies be best replies to further rational conjectures, and so on. If a strategy α were such that a sequence of such conjectures supporting it could not be extended indefinitely then every sequence of conjectures in support of a would end in a conjecture that would be inconsistent with common knowledge of the game structure, pay-off space and Bayesian rationality of all players. Accordingly Bernheim defined the *rationalizable* strategies to be just those that are supportable by such an infinitely extendible sequence of rational conjectures.

Pearce (1984) independently arrived at a formulation of rationalizability equivalent to Bernheim's. He also provided another formulation which may further illuminate the connection between rationalizability and best reply reasoning. For an n-person game a profile of strategy sets $A^1,...,A^n$ (where each A^i is included in the set of all strategies, mixed as well as pure, available to player i) *has the best response property* just in case each strategy in any of the sets A^i is a best reply to some conjecture about the strategies played from the other sets A^j. Pearce (1984, pg. 1034) showed that the *rationalizable* strategies for any player i are exactly the ones that are members of A^i for some profile $A^1,...,A^n$ of strategy sets having the best response property.

The best response property is a reasonable demand to put on any profile of strategy sets $A^1,...,A^n$ that is to count as a solution, any agent i who knows that it is common knowledge that all the players are committed to this solution concept will be able to restrict his conjectures about the strategies selected by the other players to these sets. Only strategies which are best replies to such a conjecture would be reasonable for such an agent.

If the solution concept is strong enough so that all the sets A^i are singleton's then the best response property will imply the demand that the solution be a unique Nash equilibrium. Under this assumption of uniqueness, the best response property generates the *a priori* argument for equilibrium solutions. If each strategy in any set is a best reply to any allowable profile, the best response property on the A's will ensure that each agent does his or her part to reach some Nash equilibrium or other. In this case the theory of the game will satisfy what I have called the Nash solution concept.

Rationalizability is clearly well motivated by the idea of best reply reasoning, but it is a very weak constraint. In two person normal form games the set of rationalizable strategies is equivalent to the result of iteratively eliminating strongly dominated strategies.[5] There are many games where non-equilibrium strategies are ratifiable.

In extensive form games the need to have conjectures respect the knowledge of sequential structure at information sets can give somewhat more bite to rationalizability than would be obtained simply by iteratively removing strongly dominated strategies. Pearce and Bernheim each have interesting suggestions about how to build refinements into rationalizability similar to a number of those being developed in the growing literature on refinements of the equilibrium concept. Nevertheless, rationalizability (even with these refinements) doesn't guarantee that the rationalizable strategies will be limited to Nash strategies. Unless some constraints beyond rationalizability can be put on admissible strategies even those hitherto relatively unproblematic areas of game theory such as zero-sum games where the equilibria do satisfy the Nash solution concept will be problematic.

Consider, again, Game #5. It's solution is the unique equilibrium consisting of the pair of maximin mixed strategies

$$x^* = (1/2a_1, 1/2a_2) \text{ and } y^* = (1/4b_1, 3/4b_2).$$

Where A^a is the set of all strategies (mixed as well as pure) for player a and A^b is the set of all strategies for player b the strategy set profile

$$(A^a, A^b)$$

has the best response property. Any strategy in A^a is a best response to a conjecture agreeing with the probabilities assigned in b's maximim strategy y^* and any strategy in A^b is a best response to a conjecture corresponding to a's maximin strategy x^*. Therefore, all strategies for either player are rationalizable.

This example illustrates the connection between the formulation using the best response property for profiles of sets of strategies and the formulation using infinitely extendable sequences of conjectures. For any strategy x of player **a**, a sequence starting with a point conjecture on the maximin mixed strategy y^* and then cycling between point conjectures on x^* and y^* will rationalize x. Similarly, for any strategy y for player **b**, x^* and then a cycle between x^* and y^*.

Rationalizability gives no guidance whatsoever in this zero sum game, even though the game has a unique equilibrium point. If rationalizability is all there is to rational selection of strategies then the weak equilibrium problem becomes acute. In any solvable game with a solution including at least one weak equilibrium point those non-equilibrium strategies which have expected payoffs as good as the weak equilibrium strategy relative to the conjecture that the other players play their parts in that equilibrium will be *rationalizable*. The conjecture just described followed by a cycle of point conjectures corresponding to the equilibrium will provide an infinitely extendable sequence that supports the strategy.

3.2 Ratifiability on best reply: a conditional defense

Put yourself in the point of view of a rational player in a two-person non-cooperative game. You are row chooser in the following version of game #5. Column choosers utilities have been transformed by adding a constant (+5) to the utility for each outcome.

	b1	b2
a1	(1,4)	(3,2)
a2	(4,1)	(2,3)

Your choices are causally independent. Perhaps, you each, separately, get to write down one of your strategies on a card. A referee picks up the two cards and then gives each of you the payoff assigned to you in the outcome corresponding to the two strategies. Assume your payoffs are in hundred dollar units and that your utilities are linear with these dollar amounts. You understand that the game structure and the utilities of each of you are common knowledge as is the assumption that each of you is rational, but you have not yet reasoned your way through the complexities of finding a solution. You also assume (perhaps only provisionally) that the game has a unique solution.

I think that the hypothetical *post hoc* best reply reasoning called for in von Neumann and Morgenstern's indirect argument is entirely appropriate as a way of ruling out candidates for the strategy that is your part of the solution you are trying to find. When you hypothetically assume that you will end up committing yourself to some strategy (say a1) you assume that it will turn out to be the unique strategy specified as your part of the solution. On this assumption *it is reasonable* to assume that the other player will figure this out and will choose some best reply (to a1). If each of these best replies is such that on it you would be better off doing some other strategy than the one under consideration then that strategy is ruled out. For example, the only best reply to a1 is b1, but if the other player chose b1 you would be better off playing a2 instead of a1. This is a reductio argument showing that a1 is not the strategy which is your part in any unique solution to this game.

This reasoning is cogent because (on the assumption that the game has a unique solution and the common knowledge assumptions) strategy a1 is *not ratifiable* — the hypothesis that you are going to commit youself to it gives you reason to judge that you would be better off doing another act instead. It is not reasonable to commit yourself to any unratifiable act. Indeed, it is hard to see how a rational agent could commit herself to such an act. The very act of commitment would provide rational grounds for judging that she would be better off doing another act instead.

Causal decision theory provides the resources to give a formal explication of ratifiability. The main idea is to formulate, relative to the hypothesis that you will choose a given act a (e.g. a1), the evaluation of what you would expect if you were to choose another act a' (e.g. a2) instead. We do this as follows:

$$U_a(a') = \Sigma P(a' \,\square\!\!\rightarrow bi|a) \cdot u(a',bi)^6$$

Reasoning about what would happen if I were to do a' instead is reasoning about the subjunctive conditionals

"If I were to do a' the other player would do bi"

for the various outcome determining states (pure strategies) bi. This reasoning is carried out relative to the assumption that I am going to commit myself to a by evaluating these subjunctives according to their conditional probabilities on **a**. Now we can define ratifiability.

a *is ratifiable* iff $U_a(a) \geq U_a(a')$ all a'

Act **a** is ratifiable just in acse the conditional expected utility of **a** on **a** is at least as high as the conditional expected utility on **a** of any alternative **a'**.

Here you are ready to use the indirect argument to test your various strategies as possible candidates for your part in the presumed unique solution. You can represent your commitment to the appropriate hypothetical *post hoc* best reply reasoning by using best reply priors to assess ratifiability. A best reply prior is one such that for each of your strategies a (mixed or pure) and each of the other player's strategies **b**,

$$P(b|a) = 0$$

unless **b** is some best reply to **a**.

In this game we have causal independence between the players choices of strategy. This independence is common knowledge from the game structure, therefore our best reply priors should satisfy

Causal Independence $P(a' \,\square\!\!\rightarrow bi|a) = P(bi|a)$ all a', bi and a.

This is to require that these conditional probabilities respect the causal independence built into the structure of this game. We can now show explicitly that strategy $a1$ is not ratifiable by any best reply prior. Let P be any such prior. Since $b1$ is the unique best reply to $a1$ we have $P(b1|a1) = 1$. We also have that $P(a1 \square \to b1|a1) = P(b1|a1)$; so

$$U_{a1}(a1) = 1.$$

By causal independence we have $P(a2 \square \to b1|a1) = P(b1|a1)$, as well. Therefore,

$$U_{a1}(a2) = 4,$$

so $a1$ is not ratifiable by P.

The following result (Harper 1988, pp. 35-36) ensures that our reconstruction of the indirect argument will restrict candidates for solutions to equilibrium strategies.[7]

> *Theorem 1:* In two person non-cooperative games only equilibrium strategies are best reply ratifiable.

When we apply our test to the various strategies in our game, therefore, the only possible survivors will be the mixed strategies $x^* = (1/2\ a1, 1/2\ a2)$ and $y^* = (1/4\ b1, 3/4\ b2)$ in the unique equilibrium pair.[8]

These strategies are best reply ratifiable as the following result implies (Harper 1988, pg. 37):

> *Theorem 2:* In any two-person non-cooperative game, all equilibrium strategies are best reply ratifiable, *if causal independence holds.*

This may make it look as though the best reply ratifiable strategies are exactly the equilibrium strategies; however, it is an interesting refinement of the equilibrium requirement in extensive form games where causal independence fails (Harper 1988, pp. 39-45).

Our game is an example of that narrowing down to a single possibility that von Neumann and Morgenstern characterized as the successful discovery of the solution by the indirect method (1944, pp. 147-148, see section II above). For each agent the indirect argument successfully cut down the candidates to the single strategy which is that agent's part of a unique equilibrium. That this equilibrium is weak does not undercut the solution they have achieved. For one thing, the off-equilibrium strategies have already been rejected as ones which fail the test of best reply ratifiability. For another, the fact that the indirect argument did succeed in specifying a set of alternatives (in this case a unique equilibrium) that satisfies the Nash solution concept ensures that neither agent can expect to gain by defecting to one of their non-equilibrium strategies. Our agents are trying hard to find a rational solution to their game. For such agents only a more desirable alternative solution would lead them to defect from the one they have been fortunate enough to find by the indirect argument.

The serious problem with the indirect argument is not that it sometimes specifies Nash solutions with weak equilibria. It is, rather, that in a large class of games — those with multiple non interchangeable equilibria —it fails to specify any Nash solution at all. In extensive form games causal structure can be appealed to to make the indirect argument reject inadequate equilibria. In this respect best reply ratifiability takes its place among the burgeoning collection of refinements of the Nash solution concept. Such refinements can extend the class of games for which solutions exist when the appropriately restricted set of admissible equilibrium strategies satisfies the Nash solution concept.

Ratifiability is a constraint on admissible options for rational choice, but it is not a complete theory of what to choose. The basic recommendation for causal utility theory is that you should select from among your ratifiable options one which optimizes causal utility. This suggests that the set of games for which solutions exist may be further extended by following up on the initial result of the indirect argument by reasoning about appropriate priors for using causal utility to select from among the admissible candidates.

The approach advocated here is that agents in a game retreat to rationalizability only if these methods for extending the indirect argument fail to deliver a recommendation that satisfies the Nash solution concept. The corresponding program in game theory is to attempt to develop methods that will generate such solutions for as wide a class of games as possible. The lively works on refinements of equilibrium solutions, e.g. Kreps and Wilson (1982), van Damme (1983), and Kohlberg and Mertens (1986), and the general theory of equilibrium selection developed by Harsanyi and Selten (1988) suggest that the prospects for such a program are good.

Notes

[1]In normal-form games where causal independence is assumed, one may use the set of all the other players mixed strategies as an infinite partition of alternative chance hypotheses. Your degrees of belief in outcome determining states (the pure strategies) can then be represented as expectations of their chances given some appropriate integration over these chance hypotheses. One can represent the causal independences by making each chance hypothesis y determine chances conditional on your acts so that

$$y(b_i|a) = y(b_i)$$

for each y, bi and each of your acts a. This will lead to assigning $P(a_1 \square \to b_i) = \int_y P(y)$ $y(b_i|a)dy = \int_y P(y) \, y(bi)dy = P(b_i)$, even if there are dependences in your subjective conditional probabilities so that $P(b_i|a) \neq P(b_i)$. When you move to your epistemic conditional probability you simply redistribute your subjective probability over these chance hypotheses, you don't change the causal independence built into the chance set up (see Harper 1986a).

[2]Harsanyi and Selten (1988, pg. 13) give an example with 101 non-interchangeable pure strategy equilibrium points.

[3]If in Game #4 the payoff for the outcome of playing the non-equilibrium strategy a2 against b2 were changed so that its number was less than 2, then Game #4 would be strongly solvable in Nash's sense. Game #1 is also not strongly solvable.

[4]Harsanyi has given up this view. The powerful solution concept he and Selten (1988) are developing generates a unique equilibrium solution for each game.

[5]Bernheim (pg. 1016) and Pearce (pg. 1035) suggest that for normal forms games with more than two players rationalizability is a stronger requirement than iterative elimination of strongly dominated strategies. The reason they give is that the independence among players' choices assumed in the normal form should rule out conjectures with correlated strategies. Indeed, Pearce (pg. 1035) suggest that "proofs of the equivalence of the two procedures for N = 2 could be extended to arbitrary N if a player's opponents could co-ordinate their randomized strategic actions." The independence assumed in the normal form is causal independence — independence in (and between) the chances in the objective probabilities corresponding to various mixed strategies available to the players. This independence is compatible with even extreme dependence in the subjective probabilities corresponding to conjectures over these strategies (Harper 1986). Brian Skyrms (1989) has

pointed out that this compatibility of independence in the chances with correlations in the conjectures construed as subjective expectations over those chances should allow the equivalence for two-person normal form games to be extended to n-person normal form games.

[6]On the Skyrms formulation we would have

$$U_{a1}(a3) = \Sigma \ [\Sigma P(h|a1) \cdot P(bi|h \ \& \ a2)] \ u(a3,bi)$$
$$\quad \quad i \ \ h$$

where the h's are an appropriate partition of alternative states of the relevant chance set up. If as in this game the b's are causally independent of the a's each $P(bi|h \ \& \ a3) = P(bi|h)$ so that

$$\Sigma \ P(h|a1) \cdot P(bi|h) = P(bi|a1),$$
$$h$$

just as in the Gibbard-Harper formulation.

[7]Extending this result to n-person games does not appear to be an easy task. Brian Skyrms (1989) has developed a rational for equilibrium solutions that holds for n-person games. He generates the game theoretic equilibrium from a fixed point in a dynamic deliberational process. This is a very powerful new approach to the theory of games. It departs from the classic theory in that the initial subjective probabilities assigned by each agent as well as their utilities for outcomes are included in the common knowledge assumed at the outset. It delivers impressive results about how sensitive various games are to alternative initial probability assignments (as well as to perturbations of payoffs). Moreover, some versions can be run on personal computers.

[8]This explication of the indirect argument by the test of ratifiability on best reply provides an answer to what Harsanyi and Selton (1988, pp. 14-16) call the instability problem for mixed strategy solutions. It does not require the idea of randomly fluctuating payoffs (first suggested in Harsanyi 1973, and repeated in Harsanyi and Selton 1988, pp. 14-15) in order to justify a mixed strategy solution. The solution will be justified by the indirect argument so long as the resulting recommendation satisfies the Nash solution concept.

References

Allais, M. (1953), "*Le comportment de l'homme rationnel devant le risque: Critique des postulates et axiomes de l'ecole americaine*" *Econometrica* 21: 503-46.

_ _ _ _ _ . (1979), "The so-called Allais paradox and rational decisions under uncertainty" in Allais and Hagen, eds (1979), 437-681.

_ _ _ _ _ . and Hagen (eds) (1979), *Expected Utility Hypotheses and the Allais Paradox*, Dordrecht: Reidel.

Aumann, R.J. (1976), "Agreeing to disagree", *Annals of Statistics* 4: 1236-1239.

Barwise, J. (1988), "Three views of common knowledge" in Vardi, M. Y. (ed.), 365-374.

Bernheim, D. (1984), "Rationalizable strategic behavior" *Econometrica* 52: 1007-1028.

Bicchieri, C. (1988), "Common knowledge and backward induction: a solution to the paradox" in Vardi, M. Y. (ed.), 381-393.

_ _ _ _ _ _. (1989) this volume.

Binmore, K. (1986) "Modeling rational players II", manuscript.

Brandonberger, A. and Dekel, E. (1985,) "Common knowledge with probability 1", research paper 796R, Graduate School of Business, Stanford University.

Gârdenfors, P. and Sahlins, E. (eds) (1988), *Decision, Probability and Utility: Selected Readings,* Cambridge: Cambridge University Press.

Giles, R. (1979), "Non Archimedian belief", manuscript.

Harsanyi, J. C. (1967-68), "Games with incomplete information played by 'Bayesian' players" Parts I-III, *Management Science* 14: 159-182, 320-334, and 486-502.

_ _ _ _ _ _ _. (1973) "Games with randomly-disturbed payoffs: a new rationale for mixed-strategy equilibrium points", *International Journal of Game Theory* 2: 1-23.

_ _ _ _ _ _ _. (1977), *Rational Behavior and Bargaining Equilibrium in Games and Social Situations,* Cambridge: Cambridge University Press.

_ _ _ _ _ _ _. and Selten, R. (1988), *A General Theory of Equilibrium Selection in Games,* Cambridge, Mass: The MIT Press.

Kohlberg, E. and Mertens, J. F. (1986), "On the strategic stability of equilibria" *Econometrica* 54: 1003-1037.

Kreps, D. and Wilson, R. (1982), "Sequential equilibria", *Econometrica* 50, 863-894.

Kreps, D., Milgrom, P. and Wilson, R. (1982) "Rational cooperation in the finitely repeated Prisoner's Dilemma", *Journal of Economic Theory* 27: 245-252.

Lewis, D.K. (1969), *Convention, A Philosophical Study,* Cambridge, Mass: Harvard University Press.

Luce, R.D. and Raiffa, H. (1957), *Games and Decisions,* New York: John Wiley and Sons, Inc.

McClennen, E.F. (1973), "The Minimax theory and expected utility reasoning", in Hooker et al. (eds) (1978) Vol. Ii: 337-359.

_ _ _ _ _ _ _ _. (1983), "Sure-thing doubts", in B. Stigum and F. Wenstop (eds), 117-136.

_ _ _ _ _ _ _ _. (1986), "Prisoners dilemma and resolute choice", in R. Campbell and L. Sowden (eds), 94-104.

_ _ _ _ _ _ _ _. (1988), "Ordering and independence", *Economics and Philosophy* 4, 298-308.

Pearce, D. (1984), "Rationalizable strategic behaviour and the problem of perfection", *Econometrica* 52: 1028-1050.

Popper, K.R. (1959), *The Logic of Scientific Discovery* New York: Harper Torchbook edition.

Reny, P. (1988), "Backward induction and common knowledge in games with perfect information", manuscript.

Renyi, A. (1955), "On a new axiomatic theory of probability", *Acta Mathematica* 6: 285-335.

_ _ _ _ _. (1970), *Foundations of Probability,* San Francisco: Holden-Day Inc.

Savage, L.J. (1954), *The Foundations of Statistics,* New York: John Wiley & Sons, Inc.

Selten, R. (1965), "*Spieltheoretische Behandlungeines Otigopolmodells mit Nachfragetragheit*", *Zeitschrift fur die Gesamte Straatiswissenschaft* 121: 301-324.

_ _ _ _ _ . (1975), "Re-examination of the perfectness concept for equilibrium points in extensive games", *International Journal of Game Theory* 4: 25-55.

Skyrms, B. (1982), "Causal decision theory", *The Journal of Philosophy* 79: 695-711.

_ _ _ _ _ _. (1984), *Pragmatics and Empiricism*, New Haven: Yale Press.

_ _ _ _ _ _. (1989), *The Dynamics of Rational Deliberation,* forthcoming, Cambridge, Mass: Harvard University Press.

Tan, T. and Werlang, S. (1988), "On Aumann's notion of common knowledge — an alternative approach", *CARESS* working paper #88-09, University of Pennsylvania.

_ _ _ _ _ _ _ _ _ _ _ . R. D. C. (1988), "A guide to knowledge and games", in Vardi, M.Y. (ed.), 163-177.

Vardi, M. Y. (ed.) (1988), *Proceedings of the Second Conference on Theoretical Aspects of Reasoning about Knowledge,* Los Altos, California: Morgan Kaufmann Publishers.

van Damme, E. (1983), *Refinements of the Nash Equilibrium Concept*, lecture notes in Mathematical Economics and Social Systems, No. 219, Berlin: Springer-Verlag.

von Neumann, J. and Morgenstern, O. (1944), *Theory of Games and Economic Behavior,* Princeton, N.J: Princeton University Press.

Common Knowledge and Games with Perfect Information

Philip J. Reny

The University of Western Ontario

1. Introduction

It is by now rather well understood that the notion of common knowledge (first introduced by Lewis (1969) and later formalized by Aumann (1976)) plays a central role in the theory of games. (An event E is common knowledge between two individuals, if each knows E, each knows the other knows E, etc...). Indeed, most justifications of Nash's (1951) equilibrium concept usually include (perhaps only implicitly) the assumption that it is common knowledge among the players that both the Nash equilibrium in question will be played by all and that all players are expected utility maximizers.[1] (We shall henceforth call expected utility maximizers, "rational".[2]) We hope to illustrate in an informal manner that there is in fact a large class of extensive form games, in which each of which it is not possible for rationality to be common knowledge throughout the game.

The consequences of this for many well-known extensive form refinements of Nash equilibrium are quite serious. Consider, for example Selten's (1965) notion of subgame perfect Nash equilibrium. The requirements on a solution here are not only that the strategies form a Nash equilibrium of the game as a whole, but also that the strategies induce on every proper subgame a Nash equilibrium.[3] If, however there are proper subgames beginning at (singleton) information sets at which it is not possible for rationality to be common knowledge, then Nash behavior in that subgame can no longer be justified on common knowledge grounds.[4] At best then, significant modifications are required in our explanation of Nash behavior in such subgames. At worst, Nash behavior in such subgames should not be considered as the only possibility. In either case, a re-evaluation of the subgame perfect equilibrium notion would be called for. Since extensive form refinements such as sequential equilibria (Kreps and Wilson (1982)) and perfect equilibria (Selten (1975)) involve even stronger restrictions upon behavior in subgames, the comments above apply as well to each of these notions.[5]

In addition, our result implies that arguments supporting the backward induction solution based on rationality being common knowledge at every information set, begin with a false hypothesis. Hence, the elegant argument supporting backward induction advanced by Kreps et. al. (1982) runs into difficulty. Also, the recent work of Bernheim (1984) and Pearce (1984) on "Rationalizability" involves heavy use of the common knowledge of rationality in both normal and extensive form games. If, in extensive form games, such common knowledge is not always possible, then at the very least a reinterpretation of their extensive form analyses is called for. These issues will be explored further at the end of the paper.

PSA 1988, Volume 2, pp. 363-369

Finally, others have also expressed certain difficulties with a variety of the above equilibrium concepts (See for instance Basu (1985), Binmore (1985) and Rosenthal (1981)). All of these difficulties essentially appear to involve in one way or another a problem with the assumption that rationality is common knowledge. We now illustrate by means of the simplest sort of example that there are games containing information sets at which it is not possible for rationality to be common knowledge.

2. An Example

Consider the following two-player perfect information game. A referee comes equipped with n dollars and places one in front of players one and two. Player one can take the dollar thereby ending the game, or he can leave it. If he leaves it, the referee places a second dollar in front of the players. Player two now has the opportunity to take the two dollars and end the game or not, in which case the process repeats. In general, at the kth stage of the game, the referee adds one dollar to the pot bringing its total to k dollars. If k is odd (even), player one (two) may take the k dollars and end the game, or leave it. Players' payoffs are assumed strictly increasing in dollars. Finally, should the game continue until the nth stage and the player whose turn it is decides to leave the n dollars, it is then given to the other player. Call this game TOL(n) (Take it or leave it). TOL(n) for n odd is depicted below.

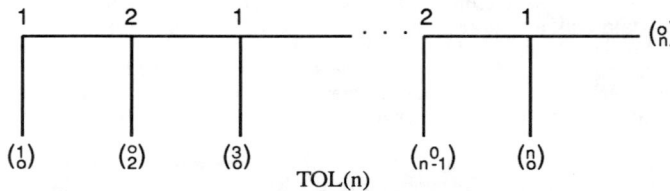

TOL(n)

In this particular game with perfect information, backward induction, and the sub-game perfect, sequential and perfect equilibrium concepts each yield the same equilibrium strategies. We proceed via backward induction. At the last stage (n odd) player one's best choice is to take the n dollars. With this in mind, two's best response at the second last stage is to take the n-1 dollars and end the game. This process continues with each player choosing to take the money and end the game on his turn if he gets the chance. In the end, backward induction (and hence the subgame perfect, sequential and perfect equilibrium concepts) yields that player one take the one dollar and end the game in the first round. This is independent of the value of n! That is, no matter how large the pot may potentially grow, the standard equilibrium notions indicate that player one will take the one dollar in the first round.[6] This sort of paradox is by no means new and is clearly reminiscent of that associated with the finitely-repeated prisoners' dilemma.

We now move to the problem of common knowledge. It is enough to consider TOL (3) (see figure below).

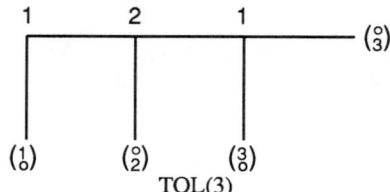

TOL(3)

We claim that if player one does not take the dollar and end the game in the first round, but instead leaves it so that player 2 must decide whether or not to take the two dollars, then it is no longer possible for rationality to be common knowledge. (i.e. At player two's information set, it is not possible for rationality to be common knowledge). The argument is really quite straightforward and proceeds by contradiction. Suppose that rationality were common knowledge at player two's information set. Player two, believing that player one is rational (i.e. an expected utility maximizer), must believe that at stage 3, player one will take the three dollars, leaving player two with nothing. A rational player two would respond to this by taking the two dollars at the second stage of the game leaving player one with zero. Hence, if at player two's information set it is the case that

(i) player two is rational, and

(ii) player two believes that player one is rational, then player two will take the two dollars leaving player one with zero.

But since rationality is common knowledge, it must be the case that in particular,

(i') player one believes that player two is rational, and

(ii') player one believes that player two believes that player one is rational.

i.e. Player one believes (i) and (ii) above. Finally, however, this implies that player one believes that player two will take the two dollars leaving player one with zero and rendering player one's choice not to take the dollar in the first round (recall that player two's information set has been reached) an irrational (non expected utility maximizing) one. Hence, player two must believe that player one is not rational which contradicts our original assumption and completes the argument.[7] A similar argument shows that in TOL(n), as soon as player one leaves the first dollar it is not possible for rationality to be common knowledge.

This observation and the work of Kreps et. al. (1982), suggests an alternative analysis of TOL(n). Recall that Kreps et. al. were interested in explaining cooperation in the finitely- repeated prisoners' dilemma. They showed that if from the start, there is a positive probability that one of the players is not rational, then cooperation could emerge as a sequential equilibrium of a long enough repeated prisoners' dilemma game suitably modified to take into account the incomplete information about the player's rationality.[8] It turns out that one can apply a generalized version of Kreps et. al.'s analysis of the prisoners' dilemma to TOL(n) and achieve similar results. That is, if from the outset there is a positive probability that one of the players is not rational, or none of the players believes the other is not rational, or . . ., and n is large enough, then allowing the pot to build for some time can emerge as a sequential equilibrium.[9] Moreover, any such sequential equilibrium must yield both players higher expected utilities than they would get if player one took the dollar at the first stage. Hence, there is a sense in which both players are better off in TOL(n) when n is large enough and rationality is not common knowledge.

Among those elements left unexplained by Kreps et. al.'s analysis and the generalized version we've applied to TOL(n), is how the positive probability that a player is not rational, or that a player believes the other is not rational, or . . . arises in the first place. What we've shown above however, is that player one can, by not taking that first dollar, create an environment in which rationality is not common knowledge. Since for n large enough, this potentially makes player one better off, this furnishes a sound explanation of where the positive probability comes from. Hence, this exogenously introduced positive probability that one of the players is not rational or that one believes the other is not rational etc..., need not be exogenously introduced at all. Our observation that it is not possible for rationality to be common knowledge once player one leaves the first dollar, supplemented by Kreps et. al's analysis once this is the environment, shows that this positive probability can arise as the result of expected utility maximizing behavior. Taken as a whole, this analysis of TOL(n) (unlike backward induction and the more traditional solution concepts) implies that two expected utility maximizing players can, acting in their own self interests, allow the pot to grow.

It is somewhat paradoxical that we can justify in terms of rational behavior player one leaving the first dollar in TOL(n), when this very justification requires that when player one does so, player two believes that player one is not rational or that two believes that one believes that two is not rational or etc.... For if an explanation based on the players' rationality is possible, won't players believe one another are rational? And won't each then believe that each believe this etc...? i.e. Won't then rationality be common knowledge?

The answer to the last question is: "absolutely not". We have already argued that once player 1 leaves the first dollar, rationality *cannot* be common knowledge. This is simply inconsistent with the structure of the game and the current position in it. The answer to each of the other questions is "not necessarily". Since rationality can not be common knowledge when player one leaves the dollar, some statement of the form: two believes that one believes that two believes that ... that one (or two) is *not* rational *must* be true. Hence the answer to at least one of the other questions *must* be no. That is, a formal proof can be constructed to demonstrate this. On the other hand, although such beliefs are possible, no formal proof can be constructed to demonstrate that after player one leaves the first dollar, two believes that one is rational and two believes that one believes that two is rational. Otherwise (assuming this proof is common knowledge) this would imply that player two believes that rationality is common knowledge. But this is impossible if players' beliefs are restricted to being consistent with the physical description of the game and the current position in it. So, although a rational explanation is available, in a formal sense it cannot be the only available explanation. And it is precisely these (necessarily) available alternatives which make a rational explanation possible at all.

As in TOL(n), one can show that by cooperating in a finitely- repeated prisoners' dilemma, the players can create an environment in which rationality can not be common knowledge. As Kreps et. al (1982) have shown, once this is the case both players may be better off. Hence one can also explain now more fully perhaps, rational cooperation in the finitely-repeated prisoners' dilemma.

3. Is TOL(n) An Isolated Example?

We shall now fulfil a promise made in the introduction and illustrate that there is a large class of games within which the problem of common knowledge of rationality arises. To do so, we shall first restrict our attention to two-person extensive form games with perfect information where no player is indifferent between any two endpoints.[10]

Instead of asking which games within this class contain information sets at which it is possible for rationality to be common knowledge, we ask a slightly different question. That is, which games in this class allow rationality to be common knowledge at every information set *simultaneously*? Some clarification is in order.

It turns out that it is possible that rationality is common knowledge at a particular information set so long as at that information set it is also common knowledge that rationality is *not* common knowledge at some other information set. Hence in such cases it may be that taken one at a time, it is possible that rationality is common knowledge at every information set, but taken together, common knowledge of rationality at one information set precludes it at another. In the latter case we say that it is not possible for rationality to be common knowledge at every information set simultaneously. Whenever it is not possible for rationality to be common knowledge at every information set simultaneously, issues such as those described for TOL(n) arise and again, the arguments supporting the traditional equilibrium concepts no longer apply (Reny (1988) contains more details). It can be shown that:[11]

Any two-person finite extensive form game with no player indifferent between any two endpoints that allows rationality common knowledge at every information set simultaneously, must be of the form:

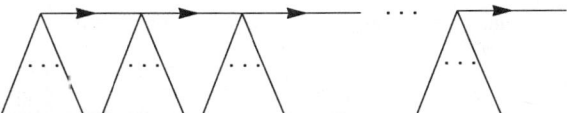

where the arrows indicate the unique subgame-perfect equilibrium (i.e. every decision node is reached in equilibrium).

In our view, this indicates that the problem of common knowledge of rationality in extensive form games occurs rather frequently.

4. Conclusion

What are we to make of all of this? We've shown that rationality cannot simply be assumed common knowledge. The physical description of the game and the current position in it may preclude this. But once rationality can no longer be common knowledge, it is not at all clear what the theory should tell the players to believe about one another. One thing however, is clear. If we insist that the theory itself is common knowledge among the players (as has implicitly been the traditional approach; and even explicitly in Bernheim (1984) and Pearce (1984)), then the theory cannot indicate that the players are rational, since this would automatically render rationality common knowledge and this is not always possible. But if the theory cannot indicate that players are rational, then the players may believe just about anything about the behavior of their opponents, since there is no obvious substitute for rational behavior. Clearly, this will lead to a plethora of possible outcomes since rational players who need not believe their opponents are rational may have many strategies which are a best response to *something* their opponents might play. Hence, we can expect a very weak (in terms of predictions) theory to result if we insist that the theory itself is common knowledge among the players at every information set. (For more on this see Reny (1988).)

The most promising avenue to pursue then, is one which explicitly allows the theory not to be common knowledge at every information set. Indeed one might reach this conclusion simply on the grounds that once a player has deviated it is impossible that the rationality of the players and the equilibrium strategies remain common knowledge. One way to consistently explain the deviation and still hold fast to the rationality of all players (including the deviator) is to postulate that the equilibrium strategies (i.e. the theory) are no longer common knowledge. We close by mentioning that this approach has been recently undertaken (Reny 1987) in a manner that yields relatively strong predictions, indicating that some of the issues raised here can be usefully taken into account.

Notes

[1]In a two-player normal form game, a pair of strategies (one for each player) is a Nash equilibrium if each player's choice maximizes his expected utility given the choice of the other player.

[2]This follows Bernheim (1982) and Pearce (1982).

[3]That is, not only must it be that (i) every player has decided what to do in every possible eventuality so that from his perspective at the beginning of the game his choices are

best given what the others have planned to do, it must also be the case that (ii) whenever any possible eventuality becomes a reality, no player will wish to change his previously decided upon choice.

[4]A physical description of chess for instance, includes the initial position, the set of all possible first moves and subsequent positions etc.... A subgame is simply the description of what the current position is and which positions are possible during subsequent play. A proper subgame must begin at a point in the game where every player has full knowledge of the past choices made by others. Hence, every subgame in chess is proper. In other games (like poker) one must take a turn in ignorance of what others have done previously. Those aspects known to a player when it is his turn to move are embodied in his "information set". Corresponding to each turn then is an information set, and so "reaching" a particular information set of player 1 say, indicates that it is a particular turn of player 1.

[5]One might argue that since Selten's (1975) theory allows players to make "mistakes", one can always preserve the common knowledge of rationality by explaining arrival at any particular information set through a sequence of independent "errors". Our response to this is that players who make "mistakes", independent or not, arbitrarily small or not, are by definition not expected utility maximizers (i.e. not rational). Expected utility maximizers always and everywhere make decisions that maximize their expected utility. Hence any explanation based on mistakes or trembles is one that embraces the lack of common knowledge of rationality.

[6]In fact, taking the first dollar is the unique Nash equilibrium outcome.

[7]A formal version is contained in Reny (1988). The formal definition given there uses infinite hierarchies of beliefs as in Mertens and Zamir (1985) and Tan and Werlang (1985). It should be pointed out that what here we've called common knowledge should more appropriately be called common belief since we never require any player to actually be rational only that players believe that one another are rational etc.... Since an event which is common knowledge (in Aumann's (1976) sense) must also be common belief (see Tan and Werlang (1985)) we have actually obtained the stronger result that at player two's information set rationality cannot be common belief.

[8]The particular kind of irrationality imposed upon the player whose behavior is not completely known is important. Kreps et. al. assume in particular that there is a positive probability that this player is one who uses the TIT-FOR-TAT strategy (i.e. cooperate in the first period and in every subsequent period copy the previous period choice of your opponent). TIT-FOR-TAT is not rational since it sometimes dictates cooperation in the last period.

[9]Again, the particular kind of irrationality involved is important.

[10]Recall that a game is one of perfect information if at every stage of the game both players know all past choices made by their opponent. A player is not indifferent between any two endpoints if whenever asked to choose which of two ways he would like the game to end, he always strictly prefers one over the other. (Chess does not satisfy the latter condition since for instance, there are many ways that the game can end with a win for white, and white is indifferent between all of these.) Note that the perfect information restriction rules out the finitely- repeated prisoners' dilemma. The no indifference between endpoints formally rules out TOL(n), but replacing the payoffs of zero at stage k by $1/k$ leaves all relevant features of TOL(n) intact and this new version is a member of the class of games described.

[11]For a proof, see Reny (1988) pp. 72-79.

References

Aumann, R. (1976), "Agreeing to Disagree", *The Annals of Statistics* 4: 1236-1239.

Basu, K. (1985), "Strategic Irrationality in Extensive Games", mimeo, Institute for Advanced Studies, Princeton.

Bernheim, D. (1984), "Rationalizable Strategic Behavior", *Econometrica* 52: 1007-1028.

Binmore, K.G. (1985), "Modelling Rational Players", mimeo, London School of Economics and University of Pennsylvania.

Kreps, D., Milgrom, P., Roberts, J. and Wilson, R. (1982), "Rational Cooperation in the Finitely Repeated Prisoner's Dilemma", *Journal of Economic Theory* 27: 245-252.

Kreps, D., and Wilson, R. (1982), "Sequential Equilibria", *Econometrica* 50: 863-894.

Lewis, D. (1969), Convention: A Philosophical Study. Cambridge: Harvard University Press.

Mertens, J.F., and Zamir, S. (1985), "Formulation of Bayesian Analysis for Games with Incomplete Information", *International Journal of Game Theory* 14: 1-29.

Nash, J. (1951), "Noncooperative Games", *Annals of Mathematics* 54: 286-295.

Pearce, D. (1984), "Rationalizable Strategic Behaviour and The Problem of Perfection", *Econometrica* 52: 1008-1050.

Reny, P.J. (1987), "Explicable Equilibria", mimeo, Princeton University and the University of Western Ontario.

_ _ _ _ _ . (1988), "Rationality, Common Knowledge and The Theory of Games", Ph.D. Dissertation, Chapter 1, Princeton University.

Rosenthal, R.W. (1981), "Games of Perfect Information, Predatory Pricing and the Chain-Store Paradox", *Journal of EconomicTheory* 25: 92-100.

Selten, R. (1965), "Spieltheoretische Behandlungeines Oligopolmodells mit Nachfragetragheit", *Zeitschrift fur die Gesamte Straatiswissenschaft* 121: 301-324.

_ _ _ _ _. (1975), "Reexamination of the Perfectness Concept for Equilibrium Points in Extensive Games", *International Journal of Game Theory* 4: 25-55.

Tan, T. and Werlang, S. (1985), "On Aumann's Notion of Common Knowledge - An Alternative Approach", mimeo, University of Chicago Business School and Princeton University.

Part XII

CONFIRMATION

Non-Bayesian Confirmation Theory, and the Principle of Explanatory Surplus[1]

Donald A. Gillies

King's College London

1. Introduction

One of the most obvious features of empirical science is that scientists attempt to relate their theories to data, that is to the results of observation and experiment. In so doing, they will find it necessary to use the term 'confirms' or 'confirmation' or at least terms with more or less the same meaning. Thus scientists may think that a theory is well- confirmed by a considerable body of observational evidence. In this case they will regard it as rational to have a high degree of confidence in the theory, and, in particular, to use it in practical applications. Alternatively a theory T may be seen as badly confirmed by the evidence. In this case it would be thought undesirable to use T in practice, and the problem would become that of modifying or replacing T so as to obtain a new theory T' which is better confirmed than T. All this goes to show that the notion of confirmation plays a central role in scientific practice, and it is therefore a most important task for philosophers of science to clarify and perhaps systematise and improve this notion, or, in other words to construct a theory of confirmation. Unfortunately it has proved surprisingly difficult to develop confirmation theory. It is not even clear how we should set about this task, or what a successful confirmation theory would be like.

In this paper our approach will be a modification of one suggested by Hesse in her (1974) book: *The Structure of Scientific Inference*. Ch.4 of this work is entitled 'The Logic of Induction as Explication', and in Section II p.97 'A more modest programme', she writes:

'I am going to take the problem of induction in its more modest version to be the problem of *explicating intuitive inductive rules*. Thus the problem falls into two parts.

(i) To formulate a set of rules which capture as far as possible the implicit rules which govern our inductive behaviour.

(ii) To formalize these in an economical postulate system.' (Hesse 1974, p.97)

I would here alter some of the terminology, preferring to speak of 'explicating intuitive judgements of confirmation' rather than 'explicating intuitive inductive rules', and of 'the implicit rules which govern our judgements of confirmation' rather than 'the implicit rules which govern our inductive behaviour'. With these changes I would

PSA 1988, Volume 2, pp. 373-380

endorse the following version of Hesse's 'modest programme'. The aim of this version of the programme is precisely to produce 'an economical postulate system'.

Our starting point or *explicandum* consists of the host of particular judgements of confirmation made by scientists. Our *explicatum,* or confirmation theory, should hopefully comprise just a few very general rules or principles which justify at least the majority of these particular judgements. Of course a few preformal judgements of confirmation may be rejected as, so to speak, whales which turn out after all not to be fishes.

Let me next give an example of a general principle of confirmation which would, I think, be accepted by most philosophers of science. This is the 'principle of severe testing' which states that the more severe the tests which an hypothesis h has passed, the greater is the confirmation of h. Popper seems to have been the first to formulate this principle, and in the *Logic of Scientific Discovery*, he puts it as follows:

> ... it is not so much the number of corroborating instances which determines the degree of corroboration as *the severity of the various tests* to which the hypothesis in question can be, and has been, subjected. (Popper 1934, Ch X, Section 82, p.267.)

My main aim in this paper is to propose a new principle for confirmation theory, which I call the principle of explanatory surplus, and which turns out to be non-Bayesian in character. I shall try to show that this principle justifies some common sense judgements of confirmation made in some parts of ordinary everyday science, and that it affords a plausible analysis of the situation regarding the confirmation of Newton's Laws and of Adler's Theory of the Inferiority Complex.

There is one other way in which general principles of confirmation theory might be justified; that is by showing that other plausible but less general principles of confirmation theory follow from them. I shall illustrate this procedure by showing that one of Hesse's proposed rules follows from the principle of explanatory surplus. The rule in question is actually her rule (h) *Simplicity* which says that:

> ... The simpler of two hypotheses supported by given evidence ought to be regarded as more highly confirmed by that evidence. (Hesse 1974, Ch.4, Section II, p.98.)

We are now in a position to formulate our suggested 'principle of explanatory surplus'. This has two sub-cases which are concerned with (a) number of parameters, and (b) number of theoretical assumptions. We will consider (a) in Section 2, and (b) in Section 3.

2. Number of Parameters

Let us suppose we have n data points $e_1, e_2, ... e_n$ say, and we are trying to explain these by some functional relationship

$$y = f(x_1, ..., x_r, \Theta_1, ... , \Theta_s) \qquad (1)$$

which depends on s parameters. A simple case would occur where each e_i has the form (x_i, y_i) and we are interested in seeing if these points lie on a line

$$y = \Theta_1 x + \Theta_2 \qquad (2)$$

Suppose first that n = 2. We can then draw a line through our two data points e_1, e_2, thereby determining Θ_1, Θ_2. Does e_1 and e_2 in this case give any support to the hypothesis that y and x are linearly related? The answer must surely be 'no', for whatever the values of e_1 and e_2, we can always draw a line joining them.

Suppose now that $n > 2$, say $n = 10$, and we are able within the limits of achievable accuracy to draw an exact line through the 10 data-points. In this case e_1 and e_2 and ... and e_{10} certainly give strong support to the hypothesis that y and x are linearly related. The parameters Θ_1, Θ_2 are fixed by only two of the data points, and the hypothesis of linearity then explains the remaining 8 data points. In other words there is here an explanatory surplus of 8 data points. My suggested *principle of explanatory surplus* is that a theory T is supported not by all the data which agrees with it, but only by that fraction of the data which can be considered as an explanatory surplus. In this case of functional relationships of the form (1), the size of this explanatory surplus can be estimated in this way. On the simple assumption that each parameter is determined by one data point, the explanatory surplus consists of n-s data points. This simple assumption will of course be too crude in general. Taking account of error and the use of statistical techniques, we may need to assume that each Θ_i requires m data-points for its determination. On this assumption, the explanatory surplus becomes n - ms. Of course I am not claiming that it would be possible to determine m with any precision, but then I am only trying to develop a *qualitative* confirmation theory. It suffices for my purpose if we can form a reasonable judgement as to whether an explanatory surplus exists, and give a crude estimate of its size. This seems to me certainly possible.

The principle of explanatory surplus is, I think, non-Bayesian in character. Let us return to our simple case of a linear relationship (2), and 10 data-points e_1, e_2, ..., e_{10}, through which we can draw an exact straight line. This straight line fixes the values of Θ_1 and Θ_2. Let h be the hypothesis of linearity with Θ_1, Θ_2 fixed in this way. Now, according to the principle of explanatory surplus, only 8 of the data points e_1, e_2, ..., e_{10} support h. Applying Bayes Theorem, we have for each ei:

$$P(h, e_i) = \frac{P(e_i, h)\, P(h)}{P(e_i)} \qquad (3)$$

Now since $P(e_i, h) = 1$, we have, taking the support of h by e $(S(h,e))$ as $P(h,e) - P(h)$,

$$S(h, e_i) = \left(\frac{1}{P(ei)} - 1\right) P(h) \qquad (4)$$

So, on the Bayesian analysis, each of the 10 data points supports h positively (and indeed to the same degree).

One aim of our confirmation theory is to give a few general confirmation rules or principles, and then, from this 'economical postulate system' to derive some subsidiary confirmation rules. I promised earlier, in accordance with this plan, to derive Hesse's rule (h) *Simplicity* from the Principle of Explanatory Surplus, and I will now show how this can be done in the special case (*number of parameters*) at present under consideration.

Hesse states her principle as follows (1974, p.98): 'the simpler of two hypotheses supported by given evidence ought to be regarded as more highly confirmed by that evidence'. To give a complete treatment of this principle, we would have to give a thorough analysis of the notion of simplicity as applied to scientific hypotheses. I am not here going to attempt this arduous task. Instead I shall make use of some helpful observations on simplicity which Hesse makes in Ch.10 of her book. Here Hesse lists a number of aspects of simplicity. We will consider just two: namely her (b_1) *Economy of parameters*, (1974, p.231) and (e) *Economy of theoretical premises* (1974, p.237).

According to (b_1), if h_1 and h_2 are otherwise comparable scientific hypotheses and h_1 has fewer arbitrary parameters than h_2, then h_1 must be considered simpler than h_2. Thus a straight line is simpler than a conic section, which is simpler than a curve of third degree, and so on. According to (e), if h_1 and h_2 are otherwise comparable scientific hypotheses and h_1 involves fewer theoretical premises than h_2, then h_1 must be considered simpler than h_2. Thus, for example, Newton's theory of gravitation would be simpler than a theory which held that gravitation was partly due to inverse square attraction and partly to the effect of Cartesian vortices in the aether.

Now it is obvious that Hesse's confirmation rule (h) *Simplicity*, with simplicity taken in the sense of economy of parameters follows from case (a) of the principle of explanatory surplus. Suppose our given evidence consists of n data points $e_1, ..., e_n$. Suppose h_1 and h_2 are otherwise comparable, but that h_1 involves r_1 parameters and h_2 r_2 parameters, where $r_1 < r_2$. Suppose further that s data points are needed to fix a parameter. Then h_1 is supported by n - r_1s of the data points, and h_2 by n - r_2s. So h_1, the simpler hypothesis, is more strongly supported by the given evidence than h_2.

It will be interesting to compare this treatment of simplicity with Jeffreys' Bayesian account. He writes:

> It is asserted, for instance, that the choice of the simplest law is purely a matter of economy of description or thought, and has nothing to do with any reason for believing the law. No reason in deductive logic, certainly; but the question is, Does deductive logic contain the whole of reason? ... Does the Nautical Almanac Office laboriously work out the positions of the planets by means of a complicated set of tables based on the law of gravitation and previous observations, merely for convenience, when it might much more easily guess them? Do sailors trust the safety of their ships to the accuracy of these predictions for the same reason? Does a town install a new tramway system, with expensive plant and much preliminary consultation with engineers, with no more reason to suppose that the trams will move than that the laws of electromagnetic induction are a saving of trouble? I do not believe for a moment that anybody will answer any of these questions in the affirmative; but an affirmative answer is implied by the assertion that is still frequently made, that the choice of the simplest law is merely a matter of convention. I say, on the contrary, that the simplest law is chosen because it is the most likely to give correct predictions; that the choice is based on a reasonable degree of belief; ... (Jeffreys 1939, I Section 1.0, pp.4-5).

I agree with what Jeffreys says here, and what he requires does indeed follow from our treatment of simplicity. Suppose h_1 and h_2 both explain some body of evidence e, and that h_1 is simpler than h_2. It then follows from the principle of explanatory surplus that e confirms h_1 more than it confirms h_2. It is thus reasonable to have a higher degree of belief in h_1 than in h_2, and, in particular, to base practical actions on h_1 rather than h_2. These results are exactly what Jeffreys wants, but it should be noted that they are obtained in a quite different way in his own Bayesian approach.

Jeffreys accepts that *ceteris paribus* the smaller the number of adjustable parameters in a law, the greater its simplicity (1939, I Section 1.6 p.48). Suppose then that h_1 has r_1 parameters, and h_2 r_2 parameters, where $r_1 < r_2$. Given a body of data e say, let us use it to fix the parameters in h_1 and h_2 as to obtain h'_1 and h'_2 say, where $h'_1 \vdash e$ and $h'_2 \vdash e$. We then have by Bayes' theorem.

$$P(h'_i, e) = \frac{P(h'_i)}{P(e)} \qquad i = 1,2 \qquad (5)$$

So, using the fundamental Bayesian assumption that confirmation is a probability function, or, in symbols, that C(h,e) = P(h,e), we have

$$C(h'_1, e) > C(h'_2, e) \text{ iff } P(h'_1) > P(h'_2) \qquad (6)$$

In other words, the simpler of two hypotheses is better confirmed by given evidence if it has higher prior probability. This is exactly what Jeffreys assumes to be the case, for he writes (1939, I para. 1.6 p.47): 'All we have to say is that the simpler laws have the greater prior probabilities'. But here Jeffreys makes the whole of scientific reasoning depend on the *a priori* postulation of a metaphysical principle concerning the simplicity of the natural world. I can see no a priori justification for supposing that nature is simple rather than complex, and thus Jeffreys' argument seems to me to fail. Hesse's confirmation rule (h) (*Simplicity*) follows, however, from the principle of explanatory surplus without there being any need to postulate *a priori* that nature is simple.

3. Number of Theoretical Assumptions

For our second example of the principle of explanatory surplus, we turn to a situation which is different from the previous one but just as characteristic of the scientific enterprise. Let us suppose that a scientist is trying to give a theoretical explanation of a set of facts which we shall denote by $f_1, ..., f_n$. To do so, he or she makes a number of theoretical assumptions which we shall denote by $T_1, ..., T_s$. The facts in question may be concerned with single events, or may be universal laws connecting observables such as 'All ravens are black' or Kepler's Laws. In either case we assume that $f_1, ..., f_n$ are well confirmed by observation and experiment, and so can be assumed to be true (at least when interpreted as approximations), while the attempt at theoretical explanation is being made.

Let us now further suppose that each of $f_1, ..., f_n$ follows logically from some subset of $T_1, ..., T_s$ together with initial conditions which are established by observation and experiment. In symbols our assumption is that

$$h_i \vdash f_i \quad 1 \leq i \leq n \qquad (7)$$

where

$$h_i = 0_i \ \& \ T_{i_1} \ \& \ ... \ \& \ T_{i_p} \qquad (8)$$

0_i is a conjunction of observation statements, and

$$T_{i_j} \in \{T_1, ..., T_s\} \quad 1 \leq j \leq p$$

The question before us is: 'given this general situation, to what extent, if any, are T_1, ... T_s supported by $f_1, ..., f_n$?' The principle of explanatory surplus states that $T_1, ..., T_s$ are supported not by all the facts which they explain but only by that fraction of the facts which can be considered an explanatory surplus. There is implicit here an economic analogy. The successful theoretician is like a successful entrepreneur.

To be successful, an entrepreneur has to choose investments $I_1, ..., I_s$ for his or her capital in such a way that he or she obtains a surplus, and the bigger this surplus the more successful the entrepreneur. Similarly to be successful, the theoretician has to choose theoretical assumptions $T_1, ..., T_s$ in such a way as to generate a surplus, and the bigger this surplus the more successful the theoretician. In the first case, the surplus is an economic one, and takes the form of an excess of receipts over outlays. In the second case, we are concerned with an explanatory surplus, which consists, roughly speaking, of an excess of facts explained over theoretical assumptions employed.

We have next got to consider how the explanatory surplus should be estimated. The simplest and most straightforward method is to subtract the number of theoretical assumptions used from the number of facts explained. So in our notation, the size of the explanatory surplus is estimated as n-s. Thus if a theoretician has to explain n facts, and needs to postulate n or more than n theoretical assumptions to do so, then the facts do not support the theoretical assumptions at all, even if the theoretical assumptions do explain the facts (in the sense of the deductive model of explanation).[2] A theoretician in such a situation is like an unsuccessful entrepreneur, who either breaks even or makes a loss, but in any case fails to make a profit.

As in the previous case, Hesse's rule (h) *Simplicity* follows from the principle of explanatory surplus, though this time simplicity must be judged in accordance with her (e) economy of theoretical premises. The derivation is exactly the same as before, and there is no need to give it in detail. A simple modification of the argument used in the case of number of parameters, suggests that, in the present case as well, the principle of explanatory surplus is a non-Bayesian principle.

Let us next consider an objection which might be made to our method of estimating the size of the explanatory surplus. It could be said that the division into separate theoretical assumptions $T_1, ..., T_s$ or separate facts $f_1, ..., f_n$ is rather arbitrary. We might, for example, consider the conjunction f_{n-1} & f_n as a single fact f'_{n-1} thereby reducing the number of facts and hence the explanatory surplus by one. It might in some cases be possible to represent f_{n-1} & f_n quite naturally as a single fact. For example,

let $\quad f_{n-1}$ = X is a sibling
$\quad\quad f_n$ = X is a male, then
$\quad\quad f'_{n-1}$ = X is a brother

There is undoubtedly a real difficulty here, but it does not, in my opinion, make the suggested method of estimating the explanatory surplus valueless. In a concrete scientific situation, where standard linguistic formulations are in use, there will generally, so I would claim, be a natural way of effecting the division into separate facts or separate theoretical assumptions. Of course this division will never be completely determinate, but then we are aiming only at a qualitative confirmation theory, and rough qualitative estimates - not at anything precise and quantitative.

There is another consideration which goes some way towards resolving the problem in hand. In practice, we often want to estimate degrees of confirmation in order to evaluate two competing theories, such as the Copernican and the Ptolemaic at some time in the period 1543-1687. Now in such a case the precise details of how we make the division into separate facts and separate theoretical assumptions does not matter too much, *provided* it is done in the same way for each of the two competing theories.

This is all I want to say about the problem in general terms. Here, as so often, the proof of the pudding is in the eating. My approach to confirmation theory in general, and the principle of explanatory surplus in particular, has been designed to enable confirmation theory to be applied to actual scientific examples - whether historical or present day. If in practice it proves possible to estimate the size of the explanatory surplus in a sensible and natural way, and if the principle of explanatory surplus leads to satisfactory results, then there is a strong case for adopting the principle. If not, not. The evaluation of the principle is thus a matter for further investigation, and I shall content myself here by showing very briefly how the principle can be applied in two different cases, one involving a successful theory, and the other an unsuccessful theory. These cases are (a) Newton's laws, and (b) Adler's theory of the inferiority complex.

4. Two examples

(a) Newton's Laws

Newton presented his theory as consisting of three laws of motion, and the law of gravity. This was a perfectly natural formulation at the time, and we may say therefore that his theory divided naturally into 4 theoretical assumptions: T_1, T_2, T_3, T_4 where T_1, T_2, T_3 are the three laws of motion and T_4 is the law of gravity.

Let us turn next to the facts which Newton sought to explain. Here again there is a natural division into Kepler's 3 laws of planetary motion, and Galileo's law of falling bodies. Applying the principle of explanatory surplus, we conclude that if Newton had explained Kepler's laws and Galileo's law and nothing else, he would not have generated an explanatory surplus, and his theory would not have been confirmed.

At first sight this result may seem surprising, but further reflection shows it to be reasonable. Newton produced a complicated theoretical system involving new concepts (force and mass), and a bold and curious assumption concerning gravitational attraction. What would have been the point in adopting such a system if it explained nothing more than the observational laws of Kepler and Galileo? In such a situation it would surely have been better to stick to the observational laws and reject the theory as a piece of superfluous metaphysics. This is exactly what the principle of explanatory surplus suggests by assigning zero support to the theory.

Here, however, we are talking merely hypothetically and not in accordance with historical reality. Newton's theory did not explain just Kepler's laws and Galileo's law but a great deal more besides. In the *Principia* Newton explained, with reasonable success, the laws of impact, the tides, the inequalities of the Moon's motions, and some planetary perturbations. He was also able to derive results concerning the figure of the Earth and comets. We clearly have here a large explanatory surplus, and Newton's theory is correspondingly strongly confirmed.

(b) Adler's theory of the inferiority complex

Popper, who worked with Adler for a while, discusses the theory of the inferiority complex briefly in *Conjectures and Refutations* (1963, Ch.1 p.35). I shall here present a variant of one of Popper's examples.

In this example we have two facts: f_1 and f_2. f_1: Mr A. is walking beside a river. He sees a child fall in. Without hesitating he jumps in, and gallantly rescues the child. f_2: At the same place, but on another occasion, Mr B. is walking beside the river. He sees a child fall in. Although he can swim as well as Mr A. he fears that he might drown if he tries to rescue the child, and so walks quietly away. (In order not to make the story too tragic, let us suppose that the child is washed ashore without drowning.)

How are the facts f_1 and f_2 explained on Adler's theory? According to that theory, everyone has an inferiority complex. Some people, however, struggle to overcome this complex by performing difficult and dangerous feats whenever possible. Mr A. falls into this class, and his behaviour is thus explained. Other people, however, are totally mastered by their inferiority complex, and will never undertake anything which appears difficult or dangerous because they feel too incapable and inferior to be able to carry out a task of that kind with success. Mr B. is such a person, and this explains why he acted as he did. Here then are two characteristic Adlerian explanations of observed human behaviour. The question before us is whether such explanations give any support to Adler's theory.

In order to give a reply, we have to analyse the number of theoretical assumptions used in these Adlerian explanations. Clearly we have the assumption T_1 that all human beings have an inferiority complex. However, in order to obtain the explanations, this general assumption must be supplemented by two particular assumptions concerning Mr A. and Mr B. respectively. These are: T_2: Mr A struggles to overcome his inferiority complex by performing difficult and dangerous feats whenever possible. T_3: Mr B is so mastered by his inferiority complex that he will avoid even attempting anything that appears difficult or dangerous. In this case, then, two facts (f_1, f_2) are explained by three theoretical assumptions (T_1, T_2, T_3). Thus no explanatory surplus is generated, and the explanations give no support to Adler's theory of the inferiority complex.

Notes

[1]I have been particularly helped in preparing this paper by discussions with Mary Hesse and Colin Howson, although my views differ from theirs on quite a number of points.

[2]In one of his early papers on logic, Frege proposes something very like this form of the principle of explanatory surplus. He says:

> The value of an explanation can be directly measured by this condensation and simplification: it is zero if the number of assumptions is as great as the number of facts to be explained. (Frege 1880/1, p.36)

References

Frege, G. (1880/1). "Boole's Logical Calculus and the Concept-script", in *Posthumous Writings*, Oxford: Basil Blackwell.

Hesse, M.B. (1974). *The Structure of Scientific Inference,* London: Macmillan.

Jeffreys, H. (1939). *Theory of Probability.* Oxford University Press.

Popper, K.R. (1934). *The Logic of Scientific Discovery English Translation*, London: Hutchinson, 1959.

_ _ _ _ _ _ _ . (1963). *Conjectures and Refutations*, London: Routledge & Kegan Paul.

Accommodation, Prediction and Bayesian Confirmation Theory[1]

Colin Howson

London School of Economics

'A hypothesis can only be received upon the ground of its having been *verified* by successful *prediction*', C.S. Peirce, *Collected Papers*, Vol. II, p.466.

1. Introduction

Hempel, in his classic [1966], raised a problem which already had a long pedigree. Balmer, having examined measurements of the wavelengths of lines in the emission spectrum of hydrogen, constructed a now well-known formula for generating these, and also other lines at that time unknown to him. Hempel asked whether Balmer's formula would have been as strongly confirmed by the new lines had they too formed part of Balmer's database. Hempel's question poses the famous and ancient accommodation-versus-prediction problem: is a hypothesis h, designed among other things to account for data e, confirmed by e as much as h would have been, had it been found to explain e without, however, having been constructed to do so? An unconditionally negative answer has been given by a numerous and influential body of scientists and philosophers; this unconditional view of the matter I shall call *predictionism*. Its most general formulation avoids counterfactuals (though it implies them) as follows: a hypothesis which explains an effect e without having been designed to do so gets more support from e than one which was so designed. The advantage of this way of putting it is that it highlights the methodological application: favour the independent explanation over the accommodating one.

Hempel cites, as an argument in support of the predictionist position, the fact that you can always fit a formula - infinitely many formulas, in fact - to any set of data. I shall discuss this point at some length in section 3, and will defer comments till then. Hempel also adduces, as a counter to that argument, that Balmer's is not any old formula: it is a strikingly *simple* formula to subsume all the data. Simplicity is the US cavalry for many people, the final saviour - of objectivity, so it is thought in this case - when all seems lost. A little reflection ought to dispel this belief. Simplicity considered as a property of *expressions,* for example, has an obvious defect as a refuge for objectivity in that it is representation-relative. The straight line formula '$y = $ constant' for plane Cartesian coordinates seems to be simpler, for example, than '$x^2+y^2 = $ constant'; but on representing the same curves in polar coordinates that order is reversed.

There are, it is true, other concepts of simplicity more robust than this one, but we need not dally longer with this elusive notion. For the invocation of simplicity cannot solve

PSA 1988, Volume 2, pp. 381-392

Hempel's problem. Suppose that a hypothesis h is proposed which predicts a highly nonsimple curve for x and y. To test h, let us suppose that a straight line is chosen more or less at random, subject to its intersecting the curve at a large number of points, and that the abscissae of the intercepts are chosen as the test values of x. The observed y values all fall on the line, as predicted by h. But h is, of course, not the simplest hypothesis to fit the data: the straight line is. Hempel would be forced, it seems, to say that the latter is the better confirmed hypothesis, not h. But intuitively, it is h which is best confirmed. But if it is not because h genuinely predicted the data, while the straight line did not, what other reason is there for preferring h? The reason cannot, of course, now be that it is the simpler hypothesis.

My brief answer to this question, which I shall elaborate in the remainder of this paper, is that theories are not usually proposed and tested without some prior reason to think that they may be true; and this will presumably be the case with h. The only evidence for the straight line, on the other hand, is, by assumption, the data: nothing in existing theory, as far as anybody can now see, tells us that the data should be collinear (of, course, it is possible that somebody may one day find something). Heuristic reasoning to theories, or to more or less powerful constraints on theories, is regarded as very important (as Nickles in his paper in this volume forcefully points out); a great deal of the material presented in physics texts, for example, consists of heuristic *derivations* of physical principles from plausible general assumptions. The difference in support between h and the ad hoc straight line in the example above is, I claim, due simply to a lack of any prior support for the straight line hypothesis: it just is ad hoc. Such a conclusion was reached earlier this century by Keynes:

> If an hypothesis is proposed a priori, this commonly means that there is some ground for it, arising out of our previous knowledge, apart from the purely inductive ground, and if such is the case the hypothesis is clearly stronger than one which reposes on inductive grounds only. ... It is the union of prior knowledge, with the inductive grounds which arise out of the immediate instances, that lends weight to an hypothesis, and not the occasion on which the hypothesis is first proposed. (Keynes, [1921], p.336)

There is one qualification that needs to be made to this statement, however, and that is to restrict its scope to scientific hypotheses. In section 2 we shall see, using a clever example of Maher's, that there certainly are situations in which the predictionist thesis is valid. But they are, I shall claim, the result essentially of witholding information, and to that extent unrepresentative of science.

The appeal to prior support, anathema to some, seems indispensable. Take the 'law' of free fall. What distinguishes this on the available evidence from Jeffreys's bogus law

$$s = (1/2)gt^2 + (t-t_1)...(t-t_n)f(t)$$

where $t_1,...,t_n$ are the times at which s and t have been co-observed, and f is some function of t nonvanishing outside $t_1,...,t_n$ (Jeffreys [1961] p. 3)? You may say, the free fall law follows as an approximation from higher-level laws in which we have reason to believe, and we have no corresponding reason to believe in the bogus law; in fact, given that the values of $t_1,...,t_n$ could have been anything we wished, we have very strong reason to disbelieve it. Well and good. The point is conceded that there is nothing in the empirical evidence itself to discriminate the real law from the bogus one: the asymmetry is prior to this, which is all that was to be shown.

How do you measure prior support? Support is not an isolated concept, and assessments of support are usually assessments of something like credibility. Now the only satisfactory theory of the credibility of hypotheses on data is, I believe, a probabilistic one, and the only satisfactory probabilistic theory seem to be the subjective Bayesian one.

Popper, of course, has denied that support is credibility, but Popper's own attempt to divorce the two notions is now widely accepted to have failed. Indeed, the fact that one can easily generate an infinity of grue-like alternatives (like the bogus free fall law) is widely, and I think correctly, regarded as fatal to his approach. There have, it is true, been many objections raised against the subjective Bayesian approach. They all seem to me misguided, but I shall say try to answer, very briefly, some of the standard arguments at the end of the paper. This is enough by way of preamble. Let us see how Bayesian confirmation theory resolves, as I believe it does, the prediction/accommodation issue.

2. Bayesian confirmation theory and the prediction/accommodation issue.

It is often suggested that Bayesian confirmation theory is *inconsistent* with the idea that observations support a hypothesis more if the latter independently predicts the observations than if it merely accommodates them. That this is not true was pointed out by Simon ([1955]), and more recently by Maher ([1988]): one and the same hypothesis may consistently be assigned different probabilities on the same data precisely because the statement that one hypothesis independently predicts the data and the other doesn't can itself be regarded as reporting relevantly different pieces of background information. But rendering Bayesian confirmation theory consistent in this way with the predictionist thesis merely accommodates that thesis, in a particularly *ad hoc* way. *Why* should the information about the differing geneses of h affect probabilities in this way? - that question simply poses again the original prediction/accommodation problem.

There are situations, however, where such information does seem to naturally induce such differences in the posterior probabilities. One such, recently noted by Maher ([1988]), is where a successful prediction of a sufficiently *a priori* improbable event serves to confirm the hypothesis that the predictor possesses a reliable prediction *procedure* (a general theory, maybe, of the data source), and this support is then passed on to the predictor's later predictions. In Maher's example a subject predicts 100 outcomes of a tossed coin. The coin is tossed 99 times and the outcomes are as predicted (assume they're pretty randomly assorted). Intuitively, the odds are now very high on the prediction of the 100th toss - much higher than they would have been on that same prediction had it been made only after the results of the 99 tosses had become known. For in the former case we believe that the subject had advance information about the outcomes of the tosses, and this belief is correspondingly confirmed by the success of the predictions - the supposition that the agreement with the facts arose by chance is so improbable that it can be safely neglected.

This informal argument can be translated into standard Bayesian terms. Let e be the prediction of the 99 tosses, and H the hypothesis that the subject has access to reliable advance information about the outcomes of the tosses. Assume that H has a nonzero prior probability. Even for very small values of this probability, H will have its probability raised to somewhere in the neighborhood of one by the successful prediction e. This is because we will identify $P(e/-H)$ with the probability of e on the assumption that the agreement was due to chance (the structure of this problem is, therefore, like a test of the null hypothesis of chance, and is very similar to Fisher's famous problem of the tea-tasting lady in Fisher [1935], Ch. 10). This can be easily calculated and is very small, and so by Bayes's Theorem $P(H/e)$ is close to 1. Let h be the conjunction of e and the subject's prediction of the outcome of the hundredth toss - head, say. H, given the background information which is that the subject has predicted e, and predicts that the next toss will be a head, implies h implies e. It is now easy to see that

$$P(h/e) = P(H/e) + P(h/-H\&e)P(-H/e) \tag{1}.$$

Since $P(H/e)$ is close to 1, so is $P(h/e)$. Now consider the accommodating case where the information relative to which P is computed is that the subject knows the outcome of the

99 tosses, and predicts that the next toss will be a head. H is as before, but since the background information now does not itself specify the outcome of the 99 tosses, H does not entail h or e, although H&e entails h. The probability of e conditional on H and on -H is the same, we can suppose, as its probability conditional on the hypothesis of chance. This being so, it is straightforward to show by Bayes's Theorem that P(H/e) = P(H). Hence, where -H is the null, chance, hypothesis,

$$P(h/e) = P(H) + P(h/-H\&e)P(-H) \tag{2},$$

which is approximately equal to P(h/chance&e) if P(H) is small.

We should note, however, that the asymmetry even in this example in favour of the independent prediction is not unconditional. It presupposes that the prior probability of H is nonzero: if it is zero there is nothing to choose between the accommodating prediction and the nonaccommodating one. Could the prior probability ever plausibly be zero? Yes. Examples are easy to construct. The following will do. Consider 2^{100} computer programs, each of which generates a distinct sequence of exactly 100 0s and 1s. These programs are set to work, and when they have stopped a coin is tossed 99 times, each head being recorded as a 1, and each tail as a 0. Consider the unique computer output agreeing with the coin data, and whose hundredth digit is a 1, say. The program generating it has, of course, correctly 'predicted' the 99 outcomes, but no-one would place any higher odds on its being correct about the hundredth than on the same prediction given the assumption that its past success was due entirely to chance. By inference, the prior probability of H is zero. Keynes again makes the salient point: 'If [the hypothesis] is a mere guess, the lucky fact of its preceding some or all of the cases which verify it adds nothing whatever to its value' ([1921] p.338).

Maher, who sets up and gives a more elaborate analysis of the coin-tossing problem in [1988} and [1989], thinks it illustrates more than a special and scientifically rather uninteresting possibility. Indeed, he seems to think that the reasoning involved in its analysis can be generalized to cover many cases of scientific prediction. His argument is that the successful prediction of a sufficiently improbable event e points to there being a reliable method by which that prediction was generated. This, we have just seen, is reasonable if the prior probability of there being such a method is nonzero. Let us grant that it is nonzero. Maher then concludes that a further prediction from the same source would be less credible had it not been preceded by the first successful prediction.

I shall now argue that this is a non sequitur, and that Maher's conclusion ignores a feature of the coin-tossing case which is peculiar to it, and is (usually) not present in science. There predictions are generated from publicly available theories. If H is such a theory then if H generates a prediction h and evidence e verifies some of h's content and e would be improbable were H not true, then H and h are correspondingly confirmed: the argument is therefore the same as that which generated equation (1) above. But there is no analogue of (2); for even if H were proposed after e became known, the entailment relations between H and h and e would still persist, and if we are entitled to assume that it is still possible that e would be an improbable event were H not true, then H is confirmed to exactly the same extent by e. It is precisely because such entailments did not hold in the situation characterized by equation (2) that the accommodating hypothesis got no support. So (1) holds for both cases, accommodation and prediction, and there is no difference in the support of h by e in either case.

We can illustrate the point nicely with the episode in the history of science which Maher (ibid.) supposes to exemplify his own thesis. This episode is Mendeleev's prediction, and the contemporary response to it, of three new elements on the basis of his theory of the Periodic Table (a rather similar episode occurred more recently, with the prediction by Gell-Mann and Okubo, of the baryon Ω^- to fill an empty place in the SU(3) classification of hadrons, and its subsequent discovery in 1964). Maher points out that as a consequence of the discovery of only the first two of those elements the Royal Society awarded the Davy Medal to Mendeleev: the implication Maher draws, which is correct as far as I can see, is that the existence of the third was by then regarded as a foregone conclusion

The diagnosis of this confidence in Mendeleev is, of course, familiar from the discussion of the coins, where the successful prediction of an otherwise improbable event makes it likely that the predictor has a reliable method up his sleeve. But the significant difference from the coin case, *is the fact that Mendeleev's method was freely divulged by him*, and was just the theory of the Periodic Table. It is *that* hypothesis which greatly increased its credibility as a result of the successful predictions and transfers its credibility to the prediction not yet confirmed. In other words, the prediction of germanium was strongly supported by the discovery of scandium and gallium because the theory of the Periodic Table itself was strongly supported by them. And the theory of the Periodic Table was supported strongly by them because, improbable though *it* was, there was no other equally plausible theory which predicted the existence and properties of those two elements.

If this is correct, and I think it is, then it follows that, in the absence of any alternative account, this support should have been equally forthcoming had Mendeleev discovered the Periodic Table after and not before the discovery of scandium and gallium. The evidential force of the discovery of these two elements resides in the fact that Mendeleev's was the only satisfactory, public theory to account for them. The fact, in other words, that they were predicted in advance has no significance in itself.

If you're inclined to doubt this, try the following thought experiment. Suppose that Mendeleev had only *privately* predicted the existence and many of the properties of the three elements scandium, gallium and germanium (as they are now called), and made public his theory of the Periodic Table and his prediction of the existence of germanium after the discovery of the first two of these. Suppose also that he waited for the scientific community to consider its attitude to the theory, and only then revealed that he had formulated the theory *before* the discovery of scandium and gallium. Would this last, purely autobiographical revelation have caused people to reassess profoundly their confidence in his theory? Surely not. (I should say, however, that Maher ([1988], [1989]) offers rather complex formal arguments for his contrary conclusion. These rest on a considerable number of assumptions, some of which seem to me just as questionable as his conclusion. For a fuller discussion see Howson and Franklin [1989].)

3. An (invalid) Argument for Predictionism

The Bayesian argument of the last section for the confirmational irrelevance of whether the theory was proposed in the light of or independently of the evidence presupposed, as I pointed out, that even when H is constructed to explain e, e may still be regarded as being very improbable if H were false. I did not defend this assumption at the time, but we are now going to consider an argument one of whose formulations explicitly contradicts it.

The argument is an old one, and at first sight very compelling. Peirce ([1932], p.462) employs it, though it has also been noted by Poincaré, Jeffreys and Goodman, and has recently been restated, with minor variations, by Giere ([1984], pp. 118 and 161), and Redhead ([1986]). It is the argument cited by Hempel whose discussion we postponed until now. It is, in full, as follows. Suppose h was constructed to explain e. Given any finite data like e, it is possible to construct not just one but infinitely many mutually exclusive hypotheses entailing those data. However, *since there is clearly no chance of e refuting any of these 'tailored' hypotheses, the truth of e can provide no evidence in their support, and a fortiori no support for h.*

We should note that a consequence of the argument is that no data, however extensive, can support at all *any* hypothesis constructed to explain it. This is strongly counterintuitive, and should at the very least kindle a suspicion that the argument is unsound. Indeed, it is not difficult to see where it breaks down. It does so because e and h are definite statements, and e either refutes h or it does not. If it does, it does so whether h was constructed to yield e or not, and chance is present in neither case. The only thing which has the chance or not of refuting h is the experimental set up E, and the chance of E's refuting h is the chance of E's yielding the outcome -e; again, this is independent of whether h was constructed to yield e or not. The argument based on chance therefore founders.

But there are probabilities other than objective chances. One way, therefore, to salvage the argument might be to eschew chances in favour of more obviously *epistemic* (i.e. degree of belief) probabilities. Some of its defenders, like Redhead, do formulate it explicitly in such terms. And there is an advantage in employing the language of epistemic probability, which is that it permits a precise characterization of support: e supports h if it raises h's probability. Anyway, the thesis that if h is constructed to explain e then there is no chance of e refuting h (if h is false) becomes: the epistemic probability of -e (i.e. of E's refuting h) on the assumption that h is false is zero; i.e. $P(-e/-h) = 0$, in an obvious notation. It follows immediately that $P(e/-h) = 1$, whence by Bayes's Theorem $P(h/e) = P(h)$. Hence e does not support h. Q.E.D.

But does $P(e/-h) = 1$ really follow from the fact that h was constructed to explain e? Redhead has recently come up with a novel argument that it does. Implicit in trying to construct an explanation e is, he claims, a decision that of the possible alternatives to any such explanation, only those which also explain e will be given positive prior probabilities (this is Redhead's so-called filter condition). It certainly follows from this filter condition that $P(e/-h)$ equals 1, where h is one such explanation, because expanding out $P(e/-h)$ into a sum over the probabilities of e relative to a partition of alternatives to h, only those h_i for which $P(e/h_i) = 1$ will have nonzero probabilities. But the filter condition does *not* follow from my constructing an explanation of e. That I have constructed h to explain e does not imply that if I consider some other hypothesis h' consistent with e to be an alternative picture of reality then h' must also explain e: h' may simply be too weak.

So we are still left with the question: does the fact that h was constructed to explain e imply that $P(e/-h) = 1$? It has been argued (e.g. by Glymour ([1980] Ch.3) that since e is, by assumption, known at the time h is formulated, its probability must be 1, so that $P(e/-h) = 1$ also, and our question is answered. But is it? If Glymour is right then $P(e/-h)$ would be 1 even if h had not been constructed to explain e, if e is a known fact, and so e would not support h in this case either. But Glymour is not right. The Bayesian (who is the only dealer in epistemic probability) does not set $P(e/-h) = 1$ just because he knows e - or at any rate not when computing the support of h by e. For were he to do so, no hypothesis would ever be regarded as being supported by any evidence at all once it became known. This would be absurd. Instead, the Bayesian interprets $P(e/-h)$ as how likely you think e *would be* were h to be false (Howson [1985] p. 307). Granted this, it not only makes perfect sense to attribute values other than unity to $P(e/-h)$ when e is known, but also when h is constructed to explain e. Indeed, on this construction the value of $P(e/-h)$ is *independent* of whether h was or was not constructed to explain e.

The fact that there is no possibility of evidence refuting a hypothesis does not then, if I am correct, imply anything at all about the level of support h may or may not attain on e. It certainly does not imply that it is zero. Scientists may not be the fount of all wisdom about the nature of scientific evidence, but their beliefs deserve some attention. And they certainly do not believe the zero support thesis. Kepler's laws, themselves considered as approximations warranted by the facts, were believed to support Newton's gravitational theory (Franklin and Howson [1985]). Also, consider the widespread use in statistics of maximum-likelihood estimators, i.e. methods of generating hypotheses which *guarantee* the least conflict with the data. Everybody regards these as reliable, at any rate within certain intervals; indeed, such intervals figure systematically in the Bayesian theory as those of asymptotic maximum posterior probability under very general conditions (Jeffreys [1961] p.193).

4. The Principle of Explanatory Surplus

The argument we have just considered purported to show that no hypothesis is supported by data it was constructed to explain. That argument was unsound, and its conclusion is, there is every reason to believe, false. There are, however, defenders of the predictionist position who are happy to acknowledge both these facts. Their position is more subtle. It is this: a hypothesis may indeed be supported by the data it is constructed to explain. But it is supported not by the whole of the data, but only that part which was not employed in its construction. Such a theory of support has been advanced by Worrall ([1978] p.48) and Zahar ([1983]). More precisely, Zahar and Worrall claim that if a part e' of the total data e is used in the construction of a hypothesis h, then e' is not to be counted as supporting h. If h entails all of e (given initial conditions etc.), then it is only the residue e-e' which supports h. It follows from this principle that if another hypothesis H explains e, and was constructed independently of e, then all of e supports H, and H receives more support from e than does h. This account entails predictionism, therefore. Is it true?

It is difficult to answer this question without being told what exactly is meant by saying that evidence is *used* in the construction of a theory, and how to identify which part of the evidence is 'used'. However, there are examples of theories where it is possible fairly unambiguously to determine which part of e, or better how much of e, was used in their construction. These are theories in which a part or the whole of e uniquely determines the value of an adjustable parameter. In this type of situation Zahar's and Worrall's proposal seems to be very similar if not identical to that of Gillies (this volume, pp.), which is to to subtract from the total of confirming data that subset of it required to determine the parameter. If, as in the case of a functional hypothesis h fitting all the data points, any subset of k points would determine the parameters of h, then on this account k units would presumably have to be subtracted from the total support for the determinate form of h by the data. Gillies calls this the Principle of Explanatory Surplus, and I shall adopt that very nice terminology.

An example supposed by Worrall (op. cit.) to bear out this principle is the following. The observed value for Mercury's annual perihelion shift can be explained by classical gravitational theory if the existence of a suitably positioned dust cloud is postulated, whose mass distribution is then determined by the magnitude of the shift. That magnitude was, however, explained straight off by General Relativity, and methodologists inform us that this fact told strongly in favour of the latter; the fact that it could also be explained by classical theory by introducing a suitable new parameter was not in itself regarded as supplying evidence in favour of the resulting theory (though independent evidence for the dust cloud certainly would have been). According to Worrall, the observed value of the perihelion shift provided no support for that theory because it served only to evaluate the new parameter.

I shall argue that this example does not confirm the principle of explanatory surplus, because there is convincing reason to suppose that principle false. I shall construct a simple counterexample and analyze it, and then come back and give an alternative explanation of the Mercury example. Here is the counterexample (Franklin and Howson [1985] p.380). A box contains, we are informed, red and black tickets in a proportion p of red to black. The value of p is, however, left unspecified. Let h(p) represent the information we are given. We now inspect the contents of the box, and discover that there are r red tickets and s black ones. Let e signify this empirical information. We conclude that p = r/s, and hence that h(r/s) is true. The e has obviously been used in evaluating the parameter p, and it seems reasonable, to put it mildly, to suppose that e supports $h(r/s)$; after all, e entails $h(r/s)$ relative to some uncontentious background information, and it is difficult to see how evidence which virtually *entails* a hypothesis fails to support it. It might be objected that $h(r/s)$ is hardly a scientific hypothesis, since it is entailed by e, but this is not a good objection. $h(r/s)$ is not entailed strictly by e; the entailment was relative to assumptions which we are disposed to think obviously true, but which may be false nonetheless. In other words, our prior probability for the hypothesis that $h(p)$ is true for some p is very close to 1. This suggests that the support for hypotheses $h(t_0)$ for some adjustable parameter t by data which just suffice to evaluate t as t_0 is large when the prior probability of $h(t)$ is large.

The Bayesian theory provides an easy demonstration of this. Let us see how it works in the context of a slightly less fanciful example. Suppose the hypothesis, $h(t)$, to be that the relationship between x and y is linear, with unspecified parameters: i.e. h says that there are a,b such that for all x, $y(x) = ax+b$, and t is the pair of parameters (a,b). Strictly, t is bound in $h(t)$ by an existential quantifier, but the notation would be unwieldy if we wrote it in explicitly. Idealizing, let us suppose that $e_1,...,e_n$ are n collinear joint readings of x and y, where $n>1$. We assume that our background information will include the x values of these points, and all entailments will henceforward be regarded as relative to this information. Then $h(t)\&e_1\&e_2$ is equivalent to $h(t)\&e_1\&..\&e_m$, for any m where $1<m\leq n$. For $m>2$ it is easily seen that

$$P(h(t)/e_1\&...\&e_m) = P(h(t)/e_1\&...\&e_{m-1})/P(e_m/e_1\&...\&e_{m-1}) \qquad (3).$$

As before, let t_0 be the value of t, i.e. of the pair (a,b), determined by the data. It follows that $h(t_0)$ is equivalent, given background information as above, to $h(t)\&e_1\&...\&e_m$ for $m>1$; so

$$P(h(t_0)/e_1\&...\&e_m) = P(h(t_0)/e_1\&...\&e_{m-1})/P(e_m/e_1\&...\&e_{m-1}) \qquad (4),$$

and where m is small, but greater than 1, it is plausible to assume that the denominator on the right hand side of (3) and (4) is very small, giving a large value to the ratio of supports of both $h(t)$ and $h(t_0)$ by $e_1\&e_2\&e_3$ and $e_1\&e_2$ respectively (Gillies claims in his paper in this volume that the Bayesian theory is incapable of assigning differing confirmatory power to the e_i; as we see, this claim is false). We also have

$$P(h(t_0)/e_1\&e_2) = P(h(t_0))/P(e_1\&e_2)$$

and $P(h(t_0)) = P(h(t)\&e_1\&e_2)$. But it is also plausible to assume that $h(t)$ and $e_1\&e_2$ are independent, on the ground that, in the absence of any other information, e_1 and e_2 convey no information about the truth of $h(t)$. But then

$$P(h(t_0)/e_1\&e_2) = P(h(t)),$$

and hence the posterior probability of $h(t_0)$ on the minimal parameter-determining data is equal to the prior probability of $h(t)$. $h(t)$ is implied by $h(t_0)$ and $P(h(t_0))$ will generally be very much smaller than $P(h(t))$, and if we define support as the incremental probability supplied by the data, it follows that the support of $h(t_0)$ by $e_1\&e_2$ is comparable to the prior probability of $h(t)$, and therefore large when that is, which was to be proved.

It might be objected that $e_1\&e_2$ ought not to support $h(t_0)$ since those data do not in general increase the probability of the underlying hypothesis $h(t)$; they only confirm the value of the parameter *given* that $h(t)$ is true for some t (Worrall (ibid.) makes a remark to this effect). But this objection seems to be founded simply on a conflation of $h(t)$ and $h(t_0)$; the data may not increase the probability of the former, but they do the latter.

The principle of explanatory surplus thus seems to be false. But there is something in it. To see what that is, go back to equation (3). As we see there, there is, for $m>2$ but not too much so, and especially for m=3, large incremental support for the pure linear hypothesis h(t) from m collinear data points, which is very intuitively appealing. But we also saw that

$$P(h(t)/e_1 \& e_2) = P(h(t)/e_1) = P(h(t)),$$

i.e. there is no incremental support for that hypothesis from one or two points. This seems to me to be the valid core of the principle of explanatory surplus: the pure linear hypothesis does not explain one or two data points, and gets no increase in probability from them. But it gets a big increase in probability from each further point observed to be on the same line as the first two (with suitably diminishing returns for large m). This is because it explains *the linearity itself* of those points, and it picks up confirmation for so doing. This is the *valid* principle of explanatory surplus, and it is just what the Bayesian formulas say.

Let us now return to the Mercury example. Formally, suppose that h_1 and h_2 are rival hypotheses, that h_1 independently predicts e, that h_2 contains an adjustable parameter which is evaluated from e, that when this value is inserted into h_2 the resulting hypothesis, h_2', entails e, and finally that the prior probability of h_1 is at least as great as that of h_2 (recall that the success of *Special* Relativity had cast severe doubt on classical theory, and that there is no independent evidence for the parameter). Now

$$P(h_2') = P(h_2'/h_2)P(h_2) + P(h_2'/-h_2)P(-h_2),$$

and the right hand summand is zero, giving $P(h_2') = P(h_2'/h_2)P(h_2)$. But we assumed that $P(h_1) \geq P(h_2)$, and if the interval within which e determines the value of the parameter is small in proportion to its total range, as it standardly is, then we may also have $P(h_2'/h_2) \ll 1$. At any rate $P(h_2') < P(h_2'/h_2)P(h_1) < P(h_1)$. But $P(h_2'/e) = P(h_2')/P(e) < P(h_1)/P(e) = P(h_1/e)$. In other words, the Bayesian theory can explain the inequality in support as a consequence of an initial distribution of probabilities, consistently with denying the principle of explanatory surplus.

Clearly, the inequality above can be achieved without requiring that the prior probability of h_1 be no smaller than h_2; I chose to impose that condition because it seemed historically justified, and is certainly sufficient for the result. Note also that the support of h_2' is not by any means necessarily zero, merely less than that of h_1. In the historical case, there was a relatively small prior probability of h_2, because of the success of Special Relativity, and because the *introduction* of the parameter was completely *ad hoc*. Nickles, a nonBayesian, has come to much the same conclusion. He points out ([1985] p. 200) that in those cases where parameter-fixing is regarded as *ad hoc*, this is not because the data were used to evaluate the parameters, but because there is no justification relative to what we feel we know about the physical set-up for introducing the parameters at all.

5. Conclusion

I have argued that the unqualified predictionist thesis is, except in scientifically unimportant cases, wrong. Nevertheless, it retains great intuitive appeal. It has, in particular, numbered among its advocates Bacon, Leibniz, Descartes, and later C.S. Peirce, Whewell, and numerous practising scientists. What is it about the doctrine which enlists such people in its support? I shall hazard a guess as to the answer. There seem to me to be two reasons. One is that all these people, as far as I am aware, took the predictionist doctrine to be a doctrine about the relative merits of theories predicting facts previously unknown and theories constructed to accommodate known facts. Now a fact e, *previously unknown* and predicted by h, is unlikely (though it is not impossible) to be explained by available theories other than h. Hence the criterion of a very small value of $P(e/-h)$ is likely to be satisfied in this case. And, as we have seen, this is a quite valid criterion of high support by e. That same criterion is, of course, also going to be satisfied when e is known but has no satisfactory explanation in current theory other than h. If I am correct, then, predictionists have simply misidentified what is often (but by no means always) a sufficient condition for a small value of P(e/-h) with that criterion itself.

As we have seen, that criterion is explicitly incorporated into the Bayesian theory. But the Bayesian theory invokes prior probabilities, and here I come to the second reason to explain the popularity of predictionism. This is that it offered an 'objectivistic' alternative to the invocation of apparently subjective prior probabilities in solving the problem of grue-type alternatives to some given hypothesis h, advanced after test-data have been collected which appear to support h. But it has been the burden of the foregoing pages that there really is no alternative to prior probabilities, and to the only satisfactory theory which sustains them, Personalist Bayesianism. To close, I shall say a little more about this Bayesian theory, and attempt to rebut the principal objection to it that in invoking subjective probabilities it constitutes a sell-out to subjectivism and 'psychologism'.

The main principles of this theory, as far as it relates to confirmation theory, are (i) that assessments of support of hypotheses h by data e are gauged by individuals in terms of the difference between their conditional degrees of belief P(h/e) in h relative to e and the unconditional degrees of belief P(h); and (ii) these degrees of belief, if collectively consistent, are formally probabilities. One's degree of belief in h is defined to be what, in the light of one's contemporary information, one feels the fair betting quotient would be on h, were anyone to bet on h and were the truth or falsity of h actually to be decidable after finite time (this is, of course, often a counterfactual condition). Note that one is *not* indicating by revealing one's degree of belief that one is actually willing to bet at these odds oneself. The famous Dutch Book Argument then proves that a necessary condition for consistency in one's degrees of belief is their satisfying the probability calculus. Let us now look at the standard objections to subjective Bayesianism, in particular at the charge of its being a capitulation to total subjectivism. I shall list the objections, and my answers. Lack of space, I am afraid, entails brevity in these; the reader who would like amplification, as well as the references to the authors of the objections, may consult Chapter 11 of the book by Howson and Urbach cited above.

(a) objectivity is abandoned. This is simply untrue in fact, and it is one of the scandals of contemporary philosophy that the charge has been allowed to persist. The consistency requirement expressed in obedience to the probability calculus imposes a quite objective constraint, and one so powerful that nobody could ever have consistent degrees of belief over sets of logically related hypotheses of more than fairly minimal complexity (further to this, see (c) below.) And the probability calculus regulates the passage from prior to posterior degrees of belief; in this, it is no less objective than the rules of first order logic itself.

(b) nobody has real-valued degrees of belief. True, and mountains do not have real-valued heights, oceans real-valued depths, or rooms real-valued lengths. This does not prevent the useful application of real arithmetic in computations involving these quantities: the results of such computations are regarded as valid subject to the uncertainties in the values of the input parameters. Similarly here; the Bayesian regards as of general methodological significance only those results which are invariant over the fairly considerable uncertainties in the values of the priors.

(c) Nobody has consistent degrees of belief over sufficiently complex sets of hypotheses. In particular, often the purely logical relationships between the hypotheses is not known completely. True, but beside the point. If we were perfect deductive logicians we should not have needed to evolve the discipline of deductive logic to guide us, to the extent that we can follow it. Similarly with probability theory.

(d) the support of h by e is always 0 if e is known. For then P(e) = 1 and so P(h/e) = P(h). But I have already argued, in section 3, that this is false.

Notes

[1]I should like to thank Peter Urbach and Don Zilversmit for reading this paper and for their very helpful suggestions for improvement.

References

Fisher, R.A. (1935). *The Design of Experiments*, 4th edition, Edinburgh: Oliver and Boyd.

Franklin, A. and Howson, C. (1985). "Newton and Kepler; A Bayesian Approach", *Studies in History and Philosophy of Science*, vol. 16: 379-385.

Gillies, D.A. (1989). "NonBayesian Confirmation Theory and the Principle of Explanatory Surplus", *PSA 1988*, vol. 2: pp. 373-380. East Lansing, MI: Philosophy of Science Association.

Glymour, C. (1980). *Theory and Evidence*, Princeton: Princeton University Press.

Hempel, C.G. (1966). *Philosophy of Natural Science*, Englewood Cliffs: Prentice-Hall.

Howson, C. (1985). "Bayesianism and Support by Novel Facts", *British Journal for the Philosophy of Science*, vol. 35: 245-251.

_ _ _ _ _ _ _ . and Franklin, A. (1989). "Maher, Mendeleev and Bayesianism", forthcoming.

_ _ _ _ _ _ _. and Urbach, P.M. (1989). *Scientific Reasoning: the Bayesian Approach*, La Salle: Open Court.

Jeffreys, H. (1961). *Theory of Probability*, Cambridge: Cambridge University Press (third edition).

Keynes, J.M. (1921). *Treatise on Probability*, London: Macmillan (1973 edition).

Maher, P. (1988). "Prediction, Accommodation and the Logic of Discovery", *PSA 1988*, vol.1.

_ _ _ _ _ . (1989). "How Prediction enhances Confirmation", *Festschrift for Nuel Belnap*, forthcoming.

Nickles, T. (1985). "Beyond Divorce: Current Status of the Discovery Debate", *Philosophy of Science*, vol. 52: 117-207.

_ _ _ _ _ _. (1989). "Truth or Consequences? Generative Versus Consequential Justification in Science." *PSA 1988* vol. 2: pp. 393-405. East Lansing, MI: Philosophy of Science Association.

Peirce, C.S. (1932). *Collected Papers*, volume II, Cambridge: Harvard University Press.

Redhead, M.L.G. (1986). "Novelty and Confirmation", *British Journal for the Philosophy of Science*, vol.37: 115-118.

Simon, H.A. (1955). "Prediction and hindsight as confirmatory evidence", *Philosophy of Science*, vol.22: 227-230.

Worrall, J. (1978). "The Ways in which the Methodology of Scientific Research Programmes improve on Popper's Methodology", in Radnitzky, G. and Anderson, G. (eds.) *Progress and Rationality in Science*, Dordrecht: Reidel.

Zahar, E.G. (1983). "Logic of Discovery or Psychology of Invention", *British Journal for the Philosophy of Science*, 34: 243-261.

Truth or Consequences?
Generative Versus Consequential Justification in Science[1]

Thomas Nickles

University of Nevada-Reno

1. Introduction

For the past century the Central Dogma of confirmation theory has been:

All empirical support (of a theoretical claim) = empirical evidence = empirical data = successful test results = successful predictions or postdictions = true empirical consequences (of the claim plus auxiliary assumptions).

According to the Dogma, *all* empirical support derives from empirical testing of predicted consequences. I shall attack this *pure consequentialism* and defend the importance of *generative* justification in science.

One question which has been widely debated through the years is whether "postdicting" or explaining already known phenomena provides as much epistemic support for an hypothesis as predicting new phenomena or, indeed, whether explanation provides any support at all. I believe that explanation is on a par with prediction, but I want to address a different question here. Both explanation and prediction (as usually construed) involve consequential reasoning. I ask whether purely consequential success of either kind can confer a determinable amount of positive support upon an hypothesis. In particular, I challenge the idea that novel prediction alone can yield a more positive conclusion about the truth status of an hypothesis than Popperian corroboration does. The complementary question is whether generative reasoning can provide epistemic support for scientific claims and, if so, whether it can do better than purely consequential reasoning in helping us select good theories. I claim that it can. This means that I must agree with weak versions of "predictivism" which say that there can be a difference between prediction and accommodation—but not for the usual reasons. Against the critics of accommodation, I insist that accommodation often does provide support and that this support can actually be stronger than prediction of the same information. In some cases accommodated information is logically sufficient for a theory, given the background knowledge, whereas predictions are at best logically necessary.

Initially at least, the generative justification idea sounds crazy to most empiricists. Their objections can be answered, I believe, but I use my limited space on this occasion to challenge pure consequentialism, which stands on shakier ground than most philosophers suppose. The failure of pure consequentialism is the "negative" route to my conclusion that an adequate confirmation theory must include a dose of generative justification.

PSA 1988, Volume 2, pp. 393-405

To begin with the familiar: consequential reasoning, whether predictive or explanatory, reasons *from* a conjectured hypothesis *to* consequences which are known or fairly readily knowable. Empirical support derives from successful predictions and perhaps postdictions as well, including explanations. A *pure* consequentialist is one who holds that consequential justification is justification enough, that generative justification is impossible or at least unnecessary to scientific work and hence unimportant.

By contrast, generative reasoning flows *from* what we already "know" (so-called background knowledge or positive science) *to* some other claim or problem solution. The methodological strategy behind generative justification is opposite that of consequential testing. The generative strategy is to provide empirical support by direct construction of the claim from what we already, fallibly know, while the consequentialist strategy (I claim) can only be indirect and eliminative.

The premises of generative arguments often include theoretical principles and results as well as phenomenal claims established by experiment. While generative reasoning includes traditional inductivist models of scientific reasoning, it is not exhausted by them. An attack on inductivist methodology is therefore not necessarily an attack on generative methodology whole hog.

The basic idea of traditional, generative justification, as found in Bacon and Newton and Herschel, for example, was that *the strongest form of justification is an idealized discovery argument.* An early exemplar is Newton's first paper, on light and color. By 'idealized' I mean that the final "discovery" argument, if achieved at all, is normally quite different from the considerations which initially suggested the hypothesis to someone. Over a period of days or decades, interested scientists reformulate and tighten up the problem, locate additional constraints on its solution, and refine or replace the arguments leading to this or that feature of the solution, which itself has typically evolved well beyond the initial conjecture. The final argument amounts to a *potential* discovery argument in the sense that it constructs the theory (largely) from what is, by now, already known. Hence the counterfactual: had scientists known then what they know now, they could have discovered the original idea in just this way. Accordingly, I distinguish original *discovery* arguments from *discoverability* arguments. The ugly term 'discoverability' is what I mean by 'idealized discovery' or 'potential discovery' (Nickles 1984, 1985). Generative justification consists in offering one or more discoverability arguments.

2. The Methodological Predicament

If generative justification, as a complement to consequential testing, is a partial solution, what is the problem? The central methodological problem is the global underdetermination problem, which I call *the methodological predicament.* Supposing that there are an infinitely or indeterminately large number of possible hypotheses for a given domain of study at most one (or a small finite number) of which are true, then what methodological strategies make it likely that a correct hypothesis will be *formulated* and eventually *selected* from among its competitors, as the most warranted among them?

Notice that there are two parts to the question. Whatever its character, the processes by which hypotheses were actually discovered (or could or would have been discovered) filter from the infinite domain of possible hypotheses the small subset of ideas actually considered. Given a suitably ample discovery stage, the "final justification" stage may be restricted to more local methods of selecting the best among the hypotheses actually on the table. My topic here is "final" justification rather than discovery.

The methodological predicament makes finding the One True Theory the goal for any domain of science. This seems unrealistic for some domains of inquiry. This sort of epistemic optimality is not the only worthy scientific goal, and I am sympathetic with more pragmatic conceptions of inquiry. However, for simplicity I shall stick to the classical methodological predicament. Some historical perspective will help to highlight key issues.

3. The Great Logical Inversion

Since it achieves a strong sort of "closure" in research, full, generative justification is often unavailable, but it remains an ideal toward which many scientists strive. I therefore reject the common view, defended by Larry Laudan (1980), that consequentialism, in the form of the method of hypothesis, defeated generativism in a series of 19th-century debates involving the likes of Herschel, Whewell, Mill, DeMorgan, Boole, and Jevons. Laudan is right that philosophers of science largely abandoned generative methodology from this time on, but scientific practice is another story.

For classical generative methodology, *evidence* in the primary sense consisted of established phenomenal claims from which a theory can be derived somehow, against the background of established principles. Hypotheses and hypothesis testing functioned mainly as heuristic devices for establishing a solid evidential basis for genuine theories. This, I think, was the point of Newton's *hypotheses non fingo* (Worrall 1983).

Later, 19th-century methodologists, culminating in Jevons, turned this conception of evidence and justification upside down in favor of a radical H-D standpoint. From a thoroughly hypotheticalist perspective, the heuristic device becomes the main method of justification. Now the relevant evidence is logically posterior rather than anterior to the theory. Previously, logical (not historical) pedigree was all-important. Now "ye shall know them by their fruits," and "old" evidence counts for naught, at least when used generatively, to construct the theory.[2] Some *novel* consequentialists went still further to deny that the (consequential) explanation of old evidence counts for anything.[3] For these strong predictivists, the term 'old evidence' was an oxymoron.

I call this methodological switch the *Great Logical Inversion*. While it is an historical fact that several methodologists from this time on have espoused strong hypotheticalist views tantamount to pure consequentialism, I am unable to find good reasons for adopting such an extreme anti-generativist position, then or now. How to logically explain the historical switch is the "mystery" of the Great Logical Inversion. In my opinion the mystery is logically unsolvable because the radical switch was unjustified.

Paralleling this ambiguity of the term 'evidence' is an ambiguity of the phrase 'inductive inference'. Few methodologists have equivocated here, perhaps, but the distinction marks a methodological divide. Today, inductive inference concerning hypotheses is a consequentialist idea. For the early inductivists, it was largely a generative idea. It reasoned from information in hand to some further claim as a conclusion. The prior probabilities of various Bayesians often seem to combine these two kinds of support. Yet, if the confirmation of h by e is conceived as a partial entailment of h by e (as by Carnap 1950), then partial generative support would seem to be a more promising explication of this than consequential support.

Natural philosophers eventually realized that the body of available information is often too thin to furnish either deductive or inductive derivations of target claims. Notably, there seemed to be no direct route from phenomenal claims to deep-structural claims couched in highly theoretical language. Enter hypothetical methods, which helped bridge the gap between low-level data and deep-structural theories *by reversing the order of reasoning*, by reasoning from the unknown hypothesis to an observational consequence that is known or fairly readily knowable. Henceforth, epistemic support was thought to accrue mainly from successful consequential testing. The more puritanical hypotheticalists abandoned all thought of generative justification in favor of the idea that a theory can be adequately supported by the smattering of successful observational predictions that we succeed in deriving from it. (Notice that the derivation of testable consequences is itself a generative task that is often more difficult than consequentialists have indicated.)

This sort of inductive support is, of course, quite different from that of generative inductive support, as there is often no question of reasoning "in the order of discovery" from a few, scattered, consequential test results plus other premises to the hypothesis as a conclusion. For a hypotheticalist to say that scientific claims are based on inductive reasoning now means only that the *truth* (or probability or some other epistemic virtue) of the claim is inferred from the predicted evidence, not that the claim itself can be constructed from the evidence. The question is whether the ardent hypotheticalist has thrown out the baby with the bath water, for can a pure consequentialist justifiably ascribe truth or positive probabilities to hypotheses at all? If not, a conjectural claim must remain forever hypothetical, no matter what its predictive success.

Incidentally, the strong empiricists who have dominated philosophy of science over the past century have made the generative difficulty more acute than it is by collapsing discoverability into original discovery and by claiming that generativism depends on the existence of strong logics of discovery. They have collapsed the broad and flexible generativist position into its specific variety, inductivism. Finally, unlike Newton and other classical methodologists, they have proceeded to reduce supporting information to finite sets of observational *data*, as opposed to lawful *phenomena* (Bogen and Woodward 1989)—not to mention other, more theoretical claims already established. Collectively, these moves have reduced the generative program to the absurdity of trying to discover deep theories by deriving them from low-level experimental data by means of a magical logic of discovery.

4. Some Arguments for Pure Consequentialism

The major arguments that have been offered for pure consequentialism and against the need for, or even the possibility of, generative justification in empirical science include the following.

1. *The argument from the failure of inductive logic.* The old inductivist methodologies failed on both logical and practical grounds. This failure of generative methodology leaves the field open to purely consequentialist methodologies, as the only alternative.

2. *The argument from the failure of logic of discovery.* Generative justification presupposes the existence of powerful, general discovery logics. But none have been found, nor is there reason to think that any can be found.

3. *The ampliative argument.* Generative justification employs a nonampliative form of proof and thus cannot handle new scientific claims, which are precisely the claims that need justification.

4. *The argument from fallibility.* Human inquirers are never in possession of an infallible body of positive scientific knowledge. Since generative justification is based on antecedent knowledge, it is both more dogmatic and less reliable than the consequential testing of hypotheses.

5. *The argument from ignorance.* Human inquirers possess no prescience. We cannot know now what we shall only know later. All growth of knowledge results from blind variation plus selective retention. This means that all justification is ultimately empirical and consequential.

6. *The argument from scientific practice.* From at least the early 19th century, successful scientific practice became pure consequentialist. It is this success of H-D methodology which shows that generative methodology is superfluous.

7. *The novel prediction argument.* Successful novel prediction permits consequential justification to be as powerful as we please and renders generative justification otiose.

8. *The Bayesian convergence argument.* Given a highly arbitrary distribution of initial degrees of belief in a hypothesis, a series of consequential tests can produce a virtual consensus in the scientific community. This "swamping of the priors" argument refutes the charge that Bayesians are paralyzed by lack of sufficient information about prior probabilities. As a probabilistic generalization of H-D methodology, Bayesianism renders generative justification pointless.

9. *The "no peeking" argument.* In a variety of instances, good methodology of experimental design and statistical inference requires a certain ignorance or blindness on the part of the experimenter or statistician, in order to define a good test or to avoid various forms of bias.

10. *The adhocness argument.* More generally, generative justification is untenable because it permits and even encourages ad hoc tinkering with hypotheses and is therefore illegitimate. No fair peeking at the test data while engaged in hypothesis construction!

11. *The "no double use" argument.* Still more generally, generative justification is illegitimate because it uses the same information twice, once to construct an hypothesis and then again to "test" it.

I have neither the space nor the skill to prove conclusively that all of these arguments against generative justification are mistaken, but that will not stop me from suggesting that none of them succeed in establishing either the nonexistence or the unimportance of generative justification. Answers to arguments 1, 2, 4, 6, 10, and 11 can be found in Nickles (1985, 1987a and b). Here my focus will be on novel prediction and related matters.

First, however, it is worth pointing out that we do have some general answers which at least cast doubt on several of the above arguments. Consider these fundamental questions, where 'justification' means 'support sufficient to resolve the methodological predicament'.

A. Is a purely generative justification of a significant empirical scientific claim possible?

B. Is a purely consequential justification of a significant empirical scientific claim possible?

C. Is it possible to combine generative and consequential elements in a single account of scientific justification?.

D. If combination is possible, what is the best recipe, when we have a choice? How can we best deploy our epistemic resources?

The answer to A is Yes, given that we already know a lot! Generative justification is possible because it actually has been accomplished for any number of scientific cases, including deep-structural theories (see Dorling 1973, or open any standard textbook of a reasonably mature science). There is no logical impossibility in justifying novel claims in this way once we clearly distinguish epistemic novelty from logical novelty (Blachowicz 1989). This means dismissing the assumption that scientists are perfectly rational agents, but that is a desirable step in the direction of "realism."

Now given a deductive argument from established scientific knowledge to a new or old theory as conclusion, pure consequentialists must evidently deny that those premises do in fact support the conclusion. This is directly contrary to the standard appraisal of deductive argument. Generative examples of an inductive variety are easy to come by as well. Most phenomenal claims are established by various kinds of generative reasoning.[4]

Generative justification often is the end product of a kind of (generic) bootstrap process, in science as in mathematics. A conjectural claim inspires sufficient empirical and theoretical research that eventually we know enough to prove or disprove the claim relative to our background knowledge. Consequential testing provides a further check on the reliability of the hypothesis and of key elements of the background knowledge on which it depends. Hence (C) is surely true as well, although it remains to show what forms the integration can take and how to get maximum epistemic punch from the information available (Question D). As Howson and Franklin (1985) point out, the Bayesian calculus suggests itself as the formal device for doing this.

Oddly enough, considering the prevalence of consequentialist thinking over the past century, it is the answer to question B, not A, that is in doubt. Do we have a single, clear case of a purely consequentialist justification of the credibility of a theoretical claim? Can consequential testing alone ever resolve an instance of the methodological predicament? Can consequential testing alone do more than Popperian corroboration, which does not indicate the true or even probable hypothesis? The burden of proof would seem to lie with pure consequentialists. Popper is one pure consequentialist who has faced squarely the limited power of consequential justification.

5. The Novel Prediction Argument

H-D consequentialists have a reply to the generativist charge that consequential inductive support is too skimpy to resolve the methodological predicament. The reply is that certain kinds of tests—namely, *novel predictions*—carry special weight. The strong (non-Popperian) version of the novel prediction argument concludes that successful novel prediction can endorse an hypothesis as highly probable. This reply not only promises to save consequentialism but also to render generativism otiose. One can have positive support without embracing a constructive methodology.

This claim that novel predictions are *logically* special, and not merely psychologically impressive, has long been challenged by generativists and others. (L. Laudan is an example of a consequentialist who nevertheless denies any special epistemic weight to novel prediction.) Mill (1843) and Keynes (1921) famously ridiculed the idea that novel predictions, as such, carry special weight. Our intuitions that hypotheses yielding novel predictions must be true, or at least on the right track, only illustrate the boundedness of human rationality, the limits of human imagination, and the shortness of human memory.

Notice that those pure consequentialists who invoke the Central Dogma without argument, simply beg the question against generative justification. Any number of writers of all stripes just take for granted that all support derives from testing. In so doing, they automatically exclude the possibility of generative justification. Although they do argue related points, I find this sort of blindness in many Popperians and Lakatosians, many Bayesians, and other prominent confirmation theorists such as Ronald Giere (1983). (Giere would reject my setup of the problem.) For example, it is almost always assumed in discussions of ad hoc theory construction that designing a theory to fit known phenomena never increases its prior probability over what it would have been otherwise. And since accommodated phenomena do not provide a genuine test of the theory, it is claimed, the posterior probability is the same as the prior. The conclusion is drawn too easily that the accommodated information provides no support. My immediate response is to ask, How could it be better to have a theory "Popper" into your head by reading tea leaves than to deliberately and rationally design it in accordance with knowledge already in hand?

6. Probabilistic Novel Consequentialism

Without much loss of generality, let us look at novel prediction in Bayesian terms. My target is those Bayesians who are pure consequentialists, those who believe that novel prediction is the way out of the methodological predicament. I can appeal to no Gallup poll, but I suspect that many Bayesians do fall into this category. Given the near invisibility of generativism in the modern discussions, it is hard to say how principled their purity is. Most practitioners see Bayesian methodology as a generalization of a simple, bivalent H-D methodology (probabilities replacing truth values), which happens to be the traditional enemy of generativism. The original appeal of subjective Bayesianism, applied to scientific hypotheses, was that the prior probability assignments made by individual scientists could be highly arbitrary, constrained only by a requirement of probabilistic coherence. For these "priors" would eventually be swamped by the results of consequential testing. Given finite priors, we can imagine a remarkably short series of novel predictive successes which will raise the posterior probability as high as anyone might wish. Or so Bayesians have claimed.

Before turning to novel prediction, let us touch on some preliminary problems with this view. First, it is difficult for critics to see what a convergence of subjective opinion has to do with truth (cf. Hesse 1974, p. 117; Glymour 1980, p. 73). Second, in scientific practice a steady convergence is unlikely. The original, strict-Bayesian idea was that, after the priors are fixed, belief change occurs only by conditionalization on new evidence, according to Bayes's rule, thus assuring long-run convergence. But if we treat Bayesian conditionalization as a kind of rational inference rule, akin to *modus ponens*, then we are free to reject the "premises" rather than to swallow the "conclusion." Consider the Howson-Urbach example (Howson, this volume). If conditionalizing on Velikovsky's surprising predictive successes yields a "posterior" of (say) .1, that is enough to convince me that my prior was set too high, however low it was to start with. Today, most Bayesians would agree that conditionalization on the evidence is not the only means of rational belief change. Third, in any event the swamping argument is, in a sense, symmetrical, as the base-rate ("Bayes rate"?) fallacy reminds us. Small priors may be swamped by consequential evidence, to be sure; but any amount of evidential success may also be swamped by picking sufficiently low priors.

Fourth, a small difference in prior assignments to two competing hypotheses can be amplified (as posterior differences) under a string of predictive successes for both. In any case the underdog can pass the leader only if the leader is eliminated by a future failure (assuming that the competing hypotheses entail the correct prediction e, or not-e, in each case). Moreover, every time we dream up a plausible alternative claim, we must alter our probability assignments, perhaps radically. Strict Bayesians avoid this problem by taking a more global approach: we begin by assigning priors to a *partition*, a set of mutually exclusive and jointly exhaustive hypotheses that includes the true hypothesis. But where did these hypotheses come from in deep-structural scientific cases? How do we know they are exclusive and exhaustive and contain the truth among them? Any sort of exhaustiveness would seem to presuppose a strong *generative* competence, so the pure consequentialist game is up already. I shall return to this obvious point. In fact, all the points to follow are well known; and many of them already appeared in 19th-century critiques of Bayesian-Laplacean inference.

It seems almost magical to generativists how something as unremarkable as Bayes's rule can amplify novel predictive success into a solution to the methodological predicament. But perhaps Laplace was right, that the Bayesian method is only "good sense reduced to calculation." Then *calculemus*! In the following, standard formulations of Bayes's theorem (here used as a belief-change rule), I have omitted the dependence of each expression on background knowledge b. When h entails e, then $p(e/h) = 1$, and the formulas are yet simpler.

(A) $p(h/e) = p(h)p(e/h) / p(e)$

(B) $p(h_i/e) = p(h_i)p(e/h_i) / \sum_{i=1}^{n} p(h_i)p(e/h_i)$

(C) $p(h/e) = p(h)p(e/h) / [p(h)p(e/h) + p(-h)p(e/-h)]$

It is surprising how many writers merely identify novel prediction with low $p(e)$ in simple formulation (A), as if $p(e)$ can be fixed independently of the other terms in the formula. The expanded forms of Bayes's rule show that the denominator can never be smaller than the numerator (which, anyway, would lead to nonsense probabilities greater than 1). On this definition of novelty, then, well-established theories cannot yield novel predictions (Redhead 1978). Recently, more methodologists have made small $p(e/-h)$ the condition of novelty, but some of the best treatments stop at this point. They typically illustrate how low $p(e/-h)$, the "catch-all" likelihood, yields a high posterior probability $p(h/e)$ without justifying the low estimate of $p(e/-h)$ for any actual case. Other writers defend a low value for $p(e/-h)$ by talking about experimental precision or by making intuitive remarks about how peculiar and unexpected the prediction e is. For the most part, these intuitive maneuvers still inspire the reaction of Mill and Keynes.[5]

Is small $p(e/-h)$ either necessary or sufficient for every kind of novel prediction or severe test of h? This formal conception of novelty can be rather sterile and need have little to do with content in the scientifically interesting sense. Consider just three cases. First, this definition does not distinguish between severe tests in point of precision and novel tests in other senses. It is easier to quantify degree of precision (in terms of significant figures) than it is $p(e/-h)$. And it may be a severe test of h to see whether it can explain some e already well explained by a competing theory. This case suggests a second. How do Bayesians handle a novel prediction e in a different domain than that for which h was designed, when that domain already has an accepted theory T that explains e? The existence of T precludes low $p(e/-h)$. Third, how do Bayesians handle the sort of case that L. Laudan (1977) termed "nonrefuting anomalies"? How do we calculate the denominator of (B) or (C) when e falls within h's domain of responsibility but, as far as we know, h entails neither e nor its negation? Bayesians can evade this difficulty by opting for Laudan's own comparative solution, whereby the failure counts against h only if a competitor already predicts e; but that seems inadequate in some cases. Similarly, by adapting a Lakatosian, historical conception of confirmation, Bayesians can deny that prediction e in a new domain is really novel if another theory for that domain already has explained e, but this solution seems unsatisfactory also. In any case most Bayesians rightly reject historical conceptions of confirmation for the reasons given by Musgrave (1974).

The key question is, How in practice can we determine the denominator of (B) or (C)? Even if the relative values of the priors of the few hypotheses under active consideration can be set by subjective preference or some other means, we are left with the problem of determining the "likelihoods" $p(e/h_i)$ or else $p(e/-h)$. In the famous case of the Poisson-Arago spot derived from Fresnel's wave theory, the catch-all term $p(-h)p(e/-h)$ is just obviously very small, most writers assure us. But how do they know? It only takes one serious, unknown, competing hypothesis to spoil the whole calculation. In the pure consequentialist problem situation, there always exist any number of theories which could have fired the prediction which hit nature's bull's eye.

There exist an infinite number of competing hypotheses (virtually all false), which predict any given body of evidence. Epistemically, all those competing hypotheses which entail a series of successful predictions do not change their relative standing one whit during this process. Their predictive successes normally will have increased their posterior support, but not to any degree that we can quantify in ignorance of the priors of those other hypotheses which failed this battery of tests and dropped out of the competition. As Jon Dorling, a subjective Bayesian, has remarked:

In analyzing any particular inference according to the Bayesian schema we will necessarily have constructed a mutually exclusive and jointly exhaustive set of hypotheses ... [and made judgments about the likelihoods of e on them]. (Dorling 1972, p. 183)

A similar point holds for orthodox methods of hypothesis testing (Rosenkrantz 1977, pp. 84, 177). Thus pure consequentialism loses either way.

There is no intuitive shortcut to somehow calculating the denominator of (B). In everyday, practical uses of inverse inference, we already know how many marksmen fired at the target, their firing rates, and what probability each had of hitting the bull's eye. We already know how many cab companies there are and the proportionate number of vehicles that each operates. But in the case of scientific prediction, we do *not* usually know the corresponding alternatives and their probabilities of success. To add psychological insult to logical injury, research shows that most people overestimate low base rates and underestimate high ones. This must be what happens in those Gee Whiz! examples that Bayesian furnish, in which a single, true prediction takes the probability of an hypothesis from (say) .01 to .85.

Words nearly as apt as the Dorling quotation can be found among 19th-century critics of the overextension of Bayesian-Laplacean inference. It is ironic that just as some 19th-century methodologists were becoming enamored of novel prediction, others, such as George Boole (1854), were undermining the probabilistic basis of such inference. These critics were working their way toward the modern view that Bayesian inference, like deductive inference, yields conclusions which are only as good as their "premises." Garbage in, garbage out. A well founded empirical conclusion must have well founded empirical premises, not *a priori* premises and certainly not premises based purely upon ignorance or indifference.

7. Conclusions

Two significant conclusions can be drawn. First, Bayesian novel consequentialism turns out to be comparative and eliminative. This is not surprising; for unlike direct construction, the completion of an optimization task by consequential means, in this case to find the One True Theory (or at least *a* true theory), requires an exhaustive, comparative search through possibility space. A Bayesian working in a local context who happens upon a new but plausible hypothesis must, for that reason alone, revise downward the priors for the small set of hypothesis under active consideration, including the prior p(e), since p(-h)p(e/-h) is one addend of p(e). This means that the mere discovery of a plausible new hypothesis can not only, by itself, disconfirm other hypotheses, but it can also reduce the novelty of a prediction (cf. Horwich 1982 on Putnam 1962; also Chihara 1987). (This sort of change falls outside of standard conditionalization, which depends on background knowledge b remaining fixed.) In order to avoid this unreliable jumping around, the strict Bayesian must require scientists to somehow construct and canvass the entire range of remotely possible alternative hypotheses compatible with current knowledge and then fix the priors and the likelihoods. Once this is done, Bayesian inference proceeds via eliminative induction of a probabilistic sort. Novel predictionism turns out not to be a distinct methodological strategy, for it collapses into the exhaustive-elimination strategy for finding the true theory. And we see that Bayesian methodology, often touted as the generalization of, and successor to, H-D methodology, is perhaps better characterized as a successor to eliminative induction. At this point we seem to be left with an exhaustive-eliminative strategy versus a directly generative strategy for finding the truth.

But, second, the need to begin from a distribution of prior probabilities and likelihoods over a partition of mutually exclusive and jointly exhaustive hypotheses, or some reasonable approximation thereof, means that the Bayesian strategy is not opposed to generative justification after all; on the contrary, it presupposes it! The good Bayesian truth seeker, like every exhaustive-eliminativist, must presuppose generative competencies of a very strong sort. The scientific community must be able to construct the true disjunction of exclusive and exhaustive candidate theories—roughly, the disjunction of every distinct hypothesis or family of hypotheses that is remotely plausible relative to current background knowledge—and *know* that this disjunction is both true and (reasonably) exhaustive. This ability does not necessarily require the existence of strong, general logics of discovery, for in sufficiently constrained contexts we can classify all possible hypothesis families by inspection. Nonetheless, it is a far cry from the pure consequentialist program. And next to this the simple, directly-constructive strategy of the generativist looks rather modest. The indirect, Bayesian strategy is *super*constructivist in requiring construction not only of the (eventually selected) true theory but of all serious competitors as well.

If we can do all this, you say, would it not be simpler just to construct the One True Theory directly, without bothering about the multitude of competitors? No. The ability to generate a true disjunction of candidate theories does not entail that we can knowingly generate the one that is true. Here 'generation' means 'generative justification' and the generative support for any particular member of the disjunction is only partial.

My general conclusion is by now obvious. *The Bayesian testing strategy is not an alternative to a generative methodology but an instance of it!* The most serious objection to generative justification has always been that its use is restricted to contexts in which we already know a good deal; but this same restriction holds—in spades—for the Bayesian novel prediction strategy. What appeared to be an unabashed philosophy of success works, it turns out, only to the extent that it is underlain by a philosophy of failure that in turn must be underlain by a powerful generative process. Bayesian arguments for novel prediction cannot justify the Great Logical Inversion, by which the possibility of successful prediction alone renders "discovery" considerations superfluous. On the contrary, the enhancements achieved by novel prediction are too great to be credible, except in precisely those cases in which strong generative achievements convert the novel prediction strategy into an eliminative strategy. The remarkable force of novel prediction should have been a source of worry rather than of comfort to pure consequentialists.

Once we appreciate that generative and consequential components are both necessary to adequate scientific justification, the question becomes how to combine them; and the Bayesian formula is the most obvious candidate for doing this. It is not a matter of truth or consequences, after all, but of truth *and* consequences. Our conclusions therefore have a bearing on the economy of research. Contrary to pure consequentialism, information used to construct a theory or model is not wasted, epistemically speaking. Prediction is not automatically superior to accommodation.[6] Not all accommodation amounts to painting the bull's eyes around the spot where the shot already has hit.

We need to ask when it is epistemically more efficient to use given information generatively rather than consequentially. I do not think there is a uniform answer here. In heavily constrained contexts, information e may entail h, given b. (E.g., from a mathematical generalization or parameterization of the Galilean transformation, the constancy of the velocity of light suffices to derive the Lorentz transformation.) In these contexts, e will have more epistemic weight when used generatively than consequentially. In other contexts the reverse will surely be true.

My position permits a redeployment of available knowledge between hypothesis generation (or reconstruction) and testing at any time, by scientists and even by philosophers, to obtain the strongest epistemic justification possible in terms of present knowledge. Ironically, a rigid Bayesian methodology (such as Carnap's) which requires us to equate the posterior at each stage with the prior of the next stage of testing—and thus views all rational inquiry as a matter of successive conditionalizations anchored in an initial prior probability distribution—is still historical in this further sense, and, indeed, too linear and cumulative. In effect, it remains tied to the original discovery arguments for h, however flimsy. It overrates original discovery and underrates discoverability. Most Bayesians today are more flexible.

If I am right about all of this, there is a kind of irony in the development of Bayesianism. It was the more hypotheticalist or consequentialist, subjective varieties of Bayesianism, based on the convergence theorem, which first brought this approach to prominence in recent philosophy of science. However, it is the fact that the Bayesian calculus is ideally suited for taking into account base rates, in the form of prior probabilities, that enables some Bayesians to handle the considerations I have just raised, in that restricted range of situations in which we do possess strong generative competence. In other words, the priors are the loci where we can factor in a degree of generative justification. This is exactly the way in which Bayesians such as Howson and Franklin (1985) have recently employed the calculus, a use that was anticipated by Salmon (1966) more than two decades ago and by Keynes (1921) long before that.

My agreement with these Bayesians on the prediction versus accommodation issue should not suggest that I am a Bayesian after all. Agnosticism suits me fine. Although it is clearly useful in certain contexts, I doubt whether Bayesianism or any simple formal scheme can capture all of the methodological distinctions and judgments we need to make. But for a moment, let me pretend that I am a Bayesian. Then I can bring out my difference from my co-symposiast, Colin Howson, by pointing out that our whole discussion of accommodation and prediction is reflexive. Boole (1854, Ch. 20) noted that, applied to reasoning about the truth of scientific hypotheses, Bayesianism is a theory with undetermined constants (the prior probabilities and the likelihoods).

Now how do scientists treat a theory with undetermined constants? They can of course conjecture values for the constants and hope to zero in on the correct member of this family of specific theories by elimination through empirical testing. Second, they can use the data together with parts of the theory to fix the parameters. Third, they can hope to theoretically derive the constants from other factors (including more fundamental constants).

In effect, pure subjective Bayesians conjecture the values of the priors and then alter their guesses as information comes in. Howson and Franklin are more constrained personalists for whom the priors represent an *intuitive* judgment about the amount of generative support. In historical contexts they empirically determine the priors and likelihoods from the statements and the other behavior of the scientists in question. My generative approach attempts to do more than this, where possible, by actually calculating the constants from more fundamental knowledge already in hand. This is what scientists themselves do with constants and free parameters, when they can.

Notes

[1] This paper is dedicated to the memory of my colleague, Willard Day, for his helpful comments and his exemplary life. I thank my co-symposiasts for a good discussion and the National Science Foundation for research support.

[2] The idea that generative information can constitute evidence is widely disregarded. In a valuable discussion of concepts of evidence, Achinstein (1978) apparently overlooks this possibility, or perhaps rules it out.

[3] Others, such as the Lakatosians, redefine 'novel' in an innocuous way that permits explanations to count as corroborative tests of hypotheses.

[4]It is true that our theories help tell us what it is significant to observe, but insofar as theoretical deductions help to justify phenomenal claims it is because we take those theories to be (sufficiently) correct; hence, this justification is also generative rather than consequential. The phenomenal claim is the target claim here.

[5]If the judgment that p(e) is low is solidly based on established background knowledge, the justification for this is generative, with the trade-off that the prior probability for any h that entails e will also be low (since the compatibility of h with established knowledge will be in doubt). In other words, the truth of e poses a threat to established "knowledge."

[6]While the epistemic virtues of accommodation are vastly underrated, the power of prediction is often overrated. Consulting a biographical dictionary, Peirce (1883, p. 163) suggested that the tongue-in-cheek generalizations about the ages of various poets which he constructed from the data would have been far better supported as predictively confirmed conjectures. This is doubtful, for as Keynes (1921, Ch. 25) already pointed out, in this case (as in the Velikovsky case discussed above), we already know an alternative explanation, radically different from the conjectures. Hence, the prior probability of the conjectures is miniscule at best and the likelihood of the evidence on an alternative theory (namely our standard theory) is so high that predictive success will count for little.

References

Achinstein, P. (1978), "Concepts of Evidence," as reprinted in *The Concept of Evidence*. Oxford: Oxford Univ. Press, 1983.

Blachowicz, J. (1989), "Discovery and Ampliative Inference," *Philosophy of Science* 56, in press.

Bogen, J., and J. Woodward (1989), "Saving the Phenomena," *Philosophical Review* 98, in press.

Boole, G. (1854), *An Investigation of the Laws of Thought*. London: Macmillan.

Carnap, R. (1950), *Logical Foundations of Probability*. Chicago: University of Chicago. 2nd ed., 1962.

Chihara, C. (1987), "Some Problems for Bayesian Confirmation Theory." *British Journal for the Philosophy of Science* 38: 551-560.

Dorling, J. (1972), "Bayesianism and the Rationality of Scientific Inference," *British Journal for the Philosophy of Science* 23: 181-190.

_ _ _ _ _ _. (1973), "Demonstrative Induction: Its Significant Role in the History of Physics," *Philosophy of Science* 40: 360-372.

Giere, R. (1983), "Testing Theoretical Hypotheses," in *Testing Scientific Theories* (Minnesota Studies in the Philosophy of Science, Vol. 10), J. Earman (ed.). Minneapolis: Univ. of Minnesota Press, pp. 269-298.

Franklin, A., and C. Howson (1985), "Newton and Kepler, A Bayesian Approach," *Studies in History and Philosophy of Science* 16: 379-385.

Glymour, C. (1980), *Theory and Evidence*. Princeton: Princeton Univ. Press.

405

Hesse, M. B. (1974), *The Structure of Scientific Inference*. Berkeley: Univ. of California Press.

Horwich, P. (1982), *Theory and Evidence*. Cambridge: Cambridge Univ. Press.

Keynes, J. M. (1921), *A Treatise on Probability*. London: Macmillan.

Laudan, L. (1977), *Progress and Its Problems*. Berkeley: Univ. of California.

_ _ _ _ _ _. (1980), "Why Was the Logic of Discovery Abandoned?", in *Scientific Discovery, Logic, and Rationality*, T. Nickles (ed.). Dordrecht: Reidel. Reprinted with changes in Laudan's, *Science and Hypothesis*. Dordrecht: Reidel, 1981, pp. 181-191.

Mill, J. S. (1843), *A System of Logic*. London: Longmans, Green.

Musgrave, A. (1974), "Logical versus Historical Theories of Confirmation," *British Journal for the Philosophy of Science* 25: 1-23.

Nickles, T. (1984), "Positive Science and Discoverability," *PSA 1984*, Vol. 1, pp. 13-27.

_ _ _ _ _ _. (1985), "Beyond Divorce: Current Status of the Discovery Debate," *Philosophy of Science* 52: 177-206.

_ _ _ _ _ _.. (1987a), "From Natural Philosophy to Metaphilosophy of Science," in *Kelvin's Baltimore Lectures and Modern Theoretical Physics*, R. Kargon and P. Achinstein (eds.). Cambridge: MIT Press, pp. 507-541.

_ _ _ _ _ _.. (1987b), "Lakatosian Heuristics and Epistemic Support," *British Journal for the Philosophy of Science* 38: 181-205.

Peirce, C. S. (1883), "A Theory of Probable Inference," in *Studies in Logic*, Baltimore: Johns Hopkins University.

Putnam, H. (1962), "Degree of Confirmation and Inductive Logic," reprinted in *Mathematics, Matter and Method, Philosophical Papers*, Vol. 1. Cambridge: Cambridge University Press, 1975.

Redhead, M. (1978), "Ad Hocness and the Appraisal of Theories," *British Journal for the Philosophy of Science* 29: 355-361.

Rosenkrantz, R. (1977), *Inference, Method, and Decision*. Dordrecht: Reidel.

Salmon, W. (1966), *The Foundations of Scientific Inference*. Pittsburgh: University of Pittsburgh Press.

Worrall, J. (1983), "Hypotheses and Mr Newton," N. R. Hanson Memorial Lecture, Indiana University (unpublished).

Part XIII

FORMAL LEARNING THEORY

Finite Axiomatizability and Scientific Discovery[1]

Daniel N. Osherson and Scott Weinstein

Massachussetts Institute of Technology and University of Pennsylvania

1. Introduction

Imagine a scientist who examines an unending sequence of data about an unknown reality and responds to each datum by announcing a first-order theory in a fixed language L. The scientist hopes to determine the truth or falsity of every sentence of L, but he does not feel obliged to reach a stage in his investigation when all such questions will have been resolved. One way to represent the scientist's project is as follows. Let the unknown reality correspond to the L-structure S. Then the scientist hopes that for every sentence $\phi \in L$, if $S \models \phi$, then all but finitely many of the theories he emits imply ϕ and do not imply $\neg \phi$. Following Kelly & Glymour (1987), we say in this case that the scientist "AE-identifies" S. The AE prefix signifies that *for all* $\phi \in L$, *there is* a stage beginning at which the scientist's theories correctly decide ϕ. If the scientist can AE-identify any potential reality (construed as alternative L-structures) drawn from a given class $\mathcal{K}$, we say that he AE-identifies $\mathcal{K}$. Section 2 below formally specifies the AE-identification paradigm of scientific discovery.

In the present paper we focus on scientists whose behavior can be simulated by a computer. Such computable scientists are conceived as emitting theories in the form of programs for enumerating axioms. The theory corresponding to a given program is the deductive closure of the axioms it generates. Now call a scientist "finite-minded" if no program that he emits generates an infinity of axioms. The question arises: Are there classes $\mathcal{K}$ such that some computable scientist AE-identifies $\mathcal{K}$ but no finite-minded, computable scientist AE-identifies $\mathcal{K}$? We provide a negative answer to this question in Section 3. Thus, with respect to AE-identification by computable scientist, no constraint is imposed on scientific inquiry by limiting hypotheses to finitely axiomatizable theories.

2. Preliminaries

2.1 Languages, Structures, and Environments

We fix a countable first-order language L with identity. The formulas of L will be used to express both the data available to scientists about their objects of investigation and the theories they frame in response to such data. A formula of L is *basic* iff it is an atomic formula or the negation of an atomic formula. The set of basic formulas is denoted *BAS*. The set of finite sequences of basic formulas is denoted *SEQ*. For $\sigma \in SEQ$, $l(\sigma)$ denotes the length of σ. For $\sigma \in SEQ$ and $n \leq l(\sigma)$, $\sigma\{n\}$ denotes the initial segment of σ of length n. The set of sentences (closed formulas) of L is denoted *SEN*. The set of individual variables of L is denoted *VAR*.

Throughout the paper, by *structure* is meant a countable structure for L. The class of structures is denoted *STR*. If $S \in STR$, $|S|$ denotes the domain of S. If $a : VAR \rightarrow |S|$ is onto, then define $Th_a(S) = \{\sigma \in BAS \mid S \models \sigma[a]\}$. If $e : N \rightarrow Th_a(S)$ is onto, then e is an *environment for S*. e is an *environment* just in case e is an environment for some structure. If e is an environment, $e\{n\}$ denotes the initial segment of e of length n. Observe that $e\{n\} \in SEQ$, for all e and n. The reader should note that if S, $S' \in STR$ and e is an environment for both S and S', then S is isomorphic to S' (see Osherson & Weinstein (1986), Lemma 3.1A). If e is an environment for S, let $Th(e) = Th(S) = \{s \in SEN \mid S \models s\}$.

2.2 Scientists and Identification

Let $\{W_i \mid i \in N\}$ be a standard indexing of the recursively enumerable subsets of *SEN*. A computable mapping of *SEQ* into N (the latter thought of as indices for $\{W_i \mid i \in N\}$) is called a *scientist*. A scientist Φ is *finite-minded* just in case for every $i \in N$, $W_{\Phi(i)}$ is finite.

(1) DEFINITION: Let environment e, scientist Φ, and $\mathcal{K} \subseteq STR$ be given.

    (a) We say that Φ *AE-identifies* e just in case

        i. for cofinitely many $i \in N$, $W_{\Phi(e\{i\})}$ is satisfiable; and
        ii. for all $s \in SEN$, if $s \in Th(e)$, then for cofinitely many $i \in N$, $W_{\Phi(e\{i\})} \models s$.

    (b) We say that Φ *AE-identifies* $\mathcal{K}$ just in case for all $S \in \mathcal{K}$ and for all environments e for S, Φ AE-identifies e.

    (c) We say that $\mathcal{K}$ is *AE-identifiable* just in case there is a scientist Φ which AE-identifies $\mathcal{K}$

Proposition 88 of Osherson & Weinstein (1988) provides an example of an AE-identifiable collection $\mathcal{K} \subseteq STR$ such that for some $S \in \mathcal{K}$, Th(S) is not finitely axiomatizable. Such examples might suggest that not every AE-identifiable collection of structures can be AE-identified by a finite-minded scientist. The following theorem shows, however, that finite-minded scientists constitute a canonical form for AE-identification.

3. Finite Axiomatizability and Scientific Discovery

(2) THEOREM: Let $\mathcal{K} \subseteq STR$ be given. If $\mathcal{K}$ is AE-identifiable, then some finite-minded scientist AE-identifies $\mathcal{K}$.

Let scientist Φ AE-identify $\mathcal{K}$. We may suppose, without loss of generality, that

(3) for every $\sigma \in SEQ$, $\{s \in SEN \mid W_{\Phi(\sigma)} \models s\} = W_{\Phi(\sigma)}$,

since any index for an r. e. set of sentences may be uniformly effectively transformed into an index for its deductive closure. We construct a finite-minded scientist Θ which AE-identifies every environment Φ does. By (1), this suffices to establish the theorem.

Given environment e and $n \in N$, Θ proceeds as follows in framing its conjecture on $e\{n\}$. Θ consults the first $n + 1$ conjectures of Φ on e, that is, $\Phi(e\{0\}), \ldots, \Phi(e\{n\})$.

Let $\{s_i \mid i \in N\}$ be a fixed, recursive enumeration of *SEN*. For each $i \leq n$, Θ then determines the longest initial segment $\tau_{i,n}$ of $< s_0,..., s_i >$, each sentence of which is decided by $\Phi(e\{i\})$ in at most n steps of computation (that is, for each $j < l(\tau_{i,n})$, either s_j or $\neg s_j$ is enumerated into $\Phi(e\{i\})$ in at most n steps of computation). Θ then picks the longest of the $\tau_{i,n}$ thus constructed, call it τ_n, and the largest j such that $\tau_n = \tau_{j,n}$, call it j_n. Observe the following:

(4) for every n, $l(\tau_n) \leq l(\tau_{n+1})$;

(5) for every n, $l(\tau_n) - 1 \leq j_n$.

Finally, Θ mimics Φ's conjecture on $e\{j_n\}$ in framing its own conjecture on $e\{n\}$. Specifically, $\Theta(e\{n\})$ is an (effectively chosen) index for the finite set of sentences $\{s_i \mid i < l(\tau_n)$ and s_i is enumerated into $W_{\Phi(e\{j_n\})}$ in at most n steps of computation $\}$. It is clear that Θ so defined is a finite-minded scientist. Note that the definition of Θ implies that

(6) for every $n \in N$, $W_{\Phi(e\{n\})} \subseteq W_{\Phi(e\{j_n\})}$.

Suppose that Φ AE-identifies e. It follows from (1) and (3) that Φ satisfies the following two conditions.

(7) For all but finitely many $k \in N$, $W_{\Phi(e\{k\})}$ is satisfiable.

(8) For all $i \in N$, if $Th(e) \models s_i$, then for all but finitely many $k \in N$, $s_i \in W_{\Phi(e\{k\})}$.

In order to show that Θ AE-identifies e, it suffices to establish that conditions (7) and (8) hold for Θ in place of Φ, that is,

(9) for all but finitely many $k \in N$, $W_{\Theta(e\{k\})}$ is satisfiable; and

(10) for all $i \in N$, if $Th(e) \models s_i$, then for all but finitely many $k \in N$, $s_i, \in W_{\Theta(e\{k\})}$.

To establish (9) and (10), we first show:

(11) for every $k \in N$, there is an $n \in N$ such that $l(\tau_n) > k$.

To prove (11), fix k and, by (8), choose $c \geq k$ large enough so that for each $d \leq k$, either $s_d \in W_{\Phi(e\{c\})}$ or $\neg s_d \in W_{\Phi(e\{c\})}$. We may choose $n \geq c$ large enough so that $\Phi(e\{c\})$ decides s_d in at most n steps of computation, for each $d \leq k$. It follows immediately that $l(\tau_n) > k$.

We proceed to establish (9). By (7), we may choose $c \in N$ such that for every $k \geq c$, $W_{\Phi(e\{k\})}$ is satisfiable. By (11), we may choose m such that $l(\tau_m) > c$. It follows by (4) that for every $n \geq m$, $l(\tau_n) > c$, and hence, by (5), that for every $n \geq m$, $j_n \geq c$. But then it follows immediately by (6) that for every $n \geq m$, $W_{\Theta(e\{n\})}$ is satisfiable.

Finally, we establish (10). Fix $i \in N$ and suppose that $Th(e) \models s_i$. We must show that

(12) for all but finitely many $k \in N$, $s_i \in W_{\Theta(e\{k\})}$.

By (7) and (8), choose $c \geq i$ such that for every $j \leq i$ and for every $k \geq c$:

(13) (a) $W_{\Phi(e\{k\})}$ is satisfiable;

(b) if $s_j \in Th(e)$, then $s_j \in W_{\Phi(e\{k\})}$;

(c) if $\neg s_j \in Th(e)$, then $\neg s_j \in W_{\Phi(e\{k\})}$

As in the immediately preceding argument for (9), we may choose m so that for every $n \geq m$, $l(\tau_n) > c$ and $j_n \geq c$. Fix $n \geq m$. Since $l(\tau_n) > i$ we have, by (13b), $s_i \in W_{\Phi(e\{j_n\})}$, and hence, by (13a), $\neg s_i \notin W_{\Phi(e\{j\})}$. But then, by the definitions of τ_n and j_n, it follows that s_i is enumerated into $W_{\Phi(e\{j\})}$ in at most n steps of computation. Since this holds for every $n \geq m$, (12) follows immediately by the definition of Θ.

4. Conclusion

The foregoing Theorem shows that a certain restriction on the form of the conjectures which a scientist can advance does not limit the class of discovery problems he can solve, up to the criterion of success imposed by AE-identification. Other issues about the ways in which the representation of scientific knowledge bear on the prospects for scientific success may be fruitfully addressed within the framework provided by the theory of machine inductive inference. Osherson & Weinstein (1988) deals with a number of these questions.[2]

Notes

[1]Research support was provided by the Office of Naval Research under contract No. N00014-87-K-0401. Correspondence to D. Osherson, E10-006, M.I.T., Cambridge, MA 02139.

[2]We would like to take this opportunity to correct a minor error in the proof of Proposition 69(a) of Osherson & Weinstein (1988). The correct proof may be extracted from Example 3.4A of Osherson & Weinstein (1986).

References

Kelly, K. and Glymour, C. (1987), "On convergence to the truth and nothing but the truth", *Philosophy of Science*, in press.

Osherson, D. and Weinstein, S. (1986), "Identification in the limit of first order structures", *Journal of Philosophical Logic*, 15, pp. 55-81.

_____. (1988), "Paradigms of truth detection", *Journal of Philosophical Logic*, 18, pp. 1-42.

Formal Learning Theory and the Philosophy of Science

Kevin T. Kelly

Carnegie-Mellon University

Consider the following collection of familiar questions in the philosophy of science.

Underdetermination: What collections of theories can be reliably distinguished from one another on a given sort of evidence presentation? How do differences in evidence presentation, background knowledge, and hypothesis vocabulary affect underdetermination?

Realism: How plausible is it that the methods employed by science are capable of arriving at the truth in a given domain of inquiry? If not, is it plausible that there exist such methods? How does the answer differ from domain to domain?

Scientific progress: Is it possible that scientific knowledge could be continually improved, even if it is never perfectly correct? Is there any sense to be made of the increasing verisimilitude of inquiry?

Methodology: How do standard methodological directives interact? Do different directives interfere with one another or do our favorite ideas about method complement one another? Is experiment really more powerful than passive observation? How do popular ideas about evidence (e.g. bootstrapping) focus inquiry toward particular kinds of possible worlds and away from others?

Bounded rationality: What methodological principles are appropriate for bounded agents (as opposed to the ideal agents frequently assumed in philosophical discussions). What does methodology look like when applying an ideal method is itself an inductive problem for a bounded agent?

The problem of induction: Are there precise levels of inductive unsolvability among inductive inference problems? If so, can the *intrinsic* level of unsolvability of an inductive problem be determined, independently of any particular choice of inductive rules? How do formal properties of the syntax and semantics of the hypothesis and evidence languages relate to the intrinsic difficulties of the resulting inductive problems?

The logic of discovery: Is there a metatheoretic setting for the comparison and assessment of inductive problems and hypothesis generation methods that aspires to the rigor and to the kinds of formal insights familiar in mathematical logic?

PSA 1988, Volume 2, pp. 413-423
Copyright © 1989 by the Philosophy of Science Association

The reliability theory of knowledge: What is the logical structure of the various doxastic attitudes corresponding to the distinct ways of making precise the notion of inductive reliability?

The philosophy of artificial intelligence: How can we assess progress in the artificial intelligence learning literature? That is, what are the problems to be solved, and how can it be determined whether one program solves one of these problems better than another?

The philosophy of psychology: How much harder is cognitive psychology than behaviorism (i.e. how much harder is it to infer the program of a computer than it is to infer its overall behavioral dispositions)? How much harder still is it to infer the programming system in which a computer's program is written than it is to infer the program or the overall behavioral dispositions?

Four threads stitch these diverse topics into a single fabric. The first is *discovery:* the process of producing rather than merely testing scientific hypotheses. The second is *reliability:* the ability to eventually arrive at or near the truth, when the truth may be any in a wide range of possible worlds. The third is *computational boundedness:* the fact that people and the computers they build cannot determine whatever logical and probabilistic relations that methodologists tell them to evaluate. The fourth is a focus on the *intrinsic difficulty* of an inductive problem rather than on the difficulties some particular method or other has with this problem.

A confluence of efforts in computer science, linguistics, psychology, logic and philosophy has led to a study that especially embodies the concerns of reliability, computational boundedness, and a focus on the intrinsic difficulties of inductive problems in the study of hypothesis generation procedures. This study is referred to variously as *formal learning theory,* and as *the mathematical theory of inductive inference.*[1] Unfortunately, neither name is properly descriptive of the subject. Perhaps "computational reliability theory" would be more accurate, but I will continue to honor historical accident by using the term "formal learning theory" in the sequel. Although formal learning theory has proceeded in relative isolation from the philosophy of science, its results and techniques promise some welcome relief from the emphasis on historical case analysis that has characterized much of the philosophical literature on methodology for the last two decades. It offers, in contrast, a fresh approach to the positivistic ambition of a clear, formal framework for the normative study of inductive methodology. In the following sections, I explain how the topics of discovery, reliability, bounded rationality, and the intrinsic difficulty of induction are woven into the fabric of formal learning theory. It also provides a forum for the comparison of Bayesian inductive techniques with other techniques in other settings.

1. Discovery and Hypothesis Generation

The logical positivists distinguished sharply between logical principles for evaluating hypotheses on given evidence and principles for generating hypotheses from evidence. The latter study was referred to as the "logic of discovery". The positivists tried hard to convince us that the study of hypothesis generation is at once hopeless and irrelevant. Popper's view (1968, pp. 31-2) was that hypothesis generation could be studied only as a psychological phenomenon, meaning that there is no such thing as a norm governing the choice of discovery methods. This view is remarkable in itself, but more remarkable is the fact that Carnap (1950) and Hempel (1956) both heartily endorsed it and added the pronouncement that there would be no such methods to study even if it were a philosophical enterprise to study them. No dogmas have been more injurious to the philosophy of science or less circumspectly embraced by it. Because of these dogmas, philosophers have, with few exceptions, ceded the logic of discovery to computer scientists and statisticians, despite the fact that the study was a common epistemological topic for two millennia.

The logic of discovery is not an afterthought in formal learning theory. It is the starting point. From this perspective, scientific inquiry is a process that produces and revises the currently accepted theory, and that aspires to approach a correct theory of its domain. Hypotheses may be tested, but the test procedures that perform them serve as cogs subordinate to the overall process of theory revision. There certainly are discovery procedures which when opened up reveal clearly separated modules for generation and for test. But there may be others in which generation and test are so smeared together that it would be fruitless to speak of separate modules. Whether or not the sharp procedural distinction between generation and test that appears in the philosophical literature (e.g. Popper 1968, pp. 32-4 and Laudan 1980, p. 182) is a good idea depends on the comparative performance of methods that are organized into separate modules and those that are not. So while the positivists might accuse us of confusing philosophy with psychology, we see them as having confused a particular way of dividing up a problem with the problem itself.

2. Reliability

Consider a process that directs the construction of hypotheses from evidence. To be concrete, consider a computer running a program, although we do not want to restrict ourselves only to computational systems. Suppose the program's name is EINSTEIN. The name gets us into the proper mood. We look up some electromagnetic data from before 1905, feed it to the program (it takes hours), and hit the return button. Lo and behold, out pops "E=mc^2"! Very impressive. While we celebrate, a colleague's daughter enters some gibberish and hits the return button. It says the same thing again. The bubble bursts. It says" E=mc^2" no matter what it is told. It was just lucky that its hypothesis was true. As a discovery method, it would have little to recommend it to anybody who didn't already know that E=mc^2. Since it wouldn't have succeeded in other circumstances (i.e. in which "E=mc^2" is false) it is an unreliable discovery procedure.

In formal learning theory, reliability is a matter of arriving at a true hypothesis over a non-trivial range of theoretically possible worlds. The set of possible worlds over which an hypothesis generator succeeds is called the generator's *inductive scope*. Ideally, a method's scope should be the set of all models of the user's background knowledge. But sometimes this is not possible, as we shall see.

Convergence: What do we mean then by "arriving at" or "converging to " the truth? Well, we don't mean any one thing.[2] Rather, we treat different convergence criteria as objects of study and refuse to propose one as the unique acceptable standard. For those with a lot of background knowledge and little time to wait for the truth, it would be sensible to demand a strict standard of convergence. This is the situation of a child learning its first language if Chomsky is correct. But where there is little background knowledge, a more liberal standard of convergence is required if reliability is to be achieved. We would prefer convergence in the short run in these cases too, but often we can't have reliable truth in the short run.

For example, a strict notion of convergence might demand that the generator stabilize to a true conjecture and report when it has done so. If a generator reliably converges to the truth in this sense, then it might be described as knowing that it knows. Another convergence criterion involves eventually stabilizing to a true conjecture, but does not require any announcement that stabilization has occurred. When a generator reliably converges to the truth in this sense, it might be described as knowing, but not as knowing that it knows. We could also repeal the requirement of stabilization to a particular sentence and demand only stabilization to a proposition, with possible flip-flops over its various representations. We can also think about defining convergence to theories that are not finitely axiomatizable. One way would be to stabilize to a recursive axiomatization of such a theory. Another would be to build up the theory piece-meal by producing a sequence of conjectures such that for each sentence, its membership in the theory is eventually settled by some stage in the sequence. That is, for each sentence in the theory there is a stage after which it is always entailed, and for each sentence in the complement of the theory there is a stage after which it is no longer entailed. Even non-axiomatizable theories can be converged to in this piece-meal sense, but only axiomatizable theories can be converged to in the all-at-once sense unless we characterize theories in some way other than by axiomatization.

Piece-meal convergence is interesting as a proposal for getting more truth or getting closer to the truth through time without developing a verisimilitude measure. So we can speak in a precise way about a process that forever approaches the truth even if each conjecture is false. The notion of piece-meal convergence is also of some historical interest, for Peirce held it to be "unphilosophical" to deny that for each clear proposition there is a time at which its truth or falsity is settled by the advance of science. That is, Peirce held that it is unphilosophical to suppose that science is not reliable in the sense of piece-meal convergence. Piece-meal convergence also turns out to yield very natural results about the scope of inquiry.

Nobody is tempted to think that something is gotten for nothing here. In a sense, the overall informational value of the method's output to its user cannot be improved by changing convergence criteria. A weaker criterion permits more reliable convergence to a more informative theory, but it becomes harder to make use of the outputs of the method in the short run. In the case of piece-meal convergence, you may always be missing most of the theory you are headed for, and each conjecture generated by the method may involve some falsehood. But depending on one's use of the outputs, this may not be so bad. For if we are interested only in a finite subset of the theory, there is a time after which this finite set's truth status is settled in the sequence of conjectures.

Of course, convergence to *some* truth is too easy a goal. The truth converged to must meet minimal standards of informativeness for a procedure to be worth bothering with. The weaker the convergence criterion, the more natural it is to beef up the informativeness of the theory sought. For example, it seems silly to require all-at-once convergence to a complete theory. In many cases such theories are not axiomatizable and hence could not be converged to for this reason alone. But piece-meal convergence does not place any inordinate demands on inquiry from one stage to the next. It demands only that some new hypothesis be explored after some finite chunk of time. So according to the piece-meal convergence criterion, we can expect a higher quality theory in the limit at the expense of possible falsehood forever in the short run. For example, we might require piece-meal convergence to the set of all true, purely universal laws. Or we might admit a certain number of quantifier alternations. Or we might demand a theory with a certain minimal weight with respect to some probability measure over sentences representing what Hempel called "explanatory strength". Many possibilities have been examined.

Data Protocol: I have not yet been explicit about the data protocol and its relationship to reliability. I use the term "protocol" generically, to cover lots of different ways for a method to interact with its environment. In the simplest setting, the method waits until it is fed a new evidence sentence by some interface to the actual world (say a laboratory that is not given any directions by the theorist).

In another protocol, the theorist asks questions to his lab, and gets the answers back. In a third, and more powerful protocol, the theorist actually helps to construct the world he is in. In this case, theoretical possibilities can be viewed as paths in a forward-branching model of tense logic. Experiments in such a setting consist of "letting a proposition be so" and then observing what happens thereafter.

Still other protocols guarantee a certain probability of observation of a given object or permit a certain probability of error or "noise" in the data (Valiant 84) and (Kearns and Li 1988).

The data protocol is crucial to the notion of reliability, for the method should succeed not only over all possible worlds, but also over all possible choices the protocol can make in ordering its data. When the method asks the questions, the protocol may get no choices because the theorist completely controls the order of the questions put to nature. At the other extreme, the protocol of the passive theorist controls all choices about data order. Some learning-theoretic studies have examined limited classes of data presentations. And there are intermediate cases in which there are some restrictions on the protocol's choices, but the restrictions are not absolute. Some studies have focused on computer-generated data presentations, on repetitive presentations, and on sets of presentations of unit probability.

When a method succeeds in converging to the truth over all possible choices of the data protocol, the method is said to identify the possible worlds, and the set of all worlds a method identifies is defined to be the method's scope. So one method is as reliable as another if its scope includes that of the other.

Since reliability requires success over all choices the data protocol can make, knowledge about the restrictions on the data presentation can be as useful as knowledge about theoretical possibility. For example, the mere fact that the data is ordered by a primitive recursive device ensures the solvability of a problem by an effective method, no matter what the subject matter.

Methodological Constraints: So far, we have focused entirely on convergence to the truth. But most methodological proposals in the philosophy of science have been short-run relations of explanation, confirmation, or test survival between hypothesis, evidence, and perhaps some other relata available to the method at the time, such as background knowledge or desires. In formal learning theory, these principles (insofar as they are sufficiently clear) can be viewed as restrictions on the set of all possible inductive methods (Osherson, Stob, and Weinstein 1986, pp. 45-95.). For example, if you think that an hypothesis should not be conjectured unless it is consistent with the data, then you thereby refuse to use methods that generate hypotheses inconsistent with the data.

To recapitulate the discussion of reliability, formal learning theory is the study of solutions to inductive inference problems. Such a problem results from a precise specification of each of the following metaphysical and normative elements:

Metaphysical elements

An hypothesis language

An evidence language

A collection of theoretically possible worlds

A protocol whereby evidence may be obtained from a given possible world

A collection of hypothesis generators that use the protocol

Normative elements

A criterion of convergence

A criterion of adequacy for hypotheses

A criterion of reliable success (identification)

Short-run restrictions on conjecturing behavior (methodological norms)

3. Computational Boundedness

There has been increasing interest among philosophers in the notion of bounded or "minimal" rationality (Cherniak 1986). Bounded rationality involves the study of how computational or methodological resource constraints can make ideally irrational behavior look like the best course of action for bounded agents. One of the most interesting applications of formal learning theory to the philosophy of science is the extensive survey by Osherson, Stob, and Weinstein (1986) of the way in which methodological principles hinder the reliability of computable discovery methods.

As we have seen, static, short-run relations of confirmation, explanation, and hypothesis test have constituted much of the methodological work in the philosophy of science. Such relations, if relied upon, are supposed to enhance our quest for knowledge. But from the point of view of formal learning theory, these principles at best fail to *restrict* the range of discovery methods we may choose among. At worst, they may throw out methods whose reliability cannot be duplicated by any procedure that obeys them. So instead of asking how these principles help in discovery, the real question is how much they *hinder* it by throwing out methods whose reliability cannot be duplicated by any method obeying the restrictions. As it turns out, many standard methodological directives are restrictive for computable scientific methods.

For example, it is easy to prove that there is a computable method that fails infinitely often to produce an hypothesis consistent with the evidence, but that is more reliable than any computable method that rationally produces only hypotheses consistent with its data. In a nutshell, the consistency requirement would allow you to turn the computable discovery method into a decision procedure for the halting problem, but the infinitely many violations of the requirement by the former procedure block the reduction. Osherson, Stob, and Weinstein have similar results for procedures that conjecture only simple hypotheses, procedures that produce maximally probable hypotheses at each stage, procedures that refuse to drop an hypothesis until it is refuted, and for procedures subject to many other restrictions proposed in the philosophy of science. Time and again, we find that bounded agents face a sharp methodological dilemma where the Greek gods of ideal epistemology see nothing but concord.

The challenge posed to the philosophy of science by such results is a serious one. How can one who does not know the epistemic cost of a methodological restriction make a competent recommendation for it? A medical doctor who doesn't know the side-effects of his prescriptions is guilty of malpractice. Is a methodologist who doesn't know the side effects of his methodological prescriptions any better? It won't do for the incompetent doctor to say that an *ideal* patient with a steel stomach would have survived the treatment. Nor will it do for the methodologist to say that a Greek god's reliability would not have been restricted by the proposed methodological principles.

Computer scientists have been working hard to examine a very bounded sense of rationality. L. Valiant (1984) and his associates have been developing a theory of "probably approximately correct" inference in time polynomial in parameters small with respect to size of the total data set. Although the problems examined may look like toys to philosophers, the character of the results is rigorous and serious. One exciting development has been to show that the form in which a definition must be cast can determine whether inference can or cannot be reliably performed in polynomial time (Pitt and Valiant 1988). Hence, in bounded rationality, norms that are invariant up to logical equivalence may be impossible to formulate, contrary to the usual practice in ideal epistemology.

4. Intrinsic Inductive Difficulty

In physics, great progress was made when we learned to distinguish effects relative to a reference frame from physical invariants; that is, when we came to distinguish what is absolute about the world from what depends upon our special perspective. In discussions of scientific methodology, our favorite inductive norms are analogous to a frame of reference: what seems hard or easy is *relative* to the methods we currently like. Although it is possible to examine the scopes of particular methods, most results in formal learning theory aspire to determine whether or not a given inductive problem is solvable, *regardless* of the method employed. As in recursion theory and the theory of computability, the focus is on invariant properties of *problems* rather than upon properties of methods for solving them.

As in recursion theory, the key to the assessment of intrinsic problem difficulty is a battery of techniques for obtaining very general negative results. The usual technique employed is the style of diagonal argument introduced by Putnam (1963) in his attack on Carnap's c-functions. The diagonal strategy is to construct an evidence presentation on which a method fails, on the assumption that it does not, thereby obtaining a contradiction. The demonic data presentation is constructed by observing what the method does on earlier data, and then confounding it later, over and over again. Other techniques include reductions of uncomputable problems, and NP-hardness reductions[3].

Evidently, the ability to prove whether or not an inductive problem can possibly be solved in the limit is of crucial relevance to thesis of convergent realism. It is also instrumental to the study of underdetermination. Consider two theories such that the same data will appear regardless of which one is true. Evidently, the theories cannot be reliably distinguished by any method (although one might be chosen over the other by methodological principles that are not reliable with respect to the two possibilities: e.g. Glymour's (1980) relation of bootstrap confirmation). But theories that would lead to the same data if true are a special case of theories that cannot be reliably distinguished. To provide an idealized illustration, imagine a Cartesian and an atomist debating whether or not matter is infinitely divisible. It can be shown that no procedure can reliably decide the question in the limit, even if we allow that they have an infinitely sharp knife capable of making any cut that can be made. So underdetermination can arise when the theories in question give rise to different total data sets. Hence, it can arise in a radical way even when no theoretical terms are involved in hypotheses. So a philosophical question arises. Why do anti-realists like Van Fraassen (1980) read so much significance into the special case of underdetermination of theories involving theoretical terms? From a learning-theoretic perspective, the special focus on underdetermination due to theoretical terms lacks motivation.

Given the ability to show that no method can solve a given inductive problem, it is possible to determine the level in a hierarchy of successively weaker notions of success at which a problem first becomes solvable. This raises the exciting prospect of actually mapping out the topography of difficulty of the problem of induction along different dimensions. One of the problems the philosophy of science has had with Hume's problem is its simplicity. It is just the observation that by definition, the premises of a properly inductive argument do not entail the conclusion. There doesn't seem much more to say. But when we look at weaker, limiting senses of inductive success, each notion of success leaves an interesting, non-trivial border between the solvable inductive problems and the unsolvable ones. And each such line is analogous to a line of elevation on a topographical chart. Hence, we can start to see in a systematic way what makes induction hard and what makes it easier, and even where exactly different parameters trade-off against one another.

No doubt such a proposal sounds naively ambitious. But several dimensions of the chart are almost completely mapped out (Kelly and Glymour 1989). In light of developments in the artificial intelligence literature on machine learning, the following question arises: how does the syntax and vocabulary of the hypothesis language affect the difficulty of the problem of inferring a complete true theory in this language? Following the lead of the literature on the decision problem, one can categorize languages by maximum function arity, maximum predicate arity, and by quantifier prefix complexity (i.e. the maximum allowable number of alternations between universal an existential quantifiers in a prenex formula). Then one may examine which classes of languages give rise to solvable problems in the all-at-once and piece-meal senses. We have also distinguished those problems solvable by computable methods or only by non-computable methods in each of these senses. Crossing the criteria yields a discrete, partially ordered scale of senses of success.

The outcome of the classification is fairly interesting. The complete, purely universal theory in a language with no function symbols can be reliably inferred by a computable method in the piece-meal sense. But it may not be reliably inferrable in the all-at-once sense, even by an uncomputable method. If we add function symbols to the vocabulary, then it is an open question whether the complete theory can be reliably inferred in the piece-meal sense by a computable method, although we know that an uncomputable method can succeed. The complete monadic theory of a structure can be reliably inferred in the all-at-once sense by a computable method, and when unary function symbols are added, the complete theory cannot be inferred all-at-once, but can be inferred reliably in the piece-meal sense. This helps to explain the popularity of monadic languages in discussions of induction from Plato through Mill and Carnap. For if we have one quantifier alternation over a binary predicate in the hypothesis language, it is possible that the resulting inductive problem is not solvable even in the piece-meal sense by an uncomputable method.

We have also completed a study that relates the quantifier complexity of a reliably inferred theory to the least quantifier complexity of the data it can be inferred from (Kelly and Glymour 1989b). It is easy to see that from data beginning with an existential quantifier and involving n alternations, it is possible to infer the complete universal theory with n+1 alternations. What is harder to show is that you cannot go to the complete universal theory with n+3 alternations. And we have not yet closed the gap to show that it is impossible to reliably jump to the complete universal theory with n+2 alternations. To prove negative results about reliable inference from quantified evidence requires careful attention to the model theory of what can be said with a certain number of quantifier alternations. The close interaction that is developing between reliable discovery and model theory shows how wrong the positivists were to fail to recognize the logic in the logic of discovery.

The ability to show that no procedure subject to a given limitation can solve an inductive problem is also essential to the investigation of inductive efficiency for artificial intelligence learning procedures. Efficiency is not just speed, but speed for a given problem. This is the issue that first drew me to formal learning theory, for the practice-oriented literature on induction in artificial intelligence is suggestive, but progress is difficult to assess. In my investigations of artificial intelligence techniques, I have found on several occasions that the (unstated) problem solved by the procedure could not have been solved at all (by an arbitrary procedure) had it been made any harder. Such results greatly enhance the interest of procedures that may appear unmotivated or simple-minded at first glance.

5. Problems and Projects

I have described some of the reasons I am interested in applications of formal learning theory to the philosophy of science. But of course there are many areas for improvement. In fact, improving the applicability and flexibility of the theory is part of the fun. For example, not much has been done yet with the difference between experimentation and passive observation. There is nothing in principle to prevent such a study, and I have already outlined how the semantics of tense logic can be adapted to provide a clear formal setting for such analyses. It is just that not much has been done with it yet.[4]

I would also like to examine more closely the relation between measure-theoretic convergence results an the non-measure theoretic results of formal learning theory. For example, our quantifier-prefix hierarchy of unsolvable inductive problems collapses if worst-case piece-meal convergence to the truth is replaced with piece-meal convergence to the truth with probability one.[5] So all of our demons can be shown by a Bayesian to live in an event of zero probability. I am currently interested in other senses (say topological) in which the sets of demons are big. A standard result in measure theory is that the real line can be exhausted by a topologically meager set and a set of measure 0 (Oxtoby 1970, Corollary 1.7, p. 5). In other words, topological and measure theoretic notions of smallness can be at odds. But whether or not one practically worries about what has zero probability, it is surely of fundamental epistemological interest to know what kinds of sets of demons Bayesians entitle themselves to ignore.

But there is no need to make Bayesianism and Learning theory appear to be competitors. Almost everywhere a measure one convergence is just another kind of criterion of success that arises when probabilities are defined over the possible worlds in the usual, learning-theoretic setting. Hence, all the questions that learning theorists examine can be asked again here. For example, Bayesian measure one convergence results usually assume either positive and negative data or likelihoods over sequences of data sentences rather than over sentences themselves (e.g. Seidenfeld and Schervish 1987). So there are learning-theory-like questions remaining in the Bayesian camp: what is the scope of conditionalization when the negative data or sequence likelihoods are not available? (i.e. when propositions describing the underlying evidence protocol are not even present in the Bayesian's conceptual scheme?) Conditionalization is just one method for updating distributions among many, and it can be compared with other methods both in the Bayesian and in other settings.

I am of the view that no matter how many times you say it, no matter how many famous people agree with you, and no matter how subtle you make your philosophy, *is does not imply ought*. So I don't worry so much about accounting for historical case examples. And I wouldn't trade theoretical insight for any historical example. But on the other hand, formal learning theory has matured to the point that the formal structure of historical inductive problems might be fruitfully analyzed. The induction of simple physical laws like Ohm's law has an interesting reconstruction as a learning-theoretic problem, but we haven't gotten around yet to deciding what, in particular, to investigate about the case.

The above shortcomings are primarily a matter of insufficient time and candle power being devoted to the questions, and hardly reflect badly on the enterprise of formal learning theory. Rather, the availability of so many clear and interesting projects is a major part of its appeal.

Other possible objections are aimed more at the formal nature of the enterprise than at its list of achievements. For example, some philosophers may worry about the fact that formal learning theory assumes a fixed evidence an hypothesis language, and that this restricts its application to "mere" normal scientific practice.

First, even if this were so, it would not justify ignoring the structure, scope, and limits of normal science. The results of formal learning theory are not trivial, so if they are about normal science, the structure of normal science is not trivial either.

Second, having a clear framework of analysis does not entail that the formal inductive agents studied in the framework cannot have any control over usage. For example, in the language learning literature, procedures are examined that "invent" such notions as "noun phrase" and "verb phrase" in order to account for observed sentences. The predicates used are arbitrary symbols that receive their entire significance from the conjectured grammars they appear in.

Third, there is no reason why the framework cannot be altered to account for meaning change both in the data and in conjectures as the theories produced by the theorist change. We can assume that an inductive problem is a set of possible *worlds-in-them-*

selves instead of a set of possible worlds. A world-in-itself can be viewed as a partial function of theories that sometimes pairs a given theory with a *world-of-experience*, which makes hypotheses and evidence sentences true or false. Convergence to the truth can be replaced with convergence to a world-of-experience and to the complete truth about this world-of-experience. Induction is not thereby trivialized unless we are radical coherentists (i.e. we assume that the inductive agent has complete control over which world-of-experience is actual). The picture of science that is implicit in this kind of formal learning theory might be called convergent *idealism* to distinguish it from *convergent realism*. We have already verified that many of the formal insights obtained in the realist version of formal learning theory also obtain in the idealist version.

Notes

[1]For a dated, but useful survey of the field, see (Angluin and Smith 82). Another useful source is (Osherson, Stob and Weinstein 86).

[2]For a detailed detailed presentation of the following discussion of convergence criteria, see (Kelly and Glymour 1989).

[3]For a lucid introduction to the theory of NP-completeness, see (Garey and Johnson 1979).

[4]Some artificial intelligence systems implicitly assume such a formal model of experimentation. For example, see (Carbonell and Gil 1987).

[5]The result follows from Theorem 2 of (Seidenfeld and Schervish 1987).

References

Angluin, D. and Smith, C. (1982), "A Survey of Inductive Inference Methods", *Technical Report* 250, Yale University, October.

Carbonell, J. and Gil, Y. (1987), "Learning by Experimentation", in *Proceedings of the Fourth International Workshop on Machine Learning"*, San Mateo: Morgan Kaufmann.

Carnap, R. (1950), *Logical Foundations of Probability*, Chicago: University of Chicago Press.

Cherniak, C. (1986), *Minimal Rationality*. Cambridge, Mass.: MIT Press.

Garey, G. and Johnson, D. (1979), *Computers and Intractability*. New York: W. H. Freeman and Company.

Glymour, C. (1980), *Theory and Evidence*. Princeton: Princeton University Press.

Hempel, C. (1956), "Studies in the Logic of Confirmation", in *Aspects of Scientific Explanation*, New York: The Free Press.

Kearns, M. and Li, M. (1988), "Learning in the Presence of Malicious Errors", *Proceedings of the 20th ACM Symposium on Theory of Computing*, Chicago, Illinois.

Kelly, K. and Glymour, C. (1989), "Convergence to the Truth and Nothing But the Truth", *Philosophy of Science*, forthcoming.

_____. (1989b), "Theory Discovery from Data with Mixed Quantifiers", *Journal of Philosophical Logic*, Forthcoming.

Laudan, L. (1980), "Why was the Logic of Discovery Abandoned?", in *Scientific Discovery, Logic, and Rationality*, T. Nickles, (ed).

Osherson, D., Stob M. and Weinstein, S. (1986), *Systems that Learn*. Cambridge, Mass.: MIT Press.

Oxtoby, J. (1970), *Measure and Category*. Berlin: Springer.

Pitt, L. and Valiant, L. (1988), "Computational Limits on Learning from Examples", *Journal of the Association for Computing Machinery*, Vol. 35, no. 4.

Popper, K.R. (1968), *The Logic of Scientific Discovery*. London: Hutchinson and Co.

Putnam, H. (1963), "'Degree of Confirmation' and Inductive Logic", in *The Philosophy of Rudolph Carnap*, A. Schilpp (ed), Lasalle, Illinois: Open Court.

Seidenfeld, T. and Schervish, M. (1987), "An Approach to Consensus and Certainty with Increasing Evidence", forthcoming in *Journal of Statistical Inference and Planning*.

Valiant, L. (1984), "A theory of the Learnable", *Communications of the Association for Computing Machinery*, vol. 27, pp. 1134-1142.

Van Fraassen, B. (1980), *The Scientific Image*. Oxford: Clarendon Press.

Learning Simple Things:
A Connectionist Learning Problem from Various Perspectives[1]

Edward P. Stabler, Jr.

University of Western Ontario

The performance of a connectionist learning system on a simple problem has been described by Hinton and is briefly reviewed here: a finite function is learned, and the system generalizes correctly from partial information by finding simple "features" of the environment. For comparison, a very similar problem is formulated in the Gold paradigm of discrete learning functions. Identification in the limit from positive text of a large class of functions including Hinton's is achievable with a trivial, conservative learning strategy. Using Valiant's approach, we place an arbitrary finite bound on function complexity and then we can guarantee text and resource efficiency relative to a probabilistic criterion of success. But the connectionist system generalizes. That is, it uses a non-conservative learning strategy. We define a simple, non-conservative strategy that also generalizes like the connectionist system, finding simple "features" of the environment. Finding the "best" features in a relevant sense is a hard problem, but it is conjectured that by placing an arbitrary finite bound on the number and complexity of the features to be found, efficient learning can be guaranteed relative to a (weaker) probabilistic criterion of success. It is possible that learners of this kind will be useful in models of human induction, but no convincing support for this view has been presented. This approach to induction has essentially the same problems as many others that have failed. By analogy, we see also that it is specific details of the connectionist learning system that account for its performance, and such systems will have little psychological interest until those details are motivated.

1. A simple learning problem and a connectionist approach

In some recent studies, Hinton has proposed that some basic, frequently repeated reasoning might be accomplished not by explicit representation and sequential reasoning, but by "microinferences" performed by a highly parallel computing device. His early work on how some the frequently used knowledge might be learned has produced some interesting results which I want to consider in some detail.

Consider the following relations: father, mother, son, daughter, brother, sister, husband, wife, uncle, aunt, nephew, and niece. We can formalize these relations in such a way that they hold between a person and a (nonempty) set of all the people that stand in the relation to the person. The family trees shown in Figure 1 name 24 individuals who play a role in 104 instances of these 12 binary relations. Hinton designed a layered "connectionist" learning system which I will call "HCLS" (Hinton, 1986), which learned the family relations shown, after being trained by "back propagation."

PSA 1988, Volume 2, pp. 424-441

We can associate each of the 24 people in the tree of Figure 1 with one of the propositions I1,...,I24, where proposition Ij is true just in case person j is the person whose relations are being computed. Similarly, each of the 12 relations can be associated with a proposition R1,...,R24, where proposition Rj is true just in case the j'th relation is the one being computed. And finally, the 24 people can be associated with "output" propositions O1,...,O24, where proposition Oj is true just in case person j is one of the members of the set of people that stand in the appropriate relation to the appropriate "input" person. In each training trial, HCLS all relations except one relation Rj are set to 0, and all input person propositions except one person Ik are set to 0, and HCLS was corrected according to the discrepancy between its output and the correct result, which would be an output representation of all and only the people Om standing in relation Rj to Ik.

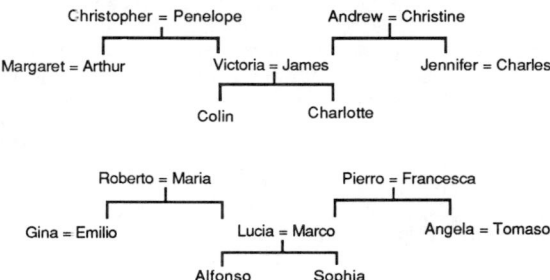

Figure 1: Family trees from (Hinton, 1986). "=" means "is married to".

HCLS was trained with 100 of the 104 different instances of the family relations, with 1000 trials per instance. After these learning trials, the system was able to compute the results for the 100 training instances reliably. As a feat of learning, this is not particularly impressive. As Hinton says,

> The fact that the network can learn the examples it is shown is not particularly surprising. The interesting questions are: Does it create sensible internal representations for the various people and relationships that make it easy to express regularities of the domain that are only implicit in the examples it is given? Does it generalize correctly to the remaining examples? Does it make use of the isomorphism between the two family trees to allow it to encode them more efficiently and to generalize relationships in one family tree by analogy to relationships in the other? (Hinton, 1986).

The striking thing is Hinton's answers to these questions are all positive. Two training and test sessions were carried out. In one of these, the system correctly computed 3 out of the 4 instances of the relationships that were not used in training. In the other session, all 4 were correctly computed. And examination of the system's performance showed that it responded in a distinctive way according to which tree the input person was in, which generation the person was in, and which branch of the tree the person was in.

How can this happen? The system is not given any specific information about family relations in advance. That is, its only knowledge specifically about these relations comes from the original exposures to instances of these relations. The architecture HCLS implicitly embodies only some very abstract ideas. For example, it requires that the input propositions group first into person features on the one hand and relation features on the other. But the original 100 instances do not suffice to determine the remaining 4. To belabor this obvious point just briefly, here is another way of putting it. We could easily define new relations which agree with the original 100 instances but differ on the remaining 4. In any such case, the very same training would lead the very same connectionist system to generalize incorrectly. This obvious point is essential to understanding the difficulty of learning problems!

There is no such thing as a general solution to the problem of how to determine missing elements of a sequence.

The interest of Hinton's remarks is in his suggestion about how the connectionist system generalizes. Apparently, the alternative relations on which the connectionist system would generalize incorrectly are, in some sense, computationally more complex than the family relations. If a learner "knows" that generalization should be done in accordance with some very restrictive complexity metric, this can guide generalization properly on functions that are, in fact, simple. This Chomskyan idea is in fact what Hinton suggests. He describes how he enforces a limit on resource complexity with the architecture of HCLS:

> The architecture is designed so that all the information about an input person must be squeezed through a narrow bottleneck of six units in the first hidden layer. This forces the network to represent people using distributed patterns of activity in this layer. The aim of the simulation is to see if the components of these distributed patterns correspond to the important underlying features of the domain...After prolonged training...it generalized correctly because it learned to represent each of the people in terms of important features such as age, nationality, and the branch of the family tree they belonged to...even though these "semantic" features were not at all explicit in the input or output vectors....The learning procedure can only discover these features by searching for a set of features that make it easy to express the associations...Thus the network constructs its own similarity metric. This is a significant advance over the simulations in which good generalization is achieved because the experimenter chooses representations that already have an appropriate similarity metric. (Hinton, 1987ms)

Hinton suggests that this discovery of an appropriate similarity metric, appropriate "features," to guide generalization is analogous to the problem of finding natural kinds in science (Hinton, 1986).

My curiosity is more piqued by these suggestions than assuaged! Does the connectionist machine's generalization really provide a model for the sorts of simplicity that human learners look for? Among the important puzzles that are left unresolved are the following:

(1) Is the set of functions that HCLS can learn big enough to be non-trivial?

(2) Does HCLS learn efficiently? For example, how much more hardware and how many more trials are needed when the number of relations being learned is doubled?

(3) Is the function HCLS learns in a significant class that is already known to be identifiable by a discrete function?

(4) Is there a resource-efficient and text-efficient discrete learner for a significant class of functions that includes the function HCLS learns?

(5) Can we provide a simple and intuitive interesting notion of "features" of an environment that can play a role in defining generalization, in the fashion suggested by Hinton?

(6) Can we design a discrete learning system such that
(i) it learns a non-trivial class of functions including the one learned by HCLS, and
(ii) it recognizes the same features found by HCLS, and uses these to generalization correctly on the family tree problem, and
(iii) it is resource and text efficient?

Unfortunately, we do not have useful mathematical characterizations of the connectionist systems, models that would allow us to find a simple characterization of such a simplicity metric by studying the constraints on the range of learnable functions embodied in the structure of HCLS. Connectionism seems to be largely an empirical business at this point. I will leave my first two questions unanswered, and provide affirmative answers to the rest. The answers to questions (3) and (4) are already well known. Questions (5) and (6) are more interesting. It turns out that there is a natural and simple notion of "feature" which can be found and exploited by a learner. These features are computationally difficult to find in general, but we can use Valiant's trick of putting a bound on the number and complexity of the features we look for to guarantee efficiency. Once we have obtained an intuitively simple account of what is going on, we consider the bearing of such approaches on models of human learning. Our answers to (3)-(6) show the importance of defining the range of problems solved by a learning system, and confirm Pylyshyn's (1988) suggestion that most significant and interesting constraints proposed for connectionist architectures have rather direct analogs in classical automata.

2. Notation for the propositional calculus formulation

In each trial, HCLS is given a person Ij and a relation Rk, and tries to guess the set $S = \{Om \mid Ij\ Rk\ Om\}$. Thus the information given to HCLS and the goal of the learning can be modeled propositionally. Let just one of propositional variables I1,...,I24 be true when the corresponding person is to be the first argument of the relation, and let one of R1,...,R12 signify the relevant relation. We can call these the input propositions. Let O1,...,O24 be the propositions signifying the second arguments of the relations. We will call these the output propositions. Altogether, we have 60 propositional variables. In advance of the trials, HCLS has no information about any connection between any of them. Similarly, our propositional representation does not provide any structure: every propositional variable is completely without structure, and we will not exploit any special ordering on these atomic propositions to sneak in an appropriate similarity metric for the family tree problem.

Let me define some terms for our propositional analysis.

Definition: The literals are the propositional variables and their negations.

Definition: A (total) vector is a truth assignment. It is convenient to think of a vector as a sequence $\{0,1\}^n$ of truth values corresponding to a sequence of the n propositional variables.

Definition: When the propositional variables are partitioned into input propositions and output propositions, then an input vector is a truth assignment to the input propositions; and an output vector is a truth assignment to the output propositions.

Definition: A Boolean function F is a mapping from the set of vectors $\{0,1\}^n$ to $\{0,1\}$.

Definition: Let F/phi be the Boolean function defined by propositional calculus formula phi.

Definition: A vector v verifies a function F if $F(v)=1$.

Definition: $F => G$ iff for all vectors v, if $F(v)=1$ then $G(v)=1$. In this case we say F implies G.

It is probably most natural to regard HCLS as computing a function from 36 inputs to 24 outputs:

h: $\{0,1\}^{36} \rightarrow \{0,1\}^{24}$.

A function f that maps binary sequences of length n into sequences of length m is most naturally regarded as m different Boolean functions, but it is more convenient for our purposes to consider the single Boolean function

$$F: \{0,1\}^{\wedge}(n+m) \to \{0,1\}$$

such that $F(v)=1$ iff $f(vI)=vO$, where v is the result of appending the m-element output vector vO to the n-element input vector vI. The function h learned by HCLS, then, can be modeled with a single Boolean function H on the 60 propositional variables introduced above.

Given a partition on the variables into input and output, we can make the following distinctions:

Definition: If there is some truth assignment for the input propositional variables which is such that no extension of this truth assignment verifies F, then F corresponds to a partially undefined i/o function.

Definition: If there is some truth assignment for the input propositional variables which is such that more than one extension of this truth assignment verifies F, then F corresponds to a nondeterministic i/o function.

3. Defining propositional learners

In Hinton's experiment, each training trial can be regarded as providing a vector that verifies the "target" Boolean function H. Actually, since HCLS makes a guess in each trial and then is corrected, it gets negative information as well, but the propositional learners we will consider here do not need this information about points that do not verify the function. The object of the learning is to determine this function H which defines the appropriate relations between our propositional variables. For example, if Ij, Rk, and Om correspond to Colin, the father relation, and James, respectively, then we will have, for example, H => F/phi when phi is the formula that says that when the when Ij and Rk are true but all other input variables are false, then Om is true and all the other output variables are false.

With this picture of learning, the following definition is natural:

Definition: A propositional learner PL is a (total) mapping from finite sequences of total vectors (corresponding to the trials) to Boolean functions (corresponding to the guesses about the target).

Definition: Given a function F, a (positive) text T for F is a sequence of total vectors that contains all and only vectors that verify F. We will call this function F the target function of the text.

As usual, given a text, we can think of a propositional learner as making a guess on each initial segment of the text. Intuitively, as the learner sees more and more evidence, its conjectures are based on longer and longer initial segments of a text. We can define the following familiar notions:

Definition: PL converges on text T iff there is a function F that is the value of PL for all but finitely many initial segments of T. In this case we say PL(T)=F.

Definition: PL identifies text T iff PL converges on T and T is a text for PL(T).

Definition: PL identifies function F iff PL identifies every text for F.

Definition: PL identifies a class of functions CLF iff it identifies every F in CLF.

Definition: A class of functions CLF is identifiable iff there is a learning function that identifies it.

4. Some trivial propositional learners for the family tree problem

Now let's consider the simple learning problem solved by HCLS. It is a trivial matter to define a learning function that will identify the appropriate Boolean function H on 60 propositional variables. Simply let the learning function conjecture H on every input. The moral of this trivial observation is: a learning system that is shown to learn only one thing is not impressive. Hinton describes the performance of his connectionist learning system on only one problem, but the suggestion that it could learn some other (unspecified) things is implicit. We will describe some learners that can learn large collections of functions that include H.

In our framework, it is easy to see that the class of Boolean functions of n variables is identifiable from positive text (for any n). Of course, H is one member of the $2^{\wedge}(2^{\wedge}60)$ functions in the class of Boolean functions on 60 variables. Since the set of Boolean functions on n variables is finite, and since each of these functions has a finite domain, this result is a special case of well-known results. (Cf. Gold, 1967 or Osherson, Stob and Weinstein, 1986, Proposition 1.4.3A and Exercise 1.4.3C.) It is perhaps enlightening, though, to consider some algorithms that will compute a suitable learning function for this class.

To formalize the matter, let's represent each vector by the conjunction of literals it verifies. The conjectures can be represented by formulas as well. One particularly straightforward learning strategy is to begin with the conjecture that the target does not verify any vector, and then add the cases demanded by the evidence. Consider the following simple but impractical algorithm that uses this conservative strategy:

Algorithm PL-DNF:

    Input: A j-element initial segment of a text T for target function F.

    Output: A DNF formula phi(j).

    * Initialize

    Let phi(0) be the (always false) empty disjunction.

    Let i=1.

    * Refine

    For i=1 to j do

        Select vector v(i) from T.

        Let phi(i) be the result of adding v(i) as a new disjunct to phi(i-1) whenever it is not already included.

It will not introduce confusion if we call the function computed by algorithm PL-DNF simply "PL-DNF." The following proposition is obvious.

Proposition 4.1: PL-DNF identifies the class of Boolean functions.

Proof: By definition, every text for F contains all of the vectors verifying F, and since there are only finitely many of them, they all must occur in some j-element initial segment of the text, for finite j. For any such j, then, it is clear that {v|F(v)=1} = {v|F/phi(j)(v)=1}, i.e. F=F/phi(j). Furthermore, it is clear that for all k > j-1, phi(k)=phi(j). QED

Exactly the same effect can be achieved using CNF rather than DNF, and we will see in a moment that the CNF learner, although rather less intuitive, is more easily made efficient.

Algorithm PL-CNF

　　Input: A j-element initial segment of a text T for target function F.

　　Output: A CNF formula phi(j).

　　* Initialize

　　Let phi(0) be the conjunction of all disjunctions of literals.

　　Let i=1.

　　* Refine

　　For i=1 to j do

　　　　Select v(i) from T.

　　　　Let phi(i) be the result of deleting every conjunct of phi(i-1) that is not verified by v(i).

Proposition 4.2: PL-CNF identifies the class of Boolean functions.

Proof: (Cf. Valiant's (1984) Thm A.) Let F be an arbitrary target Boolean function and let T be an arbitrary text for F. We must show that there is a finite j such that for all k > j-1, the k-element initial segment T(k) of T is such that PL-CNF(T(k))=phi(j) where F/phi(j)=F. This result follows if we establish the following claims:

(1) At every step, F/phi(i) => F.

(2) There is a finite j such that for all k > j-1, F => F/phi(k).

(Proof of Claim 1) Initially, for i=0, it is obvious that {v|phi(i)(v)} is the empty set, and so F/phi(i) => F is trivially true. Let B be the product of all the disjunctions c such that, for all vectors v, if F(v)=1 then F/c(v)=1. Obviously, the algorithm will never delete any clause in B from phi(i), and so {v|phi(i)(v)=1} is a subset of {v|F/B(v)=1}. Thus it suffices to prove that F/B=F. This is established by showing B => F and F => B. By definition, every clause in B is implied by every vector that implies F, so F => B. If some clause c' verified by F did not occur in B, then, by the definition of B, there would have to be a vector v such that F(v)=1 but F/c'(v)=0. But this is impossible since if c' is verified by F and if F/c'(v)=0, then F(v)=0.

(Proof of Claim 2) At each step, algorithm PL-CNF deletes all of the disjunctions in phi(i-1) that do not contain a literal that is verified by v(i). Thus, F/phi(i)(v(j))=1 for all j < i+1. By definition, every text for F contains all of the vectors verifying F, and since there are only finitely many of them, they all must occur in some j-element initial segment of the text, for finite j. For any such j, then, it is clear that {v|F(v)=1} is a subset of {v|F/phi(j)(v)=1}, i.e. F => phi(j). Furthermore, it is clear that for all k > j-1, phi(k)=phi(j). QED

Unfortunately, these learners are not feasible because of the size of the conjectures. In the case of PL-DNF, the conjectured formula contains every vector seen so far. If there are n propositional variables, there are 2n literals, and so there can be as many as 2^{2n} disjuncts in a conjecture. In the case of PL-CNF, the problem arises in the initialization step. If there are n propositional variables, there are 2n literals, and the 2^{2n} subsets of these literals correspond to the 2^{2n} conjuncts of phi(0) in PL-CNF.

More importantly, PL-DNF and PL-CNF do not shed any light on the principal concern of this paper, which is to find some constraints on the learner or the class of learnable functions such that only one conjecture is compatible with the first 100 family relationships, and this conjecture generalizes appropriately to the remaining 4. Since PL-

DNF and PL-CNF can learn any Boolean function, they still have lots of options after the first 100 trials, and so they do not know how to generalize to the next 4 cases.

5. Two efficient propositional learners with limitations

We have to restrict our learner if its conjecture is to be determined by the 100 trials of Hinton's example! Valiant notes a constraint that limits the class of learnable functions and improves efficiency. We want to consider only "simple" functions, where the relevant "simplicity" is one which will give us learning more like that of HCLS. Valiant's idea is to make use of the fact that Boolean functions have canonical CNF formulas, and so we can place restrictions on the syntactic complexity of these. He simply puts an arbitrary finite bound on the length of the conjuncts in the CNF formula. That is, he considers the class of functions that are expressed by k-CNF formulas, where these are CNF formulas whose conjuncts are disjunctions of no more than k literals. It is straightforward to modify algorithm PL-CNF so that it identifies the class of Boolean expressions defined by k-CNF formulas. The only change is that the initial conjecture, phi(0), is the conjunction of all disjunctions of at most k literals. This has the effect of keeping the initial conjecture to a size that is a polynomial function of n. When there are n propositional variables, there are less than $(2n)^{(k+1)}$ disjunctions in phi(0). Call this algorithm and the function it computes "PL-k-CNF." With a little reflection, it is clear that PL-DNF is not so easily simplified.

The algorithm "PL-k-CNF" still has an efficiency problem though. It will not converge on the target until it has seen every vector in the text. Obviously, the number of vectors that verify a function can be infeasibly large, which means that we cannot be guaranteed to converge on the correct function until an infeasibly large initial segment has been processed, even when there is no redundancy in that initial segment. Valiant has shown that a straightforward probabilistic version of this algorithm can allow us to achieve a reasonable probabilistic learning criterion in a feasible number of steps. The idea is roughly the following. The only change we need to make in the algorithm is to use a "probabilistic text:"

Definition: Let a probabilistic text be a source of vectors such that the probability of getting any one of these vectors v from the text at any step is D(v), where D is a probability distribution over the vectors that verify the target.

We can call the algorithm that results from this change PPL-k-CNF (for "Probabilistic Propositional Learner of k-CNF functions"). Now consider a the following probabilistic criterion of successful learning:

Definition: Letting epsilon be a parameter that determines how close we must get to identification of the target, we will say a learner nearly identifies a function F iff there is a probability of at least 1-(1/epsilon) that for some i, all conjectures phi(j), j > i-1 are such that

(i) F/phi(j) => F, and
(ii) the sum of the probabilities of the vectors that verify F but not the conjecture is at most 1/epsilon (i.e., if it were not for vectors that occur with probability of at most 1/epsilon, we would have F => F/phi(j)).

Valiant's result is the following:

Theorem: (Valiant, 1984) PPL-k-CNF is resource efficient and near identification of the class of functions defined by k-CNF formulas is achieved by conjecture phi(j) which is output after PPL-k-CNF has seen j elements of probabilistic text, where j is a polynomial function of epsilon and the number of propositional variables n.

In short, PPL-k-CNF is resource-efficient and text-efficient with respect to a probabilistic criterion of success. So we see that the limitation to k-CNF functions can vastly improve the efficiency of the learning process, especially when k is small relative to the number n of propositional variables. What bound k can be used in our simple learning problem? Well, k must be at least large enough to allow us to define an appropriate target function H, but we have not yet been quite clear about what that target function should be. Now consider the problem of finding the smallest k such that H can be defined by a k-CNF formula. If we want a Boolean function H which corresponds to a partially unde-fined but deterministic i/o function, where H is 0 at all points except the ones that corre-spond to the intended family relationships, then k must be at least 37. Each 37-literal con-junct can define the value of one output proposition on the basis of the 36 input proposi-tions. The number of conjuncts that our conjecture could require is now bounded by a value that is not exponential in n: $(2n)^{(k+1)}=120^{37}$. This is still a very big number! We have efficiency only in the technical sense that the bound is polynomial in n.

So let's consider the properties of PL-k-CNF and of PPL-k-CNF that are important for present purposes. In the first place, the restriction on the range of conjectures does of course imply that there are some "false generalizations" from the evidence that will never be con-sidered. This is unsurprising. Given the conservative strategy of the earlier algorithm PL-CNF, it never conjectures functions that are not 37-CNF either. The disappointment is that we have not yet made any progress toward our goal of getting a system that will always generalize correctly from 100 instances to the remaining 4. Obviously, given any 100 of the 104 vectors that verify H, there are many different Boolean functions defined by 37-CNF formulas that are verified by these vectors which are not verified by the other 4.

6. Extremely limited propositional learners

The problem with the learners considered so far is that they do not generalize the way HCLS does. After looking at 100 relations, lots of options are still open for the remaining 4. The k-CNF complexity measure used to restrict the problem did not rule out any of the spuri-ous generalizations in the final four cases of the family tree problem. We need a rather rather different limitation, one that is so severe that it rules out the spurious generalizations and forces us to abandon the conservative strategy of conjecturing that the target verifies only the evidence seen so far. Putting matters this way, it is easy to see that the problem is just to char-acterize an appropriate difference between H and functions that are like H except that they are verified by different extensions of four truth assignments to the input variables. Here it is important to remember that we want a domain-independent difference. In the passage quoted above, Hinton points out that he used tight resource "bottlenecks" to restrict the options of his system. It is much less interesting to simply cook things up so this particular case will work.

So what domain-independent measures distinguish the family relations from a similar set of relations that differ on just four cases? Here is one example of the right kind of measure. Since there are 24 output propositions, there are 2^{24} output states, each of which could be a value for some of the $2^{(24+12)}$ possible input states. But the bottlenecks in HCLS drastical-ly restrict the possible mappings between the inputs and outputs. We have already noted that the only inputs of interest are 104 in which exactly one person input and one relation input are true. The number of output states needed for these inputs is not 2^{24} but only 28! The sets of correct outputs comprise the 24 singleton sets of people, and 4 sets containing two people each — the two aunts and the two uncles of Charlotte, Colin, Alfonso and Sophia. A "bottleneck" is easily characterized in non-procedural terms as follows:

Definition: An output state of Boolean function F on input and output propositional variables is an output vector vO that is included in some total vector that verifies F.

Definition: A function F can go through a bottleneck of size p iff F has no more than 2^p output states.

Hinton's target function H can thus go through a bottleneck of size 5. In an implementation this would involve using what Hinton calls a distributed representation of the input vectors. That is, a bottleneck of size p=5 exists just in case there is a 1-1 mapping or assignment of the 28 output states into the binary states of 5 units, so we can regard the bottleneck as such a mapping:

bottleneck: outputs -> {0,1}^p,

where the mapping is 1-1. Given such a mapping, the bottleneck effectively defines a set of 5 features on the set of outputs, in the following sense:

Definition: A feature of elements of a set P is a binary partition on P.

A bottleneck of size m for a set of propositions defines a set of m features in the following way:

Definition: For $0 < i < m+1$, feature i of set P is possessed by all those p in P such that bottleneck(p) has a 1 in the i'th place.

Since the inputs determine the outputs in the cases we are considering, the bottleneck can also be regarded as determining features of the input states, in the obvious way. Hinton's idea is that if the learning system identifies these features appropriately, then it can generalize appropriately from 100 cases to the next four.

It is easy to see that the bottleneck is, in fact, a step toward the goal of avoiding spurious generalizations. This follows since spurious generalizations from the original trials can easily change the number of output states required, with the consequence that the function would not fit through a bottleneck of size 5. This metric does distinguish the family relations from some spurious generalizations, but still does not determine any particular mapping on the basis of 100 of the 104 vectors. It is easy to see that the output on each of the 4 untrained inputs could be any of the 28 output vectors.

Two points are worth noting here. First, although there is some flexibility in HCLS due to changing weights between computing units, still each computing unit is restricted to the computation of a very simple function of its inputs. The idea of our "bottleneck," on the other hand, is inspired by the idea that while we may have a restricted number of computing units, still, each unit can compute an arbitrary Boolean function. This may be the reason that our simple complexity measure, the number of output sets, is inadequate. It is well-known that on many, more sensitive complexity measures, bottleneck functions of a given size can differ greatly! A second, related point to notice is that the deviant generalizations noted in the previous proof keep the number of output sets low by violating the simple patterns of relationships in the family trees. So let's try restricting the computational complexities of our conjectures in such a way that features corresponding to our bottlenecks must be simple, in a sense we will need to make precise.

7. A learner that uses features

To make the presentation more comprehensible, let's consider just part of the family tree problem. Consider the following subtrees from Hinton's example:

Figure 2: Two subtrees from Hinton's family trees.

Five of Hinton's 12 family relationships hold in these little subtrees: mother, son, daughter, sister, and brother. We consider learning just these.

Setting up this subproblem in just the way we set up the larger one, the learner is to identify a Boolean function on the basis of vectors that verify the function. In this case, the target function is verified by only 12 vectors, so it is feasible to use them directly in the specification of the target. We assign an input and an output variable to each person (penelope, arthur, victoria, maria, emilio, and lucia). And we assign 5 input relation variables to the relations (mother, son, daughter, sister, and brother). Then the vectors given to HCLS define the target Boolean function which is verified by the vectors shown in the following truth table (we use blanks to represent the preponderant 0's):

I_p	I_a	I_v	I_m	I_e	I_l	R_m	R_{so}	R_d	R_s	R_b	O_p	O_a	O_v	O_m	O_e	O_l
1							1					1				
1								1					1			
	1					1					1					
	1								1				1			
		1				1					1					
		1								1		1				
			1				1								1	
			1					1								1
				1		1								1		
				1					1							1
					1	1								1		
					1					1					1	

Figure 3: The vectors that verify Hm.

The problem of finding simple representations of Boolean functions has been well studied, so we can just consider some of the standard techniques. Quinean minimization of a DNF formula (Quine, 1952) involves repeated application of the rule ((p q) or (~p q)) <-> q. It is clear that this process will not simplify the formula for Hm shown above, since every conjunction has the same number of positive literals. We could use the length of this minimal DNF as a complexity measure, but this measure is nowhere near sensitive enough to distinguish Hm from other functions that are like Hm on, say, 11 of the 12 verifying vectors.

There are clearly regularities in Hm that are not reflected in the length of minimal DNF representations. Notice, for example, the large blocks of 0's in the truth table shown above. For example, I have arranged the variables and vectors in such a way that (~Im ~Ie ~Il), and (~Ip ~Ia ~Iv) stand out. These conjunctions of negative literals or "cubes," each of which is present in 6 vectors, correspond to the properties "not Italian" and "not British". Let the "size" of a cube be the number of literals occurring in it times the number of verifying vectors it occurs in. The suspicion that we may be on the right track is further confirmed by the fact that the presence of large cubes of this kind can be exploited for efficient computation of the function. (Cf. Savage (1976); or Brayton and McMullin (1982).) However, just among the input person variables, there are 20 cubes as large as (~Im ~Ie ~Il), and (~Ip ~Ia ~Iv). We require a more sensitive measure of complexity that will be reduced only when the large cubes have some other property.

One measure that is commonly used is to seek cubes that determine the largest number of other values in the verifying vectors. (This roughly corresponds to the strategy that is known in switching theory as "minimizing variable interdependence" in multilevel combinational networks. See Hartmanis and Stearns (1966), de Micheli et al. (1985).) This is one of the striking features of the cubes (~Im ~Ie ~Il) and (~Ip ~Ia ~Iv). Not only are these cubes fairly large, but we can see that whenever a vector that verifies Hm also verifies one of these cubes, a large number of the output proposition truth values are also determined. For example, any vector that verifies (~Im ~Ie ~Il) also verifies (~Om ~Oe ~Ol). The idea can be formalized as follows:

Definition: The value of a cube c in F is the number of literals l not in c such that l is verified by every assignment that verifies F and c.

This measure provides the needed sensitivity. The cubes (~Im ~Ie ~Il) and (~Ip ~Ia ~Iv) have each have a value of 3. A less natural cube like (~Ie ~Ip ~Il) has a value of 1.

Hinton's discussion of HCLS indicates that 3 different bottlenecks are used in his system: one for input persons, one for input relations, and one for outputs. We can use a similar strategy on our subproblem, with smaller bottlenecks. The set of assignments corresponding to the 6 input persons in our subproblem can be defined with a bottleneck of size 3. That is, each of three binary units in a bottleneck can represent some feature of the input persons in such a way that an input person vector verifying exactly one person has a unique triple of features. In this case, we will say that the triple of features is a "decomposition" of the input person variables. This notion extends in the obvious way to n-tuple decompositions of arbitrary sets, where we call n the "size" of the decomposition.

We can now compute the most valuable decompositions, the most valuable bottlenecks. Consider the input person bottleneck first. We decompose the different input person vectors into three features, each of which is a cube, and we look for the decompositions with the most valuable cubes. In our subproblem, there is a 6-way tie among these decompositions, but all 6 of the winners are intuitive:

(1) (not Young), (not Left or Center), (not British)

(2) (not Young), (not Left or Center), (not Italian)

(3) (not Young), (not British), (not Right or Center)

(4) (not Young), (not Italian), (not Right or Center)

(5) (not Left or Center), (not British), (not Right or Center)

(6) (not Left or Center), (not Italian), (not Right or Center)

Each of these decompositions is sufficient to represent the input person propositions. For example, using the first of these decompositions, the input person vector in which penelope is the person would be represented as ((not Young) ~(not Left or Center) ~(not British)). No other input person vector verifying only one person has this same representation. Where the value of a triple is the sum of the values of its elements, these triples all have the value 13. No other decompositions of the input person assignments are as valuable.

Now we can raise the question of whether the presence of these particularly valuable decompositions distinguishes the verifying vectors of Hm from all other functions which agree on only 11 of the 12. Unfortunately, a little exploration shows that this does not suffice, but we are getting close. If we restrict our learners to functions that are i/o deterministic, and if we add weighting function which might be called a "low energy constraint," we can learn like HCLS.

Definition: The energy of a vector is the number of atomic formulas it verifies.

Now it is easy to define a measure on decompositions that increases with value and decreases with energy:

Definition: The goodness of a decomposition is the value of the decomposition divided by the average energy of the assignments in the block of variables decomposed.

Proposition 7.1: The function Hm is verified by exactly 12 vectors. The assignments to input person variables, input relation variables, and output variables can be encoded by triples with goodness 13, 20 and 13, respectively.

We can now state the key result, which was established by calculation:

Proposition 7.2: There is no function Hm' with the following properties:

(a) Hm' is i/o deterministic.

(b) Hm' is verified by more than 11 vectors but only 11 of them also verify Hm.

(c) the verifying assignments to input person variables, input relation variables, and output variables can be encoded by triples with goodness of at least 13, 20 and 13, respectively.

This proposition suggests the following learning strategy. When the learner sees n verifying vectors, if these have 3 element decompositions of each of the three blocks with goodness of at least 13, 20 and 13, respectively, then the function conjectured is one that is verified by the largest number $m > n-1$ vectors such that these can be encoded by 3 triples with goodness 13, 20 and 13, respectively. In general, we can consider the variables partitioned into k blocks, and for each block we specify a decomposition size and a lower bound on goodness. The blocks are simply the sets of variables to be decomposed into feature sets, the size of each decomposition is just the number of features to be used, and the goodness of each decomposition is a measure of its computational "simplicity."

To formalize this general perspective, we assume an ordering on the n variables. A learner for Hm can be specified then by saying that the first 6 variables have a 3 element decomposition with a value of at least 13, the next 5 variables have a 3 element decomposition with a value of at least 20, and the final 6 variables have a 3 element decomposition with a value of at least 13. For short, we will call this a ((6,3,13),(5,3,20),(6,3,13))-learner. In general, we can consider (k,s,g)-learners. Now the peculiarities of our problem begin to resolve themselves into a familiar pattern, but more on this later. Let's first provide a learning algorithm. It is convenient to introduce the following definition:

Definition: A (k,s,g) Boolean function, where (k,s,g)=((k1,s1,g1),...,(ki,si,gi)) is a function on n variables where n is the sum of k1,...,ki, such that

(1) there is a s1 element decomposition of the verifying assignments to the first k1 variables with goodness greater than or equal to g1;

(2) there is a s2 element decomposition of the verifying assignments to the next k2 variables with goodness greater than or equal to g2;

...

(i) there is an si element decomposition of the verifying assignments to the last ki variables with goodness greater than or equal to gi.

Algorithm (k,s,g)-DNF

Input: A j-element initial segment of a text T for target function F.

Output: A DNF formula phi(j).

* Initialize

Let phi(0) be the (always false) empty disjunction.

Let i=1.

* Refine

For i=1 to j do

    Select vector g(i) from T.

    Let phi(i) be the result of adding g(i) as a new disjunct to phi(i-1) whenever it is not already included.

* Generalize

For b=1 to the number of blocks specified in the parameters (k,s,g), do

    While the best decomposition of verifying assignments in block b of F/phi(i) has a goodness that is (strictly) greater than the lower bound indicated by parameters, do

        extend phi(j) with the least vector v such that: (i) v is not already included in phi(j), (ii) v provides a new assignment to variables in block b, and (iii) F/(phi(i) + v) is (k,s,g) and i/o deterministic.

Given the previous proposition, it is not hard to establish that this algorithm handles Hm as desired:

Proposition 7.3: Algorithm ((6,3,13),(5,3,20),(6,3,13))-DNF generalizes correctly from any 11 vectors in a text for Hm to the 12th.

In HCLS, generalization was not quite perfect even after 96% of the vectors had been presented. For our learner, on the simpler subproblem, we can guarantee perfect generalization from 92% of the verifying vectors. I conjecture that a larger (k,s,g)-DNF learner generalizes correctly from any 100 vectors in a text for Hinton's problem H to the remaining 4. If not, we could make slight modifications in the definition of goodness so that it does.

I have not found any beautiful characterization of the class of functions identified by (k,s,g)-DNF. Here is the obvious one. (k,s,g)-DNF does not identify the class of (k,s,g) Boolean functions, because of the fact that so many of them lack the property noted in the following definition.

Definition: A function F is (k,s,g)-maximal iff it is (k,s,g) and there is no function G such that F => G, but it is not the case that G => F, and G is (k,s,g).

It follows from this definition and results mentioned earlier that Hm is ((6,3,13),(5,3,20),(6,3,13))-maximal.

Theorem: Algorithm (k,s,g)-DNF identifies the class of (k,s,g)-maximal i/o deterministic Boolean functions.

Proof: By definition, every text for F contains all of the vectors verifying F, and since there are only finitely many of them, they all must occur in some j-element initial segment of the text, for finite j. For any such j, then, it is clear that $\{v|F(v)=1\}$ is a subset of $\{v|F/phi(j)(v)=1\}$, i.e. F => F/phi(j). Furthermore, when F is i/o deterministic and (k,s,g)-maximal, it is clear that the generalization step will have no effect after all the vectors have been seen, and so F=F/phi(j). QED

Conjecture: Algorithm (k,s,g)-DNF is resource efficient. The parameters (k,s,g) guarantee a finite bound on complexity that is independent of n.

The basic idea behind this conjecture is similar to the idea behind Valiant's theorem: although the algorithm is solving a hard problem, the difficulty of the problem is not exponential in n, but exponential in some of the parameters, and these are set to specific finite values. (See Stabler, 1989, for details.)

Now suppose we modify our algorithm so that it uses a probabilistic text, following Valiant, and call the resulting algorithm P-(k,s,g)-DNF. We cannot guarantee efficient learning relative to Valiant's probabilistic criterion of success because our learner is not conservative, it generalizes. If we eliminate part (i) of Valiant's criterion, then, of course, the text efficiency of P-(k,s,g)-DNF is a simple corollary of his result.

8. Summary

The conclusions that we have established can be informally summarized as follows. H and Hm are in significant classes of functions which can be identified in the limit from positive text using the trivial conservative strategy. They are also in significant classes of functions for which there are conservative learners that are text and resource efficient relative to a probabilistic criterion of success. A bottleneck in a Boolean network restricts the set of functions that can be realized by restricting the number of output states the function can have. The value of features represented by units in a bottleneck (a decomposition) correlates in many cases with network complexity and with intuitive "naturalness." Valuable features by themselves do not indicate appropriate generalization. More precisely, given some initial vectors, generalizing to maximize decomposition values is inappropriate for Hm. Restraining value maximization with a penalty for "energy" suffices to determine a learner that generalizes appropriately for Hm and other (k,s,g)-maximal functions, but this move seems *ad hoc*.

9. The comparison with human learners

In a critique of another connectionist learning system, Pinker and Prince (1988) say:

It is often considered a virtue of PDP models that they are powerful learners; virtually any amount of statistical correlation among [components of] a set of inputs can be soaked up by the weights on the dense set of interconnections among units. But this property is a liability if human learners are more constrained.

The idea is familiar: Unconstrained learners are not very good models of human inductive practices. The interest of Hinton's work is precisely that it is not unconstrained. Neither are the regularities it identifies merely statistical correlations in any interesting sense. It is of course true that we could treat each vector as a multidimensional observation, and a trivial "cluster analysis" would show that certain specific points in multidimensional space are represented with greater or lesser frequency. (Cf. Johnson and Wichern, 1988.) But no statistical inference technique provides a more natural basis for generalizing to unrepresented points than our algorithm does. The most interesting developments in learning theory involve the discovery of ways to restrict the power of learning systems enough so that they generalize as interesting discovery procedures must, without being so limited that they become trivial or artificial from a human perspective.

Unfortunately, on closer analysis we can see that systems like these are both too restrictive and too artificial to be of much interest.

Hinton does not argue that the behavior of HCLS on the family tree problem is human-like at either the psychological or neurophysiological level. It is clear that we do not learn family relations in anything like the way HCLS does. But the features that are discovered and used by HCLS seem to be intuitive. This suggests the possibility that learning of roughly this kind could be going on in some human situations. Hinton even suggests that learning of this kind is analogous to scientific theory formation, but I think that this is the least likely area to find HCLS-like generalization. It is a commonplace that generalization in science is largely under the influence of background knowledge and training. I would think that a better area to look for application of these simple ideas would be in domains where background knowledge might have less influence. However, I am not optimistic that the kinds of learners studied here will prove useful in any models of human induction. Our models have an ad hoc character that is all too familiar from classic discussions of induction. When we try to solve what Goodman calls "the second riddle of induction" by examining some simple examples of appropriate induction, the first thing to notice is how easy it is to rule out many spurious generalizations with ad hoc restrictions. What we find over and over is that the restrictions appropriate for one problem are not appropriate for others. It is easy to see that our (k,s,g)-learners have this same problem. Our success in this paper was the following. We discovered that a ((6,3,13),(5,3,20),(6,3,13))-DNF learner can generalize correctly from any 11 of the verifying vectors to the 12th in this case, and it does so, in effect, by discovering some "natural" features of the family environment. This is nice because the restrictions on the learner are computational, not specifically relevant to family situations. However, we had to impose an ugly set of restrictions based on i/o determinacy, number, size and goodness of the decompositions. Furthermore, notice that the same learner simply cannot learn the relations in the Figure 4.

Figure 4: Two new family trees.

The Boolean target function for these relationships is not ((6,3,13),(5,3,20),(6,3,13)). And notice that our learner cannot learn the relationships in Figure 5 either.

Figure 5: Two new family trees.

The Boolean target function for these relationships is ((6,3,13),(5,3,20),(6,3,13)), but it is not maximal, and so the ((6,3,13),(5,3,20),(6,3,13))-learner does not find appropriate features and generalizes inappropriately.

It is easy to avoid these problems with any number of ad hoc modifications in the learner. If we want results that have some mathematical interest, we had better try to find broader generalizations about learners of this kind. But if we want to find models that are useful in theories of human inductive practices, we had better try to modify the model in ways that are motivated by psychological or neurophysiological observations. As Goodman (1955) says,

"To say that valid predictions are those based on past regularities, without being able to say which regularities, is thus quite pointless. Regularities are where you find them, and you can find them anywhere." An obvious corollary of this remark applies to the attempt to define computational learners that will find regularities: there are too many ways to do it.

The connectionist work on learning all seems to have this character: a tight computational net is drawn around a particular problem, and then it is shown that generalization within the restrictions of this net on this particular problem is appropriate. The basic idea is familiar and, so far at least, notoriously unsuccessful in leading us to models of human inductive practices. It could turn out that learners that are guided by structure-specific complexity could have application in models of human induction. The structural complexity measure provides constraint on generalization. If we can discover what constraints humans respect in learning about some domain, then we can consider whether such constraints might be enforced by structural complexity bounds of some kind. But the performance of specific structured systems is of no interest if the systems handle only a narrow range of problems and if the particular constraints on the learner are not motivated by psychological or neurophysiological considerations. Contrast, for example, the various measures of the "naturalness" of a concept or feature discussed in Osherson(1978). The measures discussed there, while perhaps less sensitive than the measures used here, at least apply to a wide range of concepts in a somewhat plausible way. The question of whether connectionist architectures like Hinton's can be motivated by neurophysiological considerations is not one I want to pursue here. My point is just that, lacking any motivation for the particular constraints on the learner, and lacking any account of human learning in a particular domain to be modeled, the results can have little psychological interest.

Notes

[1]I am grateful to Geoffrey Hinton, Edward Stabler, William Demopoulos, Daniel Osherson, and Zenon Pylyshyn for helpful discussions of this material. This research was supported in part by the Canadian Institute for Advanced Research and the Natural Science and Engineering Research Council of Canada.

References

Brayton, R.K. and McMullin, C. (1982), "The Decomposition and Factorization of Boolean Expressions." *Procs. 1982 IEEE Int. Symp. on Circuits and Systems*: 49-54.

de Micheli, G., Brayton, R.K. and Sangiovanni-Vincentelli, A. (1985), "Optimal State Assignment for Finite State Machines." *IEEE Transactions on Computer-Aided Design*, CAD-4(3): 269-285.

Gold, E.M. (1967), "Language Identification in the Limit." *Information and Control* 10: 447-474.

Goodman, N. (1955), *Fact, Fiction and Forecast*. Cambridge, Massachusetts: Harvard University Press.

Hartmanis, J. and Stearns, R.E. (1966), *Algebraic Structure Theory of Sequential Machines*. Englewood Cliffs, NJ: Prentice-Hall.

Hinton, G.E. (1986), "Learning Distributed Representations of Concepts." *Procs. of the 8th Ann. Conf. of the Cognitive Science Society*: 1-12. New Jersey: Erlbaum.

Hinton, G.E. (1987ms), "Connectionist Learning Procedures." Forthcoming in *Artificial Intelligence*.

Johnson, R.A. and Wichern, D.W. (1988), *Applied Multivariate Statistical Analysis*, 2nd Edition. Englewood Cliffs: Prentice-Hall.

Osherson, D.N. (1978), "Three Conditions on Conceptual Naturalness." *Cognition* 6: 263-289.

Osherson, D.N., Stob, M. and Weinstein, S. (1986), *Systems that Learn: An Introduction to Learning Theory for Cognitive and Computer Scientists*. Cambridge, Massachusetts: MIT Press.

Pinker, S. and Prince, A. (1988), "On Language and Connectionism: Analysis of a Distributed Model of Language Acquisition." *Cognition* 28: 73-193.

Pylyshyn, Z.W. (1988ms) "The Role of Cognitive Architecture in Theories of Cognition." Forthcoming in K. VanLehn, ed., *Architectures for Intelligence*.

Quine, W.V. (1952), "The Problem of Simplifying Truth Functions." *American Mathematical Monthly*, 59: 521-531.

Savage, J.E. (1976), *The Complexity of Computing*. Toronto: Wiley.

Stabler, E.P. Jr. (1989), "Learning Simple Things: A Connectionist Learning Problem from Various Perspectives." Technical Report, Department of Computer Science, University of Western Ontario.

Valiant, L.G. (1984), "A Theory of the Learnable." *Communications of the ACM* 27:1134-1142.

Part XIV

SET THEORY

The Many Worlds Interpretation of Set Theory

Geoffrey Hellman

University of Minnesota

0. One World and (some of) its Woes

As standardly presented in axiomatic form, set theory exhibits two fundamental features which together give rise to a number of foundational and philosophical problems. These features are so basic that they are seldom even isolated. The first is that the axioms are taken as *categorical assertions* (e.g. "There exists an empty set", "There exists an infinite set", "The power set of any set exists", etc.). The second is that *actualist quantifiers* are employed in these axioms, so that they appear to be about actual objects; the domain of these quantifiers forms *the fixed universe of sets* (up to a choice of urelement basis), relata of a unique, fixed membership relation entering into the statement of the axioms.

Such presentations are associated with the philosophical positions of platonism and set-theoretic foundationalism. Set theory is viewed as the study of a unique realm of abstract objects, *the sets*, making up *Cantor's universe*; and the theory is viewed as the ultimate background theory of (virtually all) mathematics, which, as is well known, can be interpreted within it.

However, these positions confront a number of problems, some more widely recognized than others. First, there are the usual epistemic and semantic puzzles associated with objects-platonism (as I will sometimes call the view, to distinguish it from, e.g., modalism which, on some formulations, may dispense with objects entirely but still qualify as "platonist" in some more general sense). These include the problem of epistemic access, puzzles concerning the identity of the objects arising from multiple possible interpretations, and related questions of reference. (Cf. e.g. (Benacerraf 1965); also (Goodman 1977, Ch.1).)

Second, there are "problems of *imperialism*": in virtue of the two features cited at the outset, axioms of set theory are understood as true or false *simpliciter*. But there are many axiom systems and many apparent conflicts among systems, and it seems that choices must be made: we get questions such as, "Are there really no sets which are members of themselves?", "Are there really no infinite descending membership chains?", "Are there really no proper classes?", and so forth. But are all such conflicts genuine? Must we take such questions seriously? Isn't pluralism (peaceful coexistence) possible? And if so, how far does it extend? Can we, for example, be "pluralists" about the Axiom of Choice?

446

There are, moreover, problems of imperialism with respect to other, more central, branches of mathematics, such as number theory and analysis. Must these subjects be seen as embedded in set theory? Should they not be capable—even from a foundational perspective—of standing on their own? These theories, in fact, lend themselves to a structuralist interpretation; yet if we insist on carrying out such an interpretation within set theory, we face the unpalatable alternatives of adopting very strong set-theoretic axioms—far stronger than the commitments of the more elementary theories themselves—or else working within weak set theories, in which case we fail to do justice to the full generality of structuralist insights (captured in the phrase, "for number theory, any possible ω-sequence will do"). (For further development of these points, see (Hellman 1989, Ch.1).)

Finally (for our purposes here), there is what may be called the problem of *extraordinary objects*. On the fixed universe view, the domain of the actualist quantifiers would be such an object: it cannot be a set (given ZF or allied systems); but, if it is countenanced at all, it can be treated as "set-like", as the starting point of a new cumulative hierarchy (of "supersets"); and then we confront special "indiscernibility problems", problems of "apparent differences without a difference" (e.g., "Is the universe really a superset, or is it just an inaccessible level of the hierarchy of sets?", and so forth). Proper classes generally give rise to such puzzles, and this is one of the strongest reasons for resisting them. But, on the standard presentations, this is difficult to do, since "is a set" is thought of as analogous to other ordinary predicates of our language, and so why shouldn't it have an extension? (The extension cannot be a *set* in the technical sense, but it is *set-like*, and that is enough to generate the puzzle.)

Below we shall sketch some of the leading features of an alternative view of set theory, a kind of structuralist view which avoids both of the fundamental aspects of standard presentations just highlighted. But first let us look briefly at the main historical antecedents.

1. Zermelo (1930)

It is noteworthy that one of the principal founders of modern set theory, Ernst Zermelo, did not share the fixed universe view, and in his important (1930) paper, "*Über Grenzzahlen und Mengenbereiche*" ["On Boundary-Numbers and Set Domains"], he explicitly disavowed it.

Having presented (essentially) the Zermelo-Fraenkel axioms in *second-order form*—that is with a single second-order Axiom of Replacement and a single second-order Axiom of Selection (*Aussonderung*) (as opposed to infinitely many instances of first-order schemata)—Zermelo proved a series of theorems that may be called "*Quasicategoricity of ZF²*": structures satisfying these axioms ("normal domains", in Zermelo's terminology) are uniquely determined, up to isomorphism, by two (transfinite) numbers, the cardinality of the urelement basis, and the *characteristic* of the domain, that is, its ordinal height. [As a corollary: if urelements (other than the null set) are absent, any two models are, up to isomorphism, ordered by end-extension.] Moreover, the characteristic of any model is what is now called a *strongly inaccessible cardinal* (a regular limit cardinal which is a fixed point of the beth function). Thus, all models (without urelements) may be identified with the so-called "natural models", of the form $(V_\kappa, V_{\kappa+1}, \in \upharpoonright V_{\kappa+1})$, with κ strongly inaccessible and where the second item is the range of the second-order quantifiers.

Having presented these model-theoretic results, Zermelo concluded with some interesting philosophical remarks:

...Our axiom system is non-categorical, which in this case is no deficiency, but an *advantage*. For it is on just this fact that the enormous significance and boundless applicability of set theory rests... Much rather [than "*force* the desired categoricity artificially" with axioms of restriction, "at the cost of generality"] must set theory first be developed as a *science* in fullest generality, whereupon the comparative investigation of particular *models* can be undertaken as special problems. (Zermelo 1930, p. 45) (Burgess unpublished, pp.19-20)

Furthermore,

...If we make the general hypothesis that *every categorically determined domain can always be conceived of as a "set"*, i.e. can appear as an element in a (suitably chosen) normal domain, then it follows that to every normal domain corresponds a higher one with the same basis...In the same way, from every infinite sequence of different normal domains with a common basis...there arises by union and merger a categorically determined domain of sets that can again be completed to a normal domain of higher characteristic...and the series of "all" boundary numbers is just as boundless as the [ordinal] number series itself...(Zermelo 1930, p.46) (Burgess unpublished, pp. 20-21)

Finally, he concluded,

The "ultrafinite antinomies of set theory", which the scientific reactionaries and anti-mathematicians eagerly and delightedly call on in their campaign against set theory, these specious "contradictions", arise solely from a confusion between the non-categorical axioms of set theory and the various particular models of them: What in one model appears as an "ultrafinite un- or super-set" is in the next higher domain a perfectly good "set" with a cardinal number and order type of its own, which serves as the foundation stone for the construction of the new domain. The boundless series of Cantor's ordinal numbers gives rise to an equally boundless series of essentially different models of set theory, in each of which the whole classical theory can be expressed. The polar opposite tendencies of the thinking mind, creative progress and all-embracing completeness, which lie at the root of Kant's "antinomies", find their symbolic expression and resolution in the concept of the well-ordered transfinite number-series, whose unrestricted progress comes to no real conclusion, but only to relative stopping-points, the "boundary numbers" that divide the lower from the higher models. And so the "antinomies" of set theory, properly understood, lead not to a restriction and mutilation, but rather to a further development (whose scope cannot yet be taken in) and enrichment of mathematical science. (Zermelo 1930, p. 47) (Burgess unpublished, pp. 21-22)

Here then is the suggestion of a "many-worlds" interpretation of set theory, one pertaining to just a single axiom system, ZF^2. A resolution of apparent conflicts among alternative axiom systems would presumably involve still more "worlds", but Zermelo's view is that any suitably *general* axiom system should embrace many "possible models". As the above passages bring out, two informal principles underlie the view. First, there is the principle of *generality*, that any set-like object (collection, in the broadest sense) can be treated as a set, i.e. as an element of a possible model. Second, there is the principle of *extendability*, that any model can be extended to one of higher characteristic. (In addition, Zermelo stated a *strong* or *extended extendability* principle, that any sequence of models lies in a common proper extension.)

Like Zermelo, we have stated these principles informally, but could we not write them out systematically in the language of set theory itself? Of course we could, but—especially for the principle of *generality*—the natural choice of that language would be the one that Zermelo clearly preferred, namely that of second-order set theory; but then we immediately confront a serious difficulty. Our background logic is naturally taken as axiomatic second-order logic. (In fact, this is a natural choice for formalizing Zermelo's quasi-categoricity theorems. Cf. (Hellman 1989, Ch.2.) But then we have an apparent conflict between the principles and the demands of second-order logical comprehension, which gives rise to the class of all sets, the class of all ordinals, the class of all natural models, etc. For how are *these* to be treated as "sets" in accordance with *generality*, or properly extended in accordance with *extendability*? A resolution of this difficulty will be outlined below, after attending to one more historical antecedent.

2. Putnam (1967)

In quite a different context, and independently of Zermelo, Putnam in his controversial (1967) (cf. (Kreisel, 1972)) also suggested a kind of "many-worlds interpretation" of set theory. This was arrived at as part of a larger program, namely, to develop a view of "mathematics as modal logic" which would be seen to be mathematically equivalent to the standard view of "mathematics as set theory" but which would provide an alternative to the objects-platonist picture associated with that view. By taking a logico-mathematical modal operator (say, for possibility) as primitive, ordinary mathematical sentences could be represented by modal conditionals in which quantification over abstract mathematical objects is entirely absent, thereby providing a kind of "modal-nominalist" framework for mathematics. For example, any bounded set-theoretic sentence, S (with quantification over sets of rank $< \rho$, say), could be represented by a modal conditional of the form,

$$\Box \forall M (M \text{ a concrete model of Zermelo set theory of height } \rho \supset S \text{ holds in } M),$$

where a concrete model was understood as an array of "points" connected by "arrows" representing the membership relation, fulfilling the axioms of Zermelo set theory (which would be adequate for carrying out virtually any scientifically applicable mathematics, perhaps the basis for this choice). (Just what primitives were to be allowed in spelling out "model of height ρ" was not specified. From earlier parts of the paper, it appeared that the background logic was to be first-order logic, in which case it is unclear how even "model of Z"—with its infinitely many first-order *Aussonderung* axioms—is to be expressed. Presumably, "S holds in M" could be expressed by writing out S with quantifiers relativized to M, where membership as it enters in the usual relativization conditions is replaced by (say) "is an atomic part of".)

Unbounded set-theoretic sentences were to be represented by more complicated "modal translates", in which ordinary (actualist) quantifiers over sets are systematically replaced by modal quantification over concrete Z models ("graphs") and their possible extensions, e.g., a simple AE sentence of the form $\forall x \exists y A(x,y)$, with A quantifier free, would be translated as,

$$\Box \forall M \forall x [\text{Graph}(M) \ \& \ x \text{ an atom of } M \supset \Diamond \exists M' \ \exists y (\text{Graph}(M')$$

$$\& \ y \text{ an atom of } M' \ \& \ M \le M' \ \& \ A(x,y))],$$

where $\le$ is the relation of subgraph. (Again, precise details of formulation were not presented.)

Crucially, in connection with this more elaborate translation scheme, Putnam put forward a principle of *extendability*, virtually identical to Zermelo's, except that it pertained to (concrete) Z models rather than ZF² models and it explicitly employed modality: Any model *can* be extended: "Even God could not make a model...that it would be *mathematically* impossible to extend..." (1967, p. 58). And the lesson concerning proper classes was graphically drawn: No relevant structure could realize "the naive conception of the totality of all sets; for any [such] structure has a possible extension that contains more 'sets' ... we might say: it is not possible for all *possible* sets to exist in any one world!" (1967, p. 59)

As we shall see, herein lies the germ of a resolution of the tension encountered in Zermelo's principles.

3. A Modal-Structural Synthesis

It is possible to combine these interesting strands of thought on the foundations of set theory into a unified system, one which realizes a "modal-structural interpretation" ("msi") of set theory, in close analogy with such interpretations which can be developed for other fundamental mathematical theories, such as number theory (PA) and real analysis (RA). Here we present an outline of the essentials of the approach. (For details, the reader is referred to (Hellman 1989 and forthcoming).)

The approach rests on a fundamental distinction between two main purposes of axiomatization. On the one hand, there is the familiar aim of *codifying proofs*. This leads naturally to standard presentations of theories, in which, as said at the outset, axioms (usually first-order, given the success in capturing first-order logical consequence in a formal (recursively enumerable) system) are presented *simpliciter*, as if being categorically asserted. There is, of course, every reason to pursue such aims, so long as one keeps track of what one is doing. For, on the other hand, there is the distinct purpose of *expressing what type of mathematical structure(s) one intends to investigate*. Here, as is well known, second-order languages succeed in crucial cases where first-order languages fail. (For a survey, see e.g. (Shapiro 1985); also, on the topic of large cardinals and natural models, (Drake 1974).) Moreover, in such cases, it is not necessary to adopt the mathematical axioms—with their actualist quantifiers—categorically. It is quite sufficient if structures of the appropriate type are mere logical possibilities. As to their actual existence, a strict neutrality can be maintained, at least in so far as pure mathematics is concerned.

This leads to the choice of an *axiomatic second-order modal logic* as the background logic for articulating structuralism. To standard axioms of second-order logic are added those of S-5 modal logic, a choice backed up by a number of considerations. (It reflects a "non-world-relative" sense of logical possibility, appropriate for pure mathematics; iterated modalities seem not to be needed; and, on translation, the widely recognized collapse of distinctions in ordinary mathematical discourse—e.g., "if numbers are possible, then they exist, and, moreover must exist"—is sustained. For further discussion, cf. (Hellman 1989).) However, crucial decisions must be made in connection with the comprehension principles to be adopted. We have the (necessitation of the) usual scheme,

$$\exists\alpha\forall x_1...\forall x_n(\alpha(x_1,...,x_n) \equiv A), \qquad \text{(CS)}$$

where α is an *n-ary* relation variable not free in formula A, and the x_i are individual variables. However, we make the following restrictions:

(i) Only ordinary, actualist quantifiers, $\forall x_i$, are admitted, as in CS, not "possibilist" quantifiers, $\Box\forall x_i$.

(ii) The formula A is itself modal-free.

With these restrictions, CS becomes the *extensional comprehension scheme (ECS)*, and this must be understood throughout. With it, intensions are avoided; on the possible-worlds metaphor, collections and relations on objects within any given world are available, as in ordinary mathematical reasoning, but we do not allow collections or relations that span "different worlds". Moreover, the modal-structuralist analogue of (absolute) proper classes is hereby blocked: we cannot speak, for instance, of a class of all possible ω-sequences, or of all possible sets, ordinals, models, etc. (Recall the remark of Putnam's, quoted at the end of section 2.) In a relative sense, however, proper classes are perfectly legitimate possibilities. That is, within a given "possible model" or "world" there arises e.g. the class of all "sets" or all "ordinals" occurring in that model; but, as Zermelo wished, such classes can be taken as starting points for a new "possible model".

Note furthermore that only the individuals initially assumed (in a given "world") are collected or related in accordance with CS. We do not go on to collect or relate such collections or relations. No endless abstract hierarchy is recognized, but (at most) only a single abstract level. (Thus, the possibility of purely nominalistic readings of the second-order formalism is left open. In fact, in contexts where polyadic second-order quantification can be reduced to monadic—when a pairing function is available—a nominalistic interpretation along the lines of Boolos' (1985) "plural quantification" reading can be given. Other nominalist interpretations are suggested in (Hellman 1989).)

The final restriction we need is that the Barcan formula be omitted. With it, crucial distinctions we wish to maintain would collapse. We could, for example, immediately infer, from $\Diamond \exists X \exists f$("$X,f$ form an ω-sequence"), which we require as a postulate or theorem of ms number theory, that $\exists X \exists f \Diamond$("$X,f$ form an ω-sequence"), which may be false. If the actual ontology recognized is finite, it makes perfectly good sense to insist that any actual ordered sequence is necessarily finite: it—the very same sequence—could not be infinite (although it could be embedded in an infinite one). Thus, the S-5 modal logic employed must avoid the Barcan formula, in the manner of (Kripke 1963).

In this system, then, modal operators are primitive. Officially, there is no quantification over possibilia (which could be used to eliminate the modal operators). The above talk of "possible worlds" or "possible models" is heuristic only, and such talk can, in practice, always be replaced by (often more long-winded) formulations employing modal operators and the usual apparatus of quantification and variables to keep track of cross-reference. Thus, much as "the many-worlds interpretation of quantum mechanics" is a misnomer—"the one-world interpretation of quantum mechanics" serving as a better rubric (cf. (Geroch 1984))—so here perhaps "the no-worlds interpretation of set theory" would be more literally accurate. (But, since neutrality on the matter can be maintained, "the indeterminate-number-of-worlds-interpretation of set theory" would be the really accurate title, and clearly *that* must be rejected on stylistic grounds!)

Now in this framework, the msi of set theory proceeds as follows. First one explicitly defines the type of relevant structures. Following Zermelo, we take these to be the "natural models", full models of ZF^2, which can be characterized directly by

$$X,f \text{ a natural model} \equiv^{df} [\wedge ZF^2]^X(\in/f),$$

where the right side abbreviates the result of writing out the conjunction of the ZF^2 axioms with all quantifiers relativized to the second-order variable X (intuitively, the domain of the model), substituting the (two-place) relation variable f for $\in$ throughout. This is in direct analogy with the procedure for characterizing relevant structures for number theory and for real analysis. In the former case, following (Dedekind 1888), one can write,

$$X,f \text{ an } \omega\text{-sequence} \equiv^{df} [\wedge PA^2]^X(s/f),$$

in which s is the successor relation symbol occurring in the PA^2 axioms. The case of real analysis is exactly analogous. (One defines "complete, separable, ordered continuum", utilizing a second-order statement of the Continuity Axiom.)

Natural models are assumed to be *full* in the sense that the universal quantifiers—both first-order, as in the Power Set Axiom, and second-order, as in the Axiom of Replacement[2]—have their standard meanings. But, since we do not need to introduce a relation of satisfaction between natural models and sentences, we do not need an explicit *metamathematical* concept of *fullness*. Instead, remaining entirely on the level of second-order *mathematical* reasoning, employing the above definition of natural model, Zermelo's quasicategoricity theorems can be stated and proved. (Analogously, remaining entirely on the level of second-order mathematical reasoning, employing the above definition of ω-sequence, Dedekind's categoricity theorem for PA^2 can be stated and proved. Cf. (Hellman 1989, Ch. 1).)

Notice that by generalizing on the relation constant ($\in$) in the definition of "natural model", we are adopting a structuralist stance with regard to sets: sets are not thought of as absolutely identified objects, but rather as whatever might enter into a structure of the relevant type. A set is such in this purely mathematical sense only in virtue of standing in the proper relations to other constituents of a model.

In addition to the second-order logical apparatus, special axioms are to be added. In particular, we require that natural models be logically possible:

$$\Diamond\exists X\exists f[\wedge ZF^2]^X(\in/f), \tag{ME}$$

the key *modal existence postulate* for ZF set theory. Furthermore, following Zermelo and Putnam, we may adopt the *extendability* principle,

$$\Box\forall X\forall f\{[\wedge ZF^2]^X(\in/f) \supset \Diamond\exists Y\exists g([\wedge ZF^2]^Y(\in/g) \,\& $$
$$\text{"}X,f \text{ is a proper submodel of } Y,g\text{")}\}, \tag{EP}$$

in which the latter quoted clause is written out in the obvious way. (The alert reader may ask how a proper extension Y of X will "contain" X "as a member". Since the background logic is second-order, there are two "membership relations" to keep track of, that of the logic, and that of the model Y, denoted 'g' in the above formula. Indeed, although it is not even well-formed to write $Y(X)$, we can state and prove that X is *represented* (via g) by an item of Y, i.e., $\exists y(Y(y) \,\& \forall z(g(z,y) \equiv X(z)))$, and similarly for the membership relation, f, of X itself.)

In addition, strengthenings of the EP can be adopted, and, indeed, will be needed for various purposes. For example, Zermelo's *extended extendability principle* was,

"For any ordinal α, there is a sequence of natural models of order type α, ordered linearly by end-extension, and all having a common proper extension."

The ms rendering of this (letting M variables range over pairs X,f forming natural models) is,

$$\Box\forall M\forall\alpha(\alpha \text{ an ordinal of } M \supset \Diamond\exists M'\exists f(M < M' \,\& f \text{ a function in } M' \text{ such that}$$
$$f: \alpha\to |M'| \,\& f(\nu) = M_\nu \,\& (\delta < \nu \supset M_\delta <_e M_\nu)), \tag{EEP}$$

in which $<$ is the proper submodel relation and $<_e$ is (the converse of) end-extension. Note that quantification over ordinals here is always over ordinals of a (possible) model, which accords entirely with a structuralist interpretation of talk of ordinals. But there is no loss here, since no restrictions are imposed on the possibilities of models.

Now we can write down the Putnam-translate of an arbitrary sentence of the original ZF language (either first or second order). To keep things simple, consider a first-order AE sentence, S, as above, $\forall x\exists yA(x,y)$, with A quantifier free. The ms translate of S, S_{msi}, is

$$\Box\forall X\forall f\forall x\{\{[\wedge F^2]^X(\in/f) \,\& X(x) \supset \Diamond\exists Y\exists g\exists y([\wedge ZF^2]^Y (\in/g) \,\& Y(y) \,\& X \subseteq Y \,\&$$
$$f = g \restriction X \,\& A(x,y))\},$$

more colloquially, "Any element of any natural model there might be could bear A to something in a possible extension thereof." This pattern can readily be iterated to handle arbitrary sentences. (Where second-order quantifiers are involved, clauses corresponding to '$X(x)$' are replaced by clauses saying that R (say) is a relation on X (in the unary case, $R \subseteq X$).)

A useful tool for dealing with such Putnam-translates is a set-theoretic semantics, "Putnam semantics", which can be introduced inductively as a relation $\vDash_p$ holding between natural models and formulas of the original set-theoretic language. For all but the quantifier clauses, $\vDash_p$ is just like ordinary satisfaction, $\vDash$. Where we let E, E', range over evaluations of variables, the (existential) quantifier clause reads thus:

$M \vDash_p \exists x A$ [E] iff $\exists M' \geq M \exists E'$ agreeing with E except possibly at x such that $M' \vDash_p A$ [E'].

(Universal quantification is treated in the usual classical way as $\sim\exists\sim$.) Using these inductive clauses, if one simply reads off the relevant quantifier clauses and applies disquotation (substituting quantification over items in the models for quantification over evaluations of variables), one recovers the Putnam-translate, modulo the absorption of modal operators into the quantifiers with which they are uniquely associated.

Now, using Zermelo's quasicategoricity theorems, one can show that $\vDash$ is *stable* in the sense that if $M\vDash_p A$ and M' extends M, then $M'\vDash_p A$; in fact, all natural models are equivalent with respect to $\vDash_p$. Moreover, one can show that the semantics of $\vDash_p$ is *correct* in that $\vDash_p$ at (some or) any natural model agrees completely with the cumulative hierarchy, V, on the standard theory of $\vDash$, with respect to all first-order ZF sentences. (The set-theoretic proof of this requires the Axiom of Inaccessible, (AI), $\forall\alpha\exists\beta(\beta{>}\alpha$ & Inac(β)). Note that the EP implies (AI)$_{msi}$: the height of any natural model is an inaccessible occurring in any proper extension.) This is a "best possible" result, since clearly there is disagreement with respect to second-order sentences. For example, $\exists X\forall y(y \in X)$ holds in V (itself the witness, in this case), whereas the Putnam-translate of this, $\Diamond\exists M\exists X|X \subseteq |M|$ &$\Box\forall N\forall y$ $(N \geq M$ & $|N|(y) \supset y \in {}^N X)]$, fails in light of the EP. (In writing this translate, we have used the convention above on variables M, N: $|M|$ is the domain of M and $\in^M$ is the "membership" relation of M.) This disagreement is as it should be; second-order ZF, on the standard axiomatic presentation, automatically gives rise to proper classes, whereas the msi is designed to avoid them (except in the innocent, relative sense mentioned at the outset).

Finally, concerning formal developments, the ms approach, following Zermelo, provides natural methods of generating many of the so-called small, large cardinals "from below", as it were. One writes down already motivated second-order axioms, proves that natural models of them have the height of a large cardinal of a given type (occurring in any proper extension), and then appeals to the relevant extendability principle, obtaining the ms translate of the relevant large cardinal axiom (stating that every ordinal is dominated by a cardinal of that type, as in the case of AI, above). (See Hellman 1989, Ch. 2.)

It should be emphasized that the use of second-order axioms in this process is essential. It is a striking fact that, for example, ZF^1 has models of the form $(V_\alpha, \in \upharpoonright V_\alpha)$ with α a singular (and accessible) cardinal, provided it has any models of this form (as is proved by a Löwenhem-Skolem type construction, see (Drake 1974, p. 111, Ex. (4))). Once Replacement[2] replaces the first-order scheme, we have that α must be regular and (strongly) inaccessible. An exactly analogous situation obtains with respect to Mahlo cardinals and the axiom (labelled 'F' by Drake) that every normal function has a regular fixed point. (Cf. (Drake 1974, pp. 115-120).) These results highlight the superior expressive power of second-order axioms in a special way: models of the above form, e.g. V_α with α singular (called "natural models" by Drake in a slightly different sense from that of our second-order definition above), are *perfectly standard* as interpretations of the first-order formalism. The membership relation of the model is just the restriction of "the real membership relation" to the domain of the model, and the power set operation is the real, full one. Such results show that efforts to modalize set theory by taking as basic an assertion of possibility of (the truth of) all the ZF^1 axioms will spell disaster for large cardinals: perfectly standard possible models of the axioms will be too small to accommodate them.

Just how far the approach sketched here extends depends on what sorts of arguments one finds convincing for adopting the modal existence of natural models of the relevant axiom set. As we have argued, the Axioms of Infinity, Replacement, which gives rise to inaccessibles, and Axiom F, giving rise to Mahlo cardinals, can all be motivated in a similar fashion by considering fixed points of various extension-taking operations. (For details, see (Hellman 1989, Ch. 2).) Such methods are, of course, to be contrasted with the more commonly invoked (informal) principle of *reflection* on the universe. (Cf., e.g., (Reinhardt 1974) and (Kanamori and Magidor 1978).)

The game of large cardinals has been played to staggering heights, and all those mentioned here are vanishingly small compared with the truly large ones that have been defined and investigated (right up to the point of contradiction!). However, in the spirit of true climbing, we are not interested in height for height's sake, but in the challenge and beauty of the ascent. Whatever motivation can be given for even the tiniest of large cardinals can affect our view of that subject. As suggested here, and as Zermelo emphasized, they should not be viewed as esoteric; rather they are virtually inevitable given reasonable structural and iterative principles.

4. Limits of Pluralism and Second-order Logic

So far, the "many-worlds" view sketched pertains only to a single core axiom system, ZF^2. But an obvious advantage of introducing modality explicitly into set theory lies in the reconciliation afforded among apparently conflicting "intuitions about sets", over which a choice really need not be forced. A prime example is the question whether to adopt the Axiom of Regularity: "Every non-empty set has a member disjoint from it." This is intended to rule out $\in$-cycles and infinitely descending $\in$-chains. (It is symptomatic of the expressive inadequacy of first-order set theory that, by a simple compactness argument, infinitely descending $\in$-chains are not ruled out: they exist in models of ZF^1, including Regularity. However, when we move to ZF^2, with its single second-order Axiom of Replacement, the "pathologies" vanish.) On the standard presentation of set-theoretic axioms, it appears to make sense to ask, "Is it true that there are no non-well-founded sets?". If one rejoins, "True where?", the reply should be, "True in the real world of sets", or "True of the sets". There should not be the need to relativize truth to a type of structure, independently specified. Obviously, if such relativization is invoked, the answer will vary according to the specification of types of structures, and the question may appear trivialized. For example, obviously, in natural models of ZF there are only well-founded sets. Moreover, an alternative axiom system has been worked out, based on intuitive requirements on cumulative levels, from which well-foundedness of sets (along with the rest of ZF) can be derived. (See (Scott 1974).) This indeed helps show that membership relations obeying Regularity are "natural" and "well-motivated". Still, there is relativity to a type of structures. On the modal-structural view, such relativization of the question is precisely what is called for. "The real world of sets" goes by the board, and all that counts is whether structures of mathematical interest are logical possibilities (in the relevant, standard (and non-formalizable) sense of "second-order logical possibility"). If mathematicians are interested in investigating non-well-founded "membership" relations, they are perfectly free to do so without having to confront questions of "truth in the real world". It is sufficient if such structures are logically possible. One may, of course, raise questions about the bearing of such structures on ordinary usage of terms such as 'member', 'collection', and so forth. And it could turn out that at best a tenuous relation exists between such usage and the mathematical structures in question. If the relation is too tenuous, talk of "membership" in the mathematical context will appear strained. In fact, with respect to non-well-foundedness, this would seem not to be the case: every child can imagine "containers" with further "containers", *ad infinitum* (the child's Christmas morning nightmare!). Be that as it may: the point is, we have replaced an apparent question of absolute truth with one of a very different character, pertaining to a resemblance of usage between informal and mathematical language.

But are there no limits to this tolerance? Should we say, for instance, that there is no matter of fact concerning the Axiom of Choice? Again, it depends on the question being asked. If the question is, "Are choiceless universes (fulfilling, say, the ZF axioms) logically possible?", the answer is, evidently, yes, if any such universes are (in light of (Cohen 1966)). If, however, the question is, "Must Choice hold in any natural model of ZF^2?", we have quite a different situation. The quasicategoricity theorems assure us that the answer is determinate; that is, either it is redundant to add Choice to ZF^2 *for the purposes of characterizing the structures* (although it is *not* proof-theoretically redundant, cf. (Chuaqui 1972)), or else it is inconsistent, in the sense that no full model of ZFC^2 is possible! Unfortunately, we have no absolutely certain guarantees in a case like this. The best we seem capable of is a heuristic argument, obviously in favor of Choice in any full ZF model, based on the full power set operation and the consideration that a choice function on a set y of non-empty sets must occur at the next level past $y \times \cup y$. It is surprising that our axioms—those of ZF^2—which succeed in settling the question in the sense of "expressing a type of structures with a determinate answer"—an answer that, moreover, seems so obvious—nevertheless do not succeed in providing a formal proof. That is one of the deeper facts of life we must learn to live with in mathematics. Surely it does not show that our second-order concepts are "vague" in the sense of not, after all, determining a definite answer.

It is commonly said that Zermelo's second-order axioms for set theory are "vague", involving the indeterminate notion of "arbitrary (extensional) property" of sets, or "arbitrary relation" among sets (in the cases of *Aussonderung* and Replacement, respectively). The standard presentation of axioms and its associated one-world picture encourage this view, since we seem to be forced into considering arbitrary properties of (or relations on) "all sets"—i.e. "all *possible* sets"—and, since such a totality is, by its very nature, open-ended, and not even legitimate as an object ("set") of set theory, at least as much indeterminacy carries over to "arbitrary property or relation". These automatically appear to be *extraordinary objects*. From the modal-structural standpoint, however, this is highly misleading. The axioms serve to characterize a type of structures, and the second-order quantifiers are always restricted to a domain, which, as the extendability principles imply, can always be treated as a "set" (an *ordinary* object) in a more encompassing domain. Within such a domain, the second-order quantifiers are no more vague than *bounded* first-order quantification; if we are prepared to speak of all subsets of a given set, we can just as well speak of all subdomains of a given domain or all relations on a given domain. At any more embracing level, these totalities are automatically treated as "sets" within the range of the first-order quantifiers. Only at the level of modal quantifiers do we approach Cantor's "inconsistent totalities". But, really, we never reach these even here, due to the natural restrictions on comprehension, reflected in the ECP. From this perspective, it makes no sense to collect "all possible ordinals". Still, we may make a good deal of sense of the modal quantification itself, of claims as to what would hold in any natural model that there might be.

References

Benacerraf, P. (1965), "What numbers could not be", in P. Benacerraf and H. Putnam, eds., *Philosophy of Mathematics*, second edition, Cambridge: Cambridge University Press, 1983, pp. 272-294.

Boolos, G (1985), "Nominalist Platonism", *Philosophical Review*, 94: 327-44.

Burgess, J (unpublished), "Sources on the Foundations of Set Theory".

Chuaqui, R. (1972), "Forcing and the Impredicative Theory of Classes", *Journal of Symbolic Logic*, 37, 1: 1-18.

455

Cohen, P.J. (1966), *Set Theory and the Continuum Hypothesis*, New York: Benjamin.

Dedekind, R. (1888), *Was Sind und Was Sollen die Zahlen*, Brunswick: Vieweg, translated as "The Nature and Meaning of Numbers", in *Essays on the Theory of Numbers*, New York: Dover, 1963, pp. 31-115.

Drake, F.R. (1974), *Set Theory: An Introduction to Large Cardinals*, Amsterdam: North Holland.

Geroch, R. (1984), "The Everett Interpretation", *Noûs*, 18, 4: 617-633.

Goodman, N. (1977), *The Structure of Appearance*, third edition, Dordrecht: Reidel.

Hellman, G. (1989), *Mathematics without Numbers*, Oxford: Oxford University Press.

_ _ _ _ _ _ . (forthcoming), "Toward a Modal-Structuralist Interpretation of Set Theory", *Synthese*.

Kanamori, A., and Magidor, M. (1978), "The Evolution of Large Cardinal Axioms in Set Theory", in G.H. Muller and D.S. Scott, eds., *Higher Set Theory, Springer Lecture Notes in Mathematics*, vol. 669, Berlin, Springer.

Kreisel, G. (1972), *Review of Putnam* (1967), Journal of Symbolic Logic, 37: 402-4.

Kripke, S. (1963), "Semantical Considerations on Modal Logics", *Acta Philosophica Fennica*, pp. 83-94.

Putnam, H. (1967), "Mathematics without Foundations", in Benacerraf and Putnam (1983), pp. 295-311.

Reinhardt, W. (1974), "Remarks on Reflection Principles, Large Cardinals, and Elementary Embeddings", in T. Jech, ed., *Axiomatic Set Theory, Proceedings of Symposia in Pure Mathematics*, Vol. 13, Part 2, pp. 189-205, Providence, R.I.: American Mathematical Society.

Scott, D. (1974), "Axiomatizing Set Theory", in T. Jech, ed., op. cit Reinhardt (1974), pp. 207-214.

Shapiro, S. (1985), "Second Order Languages and Mathematical Practice", *Journal of Symbolic Logic*, 50: 714-742.

Zermelo, E. (1930), "Über Grenzzahlen und Mengenbereiche: Neue Untersuchungen über die Grundlagen der Mengenlehre", *Fundamenta Mathematicae*, 16: 29-47.

Sets and Point-Sets:
Five Grades of Set-Theoretic Involvement in Geometry

John P. Burgess

Princeton University

Cantor was the founder of not one but two theories: an earlier theory of point-sets and a later theory of sets. How are they related? How much more about points and sets of points can be proved with the assumption of sets of sets of points, sets of sets of sets of points, and so on, than without? The present paper is a semi-popular (nontechnical except for presupposing some familiarity with first-order logic) survey of partial answers obtained by logicians during the last decades.

Zero. Point Theory

Tarski (1959) has shown how classical geometry (of any number of dimensions, say for definiteness three) can be formalized in a first-order language, here to be called L_0, as a first-order theory, here to be called G_0. L_0 has variables x,y,z, and so on for points, and predicates for a couple of geometric relations among points. G_0 has a dozen axioms and one scheme.

Traditional geometry was a theory not just about points, but also about lines and other kinds of sets of points. Nevertheless, Tarski notes that one is able to express in his symbolism all the results that can be found in traditional textbooks, primarily as a consequence of the fact that each set of points of any kind mentioned in such results is determined by a fixed finite number of points. For example, a line is determined by two points, so that quantification over lines can be replaced by double quantification over points. The extension of G_0 formalizing in the natural way the assumption of lines would be interpretable in G_0.

Modern geometry uses coordinates, and is a theory not just about points, but also about real numbers. Nevertheless, one is able to express in Tarski's symbolism all the results that can be found in modern textbooks, primarily as a consequence of the fact that any real number is determined by three points: namely, by any x_0, x_1,x such that x_0 and x_1 are distinct, and x is collinear with them, and the ratio of the distance from x_0 to x to the distance from x_0 to x_1 equals the real number in question. Modern geometry would be formalized in a natural way as a theory AG_0 extending both the pure classical geometry G_0 and a pure classical algebra A_0; but AG_0 (and hence A_0) is interpretable in G_0. More details about this interpretation can be found in the paper of Burgess (1984). In discussing set-theoretic geometry below, it will be assumed that this interpretation has been carried out, and the usual identification of the line with the set R of real numbers, the plane with the set R^2 of pairs thereof, and the space with the set R^3 of triples thereof will be adopted.

PSA 1988, Volume 2, pp. 456-463

Conversely, AG_0 (and hence G_0) can be interpreted in A_0, primarily as a consequence of the fact that every point of the space is determined (relative to an arbitrarily fixed frame) by three real numbers, its coordinates. This interpretation allows Tarski to derive from old results about A_0 new results about G_0, notably:

Completeness: For any sentence P of L_0 either P is provable in G_0 or not-P in provable in G_0

Conservativeness: For any consistent extension G of G_0 in any extension L of L_0, and for any sentence P of L_0, if P is provable in G, then P is provable in G_0.

No more about points can be proved with the assumption of sets of points, sets of sets of points, and so on, than without.

(To enter briefly into more technical matters: As regards completeness, the length of the shortest proof of P or proof of not-P grows worse-than-exponentially as a function of the length of P; and there are conjectures P for which no proof of P or proof of not-P has been found and exhibited. As regards conservativeness, for some G and P, the length of the shortest proof of P in G_0 may be worse-than-exponentially longer than the length of the shortest proof of P in G; and there are theorems P for which a proof in an extension G has been found and exhibited, but none in G_0. This cannot happen for those G that are interpretable in G_0 as in the case mentioned above of AG_0: Use of coordinates makes for more perspicuous proofs, but not for shorter proofs. This matter may be discussed in more detail in a survey in preparation by Manders.)

One-Half. Finite Point-Set Theory.

As Tarski notes, though results about 3-sided, about 4-sided, and so on, polygons can be expressed in Lo and proved in G_0, results about arbitrary polygons cannot. They can be in an extension $G_{1/2}$ of G_0 in an extension L_1 of L_0. L_1 has variables X,Y,Z, and so on, for sets of points, and a predicate for the elementhood relation. The uniqueness axiom of *extensionality* asserts that sets X,Y having exactly the same points x as elements are identical. The instance of the existence scheme of *comprehension* for a formula Q asserts that there is a set X whose elements are all and only those points x such that Q(x). $G_{1/2}$ has the extensionality axiom and enough instances of the comprehension scheme to formalize in a natural way the assumption of *finite* sets of points. (Questions and results about certain special infinite sets of points to be discussed in the next section can be coded in an artificial way in $G_{1/2}$.) In this extension one can define what it is for a point on the line or element of the set R of real numbers to be an element of the set N of natural numbers. This definition allows one to derive from the limitative results of Gödel and others about arithmetic similar results about $G_{1/2}$ and any consistent, axiomatizable extension thereof. For any such theory, completeness fails (and conservativeness fails), as Tarski notes.

One. Point-Set Theory.

Historically, point-set theory was developed by Cantor as a foundation for classical analysis. Alternatively, it can be regarded as an extension of classical geometry, formalizable in a natural way as the theory G_1 in L_1 having the extensionality axiom and *(all* instances of) the comprehension scheme. (A weak axiom of *dependent* choice DC will also be taken to be an axiom of G_1 and used without mention, but stronger hypotheses of independent choice AC will not be taken to be axioms of G_1 and will be mentioned when used.) As noted in the last section, completeness fails for G_1. Many hypotheses about arbitrary point-sets can be expressed in L_1 but cannot be settled by proof or disproof in G_1, including all the hypotheses given names below. What can be proved are certain implications among these hypotheses, and certain restrictions of these hypotheses to special classes of point-sets. What follows is an outline survey of such results, with references to more detailed surveys where attributions of results and references to the original technical research literature can be found.

Special Classes of Point-Sets: As *basic* sets in dimension one (two) (three) we take intervals-minus-endpoints (squares-minus-edges) (cubes-minus-faces). The most important operations for obtaining other sets from basic sets are *complementation, join*, and *projection*. Here join means forming the union of sets $E(n)$ for n an element of N; projection means forming thc image $F/A/$, the set of all $F(x)$ for x an element of A, of a set A under a continuous function F, where a function F is continuous if thc inverse image $F^{-1} /V/$, the set of all x such that $F(x)$ is an element of V, for any basic set V is itself a join of basic sets. The most important special classes of point-sets are the following:

(1a)	open set:	joins of basic sets
(1b)	closed sets:	complements of open sets
(2a)	F-sigma sets:	joins of closed sets
(2b)	G-delta sets:	complements of F-sigma sets
(2)	simple sets:	sets either F-sigma or G-delta
(3)	Borel sets:	sets obtainable from basic sets by iteration of complementation and join
(4a)	analytic sets:	projections of Borel sets
(4b)	co-analytic sets:	complements of analytic sets
(5a)	PCA sets:	projections of co-analytic sets
(5b)	CPCA sets:	complements of PCA sets
(6)	projective sets:	sets obtainable from Borel sets by finite iteration of complementation and projection

For the rest of this section, where not otherwise specified a *function* (a *multifunction*) will be understood to be a function whose domain is a subset of the line and whose values are elements of (non-empty subsets of) the line. A function F (multifunction E) may be identified with its *graph*, the subset of the plane having as elements the pairs (x,y) such that $y = F(x)$ (y is an element of $E(x)$). So identified, functions and multifunctions may be classified as above. Note that classifications above with higher numbers include those with lower numbers.

Axioms of Choice: There are several inequivalent axioms of choice, the weaker ones being more "intuitive" than the stronger ones, in the sense that the pioneers of the theory were, and novices in the theory still are, more likely to use the former than the latter unconsciously. Let E be a multifunction, F a function with the same domain; then F is called a *selector* for E if $F(x)$ is an element of $E(x)$ for all x, and a *superselector* for E if also $F(x) = F(y)$ for all x,y such that $E(x) = E(y)$. AC^- (AC) is the assertion that every multifunction has a selector (a superselector). AC^+ or WO is the assertion that there is a *well-ordering* of the line, an ordering of its points—necessarily differing from the usual ordering—in which every non-empty set has a least element. AC^+ implies AC (which trivially implies AC^-), for given a well-ordering and a multifunction E, one obtains a super-selector F by setting $F(x)$ = the least element of $E(x)$.

Smallness and Niceness Properties of Point-Sets: There are are several distinct senses in which a point-set may be "small", and several distinct senses in which a point-set may be "nice". The three most important are the following:

Cardinality Properties: Cantor defined two sets to have the same *cardinal number* if they can be put in one-to-one correspondence. He also proved the line R, the plane R^2, and the space R^3 to have the same cardinal number, making dimension irrelevant here. He also defined an order on cardinal numbers and proved that that of N, which he called *denumerable*, is less than that of R, which he called *continuum*. Thus being finite-or-denumerable, which he called *countable*, is a "smallness" property. Not having *interme-

diate cardinality between denumerable and continuum is a "niceness" property. Cantor conjectured that no set has intermediate cardinality. This conjecture is the *continuum hypothesis* CH. If it fails, the number of intermediate cardinals might be anything from one (a hypothesis to be called CH') to continuum (a hypothesis to be called CH*).

Measure Properties: Only the case of dimension one will be considered here. (For the cases of two or three, substitute "square" or "cube" for "interval", and "area" or "volume" for "length".) A *measure* here is a function whose arguments are subsets of the unit interval and whose values are real numbers between zero and one, the value for the singleton of a point as argument being zero, and the value for the whole unit interval as argument being one. The more of the following conditions a measure satisfies, the better a generalization of length it will be:

Weak Additivity (Additivity) (Strong Additivity): The measure of the union of finitely (or denumerably) (or intermediately) many disjoint sets is the sum of their measures.

Invariance: The measures of geometrically congruent sets are equal.

A set X is *null* if for every positive real number e there are denumerably many intervals such that X is contained in their union and the sum of their lengths is less than e. This is another "smallness" property. A set X has the *measurability* property of Lebesque if there is a simple set S such that the difference between X and S, the set of points that are elements of one or the other but not both, is null. This is a "niceness" property: Lebesque showed that there is a *unique* additive, invariant measure defined on the measurable sets, which thus have a well-defined "size".

Category Properties: A set X is *rare* if for every basic set V there is a basic set W contained in V and disjoint from X, and is *meager* if it is a union of countably many rare sets. Meagerness is a "smallness" property: Baire proved that no basic set is meager. A set X has the *categorizability* property of Baire if there is a simple set S such that the difference between X and S is meager. This is a "niceness" property, analogous to but not identical with measurability.

Sets that are "small" in one of the above senses need not be so in another of the above senses. The most famous counterexample is due to Cantor: The set obtained by discarding the middle third of the unit interval, the middle thirds of both the right and the left remaining intervals, and so on, has cardinal number continuum, but is both null and meager.

The Determinacy Property: There is a fourth important "niceness" property. Given a subset X of the unit interval, consider the following game for two players: I discards either the right or the left half of the interval, then II discards either the right or the left half of the remaining interval, then I discards either the right or the left half of the remaining interval, and so on, until after an infinite sequence of moves by the two players alternately there is only a single point x left. Then I wins if x is in, and II wins if x is not in, the given set X. If I (II) can so play as to win no matter how II (I) plays, X is said to be *determinate* in favor of I (II). The hypothesis that every set is determinate (in favor of one or the other player) is called the *axiom of determinateness* AD. Without assuming AC, AD cannot be disproved, and it implies CH and implies that all sets are Lebesque measurable and Baire categorizable.

Consequences of AC: AC will be assumed for the rest of this section. It implies that there are non-measurable and non-categorizable sets (and hence implies not-AD). The most famous counterexample occurs in dimension three, and is due to Banach and Tarski: There are finitely many disjoint sets $A_1,...,A_n, A_1',...,A_n'$ such that A_i is geometrically congruent to A_i', the union of the A_i (A_i') is a sphere S (S'), but S is not geometrically congruent to S'. This implies that in dimension three there is no weakly additive, invariant measure defined on all sets.

Results about CH: AC and CH together imply the existence of a well-ordering of the line in which any point has only finitely or denumerably many predecessors, which might be called a *very-well-ordering*. It allows many examples, as surprising as those of Cantor or of Banach and Tarski, to be produced, a hundred or so of which have been collected by Sierpinski. The hypothesis, equivalent to the conjunction of AC and CH, that there is a *very-well-ordering* of the line might be called WO^+. Even stronger hypotheses than WO^+, implying even more surprising consequences than those of Sierpinski, have been considered in the literature, under such names as "diamond", "box", "morass".

A hypothesis called after its author *Martin's axiom* MA is implied by, but does not imply, CH. About half the consequences of CH considered by Sierpinski are provable assuming MA, and about half are disprovable assuming MA and not-CH. Another hypothesis of the same author, called *Martin's maximum* MM implies MA and not-CH, indeed implies CH' among other surprising consequences.

Results About Special Classes of Point-Sets: Perhaps the two most important question areas are (i) measurability and categorizability, (ii) existence of selectors (superselectors) for multifunctions. Without special hypotheses, the measurability and categorizability of analytic sets, and the existence of a PCA selector for any PCA multifunction can be proved. Perhaps the two most important special hypotheses are (a) the existence of a PCA very-well-ordering, (b) the determinacy of all projective sets. Each implies the existence of a projective selector for any projective multifunction, but otherwise the two are incompatible. Indeed, (a) implies there are non-measurable and noncategorizable PCA sets, and that there exists a projective superselector for any projective multifunction; while (b) implies the measurability and categorizability of all projective sets, but that there exist no projective superselectors for certain simple multifunctions.

References: more detailed surveys with limited prerequisites include several chapters in the handbook of Barwise (1977): Kunen's covers question related to AC, CH, and "diamond"; while Rudin's covers MA. Martin's covers results about special classes of point-sets. Also to be mentioned is the symposium contribution of Martin (1976), covering questions related to CH (and so overlapping with the surveys of Kunen and Martin). The original monograph of Sierpinski (1956) on consequences of CH is very readable, as is the textbook of Oxtoby (1971) on measure and category. A very recent and readable popularization of the example of Banach and Tarski on non-measurability is provided by the magazine article of French (1988). Illustrations (by computer graphics) of the example of Cantor on the distinction between different "smallness" properties, and of many other examples, are provided in the essay of Mandelbrot (1977).

Two.   Point-Set-Set Theory.

So far we have considered objects of *rank zero*, points, and objects of *rank one*, sets whose elements are objects of rank zero. In considering higher ranks, there are two options: We may take objects of rank r to be (i) sets whose elements are objects of rank (r - 1), or else (ii) sets whose elements are objects of rank less than r. Iterated, option (i) leads to a *stratified* hierarchy of sets like that of Russell and Whitehead; while option (ii) leads to a *cumulative* hierarchy of sets like that of Zermelo and Fraenkel. The stratified hierarchy has only finite ranks, while the cumulative hierarchy has also transfinite ranks, so there is an important difference between them when iterated. The difference between them is less important if one stops at rank two: The theory based on either option can easily be interpreted in the theory based on the other option. Either can be formalized in a natural way as an extension G_2 of G_1 in an extension L_2 of L_1, with variables for objects of rank two, and the rank two extensionality axiom and comprehension scheme.

Two indicate how the assumption of sets of higher rank may have implications about sets of lower rank, two examples may be mentioned of hypotheses expressible in L_2 but not in L_1 that have important consequences expressible in L_1. The first, which goes back to Zermelo, is the rank two analogue of AC^-, the existence of selectors for higher rank sets. It implies the rank one original AC^+, the existence of a well-ordering for lower rank sets. The second, which goes back to Ulam, is the assumption of the existence of a strongly additive (but ncn-invariant) measure defined on all subsets of the unit interval. It implies CH* among other surprising consequences. Examples multiply as we proceed to consider ranks higher than two.

Infinity. Set Theory.

The set theory now generally accepted derives from Zermelo and Fraenkel. Their set theory may be formalized in a natural way as a first-order theory ZFU in a first-order language L. L has variables x,y,z, and so on for objects of rank zero or *urelements*, and u,v,w, and so on for objects of higher rank or sets; it has predicates for relations among urelements or *urrelations*, and for elementhood. ZFU has a half-dozen or so axioms and schemes, the "intuition" behind which is explained in the handbook chapter of Shoenfield (1977). Taking the urelements to be points, and the urrelations to be geometric, and adding the axioms of G_0, produces a theory $G_0 + ZFU$. It is not literally an extension of G_1 or G_2, but G_1 and G_2 have such an obvious interpretation in it that, by a slight abuse of language, it will be spoken of as if it were.

Within ZFU one can define the *pure* sets as those none of whose elements, elements of elements, and so on, is an urelement or non-set. Thus there is only one pure set of rank one, the empty one, and only one pure set of rank two, its singleton. But starting with the empty set as a surrogate for zero and its singleton as a surrogate for one, it is possible to find among the pure sets surrogates for all the objects of classical mathematics, including points of classical geometry. Within $G_0 + ZFU$ it can be proved that the system of point-surrogates (and sets thereof, and sets of sets thereof) is *isomorphic* to the system of points (and sets thereof, and sets of sets thereof). Alternatively, rather than start with the mixed theory (of urelements and sets) ZFU, one can start with a pure theory ZF (of sets), and within it prove the surrogate-interpretations of the axioms of G_0 (and G_1 and G_2). The alternative approach of working in ZF is equivalent to the original approach of working in $G_0 + ZFU$, since either theory is interpretable in the other. The alternative approach is now generally accepted in the literature, including most of the references cited above, where points are identified with their surrogates.

Also now generally accepted is the axiom of choice—it is not necessary to distinguish weaker, selector versions from stronger, well-ordering versions, since as noted in the last section, the former applied to higher ranks imply the latter applied to lower ranks—addition of which to ZF produces the theory ZFC. Certain additional axioms have been proposed, but not yet generally accepted. Of these the most important are perhaps (1) the *axiom of constructibility*, addition of which to ZFC produces ZFL, and (2) the *axiom of supercompactness* (a "large cardinal axiom"), addition of which to ZFC produces ZFS. The two axioms are incompatible. Opponents of ZFL and proponents of ZFS have argued that the "intuition" behind many axioms of ZFC is that the cumulative hierarchy of sets is as "wide" as conceivable, containing as sets of rank r all conceivable sets whose elements are of rank less than r, and that the denial of constructibility is a further expression of this "intuition"; also that the "intuition" behind other axioms of ZFC is that the cumulative hierarchy of sets is as "high" as conceivable, containing sets of all conceivable ranks r, and that the affirmation of supercompactness is another expression of this "intuition". For more on the "intuitions" behind proposed additional axioms, and details on the results outlined in the rest of this section, see the recent two-part paper of Maddy (1988a,b).

The question with which this paper began may now be reformulated and supplemented with another question:

(Q1) Are there geometric-looking hypotheses about arbitrary (special) point-sets that can be expressed in L_1 but not settled in G_1 that can be settled in higher set theory (plus the axiom of constructibility) (plus the axiom of supercompactness)?

(Q2) If so, is there a single such hypothesis that, added as an axiom or scheme to G_1 has as consequences all other such hypotheses about arbitrary (special) point-sets that are consequences of higher set theory (plus the axiom of constructibility) (plus the axiom of supercompactness)?

Each of these is indeed a family of a half-dozen questions. The notion "geometric-looking" is an intuitive, not a rigorous one. If it is omitted from the questions, then they can be answered affirmatively by general results of logic due to Gödel and others. But the questions are perhaps most interesting when it is not omitted. In outline, the status of the questions is as follows.

ZFC has no known geometric-looking consequences (beyond those of the axiom of choice) about arbitrary point-sets, but it does have one known consequence about special point-sets: the determinacy of all Borel sets.

ZFL has as consequences CH, "diamond", "box", "morass", and others, that can be expressed as hypotheses about arbitrary point-sets; as the consequences get stronger and stronger, expressing them in L_1 becomes less and less a matter of natural formalization, more and more a matter of artificial coding: They become less and less "geometric-looking". The axiom of constructibility implies the existence of a PCA very-well-ordering, which as noted in an earlier section has many consequences about special sets; and it is known that all the consequences of constructibility about special sets are consequences of the existence of a PCA very-well-ordering.

ZFS has no geometric-looking consequences about arbitrary point-sets: In particular, neither the axiom of supercompactness nor any other large cardinal axiom implies anything about CH. The axiom of supercompactness does imply the determinacy of all projective sets, which as noted in an earlier section has many consequences about special sets; and it is known that all the consequences of supercompactness about such sets are consequences of the determinacy of such sets.

While it is hoped that the above results will have been found of interest for their own sake, it may be mentioned that they are also of interest for the sake of certain connections with the problem of the nominalistic reconstrual or reconstruction of the mathematics applied in physics, as will now be briefly indicated. In connection with the very closely related question of the nominalistic reconstrual or reconstruction of the physics in which is applied mathematics, Field, in his well-known book (1980) (and in a half-dozen subsequent papers, to appear together in a collection in preparation), has claimed (1) that a nominalist can accept geometry if it is taken to be a theory not about ideal mathematical points in an ideal mathematical space, but rather about real physical points in a real physical space. While there are many consistent mathematical geometries, presumably there is only one true physical geometry. For brevity, claim (1) will be discussed on the (contrary-to-fact) assumption that the one true physical geometry is G_0.

Field has also claimed (2) that a nominalist can accept certain strong and rich logical apparatus, including (a) the *mereological* apparatus of Lesniewski, and/or (b) the *plural* apparatus of Boolos. Acceptance of G_0 and either (a) or (b) would enable a nominalist to accept a theory in which G_1 is trivially interpretable by replacing quantifications "there exists a set of points ..." by quantifications (a) "there exists a sum of points ..." or (b) "there exist some points ...". Acceptance of G_0 and both (a) and (b) would enable a nominalist to accept a theory in which G_2 is trivially interpretable by replacing quantifications "there exists a set of sets of points ..." by quantifications "there exist some sums of points ...".

As noted in earlier sections, classical algebra is interpretable in G_0, and classical analysis is interpretable in G_1. So acceptance of claims (l) and (2) would enable a nominalist to reconstrue all the mathematics applied in physics at present. The only residual worry would concern conceivable future applications of higher set theory as now generally accepted (ZFC) or some proposed extension thereof (ZFL, ZFS). This residual worry will be discussed under the (almost certain) assumption that such applications would proceed by way of geometric-looking consequences of higher set theory expressible in L_1, as in question (Q1) above. Question (Q2) above becomes relevant in the following way: If all those geometric-looking consequences are implied by a single one of them P, then the very fact of its having applicable implications could be cited by a nominalist as quasi-inductive evidence for the truth of P, or more precisely of its trivial nominalistic interpretation: This, rather than higher set theory, could be accepted by a nominalist, and would suffice for applications. More detailed discussion must be left for another occasion.

References

Barwise, J.(ed.) (1977). *Handbook of Mathematical Logic*. Amsterdam: North Holland.

Burgess, J.P. (1984). "Synthetic Mechanics," *Journal of Philosophical Logic* 13: 379-395.

Field, H.H. (1980). *Science Without Numbers: A Defence of Nominalism*. Princeton: Princeton University Press.

French, R.M. (1988). "The Banach-Tarski Theorem," *Mathematical Intelligencer* 10.4: 21-29

Kunen, K. (1977). "Combinatorics," in Barwise (1977).

Maddy, P. (1988a). "Believing the Axioms I," *Journal of Symbolic Logic* 53: 481-511.

_ _ _ _ _ _. (1988b). "Believing the Axioms II," *Journal of Symbolic Logic* 53: 736-764.

Mandelbrot, B.B. (1977). *Fractals: Form, Chance, and Dimension*. San Fransisco: Freeman.

Martin, D.A. (1976). "Hilbert's First Problem: The Continuum Hypothesis," in F. Browder, ed., *Mathematical Developments Arising from Hilbert Problems*. Providence: American Mathematical Society.

_ _ _ _ _ _ _ . (1977). "Descriptive Set Theory," in Barwise (1977).

Oxtoby, J.C. (1977), *Measure and Category*. Berlin: Springer.

Rudin, M.E. (1977), "Martin's Axiom", in Barwise (1977).

Shoenfield, J.R. (1977C, "The Axioms of Set Theory", in Barwise (1977).

Sierpinski, W. (1956), *Hypothèse du Continu*, 2nd. ed. New York: Chelsea.

Tarski, A. (1959), "What is Elementary Geometry?", in L. Henkin, P. Suppes, A. Tarski, eds., *The Axiomatic Method with Special Reference to Geometry and Physics*. Amsterdam: North Holland.

Part XV

RISK ASSESSMENT

Some Public Policy Problems with the Science of Carcinogen Risk Assessment[1]

Carl F. Cranor

University of California

Government agencies and private risk assessors use (quasi) scientific "risk assessment" procedures to try to estimate or predict risks to human health or the environment that might result from exposure to toxic substances in order to take steps to prevent such risks from arising or to eliminate the risks if they already exist[2] We might think of this as an aspect of "regulatory science".

Using carcinogen risk assessment as a model I consider several aspects of risk assessment—use of epidemiology, animal bioassays, and extrapolation models in predicting risks—to show how scientific procedures and uncertainties may determine (in a way often unbeknownst to the practitioners) the public policy debates concerning the estimation of risks from carcinogens.

For one thing, actual and possible scientific uncertainties are large enough that two different researchers using exactly the same data could come to quite different conclusions; such uncertainties permit risk assessors in large measure to determine regulatory decisions even though this exceeds their authority. In addition, both the uncertainties and science policies used to overcome them permit the infection of regulatory science with public policy, moral considerations if you will, thus making regulatory science much less like ordinary science.

Also, there is an assumption that risk assessment is independent of the risk management—of the important social policy—decisions and this seems mistaken. Because of the numerous uncertainties in risk assessment, scientific and nonscientific policy considerations bridge these gaps, infecting risk assessment with policy considerations. Thus, regulatory science, for the above and other reasons is much less like ordinary science than is normally supposed.

Furthermore, in human or animal statistical studies that are the foundation of standard setting scientists may be forced, for reasons of economics or circumstances, to study very small samples of experimental and control groups. This in turn in many circumstances forces either them or risk managers who use their data into a dilemma whether they recognize it or not: they are forced to choose between adhering to the evidentiary standards of good science as conventionally conceived—using cautious inferences before drawing scientific conclusions and avoiding false positives—and doing "good regulation"—providing results (estimates or predictions) protective of human health for purposes of regu-

PSA 1988, Volume 2, pp. 467-488
Copyright © 1989 by the Philosophy of Science Association

lation and avoiding false negatives. Because of these effects, I argue that scientists or policy makers (whichever is appropriate) should approach traditional avoidance of false positives as moral decisions appropriate to the context in question, especially when they are engaged in regulatory science.

Finally, the above results raise several questions scientists, philosophers of science, moral philosophers and policy makers should address forthrightly in order to serve better the aims of science and regulation.

1. Background

Risk assessment can be divided into toxicological and environmental risk assessment. Toxicological risk assessment is the assessment of risks to human health from particular substances making inferences from animal bioassays and human epidemiological studies, while environmental risk assessment estimates risks to human beings from exposure to carcinogens when they are released into the environment, into the soil, the ground water and air.[3]

Contrasted with *risk assessment* is *risk management*.[4] Risk assessment, a seemingly scientific enterprise, when used in regulatory law, aims at providing accurate information about risks to human beings so administrative agencies can regulate exposure to potentially cancerous substances in fulfillment of their respective statutory mandates. After scientists in the technical, scientific part of the federal agencies have provided an estimate of risks to human beings from exposure to toxic substances, they then give this information to the risk managers. *Risk management* is concerned then with managing the risks in accordance with statutory requirements and other economic, political and normative considerations.

At the outset, however, risk assessment in the present state of knowledge is a third best solution to the problem of estimating harms to human beings from exposure to toxic substances. The ideal is 'harm assessment'; if we had perfect information we would provide an accurate assessment of the harmful effects to people and the environment from exposure to toxic substances. This would provide us with exact numbers of deaths and diseases, thus we would not overestimate or underestimate the kinds and amounts of harms resulting from toxic exposures.

If we distinguish between risks and uncertainties, a "risk" is a *probability* of an unfortunate or undesirable outcome (Rescher 1983, p. 5), when one can assign such probabilities to outcomes. Thus, a "risk assessment" properly speaking aims at estimating the *probabilities* of harms from toxic exposures, and is a second best solution to the problem of estimating harms to human beings from toxic exposure.

At present the task of regulators is much worse than this, for great uncertainties obtain in trying to predict such harms. For example, some have argued that high dose to low dose extrapolation models used in animal bioassays can vary by six orders of magnitude (10^6), (Cothern, et. al. 1986) and many believe there is little biological basis for choosing between these models. (Freeman and Zeisel) Thus, we should think of risk assessments at present, not as risk assessments properly speaking, but as "risk and uncertainty assessments," the third best solution to identifying harm to human beings from toxic exposures.

In addition, matters of considerable moment depend upon the products of risk assessment, for in many cases one answer (a projection of high enough risks to require regulation) will impose considerable costs on the affected industry and perhaps the larger public, while another answer (a projection of risk low enough so that regulation is not a required) may well leave innocent people at risk from exposure to dangerous substances.

Because of substantial uncertainties, regulatory agencies run risks of making mistakes, risks of regulatory false positives and false negatives. A regulatory "false positive" would occur when a substance is inappropriately regulated or regulated too stringently under a particular statute for the degree of harm that it causes. The substance might cause no harm at all, or much less harm than an agency believed it caused at the time of regulation. A regulatory "false negative" by contrast would be an outcome of a regulatory activity that resulted in a substance not being regulated at all when it should be, or regulated to a much lesser degree than it should be commensurate with the kind and degree of harm it causes and commensurate with the statute under which it was regulated. In a statistical study, because probabilities are involved, *it is certain* that if one were regulating large numbers of substances there would be both false positives and false negatives, for by chance alone, mistakes would be made. By analogy it is likely that agencies regulating large numbers of substances will make mistakes, either as a consequence of the underlying statistical studies or through other errors, which will result in both false positives and false negatives. Because of the possibility of regulatory mistakes it is a normative question concerning the parties on whom the costs of regulation or its absence should fall.

There are a number of ways to estimate risks to human beings from chemical substances—molecular structure analysis, short-term tests, long-term chronic bioassays in laboratory animals, and human epidemiology. In the following sections I summarize some results from the latter two to indicate some of the public policy problems that arise from the "science" in these fields.

2. Regulatory Science and Policy Choices

One method for estimating toxicological risks to human beings is to study the carcinogenic effects of substances on animals, and then to project risks to human beings based upon this information. An experimental group of rodents is fed high doses of a substance to see whether the cancer rate in the experimental group is significantly greater than the cancer rate in a control group. If it is, then scientists extrapolate from the high dose response rates in the rodents to project a low dose response rate in rodents (an exposure level much closer to the typical human exposure dose). Using this low dose response rate, based on principles of biology, toxicology and pharmacology (if such information is used), they then estimate, on the basis of rodent-to-human models, the likely risks which human beings would face at their levels of exposure. This risk information is then combined with exposure information in the workplace or in society at large in order to estimate the magnitude and extent of the risks to human beings. Finally, the risk information is combined with economic, policy, statutory, and technological feasibility information so that regulatory agencies can then decide how properly to manage the risks in question.

Such procedures require a number of inferences from the established experimental data from laboratory animals to the projection of end point risks to human beings, and these inferences have a number of uncertainties and gaps which must be bridged in order to produce the risk numbers. The uncertainties arise because there is insufficient information (in both theories and data) available to settle the scientific questions at issue. For example, the choice of different high dose to low dose extrapolation models (all of which have been advocated by respectable researchers) can produce variations in endpoint risk estimates that vary by a factor of 10^6. (Cothern, et. al. 1986) Use of different rodent to human extrapolation models can produce risk estimates that differ by a factor of thirteen. Use of different parameters in environmental fate models can produce results that vary by a factor of 500.[5] Some of these possible variations of estimates have not been exhibited in actual governmental risk assessments, for often different regulatory agencies have tended to agree on the same models, even though scientific data and theories are insufficient to support such choices. However, we have found in our research that actual risk assessments done by agencies may well differ by two orders of magnitude.[6]

In order to bridge the gaps created by uncertainties, a National Academy of Sciences Study has suggested that regulatory agencies adopt policies (inference guidelines) for some fifty substantial "inferential gaps" that exist in the procedures for estimating risks to human beings from basic toxicological information. (National Research Council 1983)[7] These may not always be well grounded in biological fact.

These policies may govern decisions from choice of data, e.g., should toxicologists count benign tumors in experimental animals as providing evidence that a substance causes cancer, to choice of high dose to low dose extrapolation models, which aims at modelling statistically the mechanism of cancer, to choice of statistical procedures, e.g., whether one uses upper confidence limits or maximum likelihoods in the high dose to low dose extrapolation models, to whether certain kinds of data should be permitted or required to be offered in evidence in regulatory proceedings, e.g., whether agencies should consider pharmacokinetic mechanisms in calculating the dose to which a person is exposed.[8] Some of the main areas of disagreement between agencies and litigants to cancer standard setting are summarized in Appendix B.

The policies adopted to bridge such gaps can lead to considerable controversy, for they can make large differences in the estimation of risks to human beings, and there may be little or no biological plausibility to any of the models. (Freedman and Zeisel)

Thus, since science policies may have little justification in biological fact, they are chosen at least in part on public policy considerations. For example, most agencies have adopted the linearized multistage high dose to low dose extrapolation model in conjunction with a 95% upper confidence interpretation in order to avoid underestimating risks to human beings. (OTA 1987) The public policy outcome explicitly guides the choice of model. Thus, the "science" of risk assessment at the very outset, is widely infected with extra-scientific policy judgments about the matters at issue, and the "science" of risk assessment is less than fully scientific in the ordinary sense.

If the above view is correct this poses a dilemma for risk assessment science. On the one hand, an attempt to make it more scientific in the sense of basing component models on established biological facts will result in little regulation because the knowledge is not available and because current regulatory (and tort) laws tend to preserve the status quo until evidence for changing it is provided. On the other hand, if agencies are to expedite risk assessments and not wait for answers to these scientific questions, then the science of risk assessment will be substantially permeated with nonscientific policies and thus be quite different from ordinary science.

A larger point is that the many uncertainties which at present exist in cancer risk assessment may make it difficult, if not impossible, for scientists to remain wholly faithful to scientific traditions while providing data which will permit timely and morally justifiable regulations. More likely, fidelity to scientific tradition will paralyze regulatory activity. In addition, however, even if there were not the above problems, there remain more fundamental problems with aspects of carcinogen risk assessment that should be of concern to scientists, philosophers of science and policy makers. These are considered in the remainder of the paper.

3. Hidden Policy Tradeoffs

(A) I have argued elsewhere that a wise and conscientious epidemiologist (or a risk manager using the epidemiologist's results) doing a cohort epidemiological study (OTA 1981, p. 137) with perfect evidence, but with constrained sample sizes for the disease being studied, *cannot* but face potentially controversial moral and social policy decisions in order to design and interpret an epidemiological study and to produce the risk numbers that are the outcome of such work. I summarize some of the main points of that argument here.[9]

Consider the theory of hypothesis acceptance and rejection. This provides appropriate terminology to characterize the main risk and proof variables with which epidemiologists must work, and to understand the logic of scientific proof available in this area.

Consider for regulatory purposes whether benzene causes serious disease, e.g. leukemia or aplastic anemia. The null hypothesis would predicate, e.g., that exposure to benzene is *not* associated with greater incidence of a certain disease than that found in a nonexposed population, while the alternative hypothesis would then indicate that exposure to benzene *is* associated with a greater incidence of such diseases. (Feinstein 1977, pp. 320-321)

By chance alone a researcher investigating these hypotheses risks false positives [type I errors], or false negatives [type II errors]. The probability of committing a type I error is normally designated **A** and the probability of committing a type II error is designated **B**.[10] (Feinstein 1977, pp. 320-321) The "power" of a statistical test is 1 - **B**. Conventionally, A is set at .05 so that there is only a one in twenty chance of rejecting the null hypothesis when it is true. (Walter 1977, p. 391)[11]

The low value for **A** probably reflects a *philosophical view* about scientific progress and may constitute part of its justification.[12] In building the edifice of science, by keeping the odds of false positives low, one ensures that each brick of knowledge added to the structure is solid and well-cemented to existing bricks of knowledge. Were one to tolerate higher risks of false positives, take greater chances of new knowledge being false by chance alone, the edifice would be much less secure. A secure edifice of science, however, is not the only important social value at stake in environmental health regulation.

One can think of **A**, **B**, and 1 - **B** as measures of the "risk of error" or "standards of proof." What chance of error is a researcher willing to take? Is a twenty percent (**B** = .20) chance of a study showing benzene does not cause cancer, when in fact it does, an acceptable risk? When workers or the general public may be contracting cancer (unbeknownst to all) even though a study (with one kind of high epistemic probability) shows they are not, is a risk to their good health worth a twenty percent gamble?

Alternatively, we might think of **A**, **B**, and 1 - **B** as standards of proof. How much proof do we demand of researchers and for what purposes? Must potential carcinogens be condemned by mere preponderance of evidence, somewhat more than fifty percent of the evidence (e. 1 - **B** = .51+)? That is to say, must researchers be more than fifty percent sure that benzene is a carcinogen presenting a risk to employees in the workplace before regulating it? Should scientists in agencies be permitted to take a forty-nine percent chance (**B** = .49) that substances are not high risk carcinogens to the populace, when in fact they might be?

These trade-offs are also a function of two other variables: N, the total sample of people in the exposed and unexposed groups, and **D**, the relative risk one wants to detect. (Feinstein 1977, pp. 320-324) With **A** and **B** fixed, the relative risk one can detect is inversely related to sample size: the smaller the risk to be detected the larger the sample must be. The value of **D** for which a study might be designed to detect depends upon many factors, including the seriousness of the disease, its incidence in the general population, and how great a risk, if any, the exposed group justifiably should be expected to run.

Furthermore, **A**, **B**, **D** and N are mathematically interrelated. If any three of them are known the fourth can be determined. Because the variables are interdependent, crucial trade-offs may be forced by the logic of the statistical relations. Consider the following hypothetical decision tree (summarized in Table I) which presents five related alternatives.

The first thing to notice is that if one wished to have the most *accurate* study with equally small chances of false positives and false negatives, one should adopt alternative

472

(1). This choice is likely to be very expensive or such large samples will be unavailable for study. Similar conclusions follow for alternative (2).

If one rejected these alternatives, however, and were forced to study a much smaller sample, this poses a dilemma. Alternatives (3) and (4) may leave those exposed to toxic substances at considerable risk, because by chance alone they run substantial risks of false negatives (or of failing to detect a relative risk that is of concern). Alternative (5) risks undermining the credibility of the research because it is inconsistent with scientific practice. Thus, the mathematics of epidemiology together with small sample sizes and a low background disease rate impose difficult moral choices on researchers and regulators alike when such studies are used in regulatory contexts to estimate risks to people.

Table I

Cohort Study

Assume that the prevalence of disease L in the general population is 8/10,000.

The study seeks to detect a relative risk of 3 ($D=3$), provided such risk exists.1

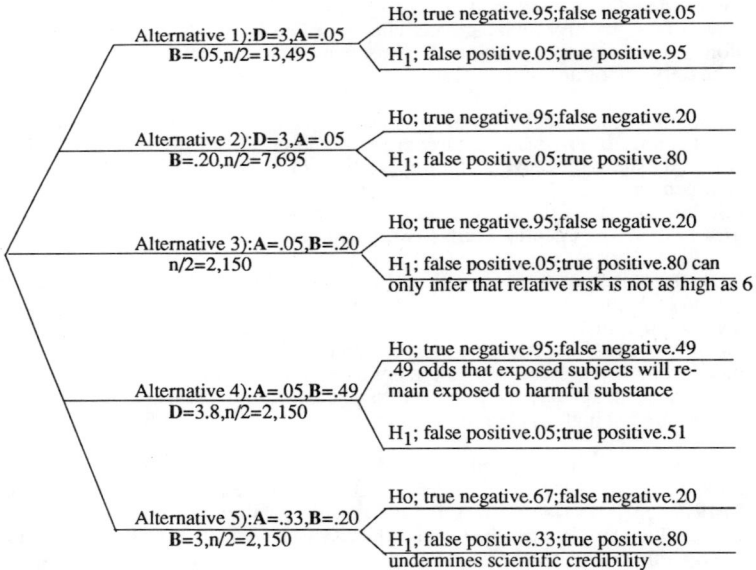

Nearly all the figures used in these examples are taken from my "Epidemiology and Procedureal Safeguards for Workplace Health in the Aftermath of the Benzene Decision," which figures were originally supplied to me by Dr. Helaine Pleet and the computers at the Centers for Disease Control in Atlanta, Georgia. That information is summarized in Appendix A. Similar numbers can be calculated from the equation (4) Walter (1977) in note 47 infra. The particular numbers in alternative (1) come from line 3 (c) of Appendix E.

Alternative (3) suggests some interesting results for "negative" or "no effect" studies. Assume a study is run on 2,150 exposed workers with **A** at .05 and **B** at .20, when the prevalence of the underlying disease is 8/10,000. With these values, we only could be confident of detecting a relative risk of six with a power of .80. But suppose none were detected, that is, the study was "negative" or showed "no effect" between some chemical C and the disease L. What could we infer? At most we would be justified in concluding that the relative risk was less than 6. It might be 5.9 or 1, but given the constraints on the study, we could not conclude so statistically. Thus, for "no effect" studies the most that can be inferred is that the relative risk to people in the exposed group is not as high as the relative risk tested for in the study. Regulatory agencies regard such results as useful mainly for setting upper bounds on risks to people. (OTA 1987)

Furthermore, negative studies based on small samples and low A can have quite unacceptable type II errors for public policy purposes. Needleman and Bellinger, surveying fourteen epidemiological studies of lead in children, with low A (< .05) found that type II error rates ran as high as 82%. Thus, by chance alone researchers had up to 82% odds of failing to detect adverse effects on children from lead exposure even when they may in fact have existed. (Needleman and Bellinger)[13]

In other cases it may be statistically impossible to detect a significant health risk when one exists. Suppose that it is thought important to detect a relative risk of three among workers exposed to toxic substances for a disease that occurs in eight people of every 10,000. If there were only 1,000 workers to study (with A at ,05 and B at .20), a relative risk could not be detected below 10 with a power of .80, even if it turned out the substance in question caused a threefold increase in mortality. (Walter 1977, p. 39)

Table II

Cohort Study

Assume that the prevalence of disease L in the general population is 8/100,000.

The study seeks to detect a relative risk of 3 (D=3), provided such risk exists.

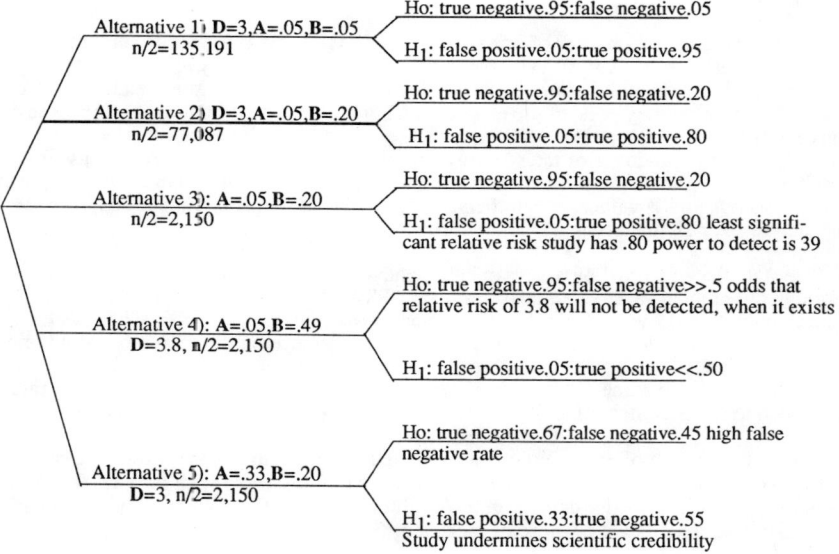

Alternative 1) D=3,A=.05,B=.05
n/2=135.191
Ho: true negative.95:false negative.05
H₁: false positive.05:true positive.95

Alternative 2) D=3,A=.05,B=.20
n/2=77,087
Ho: true negative.95:false negative.20
H₁: false positive.05:true positive.80

Alternative 3): A=.05,B=.20
n/2=2,150
Ho: true negative.95:false negative.20
H₁: false positive.05:true positive.80 least significant relative risk study has .80 power to detect is 39

Alternative 4): A=.05,B=.49
D=3.8, n/2=2,150
Ho: true negative.95:false negative>>.5 odds that relative risk of 3.8 will not be detected, when it exists
H₁: false positive.05:true positive<<.50

Alternative 5): A=.33,B=.20
D=3, n/2=2,150
Ho: true negative.67:false negative.45 high false negative rate
H₁: false positive.33:true negative.55
Study undermines scientific credibility

As striking as the preceding examples are, they only suggest the statistical problems a cohort study of a typical environmentally caused disease (e.g., benzene-induced leukemia) might pose. If the prevalence of the disease subject to study were rarer by a factor of 10, a more realistic number for leukemia, (National Cancer Institute 1981, pp. 662-63, Table 51) then the decision tree exhibits even more surprising results (summarized in Table II). In an analogue to alternative (2), holding **A, B** and **D** constant and changing the incidence of L to 8/100,000 for leukemia, one would have to study 77,087 people exposed to benzene and an equal number not exposed to obtain statistically significant results.[14] In an analogue to alternative (3), one would have to study 21,580 subjects, one could detect a relative risk of six.[15] Forced to study 2,150 subjects, one could detect a statistically significant relative risk no lower than thirty-nine.[16] If one could only study as few as 1,000 people, one could reliably detect a relative risk no lower than fifty.[17] Studying as many as 10,000, one could detect a relative risk of no less than nine.[18] In alternative (4), holding **A, D** and **N/2** constant, one would lack even fifty percent confidence in one's results.[19] Even increasing **A** to 16% would not decrease the chance of a type II error below fifty percent.[20] Analogues of alternative 5) lead to similarly unsatisfactory results. The point: the rarer a disease, the greater the problems faced by epidemiologists (or policy makers interpreting a study).

The point of the above examples is that epidemiologists even with no evidence gathering problems, when faced with the relatively rare diseases and small samples available for study typical of environmentally induced diseases, must take into account cost considerations, samples available for study, and some of the objectives of the study in order to *design* and *conduct* the study in question. The most important of these questions concerns the odds of false positives and false negatives researchers are willing to take. In regulatory contexts this will make a difference on whom the costs of possible statistical mistakes will fall. *Thus someone*, either the *scientists* designing and conducting the study or the *policy makers* relying on the statistical evidence and applying the study results in a regulatory context, must face up to some crucial social issues. Who should bear the costs of possible statistical mistakes: The company making the substance which may be the object of regulation or the public or employees of a firm who might be subject to health harms, if the substance is toxic?

Furthermore, if, as a matter of scientific practice scientists uncritically remain committed to the 95% rule (A = .05) and forward the results uncritically to risk managers, they risk deciding the public policy question of whether a risk to human health exists or not merely by how they present the evidence. *Uncritical* commitment to the 95% rule convention may beg the public policy question at issue. Whether or not this occurs depends upon the practices of the scientists involved; this is a matter of sociology. It is possible for statisticians to present their results objectively without commitment to preset or automatic **A** and **B** values, [21] but then the **A, B** value tradeoff is merely pushed back one step to the risk managers.

The statistics of animal bioassays exhibit behavior similar to that of epidemiology although the numbers are not quite as dramatic. Talbot Page has shown that if one has fifty control animals and five of these develop tumors at one site, e.g., the liver, while twelve of fifty treated animals develop tumors at that same site, reliance on the 95% rule would reject this as statistically significant evidence of a difference in tumor rates.[22] Nonetheless, use of Bayes' Theorem and some plausible background assumptions would show the tumor rate in the treated animals compared to the controls to be a matter of considerable concern.[23]

Just as in the case of epidemiological studies, in animal studies the rarer the disease rate is in control animals and the fewer the controls in the sample with tumors, the more researchers should consider using higher A values to ensure that existing diseases in animals do not go undetected because of the scientific conventions of the statistics of the studies.[24]

(B) One can see additional problems with understanding epidemiological studies for regulatory use by looking at how fixed data might be interpreted once a study has been finished. The fixed data would consist of the background disease rate, sample size, and observed relative risks. However, for purposes of *interpreting* this information epidemiologists or policy makers could vary the values of **A** and **B**. Consider one possibility.

Suppose the study of 2,150 exposed individuals produced an observed relative risk of about three, because there were five deaths compared with 1.72 (one or two) in the control group. Is this a positive result or not? The following table shows that one could interpret the study as a positive or as a negative study for any of several pairwise choices of **A** and **B** values.

Positive Results			**Negative Results**
A	**B**	**D** (Least significant relative risk which test has a power of .51 or higher to detect)	
.10	.49	2.0	When A<.10 (with B constant) or B<.49 (with A constant)
.15	.40	3.0	When A<.15 (with B constant) or B<.40 (with A constant)
.20	.30	3.1.	When A<.20 (with B constant) or B<.30 (with A constant)
.25	.25	3.1	When A<.25 (with B constant) or B<.25 (with A constant)

Any pairwise combinations of **A** and **B** in the left hand column will show that the study outcome is positive. Changing the variables slightly as indicated in the right column will produce a negative study.[25] This example shows that risk assessors or policy makers have considerable flexibility in *interpreting* the data of a study. How they *interpret* and *use* the evidence in certain regulatory and legal contexts will have important consequences for protecting human health. Similar results can be reproduced for interpreting the results of animal bioassays; researchers and policy makers should be given advice similar to that given for interpreting epidemiological studies: adherence to low **A** values (e.g., use of the 95% rule) may well bias the normative, regulatory issues in unfortunate ways. Thus, the interpretation of the tumor rates of animal bioassays should be approached as a normative, or moral question.

(C) An additional problem concerns the relative values of **A** and **B** when evaluating the health effects of relatively large numbers of substances. As long as **A**<**B** and **A** is in the neighborhood of .05, we are doing "better" science conventionally conceived, but we are also protecting possibly harmful chemicals better than human health. Suppose that we have twenty-four hundred substances to test. Assume also, that 40% of those are carcinogens and 36% of them are not, with the remainder equivocal or inconclusive (realistic numbers according to an OSHA study). Now if epidemiologists set **A** at .05 and **B** at .20 (typical values), assuming this is large enough sample, we will have one-hundred and ninety-two false negatives and forty-three false positives. With one-hundred and ninety-two false negatives, this means that one-hundred and ninety-two substances will pose some risk of cancer to the populace (and how large a risk this is will depend upon the prevalence of the disease, the relative risk associated with the substance, the substance's potency and the number of people exposed), but the test will not show it. On the other hand, forty-three false positives mean that forty-three substances will be wrongly regulated (or possibly banned altogether), depending upon the statutory authority in question. If substances are banned, the products into which they are incorporated will be more expensive to produce and market, or we will be deprived of their use and benefits altogether. If the substances are merely regulated, they likely will be more expensive to produce and market.

4. Philosophical Decisions in Regulatory Science

In certain common circumstances, described above, in which we use or need to use the tool of epidemiology, or other statistical studies in estimating risks to human beings from toxic substances, and in which scientific tradition would ordinarily rely upon low **A** values, e.g., **A** = .05, the study of relatively rare events by means of small samples forces a tension between the use of this rule and other public policy and moral concerns we might have. Roughly the tension is between a commitment to traditional scientific caution in pursuit of the truth (represented in the 95% rule), and a commitment to protecting people's health, or at least not taking chances with their health. However, the same examples show that in the circumstances described there is no necessity to the received scientific practice—it could be done differently. *Whether epidemiologists or policy makers should adhere to the 95% rule in certain contexts is a normative, a philosophical question.* Moral philosophers, philosophers of science, lawyers and those in public institutions with the authority to protect our health should explicitly acknowledge and address these issues. Those charged with regulating our exposure to toxic substances should consider such policy problems in the design of the study.

Second, the use of epidemiological data is not obviously a neutral and objective project. In subsection 3(B) above, sample size and number of deaths are fixed data in the study, but *whether a risk to human health is judged to exist depends upon the choice of values for* **A** *and* **B**. How the fixed data gets used in subsequent regulatory or legal proceedings will also depend upon these variables and may have important consequences for our health.

More importantly, in order to design and perform the studies in question, the choice of values for **A** and B commits scientists or the policy makers who use the studies implicitly, if not explicitly, to making judgments that are the equivalent of moral or social policy considerations.

Thus, since the design, reporting, interpretation and use of epidemiological data are dependent upon such judgments, and since we could change traditional practices for interpreting and using scientific evidence, we should face the use of the 95% rule in risk assessment and risk management proceedings as a normative question.

Consider an analogy in the law. In criminal trials, avoiding wrongful damage to someone's reputation and well-being is so important that we spend considerable sums of money and deliberately make proof of guilt very difficult in order to avoid wrongly inflicting harsh treatment and condemnation on the defendant. We could save money and have more unjust outcomes if we thought it worth the human costs, but we do not. Clearly a number of moral and cost considerations have influenced the institution of the criminal law. With regard to criminal trials, we have been quite self-conscious in debating the moral considerations that bear on the design and workings of such institutions. Similar problems arise in the interpretation and use of epidemiological studies, and I am suggesting that similar debates should attend the use of scientific results that may have such profound influences on our health.

A further problem is that the conventional evidentiary practices of science (use of the 95% rule) may well be much more demanding than the evidentiary requirements and aims of the tort and regulatory law. Tort law requires a plaintiff to establish his case by a "preponderance of the evidence." If this standard were expressed in quantitative terms it would require that somewhat more than 50-55% of the evidence favor plaintiff's case. In regulatory proceedings for the most part a lesser evidentiary standard is required than in the tort law to establish a case as long as the agency does not act arbitrarily or capriciously.

Thus the problem: the evidentiary standards of science as exhibited in the 95% rule are much more demanding that the legal standards of evidence where the regulatory science evidence will be used. Thus by default if regulators use conventional scientific standards their science will in many cases beg the regulatory question against regulation. Using epidemiology, as an example, when relatively rare diseases and small samples force regulatory scientists (or policy makers) to choose between high false negatives or high false positive rates that would be intolerable for normal scientific work, if they choose to tolerate high false negatives and protect the integrity of their scientific work, they thereby favor non-regulation or less regulation by their choice of evidentiary standard.

Furthermore, given the wider aims of both the regulatory and tort law and the weaker evidentiary standards that must be met in the law, there may not be good reasons *in these legal contexts* to require regulatory science to meet the same evidentiary burdens as normal scientific work aimed only at discovering the truth and adding to our stock of scientific knowledge about the world. Thus, I would urge that for legal purposes regulatory science adopt evidentiary standards much closer to that of the legal institutions it is meant to serve (although this argument must be prosecuted elsewhere).

5. Conclusions

Several conclusions emerge from the arguments presented above. First, because of substantial uncertainties in the risk assessment process, in nearly all cases it is difficult, if not impossible, to obtain *scientifically* respectable and accurate estimates of health risk to human beings. A number of the crucial inferences at present are not grounded in biological fact.

Second, in order to overcome the absence of scientific information, regulatory agencies have had to resort to extra - scientific policy judgments, resulting in subversion of the policy neutral status of regulatory science as it is presently practiced. Thus, the risk assessment policies as well as particular assessments of the risks posed by particular substances are not objective in the sense in which mechanics or electromagnetism is thought to be objective, and free from moral considerations.[26]

Third, and most important, however, is that even in circumstances in which there are no problems with scientific uncertainties or with extra-scientific policy considerations infecting regulatory science, substantial nonscientific policy judgments enter into interpreting and using the scientific evidence. Further, researchers or risk managers in interpreting and reporting the results of epidemiological studies face the equivalent of social policy decisions. Choice of **A** and **B** variables will influence how the study gets reported and may subsequently affect the regulation of toxic substances.

I emphasize these results in order to occasion debate among scientists, philosophers of science, moral philosophers, and lawyers concerning the best set of procedures to adopt in the face of such issues. It raises at least the following questions: Should the intellectual and institutional integrity of the relevant areas of science remain free from the nonscientific or extra-scientific policy judgments, if doing so would mean no regulation or would substantially slow the risk assessment process against regulation of toxic substances? Since scientific truth is not the only social goal in risk assessment, to what extent should pursuit of other aims modify that goal; in particular to what extent should the protective health aims of regulation or the need for expeditious risk assessments modify or shape pursuit of scientific truth? To what extent should the "institution" or practices of science accommodate the demands of other institutions in our society, e.g., regulatory institutions as well as the tort law?

I have only provided the background for posing these larger questions, but they should be faced by appropriate parties interested in the interface of science, law and public policy concerns in regulatory science. Although I have not answered such questions in this paper, they are being pursued in forthcoming papers. (Meetings/PSA 03/21/89)

Notes

[1]Research on the paper has been supported by a grant from the University of California Toxic Substances Research and Teaching Program for the UC Riverside Carcinogen Risk Assessment Project of which the author is the principle investigator. I am indebted to Deborah Mayo, D.V. Gokhale, and William Kemple and Kenneth Dickey for comments and criticisms on the ideas presented in this paper.

[2]When risk assessment is used in toxic tort suits, the aim is to provide evidence of harm to provide a basis compensation for injuries caused by such substances.

[3]This distinction is adapted from work done by the University of California, Riverside, Carcinogen Risk Assessment Group.

[4]The National Research Council. (1983), *Risk Assessment in the Federal Government: Managing the Process*. Washington D.C.: U.S. Gov't Printing Office pp. 18-19.

We use *risk assessment* to mean *the characterization of the potential adverse health effects of human exposures to environmental hazards*. Risk assessments include several elements: 1) *description of the potential adverse health effects* based on an evaluation of results of epidemiologic, clinical, toxicologic, and environmental research; 2) *extrapolation from those results to* predict the type and estimate the extent of health effects in humans under given conditions of exposure; judgments as to the 3) *number of characteristics of persons* exposed at various intensities and durations; and 4) *summary judgments* on the existence and overall magnitude of the public-health problem. Risk assessment also includes characterization of the uncertainties inherent in the process of inferring risk...

The Committee uses the term *risk management* to describe the *process of evaluating alternative regulatory actions and selecting among them*. Risk management, which is carried out by regulatory agencies under various legislative mandates, is an agency decision-making process that entails consideration of political, social, economic, and engineering information with risk-related information to develop, analyze, and compare regulatory options and to select the appropriate regulatory response to a potential chronic health hazard. The selection process necessarily requires the use of value judgments on such issues as the acceptability of risk and the reasonableness of the costs of control.

[5]Research done under the UCR Carcinogen Risk Assessment Project shows that the predicted movement of benzene escaping from leaking gasoline tanks can vary by a factor of 500. (Lee and Chang)

[6]See appendix A for a summary of several risk assessments done on each of benzene and ethylene dibromide and their differing conclusions. These results were compiled by Kenneth Dickey.

[7]Agencies have adopted four kinds of assumptions:
1. assumptions used when data are not available in a particular case;
2. assumptions potentially testable, but not yet tested;
3. assumptions that probably cannot be tested because of experimental limitations; and
4. assumptions that cannot be tested because of ethical considerations.
(The Office of Technology Assessment 1987).

[8]If the information were available and understanding of mechanisms were adequate (both of which appear to be doubtful at present for most substances), this should enable scientists to distinguish between the dose administered to a rodent or person external to the body from the dose which reaches the target organ where it would then do some damage.

[9]The conclusions of that earlier paper have been modified somewhat because of arguments presented by Deborah Mayo, a cosymposiast at the PSA meetings where this paper was read. See (Mayo 1989).

[10]Throughout I use the bold face **A**, **B** & **D** for Greek alphabet letters Alpha, Beta and Delta because the computer lacks Greek capabilities.

[11]This "conventional" value varies even within research science. For some fields a 5% chance of false positives is unacceptably *high*, e.g., in microbiology or in subatomic physics. For regulatory purposes, however, I suggest (below) this value may be unacceptably *low*.

[12]The low value for **A** may also be a mathematical artifice explained historically. As Giere puts it "The reason [for the practice of having a 95% confidence level to guard against false positives] has something to do with the purely historical fact that the first probability distribution that was studied extensively was the normal distribution." (Giere 1981, pp. 212-213) Two standard deviations on either side of the mean of a normal distribution encompasses 95 percent of the entire distribution.

He adds "95 percent is a comfortably high probability to take as the standard for a good inductive argument. Most scientists seem to think that science can get along with one mistake in 20, but not with too many more." (p. 213)

[13]Needleman's and Bellinger's results of type II errors are summarized in Appendix C.

[14]This figure is from line 3(d) of Appendix E.

[15]This figure is from line 5(d) of Appendix E.

[16]This figure is from Walter 1977, p. 388 (equation (4)).

[17]This figure is from Walter 1977, p. 391 (Table 2).

[18]Id.

[19]I put this point in a general way because of statistical problems. A sample of 2,150 people is so small compared to the disease rate of leukemia that the assumptions underlying the usual epidemiological equations no longer apply. The most important of these assumptions is that there is a normal distribution of diseased individuals through the population. In a group of 2,150 people on the average only .17 people would contract leukemia. Since people come in multiples of one, a probability procedure is needed to estimate how many times a coin will come up heads if it is flipped twenty-five times. The chance of k individuals contracting leukemia (with a prevalence of 8/100,000) is given by the formula for a Poisson distribution:

$$p(k)=e\frac{-M(M)^K}{K!}$$

where e is the mathematical constant and M is the mean of the sample population being studied, when A=1.5%, D=3, and **B**=90.5%. Thus, the chances of making a type II error are nearly one hundred percent. When A=16% and **D**=3, **B**=59.7%, one could not be more than forty percent confident that one could detect a relative risk of three, even when one existed. Professor David J. Strauss of the Department of Statistics, University of California at Riverside, pointed this out to me.

[20]Id.

[21]See (Mayo 1989) who argues this point.

[22]Page, T. 18.

[23]Our research shows Page is seriously mistaken about some aspects of hypothesis testing and use of the 95% rule, for the power of appropriate tests in his case is quite high, even though the test is not statistically significant by the 95% rule. The power for seventeen control and treatment tumors when A=.0542 is about .9468, while the power of the test for sixteen treatment and control tumors is .9630 when A=.0857. This point was established by the author and Bill Kemple working on the UCR Carcinogen Risk Assessment Project.

[24]Results of representative animal bioassays are taken from research under the UCR Carcinogen Risk Assessment Project by Mr. Bill Kemple and the author, and summarized in Appendix C.

[25]The data for this table are from Appendix F.

[26]In addition, particular science policy judgments may favor one set of parties to regulation over another. For example, several of the science policies are chosen to avoid false negative regulatory mistakes, thus risks to human beings tend to be overestimated leading to greater rather than lesser regulation of the substance in question. Risk assessment policies, however, might be designed to be more nearly procedurally neutral in that if the science policies have all been adopted after sufficient opportunity for notice and comment and in accordance with the Administrative Procedure Act, all the parties to the regulation are playing by the same set of rules.

Appendix A

Estimates of Lifetime Excess Cancer Risk from Exposure to Ethylene Dibromide

Risk Assessment	Excess deaths per 1,000 (95 pct upper confidence limits)		Model
	20 PPM	0.1 PPM	
EPA CAG:			
1976............	999	67	"Upper bound" on onehit
1980............	999	45	Onehit.
SRI..............	990-1000	117	Multihit, onehit.
OSHA In-House....	160-437(251-516)	0.06-3(1.4-3.6)	Onehit, multistage.
Busch............	190-490(293-588)	..........................Probit.	
Brown:			
Hemangiosarcoma.	70-110(134-148)	0.2-0.6(0.7-0.8)	Onehit, multistage.
Nasal Tumors....	725(785)	6 (8)	Onehit.
CAL/OSHA.........	400-996	0.5-2	Onehit, linear.

Benzene Risk Estimates by Agency (a)

	EPA	OSHA	White	NIOSH	IARC	API
10 ppm	34	95	44-152	634	14-140	8
1 ppm	3.4	10	5-16	5	1.4-14	0.6

(a) Lifetime cancer per 1000 people exposed.

Compiled by Kenneth Dickey, graduate student in Philosophy and
Research Assistant in the UCR Carcinogen Risk Assessment Project.

Appendix B

Decision Points in Quantitative Risk Assessment:
Alternative Data Choices and their Impact on Risk Estimates

Issue	More safety-oriented risk estimate	Less safety-oriented risk estimate
Input Data		
1. "Benign" tumors*	Data used	Data not used
2. Exposure (latent period)*	Allowance for "wasted exposure"	No allowance
3. Use of epidemiologic data*	Not used	Used, or suggestion that one wait for, epidemiologic data
4. Controls in epidemiology*	Detailed attempts	National data: (at times) "locally adjusted" controls
5. Risk groups in epidemiology*	Narrow - clearly exposed	Broad - doubts about exposure
6. Exposure estimate (actual population)	Upper 5% (or 1%) of population	Average in population
7. Use of "negative" studies	Not used	1. Used 2. Equated with positive studies (1=1)
8. NOGL (LOGL, etc.)	Concern with the experimental sample size	No attention is devoted to sample size
9. Adjustment for background exposures	Additivity in dose	Additivity in the response. Abbott's correction
10. Species (animal) extrapolation	"most sensitive"	Some average (usually geometric mean)

Issue	More safety-oriented risk estimate	Less safety-oriented risk estimate
Model		
1. Threshold*	No	Yes
2. Model	One hit	"Sensitivity" model (eg., Probit/Logit)
3. Concern with differentiating "genotoxicity" in animal bioassays*	No	Yes
4. Species (animal) for extrapolation	"Most sensitive"	Some average (usually geometric mean)
5. NOGL (LOGL, etc.)	Concern with the sample size	No attention is devoted to sample size
6. Adjustment for background response	Additivity in dose	Additivity in the response. Abbott's correction
Interpretation of Results		
1. Safety factor	Large, eg., 1000 or more	Small, eg., 10-100
2. Who is to be protected?	"Sensitive" (eg., pregnant women, children, old people)	Average
3. False positives/ false negatives	"Cost" of false negative is greater	"Cost" of false positive is greater
4. Concern for multiple sources of exposure	Yes	No
5. "Acceptable" risk	1 x 10-6 or less (lifetime)	Greater than 1×10^{-6} (lifetime; eg., 1×10^{-6} per year)

*For computations on carcinogens.

Source: Adapted from Table: Issues and assumptions in risk assessment computations, in Schneiderman, M., "Quantitation and Interpretation in toxicology: What Can We Do with the Numbers?", 1986, unpublished.

Appendix C

Type II Fallacies in the Study of Childhood Exposure to Lead at Low Dose:
A Critical and Quantitative Review

Herbert L. Needleman
David C. Bellinger

Amended Metanalysis - 9/30/86

Author	Year	N	Effect Size	Power Small Effect[*]	PBAL(IT)	2LOGeP
Ernhart Et	1974	80	0.6	0.2	0.025	7.38
Needleman	1979	73	0.35	0.47	0.015	8.4
Yule, et.al.	1981	82	0.573	0.42	0.021	7.73
Winneke, et.	1982	26	0.26	0.18	0.15	3.7
Smith, et.al.	1983	185	0.17	0.7	0.12	4.24
Winneke Et	1983	115	0.351	0.25	0.4	1.83
Harvey	1984	47		0		
Shapiro	1984	193	0.46	0.48	0.025	7.38
Lansdown	1986	162	0.07	0.48	0.66	0.83
Hansen	1985	82	0.5	0.34	0.0005	15.2
Hawk Schro	1985	75	0.64	0.25	0.0004	15.64
Schroeder	1985	104	0.5	0.33	0.005	10.6
Fulton	1986	501	0.4	0.52	0.003	11.6
Hatzakis	1986	509	0.4	0.52	0.00065	14.6

$$dx = 109.13$$
$$df = 26$$
$$P = 3 \times 10^{-12}$$

[*] Type II error rates are 1- (the number in the column "Power of small effect"). Thus, the Winneke, et.al. study had a Type II error rate of .82.

Appendix D

Representative Power Values for Fixed Alphas and Disease
Rates in Groups of Fifty Control and Fifty Treatment Animals
In Animal Bioassays Using Fisher's Exact Test*

(Type I Error)	P1 (Tumor Rate In Control Group Of Fifty Animals)	P2 (Tumor Rate In Treatment Groups Of Fifty Animals)	Power of Test
.05	.01	.03	.0104
	.025	.075	.126
	.05 (2 or 3 animals)	.15 (7 or 8 animals)	.3878
	.075 (3 or 4 animals)	.225 (10 animals)	.5927
	.1	.3	.7469
.10	.01	.03	.0391
	.025 (1 animal)	.075 (3 or 4 animals)	.2295
	.05 (2 or 3 animals)	.15 (7 or 8 animals)	.542
	.075	.225	.7359
	.1	.3	.8535
.15	.01	.03	.1195
	.025	.075	.3646
	.05	.15	.6275
	.075	.225	.8017
	.1	.3	.8995
.20	.01	.03	.1338
	.025	.075	.4426
	.05	.15	.7042
	.075	.225	.8537
	.1	.3	.9290
.25	.01 (0 or 1 animals)	.03 (1 or 2 animals)	.2894
	.025 (1 animal)	.075 (3 or 4 animals)	.5314
	.05	.15	.7672
	.075	.225	.883
	.1	.3	.9532
.3	.01	.03	(?).2894
	.025	.075	.5328
	.05	.15	.8024
	.075	.225	.9212
	.1	.3	.9641

*Compiled by Bill Kenple, graduate student, Department of Statistics, and Research Assistant, University of California, Riverside, Carcinogen Risk Assessment Project.

Appendix E

Representative Statistical Values for Prospective Epidemiological Studues of
Diseases at Various Incidence Rates

Incidence of disease in general population:		Numbers of Subjects to be Studied		
		N/2 exposed	N/2 unexposed	N total
1) δ = 1/5 (relative risk)				
α = .05 (risk of type I error)				
β = .20 (risk of type II error)				
a) 8/100	a)	691	691	1,382
b) 8/1,000	b)	7,631	7,631	15,262
c) 8/10,000	c)	77,022	77,022	154,044
d) 8/100,000	d)	770,939	770,939	1,541,878
2) δ = 1.5				
α = .05				
β = .05				
a) 8/100	a)	1,212	1,212	2,424
b) 8/1,000	b)	13,382	13,382	26,764
c) 8/10,000	c)	135,078	135,078	270,156
d) 8/100,000	d)	1,332,037	1,332,037	2,664,074
3) δ = 3				
α = .05				
β = .20				
a) 8/100	a)	62	62	124
b) 8/1,000	b)	756	756	1,512
c) 8/10,000	c)	7,695	7,695	15,390
d) 8/1000,000	d)	77,087	77,087	154,174
4) δ = 3				
α = .05				
β = .05				
a) 8/100	a)	109	109	218
b) 8/1,000	b)	1,326	1,326	2,652
c) 8/10,000	c)	13,495	13,495	26,990
d) 8/100,000	d)	135,191	135,191	270,382
5) δ = 6				
α - .05				
β = .20				
a) 8/100	a)	13	13	26
b) 8/1,000	b)	207	207	414
c) 8/10,000	c)	2,150	2,150	4,300
d) 8/100,000	d)	21,580	21,580	43,160
6) δ = 6				
α = .05				
β = .05				
a) 8/100	a)	22	22	44
b) 8/1,000	b)	363	363	726
c) 8/10,000	c)	3,771	3,771	7,542
d) 8/100,000	d)	37,845	37,845	75,690

From Cranor (1983), pp. 400-401

Appendix F

Relative Risk as a Function of Alpha Beta Values
I. Relative Risk When Disease Rate is 8/10,000

Alpha	Beta	Dis.Rate	Rel.Risk	Sample Size
0.05	0.05	8/10,000	8.9	2150
	0.10		7.5	
	0.15		6.7	
	0.20		6.0	
	0.25		5.5	
	0.30		5.1	
	0.35		4.7	
	0.40		4.3	
	0.45		4.0	
	0.49		3.8	
0.10	0.05	8/10,000	7.5	2150
	0.10		6.2	
	0.15		5.5	
	0.20		4.9	
	0.25		4.5	
	0.30		4.1	
	0.35		3.8	
	0.40		3.5	
	0.45		3.2	
	0.49		3.0	
0.15	0.05	8/10,000	6.7	2150
	0.10		5.5	
	0.15		4.8	
	0.20		4.3	
	0.25		3.9	
	0.30		3.6	
	0.35		3.2	
	0.40		3.0	
	0.45		2.7	
0.20	0.05	8/10,000	6.0	2150
	0.10		4.9	
	0.15		4.3	
	0.20		3.8	
	0.25		3.4	
	0.30		3.1	
	0.35		2.8	
	0.40		2.6	
	0.45		2.4	
0.25	0.05	8/10,000	5.5	2150
	0.10		4.5	
	0.15		3.9	
	0.20		3.4	
	0.25		3.1	
	0.30		2.8	
	0.35		2.5	
	0.40		2.3	
	0.45		2.1	
0.30	0.05	8/10,000	5.1	2150
	0.10		4.1	
	0.15		3.5	
	0.20		3.1	
	0.25		2.8	
	0.30		2.5	
	0.35		2.2	
	0.40		2.0	
	0.45		1.8	
0.33	0.05	8/10,000	4.9	2150
	0.10		3.9	
	0.15		3.4	
	0.20		2.9	
	0.25		2.6	
	0.30		2.4	
	0.35		2.1	
	0.40		1.9	
	0.45		1.7	

II. Relative Risk When Disease Rate is 8/100,000.

Alpha	Beta	Dis. Rate	Rel.Risk	Sample Size
0.05	0.05	8/100,000	65.9	2150
	0.10		52.6	
	0.15		44.8	
	0.20		38.8	
	0.25		34.1	
	0.30		30.4	
	0.35		26.9	
	0.40		23.8	
	0.45		21.2	
0.10	0.05	8/100,000	52.6	2150
	0.10		40.9	
	0.15		34.1	
	0.20		28.9	
	0.25		24.9	
	0.30		21.8	
	0.35		18.9	
	0.40		16.4	
	0.45		14.3	
0.15	0.05	8/100,000	44.8	2150
	0.10		34.1	
	0.15		28.0	
	0.20		23.4	
	0.25		19.8	
	0.30		17.1	
	0.35		14.5	
	0.40		12.4	
	0.45		10.6	
0.20	0.05	8/100,000	38.8	2150
	0.10		28.9	
	0.15		23.4	
	0.20		19.2	
	0.25		16.0	
	0.30		13.6	
	0.35		11.4	
	0.40		9.5	
0.33	0.05	8/100,000	28.2	2150
	0.10		20.0	
	0.15		15.5	
	0.20		12.2	
	0.25		9.8	
	0.30		8.0	
	0.35		6.4	
	0.40		5.1	
	0.45		4.1	

References

Cothern, Coniglio and Marcus (1986), "Estimating Risks to Health," *Environmental Science and Technology* III 20.

Cranor, C. (1983), "Epidemiology and Procedural Protections for Workplace Health in the Aftermath of the Benzene Case." *Industrial Relations Law Journal.*

_ _ _ _ _. (1987) "Some Public Policy Problems with Risk Assessment: How Good is the Use of the 95% Rule in Epidemiology?" Invited Symposium Paper, Pacific Division Meetings American Philosophical Association.

Feinstein, A. (1977), *Clinical Biostatistics.*

Freedman, D. and Zeisel, H. (forthcoming). "From Mouse to Man: The Quantitative Assessment of Cancer Risks," *Statistical Science.*

Lee and Chang, A.C. (forthcoming), "An Evaluation of Transport Modeling of Organic Chemicals in Soil from Underground Fuel Tank Leaks."

Giere, R. (1981) *Understanding Scientific Reasoning.*

Mayo, D. (1989), "Towards a More Objective Understanding of the Evidence of Carcinogenic Risk." *PSA 1988*, Vol. 2, pp. 485-503..

National Cancer Institute, Demographic Analysis Section, Division of Cancer Cause and Prevention, (1981). Monograph No. 57. Surveillance, *Epidemiology and End Results: Incidence and Mortality Data* 1973-1977, pp. 662-63 and Table 51.

National Research Council, (1983), *Risk Assessment in the Federal Government: Managing the Process.* Washington, D.C.: U.S. Gov't Printing Office.

Needleman and Billinger (forthcoming), "Type II Fallacies in the Study of Childhood Exposure to Lead at Low Dose: A Critical and Quantitative Review."

Page, T. (forthcoming). *The Economics of Risk Assessment.*

Rescher, N. (1983), *Risk: A Philosophical Introduction to the Theory of Risk Evaluation and Management.* Washington, D.C.: University Press of America.

U.S. Congress, Office of Technology Assessment (1987), *Identifying and Regulating Carcinogens.* Washington, D.C.: U.S. Government Printing Office.

U.S. Congress, *Office of Technology Assessment* (1981),

Walter (1977), "Determination of Significant Relevant Risks and Optimal Sampling Procedures in Prospective and Retrospective Studies of Various Sizes." *American Journal of Epidemiology* 105: 387-91.

Toward a More Objective Understanding of the Evidence of Carcinogenic Risk[1]

Deborah G. Mayo

Virginia Polytechnic Institute and State University

The field of quantified risk assessment is a new field, only about 20 years old, and already it is considered to be in a crisis. As Funtowicz and J.R. Ravetz (1985) put it:

> The concept of risk in terms of probability has proved to be so elusive, and statistical inference so problematic, that many experts in the field have recently either lost hope of finding a scientific solution or lost faith in Risk Analysis as a tool for decisionmaking. (p.219)

> Thus the 'art' of the assessment of risks ... is at an impasse. The early hopes that it could be reduced to a science are frustrated. ...[O]thers are tending to introduce the 'human' and 'cultural' factors. The question now becomes, to what extent should these predominate? Would it be to the reduction or exclusion of the 'scientific' aspects? For, ...if the perceived phenomena of 'risks' are interpreted as lacking all objective content or being merely a small part of some total cultural configuration, then there is no basis for dialogue between opposed positions on such problems. (pp.220-221)

The crisis of confidence in this new field comes from two directions: on the one hand it comes from the general challenge of philosophers and others as to whether there exist any objective, rational rules in science; and on the other hand there are many real cases where conflicting risk assessments are reached on the basis of the same data. It will be useful to consider throughout an example of such a risk assessment conflict. I take a recent case from the Environmental Protection Agency (EPA) as to the carcinogenic potential of the substance formaldehyde. On the basis of the very same data, the EPA in May of 1981 reached a different and opposite assessment from the one it had reached in September of 1981. My aim is to suggest how a more objective understanding of the evidence would help in resolving such a conflict.

I want to emphasize at the start that my approach is distinct from those appeals to "objective science" that deny the entry of value judgments in reaching risk assessments. Rather, my approach will be to show that despite the entry of these value judgments, it is possible to unearth what the data do and do not say about the actual extent of the risk involved.

1. Risk Assessment in the Case of Formaldehyde

The term risk assessment, as I am using it, covers the generation and analysis of data in order to characterize the extent to which an agent causes an increase in the incidence of a

PSA 1988, Volume 2, pp. 489-503

health condition, in our case, cancer, in humans, lab animals or other test systems. Data generation in arriving at risk assessments includes prospective randomized treatment-control experiments, and retrospective case-control studies. To examine whether formaldehyde increases the risk of cancer, prospective experiments on rats were conducted. Epidemiological studies on humans, in contrast, only allowed a retrospective analysis of cancer rates in various occupations. On the basis of the statistically significant increases in (nasal) cancer among formaldehyde-treated rats, the Chemical Industry Institute of Toxicology (CIIT) reached the assessment that formaldehyde is carcinogenic in laboratory rats and reported this to the EPA in November of 1980. A panel of eminent scientists convened by the National Toxicology Program confirmed this risk assessment, and concluded that "formaldehyde should be presumed to pose a risk of cancer to humans". (See *Hearing*[2], p. 191.) The lengthy document detailing the formaldehyde risk assessment was entitled the "Priority Review Level 1" (PRL-1) dated February, 1981.

Risk assessments form the basis of *risk evaluations* and *management*. These involve assessing how substantively important a risk is and what should be done about it.[3] This requires an explicit consideration of social, ethical, and economic considerations (e.g., against what level of risk should the public be protected? and what form should this control take?) The evaluation the EPA staff reached was that formaldehyde should be designated as a priority chemical under the EPA provision known as 4(f). To quote from the *Federal Register* notice:

> [T]he Agency finds that there may be a reasonable basis to conclude that formaldehyde presents a significant risk of widespread harm to humans from cancer. (Federal Register, May 1981, pp.5-6.)

This last sentence is important because triggering statute 4(f) requires only that *there may be a reasonable basis* to conclude that a significant risk exists, and not that there is a reasonable basis for such a conclusion. In itself 4(f) does not call for any regulation. It is simply a call for closer scrutiny based on an indication that there *may* be a significant cancer risk.

All of this was in 1980 and early 1981. Then there was a change in Administration; the Reagan administration entered, and along with it a new EPA Administrator (Ms. Gorsuch) and some new staff. In fact, formaldehyde was the first 4(f) recommendation brought before the new Administrator for signing. Instead of signing it members of the new EPA staff carried out a reassessment of the hazard data in the PRL-1. The new and revised version of the data became the Todhunter Memorandum, Dr. Todhunter being a new EPA Assistant Administrator. Some of the changes included blatant erasures of the highest risk estimates that had been given in the PRL-1. (See *Hearing*, pp. 349-365.) There are other, less blatant changes in the reassessment. Most significant was Todhunter's deemphasis of the positive rat studies and his emphasis of the negative epidemiological studies. Todhunter concludes, for example,

> There does not appear to be any relationship, based on the existing data base on humans, between exposure [to formaldehyde] and cancer. Real human risk could be considered to be low on such a basis. *(Hearing, p. 260)*

This hazard reassessment was then given as the basis of a changed hazard evaluation. On Sept. 11, 1981 the EPA staff recommendation to designate formaldehyde as a 4(f) priority chemical was *reversed* and the opposite hazard evaluation was made. (See *Hearing*, pp.192-193.) Whether or not this shift in hazard assessment was justified was the subject of enormous controversy. It led to a congressional hearing on formaldehyde, which I refer to throughout as *Hearing*.[4]

The suspicion which led to these hearings was that the agency was altering the widely accepted standards for carcinogenic risk assessment, and that it was doing so in order to garner "expert scientific support" for furthering the aim of anti-regulation. One of the main reasons for this suspicion was that the new Administration did not base its decision against a 4(f) designation on any new data beyond the PRL-1 document which had been the basis for the original, and opposite, recommendation (though, as mentioned, it did *conceal* some evidence supporting a 4(f) designation). Rather, the new Administration proceeded to hold a series of secret meetings restricted only to certain scientists and lawyers from the Formaldehyde Institute, the Formaldehyde Trade Association and EPA staff. In these meetings they reinterpreted the data and, without the usual peer review, came to the opposite conclusion than that endorsed by numerous eminent scientists and agencies. As one attorney with the Natural Resources Defense Council (Dr. Warren) put it:

> There are no new data to support the reversal, only a reinterpretation which has been advocated by and is quite favorable to the interests of the formaldehyde industry. Those new assumptions, as we have heard, depart radically from accepted principles of cancer risk assessment... . In our view, this has been an effort to get the Government off the back of the formaldehyde industry. (*Hearing*, p.188)

The disagreement was not about the level of risk required before triggering 4(f) i.e., for judging that a substance may pose a significant or serious human risk. Todhunter maintains that he was holding the same range of risk which agencies have tended to deem of public concern.[5] Nevertheless, on the same evidence, different conclusions are reached. If going from data to risk assessments was a matter of applying a single universally accepted best method, then this difference in resulting assessment could not occur. That disagreement does occur shows there is no such algorithm for risk assessment. The fact that assessments are nevertheless reached is typically taken as grounds to conclude that extra-scientific, cultural, social, ethical or other contextual values must be entering. For some, this conclusion shows that something is wrong and that we need to try to avoid or somehow neutralize the entry of non-scientific interests. In the formaldehyde case—which came at a time during which such politicization at the EPA was rampant—this attitude resulted in the above mentioned hearing to determine if the EPA was altering the standards for carcinogenic risk assessment. Later, the National Academy of Sciences issued a report stressing the need to separate the science of assessment from the social and policy values that enter at the level of risk management (National Research Council). But a growing body of risk literature questions the possibility that scientific risk assessment can ever be free of the policy values appropriate to risk management. To this group, policy in science is not a violation, but rather is inevitable. However, there are very different grounds for reaching such a view, and it is important to separate them.

2. The Sociological Relativist View

On one set of views, which I will call the *sociological relativist view*, scientific risk assessment—indeed science generally—is inevitably a product of, if not entirely constructed from—socio-cultural values. An example is the influential socio-cultural theory of risk assessment of Douglas and Waldavsky. According to Douglas and Waldavsky:

> The risk assessors offer an objective analysis. We know that it is not objective so long as they are dealing with uncertainties and operating on big guesses. They slide their personal bias into the calculations unobserved. (Douglas and Waldavsky 1980, p.80)

Risk assessment, on their view, is totally determined by socially constructed methods and judgments; they are social constructs. This view provides an explanation of conflicting risk assessments in terms of the different policy judgments and competing 'world-views' of different assessors.

Granted one can find competing political interests to explain the conflicting risk assessments in the case of formaldehyde—as searching the fascinating testimony shows.

The EPA and Todhunter were influenced by the political commitment to anti-regulation, and one can explain the Todhunter Memo, as one attorney argued and as we have already noticed, as "an effort to get the Government off the back of the formaldehyde industry" (*Hearing*, p.188). Likewise, in defense of Todhunter one witness (Mr. Walker) maintained that the opposition were "a few disgruntled employees of the EPA who [simply because they want to place formaldehyde under 4(f)] feel justified in waging guerrilla warfare against the Agency and those in positions of authority" (*Hearing*, p.4).

But however well a story about background interests, social values, and negotiation may explain a risk assessment, and however much our assessment tools are products of social beings and institutions, the question whether or not a risk assessment is warranted is not a matter of social values; it is a matter of what the risk actually is. Sociological relativists are led to consider "objective" physical risks either unattainable or unimportant. However, they hold an overly stringent conception of objectivity—one that is precluded by the need to make inferences under uncertainty without algorithms. Such relativists consider the entry of any and all judgments subjective and biased; as we saw in the quote from Douglas and Waldavsky.

3. Risk Assessment Policy (RAP) and RAP Relativism

While the sociological challenge to objectivity may be countered by denying that objectivity requires neutral algorithms or freedom from uncertainty, there is a different basis for challenging the objectivity of risk assessments—one which does not turn on an overly narrow conception of objectivity and does not deny the importance or possibility of measuring physical risks. This second view stems from the nature of the judgments and decisions that are required in order to carry out risk assessment estimates. For example, one must decide what data to collect, how large a sample size to take, what levels of reliability (e.g., statistical significance) to use, how to weigh studies with different results (e.g. whether they should be weighed according to statistical power), what models should be used to extrapolate from animals to humans, etc. Because these judgments involve choices with no unequivocal scientific answers—at least at present—and since these choices have policy implications, they are intertwined with policy. Thus they are not just a matter of objective science. These judgments may be called risk assessment policy (RAP) judgments.

The view that risk assessment is necessarily entwined with policy because of the inevitability of making RAP judgments may be called *RAP relativism*—to distinguish it from other more extreme relativistic views (e.g., sociological relativism). Among the first to articulate a version of RAP relativism was Alvin Weinberg, who placed what I am calling RAP judgments under his term *trans-science*—"questions which can be asked of science and yet *which cannot be answered by science*" (1972, p.209). Now the view is fairly widespread. A congressional report from the National Academy of Science in 1983, which resulted from the suspicion that agency science was being politicized (e.g., as represented by the formaldehyde case), offers a very useful delineation of over 50 junctures at which RAP judgments enter in the course of making risk assessments.[6] Carl Cranor has recently provided a clear statement of what I am calling RAP relativism in regards to the judgments required in specifying methods of risk estimation:

[T]he supposedly objective scientific studies used for estimating risks to human health...are considerably more controversial and political than most people think. ...a wise and conscientious scientist with perfect test data *cannot* help but make moral and policy judgments in order to interpret an epidemiology study and to produce the risk numbers that are the outcome... . ...the moral and policy judgments are forced by the statistical equations themselves and the choice of variables employed in them.[7] (Cranor 1987)

The basis for Cranor's allegation is that each RAP choice has policy implications. That is, each choice influences the chance that the substance will be considered a significant risk to humans; in other words, it affects the protectiveness of the risk assessment.

The conflicting assessments in the formaldehyde case stem from different choices of RAP options. In particular, while the CIIT report and the PRL-1 accorded high weight to the positive animal results and denied that negative epidemiological studies warranted concluding no increased human risk, the Todhunter Memo did just the reverse. The Todhunter Memo deemphasized the positive rat studies and—most importantly—took the negative epidemiological studies to indicate low increased human risk or none.[8] And the reason a result was considered negative was itself a result of a particular choice of statistical analysis—also a RAP judgment. With an EPA purged of scientists save those tending to favor anti-regulation, there was plenty of leeway for the Agency to consistently choose the inference option least likely to have a protective outcome.[9] Under the guise of demanding stringent scientific evidence, these policy choices made it extremely unlikely that a substance would be claimed to pose a significant human risk. In contrast, those who endorsed the original PRL-1 made a more protective RAP choice. On the RAP relativist account, therefore, the conflicting risk assessment results from the difference in the view of those concerned as to how protective regulations should be; it is a policy conflict.

Although RAP relativism is less threatening to objectivity than sociological relativism and sociological reductionism, it shares some of the same implications for risk assessment. First, it implies that risk assessment judgments (at least where there is uncertainty) inevitably reflect policy judgments, and risk assessment disagreements largely reflect disagreements about policy —including moral, social, economic or other values typically considered "non-scientific". Thus, science is given little role in an unbiased adjudication of disagreements over risk assessments. If interpreting scientific results is necessarily colored by social and political contexts, it is impossible for science to provide risk assessment oversight that is fully objective.

Second, if it is true that in reaching a risk assessment (based on RAP judgments) one cannot help making an ethical choice about how protective one should be, then it does not seem that risk assessment is an appropriate business for scientists. After all, scientists are not elected to make social policy choices about acceptable risk. (It was precisely the fact that EPA scientists were guilty of politically motivated assessments that led many to decry the politicization of EPA science during the time of the formaldehyde reassessment.) If risk assessment judgments are policy judgments, then dealing with the uncertainties involved in RAP choices should be performed by policy makers and ethicists, not scientists.

However, such a practice allows policymakers to fall into all manner of misinterpretations of the assessment evidence. (See, for example, Silbergeld.) If RAP judgments are made by non-scientist policymakers, they are divorced from the original issues and uncertainties underlying the different risk estimates. At the same time, the scientist is limited to presenting possible RAP choices, but is involved neither in making them nor in bringing out the implications for protectiveness. For example, if the scientific work ends after reporting two possible estimates that may be used, say, a maximum likelihood estimate and an upper 95% confidence bound, then the scientist will not be around to explain how far off each of these estimates is likely to be from the actual risk and why. This permits the assessor to make the final choice (e.g.,about which estimate to use) without acknowledging what standard of protectiveness he is effectively requiring in choosing a given option (e.g., that the maximum likelihood estimate is less protective than the upper 95% confidence bound).

Without investigating further the consequences of RAP relativism here,[10] I want to consider whether the need to make decisions in applying statistical risk assessment methods really does have this relativistic consequence—i.e., the consequence that interpreting results necessarily requires policy judgments. I shall argue that this conclusion may be avoided.

It may be admitted that the conflicting assessment of formaldehyde is explainable by different (more and less protective) choices of RAP options (i.e., whether or not to accord higher weight to the positive animal results than to the negative epidemiological studies). But this would not lead to RAP relativism unless adjudicating between them is itself relative to a stance on how protective we should be. I shall argue it is not so relative. Granted, there will be latitude for choice among possible RAP options. Granted also that each choice has a policy implication—influencing the likelihood that a substance will be judged to pose a significant hazard to human health. Nevertheless, this does not preclude objective scrutiny as to whether a given assessment is warranted by the evidence. Whether given evidence warrants a given risk estimate is a matter of scientific not social acceptability—that is, it is a question of how well the inferred estimate reflects what is really the case about the causal effect of the substance in question. The latitude in choosing RAP options does not preclude the objective scrutiny needed to answer this question.

Focusing on the RAP judgments involved in interpreting statistical tests, I shall argue that risk estimates are necessarily policy judgments only under misuses of the statistical tests involved—ones which, unfortunately, are encouraged by the manner in which tests are often formulated and taught.

4. Neyman-Pearson (NP) Tests

The type of statistical test standardly used in reaching risk assessments is the Orthodox or Neyman-Pearson test (NP test), often in combination with Fisherian significance tests. The test considers hypotheses, typically assertions about a property of some population: a *parameter*. In the formaldehyde case, the hypotheses are assertions about the parameter which I will call Δ, the increased cancer risk in the population—humans. The NP test splits the possible parameter values into two: one representing the *test (or null) hypothesis* H, the other the set of *alternative hypotheses* J. The test hypothesis H in the formaldehyde case asserts that formaldehyde does not cause an increase in a person's risk of dying from cancer of a give type. That is, it asserts that there is a 0-increase in the hazard rate, i.e., $\Delta = 0$. The alternative hypotheses assert that formaldehyde causes a positive increase, i.e., $\Delta > 0$. Since here one is looking for positive discrepancies from 0, this is a *one-sided* test, which I call test T+:

Test T +: Test (null) hypothesis H asserts $\Delta = 0$ (no increased risk)[11]
   Alternative hypothesis J asserts $\Delta > 0$ (a positive increased risk).

The test considers a *test statistic* that describes an aspect of the outcome of interest. One statistic in testing formaldehyde is D, the difference in cancer rates between the subjects exposed and those unexposed to formaldehyde:

Test Statistic D: the difference in cancer rates between the subjects exposed and those unexposed to formaldehyde.

Corresponding to each observed difference is its level of statistical significance, defined as follows:

The Statistical Significance Level of an Observed Difference D_{obs} is the probability with which so large a difference arises assuming the null hypothesis H is true, i.e.,
   $Prob(D \geq D_{obs}$, given that H is true).

A good way to see significance levels is as standard measures of distance from H, except with this inversion: the larger (and more significant) the distance, the smaller its significance level.

An NP test consists of a rule which specifies, before the observation is made, how statistically significant (i.e., how improbably far) an observed difference must be before it should be taken to reject H. The maximum significance level chosen beyond which D_{obs} is taken to reject H is called the *size* of the test, and is denoted by α. Thus, test T+ with size α consists of the following rule:

Test T+ with size α: Reject H if and only if observed difference D_{obs} is statistically significant at level α.

Observed differences which are not large enough to reach this preset size are taken to accept H. In this way the test maps the possible outcomes—the *sample space*—into either reject H (and accept J) or accept H. The partitioning that results from the test is illustrated

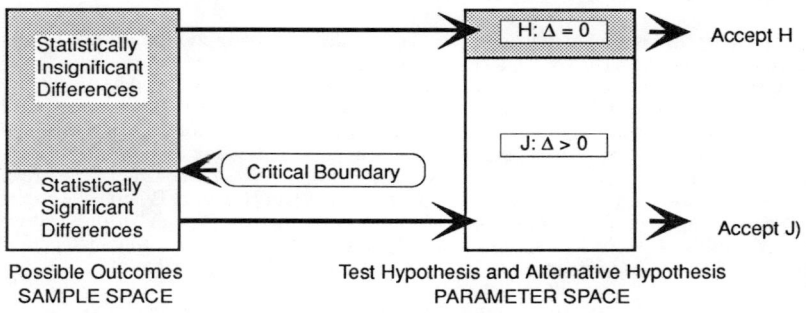

Figure 1. Neyman-Pearson Test T+ as a Mapping rule

below:

As long as there is variability in the effect (e.g., not all who are exposed get cancer, and not all who get cancer are exposed), and as long as only a sample from the population is observed, there is a chance the test will make an error. Two types of errors are considered: First, the test leads to reject H (accept J) even though H is true (the Type I error); second, the test leads to accept H although H is false (the Type II error). A test with size α rejects H when in fact H is true—i.e., it commits a Type I error—with probability no more than α. The smaller the test's size α, the less frequent the Type I error. But by making α smaller the test suffers an increase in the frequency with which it accepts H even when in fact H is false (and so should be rejected)—i.e., an increase in the frequency of a Type II error. The probability of a Type II error is denoted by ß. α and ß are the test's error probabilities:

Error Probabilities: α is the probability of an erroneous rejection of H (Type I error); ß, the probability of an erroneous acceptance of H (Type II error).

Since these two error probabilities cannot be simultaneously minimized, the NP model instructs one to first fix α, the size of the test, at some small number, such as .05 or .01. (In other words, the test is specified so as to ensure it is very improbable for the test to reject H when the hypothesis H is true.) One then seeks out the test which at the same time has a small ß. 1 - ß is the corresponding power of the test. Because in our case alternative hypothesis J contains more than a single value of the parameter, i.e, it is composite, the value of ß varies according to which alternative in J is assumed true. The "best" NP test of a given size α (if it exists) is the one which at the same time minimizes the value of ß (i.e., the probability of type II errors) for all possible values of Δ under the alternative J. I shall refer to a specific simple alternative hypothesis as $J:\Delta = \Delta'$.

We have now to ask: Are test specifications matters of policy values?

The test's error probabilities, α and ß, are objective in that they refer, not to subjective degrees of belief, but to relative frequencies of error in sequences of applications (whether similar or dissimilar) of a given experimental test.[12] And NP tests are objective in the sense that they control the error probabilities of tests regardless of what the true, but unknown, value of Δ is. However, the error probability specifications themselves go beyond the objective formalism of NP tests. As Neyman and Pearson note:

> From the point of view of mathematical theory all that we can do is to show how the risk of the errors may be controlled and minimized. The use of these statistical tools in any given case, in determining just how the balance should be struck, must be left to the investigator. (Neyman and Pearson 1933, p.146)

As a result, NP tests were developed in a certain decision theoretic framework where there would be a clear basis for the test specifications. Neyman called the resulting theory of tests an objective theory of *inductive behavior*. Here tests are formulated as mechanical rules or "recipes" for reaching one of two possible decisions: "act as if H were true" or "act as if H were false", according to whether H is accepted or rejected. Such "machinery" produces automatic acceptance or rejection of H. For example, rejecting H in our formaldehyde case may be associated with a decision to trigger 4(f). For each decision there are certain losses and costs associated with acting on it when in fact H is true. By considering such consequences the scientist is, presumably, able to specify the risks he can "afford". (It is imagined that the scientist first specifies α as the maximum frequency with which he feels he can afford to reject H erroneously, and then seeks to minimize the value of ß.)

However, this opens the door to the RAP relativist's concern. For such considerations of consequences—in our case social, ethical,and economic—are clearly policy matters; so it appears that specifying a test *is* tantamount to making a policy decision—just as the RAP relativists contend. But if the domain of a scientist is objectively finding out what is the case as opposed to setting policy goals, then he does not seem to be in the position of making the needed test specifications. And if he is left to make these value judgments, the results necessarily reflect, not just what is the case about the cause, but what he believes about when a substance ought to be regulated—e.g., placed under 4(f)—and that is precisely the RAP relativist allegation. If this is true, then it is impossible for an assessment to be wholly objective—where by "objective" I mean reflecting what is the case about the risk and not one's policy preferences.

How are we to avoid this conclusion and with it the charge that risk assessors necessarily make ethical and policy decisions in reaching and interpreting risk estimates? The answer is to be found in rejecting the automatic use of tests where the null hypothesis is accepted or rejected according to whether the preset significance level is reached, without any reflection on the evidential meaning of the specific observed result.

It is worth noting that the threat to objectivity caused by the automatic use of NP tests has been recognized since the tests were first advanced in the 1930's—most notably by R.A. Fisher. While Fisher's ideas formed the basis of NP tests, by couching them in a behavioral-decision framework he felt Neyman and Pearson had given up the ideal of objectivity in science. Although Egon Pearson, one of the co-founders of NP tests, responded to Fisher, as did others who reject the automatic use of statistical tests,[13] these automatic uses and the misinterpretations they encourage are still problematic enough in epidemiology to have given rise to a recent movement in that field to "cleanse its literature" of statistical tests and α values altogether. (Fleiss 1986, p.559. See also Walker 1986.) The misinterpretations which have led to this are: automatically equating rejections of H (statistically significant differences) with finding substantively important discrepancies from H, and failures to reject H with finding 0 or unimportant discrepancies. However, the alternative methods recommended (confidence intervals) are open to analogous automatic uses. To avoid misinterpretations we need a more objective understanding of the statistical results, which I shall now consider.

4. A More Objective Understanding of NP Statistical Tests

a. Understanding Rejections of H:

Ideally, the policy question of what counts as a substantively important increase in cancer risk is answered at the start, so that the test may be specified in order to have appropriately high probabilities of detecting all and only those increases. Substantively important increases in the formaldehyde case are those increases in cancer risk deemed serious enough to trigger 4(f)—a policy judgment. However, regardless of how the test has been specified, whether based on policy or other values, knowledge of the test's error probabilities, I claim, allows interpreting objectively what the data do and do not indicate about the increased risk—i.e, about Δ.

What is the objective import of a rejection of hypothesis H (a positive result) in the context we are considering? The proper construal of a rejection of our test hypothesis H of 0-increase is an assertion to the effect that: "This test detected an increase Δ of at least such and such". The task is to determine the approximate *lower bound* for Δ.

To accomplish this, consider how one interprets a failing test score on an (academic or physical) exam. If it is known that such a score frequently arises when a subject's knowledge is deficient only to a given degree, say δ, then one would deny that such a rejection indicated the existence of a deficiency *in excess of* δ. For example, suppose a student obtains a failing score on an examination. But suppose such a failing score arises very frequently among students who have a deficiency δ of only 10% of the material being tested (i.e.,they know 90% of the material). Then such a failing score is not good grounds to infer that the student has a deficiency, say, of 40% (i.e., that he or she knows only 60% of the material). Such a test is *too severe* for that inference. The situation is analogous in interpreting statistical tests, and the reasoning can be made precise in what we may term the *severity function*.

Let us focus on the test T+ used in our formaldehyde example. The test result is a difference in risk rates, D_{obs}. For any hypothesized value of the increased risk one can ask: what is the probability of a difference as large as D_{obs}, if in fact some hypothesized value Δ' were the true increased risk? I call the answer to this question the severity of observed difference D_{obs} against an increase Δ':

The severity of observed difference D_{obs} against the alternative hypothesis that $\Delta = \Delta'$
    equals the Prob(such a large difference, given that $\Delta = \Delta'$)
    i.e.,Prob($D \geq D_{obs}$, given that $\Delta = \Delta'$)

for Δ' ranging ranging over the possible values of the parameter Δ. By "such a large difference" I mean one as large as or larger than the one observed, D_{obs}. (Note that for the case where $\Delta' = 0$, the severity of the difference equals its significance level.) The higher the test's severity against positive values of Δ', the higher its chance to detect Δ' (by rejecting H).

We can discriminate between legitimate and illegitimate construals of a statistical result by considering the values of the severity function for various values of Δ in the parameter space. For the same reasons we noted in our examination analogy above, a rejection of H only indicates that the increase Δ exceeds some value Δ', if it is improbable that Δ' brought about so large an observed difference. This may be formulated as the following rule for understanding rejections, [RR]:

[RR]: *A* T+ *rejection with observed difference* D_{obs} *indicates that* $\Delta > \Delta'$ to the extent that so large a difference is improbable were Δ no greater than Δ'—i.e.,to the extent that the severity of D_{obs} against the hypothesis that $\Delta = \Delta'$ is *low*.

So reasonable lower bounds are alternative hypotheses (i.e., positive Δ-values) against which the observed difference has low severity.

Out of a desire to obtain a test with high severity against an alternative of interest, say Δ', it has sometimes been suggested (e.g., by Cranor) that the test's size (i.e., the α level required to reject H) be raised. The above reasoning should make clear why such a suggestion fails to accomplish its aim. While the resulting test (with raised α) may now classify a previously negative result as one significant enough to reject H, that rejection will not indicate the existence of the increase of interest. The reason stems from the following consequence of rejection rule (RR):

(From [RR]) D_{obs} is a poor indication that Δ exceeds Δ' if a difference as large as D_{obs} would occur frequently even if Δ were no greater than Δ'.

Note that if we choose a small size, the test's severity against 0 is low. So the reason one wants a small size (in a non-automatic use of tests) is not the desire for low long-run frequency of error, but the desire that each *particular* rejection of the null hypothesis warrants inferring that the increase exceeds 0. Otherwise the result is misleading.

b. Understanding Failures to Reject H (i.e., Acceptances of H):

The problem in the formaldehyde conflict was not the interpretation of the positive rat studies (where hypothesis H was rejected), but the interpretation of the negative epidemiological ones (where H was not rejected). This problem—how to interpret negative statistical results—is one that particularly plagues epidemiological studies for estimating cancer risk. For here sample sizes are typically small relative to incidence rates of the cancer in question. Even the best epidemiological studies can rarely detect increases in cancer risk of less than 1 in 10 (one additional cancer per 10 individuals), while smaller increases are typically of interest. (See, *Hearing*, p.763 and the study reported in Freiman *et. al.*, 1978.)

One of the main questions that was raised at the formaldehyde hearings was this: Do the failures to reject the hypothesis H of 0-increase indicate that there is little or no risk? The many scientists and organizations endorsing the PRL-1 document say no. Indeed, Todhunter's own epidemiologist on the staff responsible for this work wrote:

Before leaving [the EPA], I would again like to emphasize that the available epidemiologic data from studies on formaldehyde exposure are inconclusive and *not supportive of no association*, as purported by the formaldehyde Institute. (*Hearing*, p.137, emphasis added)

But Todhunter and the Formaldehyde Institute say yes, claiming:

There is a limited but suggestive epidemiological base which supports the notion that any human problems with formaldehyde carcinogenicity may be of low incidence or undetectable. ...[The ranges of risks] are of from low priority to no concern... .(*Hearing*, p.253)

Consider a study that was cited in defense of the Todhunter interpretation (*Hearing*, pp.137-138). In a mortality study of Du Pont workers, the relative risk of dying from cancer among those in the study exposed to formaldehyde was not statistically significantly greater than among those not so exposed: the null hypothesis H was not rejected.[14] Du Pont concluded that

...the data suggested that cancer mortality rates in the company's formaldehyde exposed workers were no higher than the rates among nonexposed workers. (*Hearing*, p.284)

They are inferring, in other words, that the increased risk Δ equals 0. The error in such an interpretation is this: Failure to reject the null hypothesis of 0-increased risk is not the same as having positive evidence that the increased risk is 0. For such negative results may be common (i.e., probable) even if the underlying increase in risk is greater than 0. In fact, the Du Pont study had a very small chance of rejecting null hypothesis H (i.e., of having H "fail the test") even were the actual increase in risk to exceed 0 by substantial amounts. For example, the Du Pont study had only a 4% chance of rejecting H, even if there were a twofold increase in cancer of the pharynx or of the larynx in those exposed to formaldehyde. Thus, failing to reject H does not rule out twofold increases in these types of cancers. The situation was even worse with nasal cancers and not much better with the others. This is indicated in the following chart adapted from a review of the Du Pont study by the National Institute for Occupational Safety and Health (NIOSH) (*Hearing*, p.549):

	Lung	Pharynx	Larynx
# of cases	181	7	8
Power to Detect Odds Ratio = 2*	37%	4%	4%
Least Significant Odds Ratio Detectable**	2.9	57.5	42.5

* Assumes α = .05 (1 tail)
** Assumes α = .05 (1 tail) and Power (1-ß) = .80

Although I recommend interpreting tests by considering the severity function rather than the usual NP power function as employed in this chart, this does not alter the present point because high power entails high severity.[15]

More generally, failing to reject H does not rule out increases as large as Δ', if there is a small probability of rejecting H even if the increased risk were as large as Δ' (i.e., even if the severity against Δ' is low). All of this follows very familiar reasoning. If we are unlikely to hear a fire alarm (to get a rejection of H) even if there is a bad fire, then not hearing the alarm is not grounds for thinking there is no fire.

While a failure to reject does not indicate that the increase is 0, it does permit an inference about the likely *upper bound* of the unknown increase Δ. That is to say, a failure to reject H provides reason to say "the data provide good grounds that the increased risk is no greater than such and such (upper bound)". To find plausible upper bounds requires determining the extent of risk increase that with high probability would have resulted in a rejection of H, i.e, the increase against which the observed difference had *high severity*. In the Du Pont study, the test had a high probability (.8) of rejecting H, if the risk of cancer of the larynx were 42 times higher among exposed than unexposed workers. Hence a failure to reject H *does* indicate that the actual increased risk is not as high as 42-fold. (This follows from the fact that if the test has high power against an alternative, it has high severity against it.)

This leads to a general rule for understanding an acceptance of H with respect to test T+ (converting talk of ratios to differences) rule [RA]:

[RA]: A T+ acceptance with observed difference D_{obs} indicates that the actual increased risk Δ is less than Δ' to the extent that a larger difference would be probable, were the increase as great as Δ' —i.e., to the extent that the *severity* of D_{obs} against the hypothesis that $\Delta = \Delta'$ is *high*.

So reasonable upper bounds are risk increases (Δ-values) that yield high severity values. Correspondingly, the smaller the test's severity against Δ', the *less well* a T+ acceptance indicates that $\Delta < \Delta'$.

As indicated on the chart above, the Du Pont study had a fairly good chance (80%) of detecting a 3-fold increase in the relative risk of lung cancer, and a 57-fold increase in the risk of cancer of the pharynx. So, even ignoring some methodological difficulties with the study, its negative statistical results at most indicate that the various cancers are no more than 3 or 57, etc. times as likely among workers exposed to formaldehyde. They clearly do not warrant the conclusion reached by Du Pont and others, that the study supports the claim of no increase in (relative) cancer risk among formaldehyde workers. The study does not even support the claim of low increased risk, given what Todhunter himself claimed the EPA counted as low.

We can summarize informally the interpretation of both positive and negative results in terms of what a difference D_{obs} indicates:

[RR]: An observed difference (in incidence rates) D_{obs} only indicates that the increased risk (in the population), i.e., Δ, exceeds those values that would rarely have resulted in so large an observed difference.

[RA]: An observed difference (in incidence rates) D_{obs} only indicates the nonexistence of those population risk increases (Δ values) that would frequently have resulted in a *larger* observed difference.

But how, it might be asked, should "rarely" and "frequently" be specified? We can get a feel for the increase indicated by using benchmarks such as .9 or .95 for very frequent, and .1 or .05 for rare. But by interpreting tests along the lines suggested in [RR] and [RA] the use of statistical tests should no longer be a matter of pre-specified error probabilities altogether. Instead we can understand what the actual result D_{obs} indicates more or less well by calculating all (or several) of the upper and lower bounds for different degrees of severity. This would yield *severity curves*. (While each pair of upper and lower bounds of a given degree of severity is mathematically equivalent to formulating the confidence interval at the corresponding level of confidence, the difference is that not all values within the interval are treated on par. It most closely corresponds to forming a series of confidence intervals, one for each confidence level.[16])

The criticism of the EPA assessment was based on the reasoning incorporated in the rule for interpreting acceptances, i.e., rule [RA]. A number of scientists concluded that "in order to justify its failure to address formaldehyde under 4(f)...EPA has rewritten both the science and the law." (*Hearing*, p.195). (See also Ashford, *et.al.*, 1983). Because of the criticisms of the science underlying the EPA risk assessment, the EPA ultimately did place formaldehyde under the 4(f) category in 1985.

5. Conclusion

The conflict in the formaldehyde example, we said, arose from a difference in RAP judgment. By means of the above understanding of the actual extent of risk indicated by a statistical test result, one can determine the protectiveness of a given RAP judgment. It allows one to answer the question: according to the standard being required, what extent of risk must be fairly clearly indicated before it is taken as grounds that there may be a significant human risk? It allowed critics to ascertain that—contrary to what Todhunter maintained—the Todhunter assessment reflected a change in the standard required for triggering 4(f). Thus, even granting that the judgments required in reaching statistical assessments may reflect policy values, conventions, etc., I have argued, it does not follow that the task of evaluating *whether a given risk assessment is warranted by the evidence* need also be infected with subjective policy values. This task is an empirical one that may often be accomplished objectively, in the sense of reflecting what is actually the case regarding the risk, regardless of what anyone thinks is or ought to be the case.

Notes

[1]A portion of this research was carried out during tenure of a National Endowment for the Humanities Fellowship for College Teachers; I gratefully acknowledge that support. I would like to thank Marjorie Grene for numerous useful comments on earlier drafts.

[2]*Formaldehyde: Review of the Scientific Basis of EPA's Carcinogenic Risk Assessment.* Hearing Before the Subcommittee on Investigations and Oversight of the Committee on Science and Technology, U.S. House of Representatives, May 20, 1982. All pages in parentheses following *Hearing* refer to this report.

[3]This delineation of risk assessment and risk management is in accordance with such documents as National Research Council, *Risk Assessment in the Federal Government: Managing the Process.* Other uses of "risk assessment", in contrast, take it to include risk management.

[4]See Note 2.

[5]Carcinogens may be considered problematic if they increase the risk of cancer by 1 case in 10,000 or 1 case in 1,000.

[6]National Research Council (pp.29-33). My use of the term "risk assessment policy" comes from this report.

[7]He expressed essentially the same point in his oral presentation at this session of the PSA 1988.

[8]The Todhunter formaldehyde assessment endorsed other less protective choices, such as holding to the existence of a threshold for carcinogenicity of formaldehyde, discounting benign tumors, and preferring maximum likelihood estimates over upper confidence level estimates.

[9]For a discussion of the blacklisting of scientists and "hit lists" at the EPA during this period, see Lash, et. al.,(1984).

[10]I do so in "Sociological vs. Metascientific Philosophies of Risk Assessment".

[11]The same test would be used were the null hypothesis to assert that $\Delta \leq 0$.

[12]Since within the context of NP tests parameter Δ is viewed as fixed, hypotheses about it are viewed as either true or false. Thus, since a probability is interpreted as a relative frequency, it makes no sense to assign such hypotheses any probabilities other than 0 or 1.

[13]A discussion of these responses occurs in Mayo (1985).

[14]In the Du Pont study, 481 cases of male cancer deaths among employees between 1957-1959 constituted the cases. These were matched on relevant factors with controls

who did not die of cancer. The statistic observed was the relative odds ratio, the ratio of the odds of having been exposed to formaldehyde among cases and controls. For simplicity, I refer here to the risk rather than the relative risk.

[15]The difference between power and severity is that while severity is a function of the particular observed difference D_{obs}, the power is a function of the smallest difference judged significant by a given test. Let D^* be the smallest difference test T+ judges significant. (i.e., D^* is the *critical boundary* shown in Fig. 1 beyond which the result is taken to reject H.) Then, power is defined as follows:

The power of test T+ against alternative $\Delta = \Delta'$ equals the probability of a difference as large as D^*, given that $\Delta = \Delta'$.

The severity, in contrast, substitutes D_{obs} in for D^*. The advantage of the severity function, I claim, is that it affords an understanding that reflects the difference that has actually been observed.

[16]For further discussion of this relationship, see Mayo (1985). Poole (1987).makes use of what are essentially severity curves in interpreting statistical results. Such curves are also employed by Kempthorne and Folks (1971).

References

Ashford, N.A., Ryan, C.W. and Caldart, C.C.(1983), "A Hard Look at Federal Regulation of Formaldehyde: A Departure from Reasoned Decisionmaking", *Harvard Environmental Law Review* 7:297-370.

Cranor, C. (1987), "Some Public Policy Problems with Epidemiology: How Good is the 95% Rule?". Paper presented at the Pacific Division meeting of the American Philosophical Association, March 1987.

Douglas, M. and Wildavsky, A. (1982), *Risk and Culture*. Berkeley: University of California Press.

Fisher, R.A. (1955), "Statistical Methods and Scientific Induction", *Journal of the Royal Statistical Society* (B) 17:69-78.

Fleiss, J.L. (1986), "Significance Tests Have a Role in Epidemiologic Research: Reactions to A.M. Walker", *American Journal of Public Health* 76 (No.5, May 1986): 559-560.

Formaldehyde Federal Register Notice, May 1981.

Freiman, J.A., Chalmers, T.C., Smith Jr., H. and Kuebler, R.R. (1978), "The Importance of Beta, the Type II Error and Sample Size in the Design and Interpretation of the Randomized Control Trial, Survey of 71 'Negative' Trials", *The New England Journal of Medicine* 299 (No.13):690-694.

Funtowicz, S.O. and Ravetz, J.R. (1985), "Three Types of Risk Assessment: A Methodological Analysis", in *Risk Analysis in the Private Sector*, C. Whipple and V.T. Covello (eds.). New York: Plenum Press, pp.217-231.

Kempthorne, O. and Folks,L. (1971), *Probability, Statistics, and Data Analysis*. Ames: Iowa State University Press.

Lash, J., Gillman, K. and Sheridan, D. (1984), *A Season of Spoils: The Reagan Administration's Attack on the Environment*. New York: Pantheon Books.

Mayo, D. (1985), "Behavioristic, Evidentialist, and Learning Models of Statistical Testing", *Philosophy of Science* 52:493-516.

_ _ _ _ _ ., "Sociological vs. Metascientific Philosophies of Risk Assessment", in *Acceptable Evidence: Science and Values in Risk Management*, D. Mayo and R. Hollander (eds.). Forthcoming, Oxford.

National Research Council, *Risk Assessment in the Federal Government: Managing the Process*. Washington, D.C.: National Academy Press, 1983.

Neyman, J. and Pearson, E. S. (1933), On the Problem of the Most Efficient Tests of Statistical Hypothesis", *Philosophical Transactions of the Royal Society* A 231: 289-337. (Reprinted in Joint Statistical Papers, Berkeley: University of California Press, 1967, pp.276-283).

Pearson, E.S. (1955), "Statistical Concepts in Their Relation to Reality", *Journal of the Royal Statistical Society* (B) 17:204-207.

Poole, C. (1987), "Beyond the Confidence Interval", *American Journal of Public Health* 77 (No.2, Feb. 1987):195-199.

Silbergeld, E.K., "Risk Assessment and Risk Management—An Uneasy Divorce", in *Acceptable Evidence: Science and Values in Hazard Management*, D. Mayo and R. Hollander (eds.). Forthcoming, Oxford.

U.S. House of Representatives, *Formaldehyde: Review of the Scientific Basis of EPA's Carcinogenic Risk Assessment*. Hearing Before the Subcommittee on Investigations and Oversight of the Committee on Science and Technology, 97th Congress (second session), May 20, 1982.

Walker, A.M. (1986), "Reporting the Results of Epidemiologic Studies", *American Journal of Public Health* 76 (No.5, May 1986):556-558.

Weinberg, A. (1972), "Science and Trans-Science", *Minerva* 10:209-222.

Risk Assessment and Uncertainty

Kristin Shrader-Frechette

University of South Florida

Risk assessments done by the Ford Foundation-Mitre Corporation and the Union of Concerned Scientists agree on the hazard probabilities and consequences associated with commercial nuclear fission, but make opposed recommendations regarding using atomic energy to generate electricity. The UCS risk analysis decided against use of the technology; the Ford-Mitre study advised in favor of it (Union of Concerned Scientists 1977; Nuclear Energy Policy Study Group 1977; Cooke 1982, p. 334).

The two studies agreed on the data but made different policy recommendations because they followed different methodological rules at the third, risk-evaluation, stage of assessment. (The first two stages are risk identification and risk estimation.) The Ford-Mitre research was based on the widely accepted Bayesian decision criterion that it is rational to choose the action with the best expected value or utility, where 'expected value' or 'expected utility' is defined as the weighted sum of all possible consequences of the action, and where the weights are given by the probability associated with each consequence. The UCS recommendation followed the rule that it is rational to avoid the worst possible consequence of all options (Cooke 1982).

The "prevailing opinion" among decision theorists, according to John Harsanyi, is to use the Bayesian rule (Harsanyi 1975, p. 594), even in situations of uncertainty (Harsanyi 1986, p. 88; 1977a, p. 322; see 1976, pp. 38, 95; and 1977b, p. 94; see also Tversky and Kahneman 1986, p. 125). Ron Giere, making a more modest point, claims that philosophers and decision theorists do not believe that either maximax or maximin is more rational in cases of uncertainty (Giere 1979, pp. 317-318).

1. Introduction and Overview

I want to argue that the prevailing opinion is wrong, at least in the case of *societal* risks under *uncertainty*. Admittedly Bayesian rules are better in many cases of *individual* risk or certainty. (Both Bayesian and maximin strategies are sometimes needed.) Although I shall not take the time to defend all these points in detail, I shall argue (1) that there are compelling reasons for rejecting Harsanyi's defense of the Bayesian strategy under uncertainty; (2) that it is more rational, in specific types of situations, to prefer the maximin strategy; and (3) that calibrating expert opinions is superior to using the equiprobability assumption or subjective probabilities. Let's look first at Harsanyi's arguments.[1]

PSA 1988, Volume 2, pp. 504-517

2. Harsanyi's Arguments

I don't want to go into many details regarding Harsanyi's main *ethical* arguments about Bayesian and maximin strategies under uncertainty.[2] Likewise I don't want to concern myself with *general epistemological* criticisms of Bayesianism versus maximin.[3] Since uncertainty is the most problematic case, for the Bayesian, let's focus on Harsanyi's main epistemological arguments in favor of the Bayesian rule, and against maximin, in cases of uncertainty. He claims that those who reject the Bayesian strategy are guilty of (1) making irrational decisions; (2) sanctioning irrational consequences; and (3) irrationally presupposing that the worst outcome has a probability of one.

2.1 A First Argument for the Bayesian Strategy

The main assumption underlying Harsanyi's first argument is, in his words: "It is extremely irrational to make your behavior wholly dependent on some highly unlikely unfavorable contingencies, regardless of how little probability you are willing to assign to them." (Harsanyi 1975, p. 595)

To substantiate his argument, Harsanyi gives an example: Suppose you live in New York City and are offered two jobs, in different cities, at the same time. The New-York-City job is tedious and badly paid, but the Chicago job is interesting and well paid. However, to take the Chicago job, which begins immediately, you have to take a plane, and the plane travel has a small, positive, associated probability of fatality. This means, says Harsanyi, that following the maximin principle would cause you to accept the New York job. The situation can be represented, he claims, on the following table:

	If the Chicago plane crashes	If the Chicago plane does not crash
If you choose New York job	Then you have a poor job but will be alive.	Then you have a poor job but will be alive.
If you choose Chicago job	Then you will die.	Then you have a good job but will not die.

In the example, Harsanyi assumes that your chances of dying in the near future from reasons other than a plane crash are zero. Hence he concludes that, because maximin directs choosing so as to avoid the worst possibility, therefore it is irrational; a rational person, using expected-utility, would choose the Chicago job because of its desirability and the low probability of a plane crash.

2.11 A Problematic Example

Harsanyi's example, however, fails to establish the superiority of Bayesian rules under uncertainty. *First*, the example is highly counterintuitive; hazard assessors have repeatedly confirmed that the average annual probability of fatality associated with many other activities, e.g., driving an automobile, is greater, by an order of magnitude, than the risk of dying in a plane (Starr and Whipple 1980, p. 1118; Hushon 1979, p. 748; Okrent 1980, p. 374; Maxey 1979, p. 401; Graham 1982, pp. 692-704; Comar 1979, p. 319; Cohen and Lee 1979, p. 707).

Second, in assuming that one's chances of dying any way but in a plane crash are zero, Harsanyi has not given us a case of uncertainty and has therefore begged the question of using Bayesian rules under uncertainty. Moreover, he can hardly claim that people behave *as if* they maximized utility in situations of uncertainty since, by definition, in cases of uncertainty we do not know the relevant probabilities.

Third, even if the example in this first argument were plausible, it would prove little about the undesirability of using maximin in situations of *societal* risk under uncertainty, like deciding whether to build a nuclear reactor or to open a liquefied natural gas facility. In the *individual* case, the risk is typically freely chosen by one person, but in the *societal* instance, it is often involuntarily imposed on a group, without consent. Moreover, democratic *process* is probably *more important* in cases of societal risk under uncertainty, since it would be more difficult to insure informed consent (March 1986, p. 148).

Harsanyi, however, disagrees that the individual and societal cases are assymetrical; he claims that although his counterexamples refer to individual situations, "it is very easy to adapt them to large-scale situations," because scale is not "a fundamental variable in moral philosophy." (Harsanyi 1975, p. 605) However, if I make a decision regarding *my own* risk, I can ask "how safe is rational enough?" and I can be termed "irrational" if I have a fear of flying. But if I make a decision regarding risks to *others* in society, I do not have the *right* to ask, where their interests are concerned, "how safe is rational enough?" Because I am bound by moral and legal obligation to others, instead I must ask "how safe is free enough?" or "how safe is fair enough?" or "how safe is voluntary enough?[4]

If rational societal decisionmaking requires an ethical rule that takes account of the fairness of the allocational *process*, not merely the *outcomes*, then there are strong reasons to doubt that the individual and societal cases of risk assessment are symmetrical.[5] Moreover, if there are grounds for doubting the sure-thing principle, in cases of societal decisions, because it ignores ethical process and focuses only on outcomes, then Harsanyi needs a societal example, which he has not given.

2.12 Why Not Base Some Decisions on Consequences?

Harsanyi's first argument is also problematic because it is built on the supposition that "it is extremely irrational to make your behavior wholly dependent on some highly unlikely unfavorable contingencies regardless of how little probability you are willing to assign to them." (Harsanyi 1975, p. 595) I don't think so. Suppose one has the choice between buying organically grown vegetables and those treated with pesticides, and that the price difference between the two is very small. And suppose, third, that the probability of getting cancer from the vegetables treated with pesticide is "highly unlikely," to use Harsanyi's own words. It is not irrational to avoid this cancer risk, even if it is small, particularly if one can do so at no great cost. Some consequences are so undesirable that it makes sense to try to avoid them, regardless of their probability, and especially if one can do so at no great cost. Many rational people do not wish to gamble, if their lives are at stake, except to obtain a great benefit (Watkins 1977, p. 351). Moreover, a rational person might decide to base a decision solely on consequences since subjective probabilities are often flawed. If Kahneman and Tversky are right, that the same type of systematic heuristic errors as committed by laypersons, biases of representativeness, anchoring, etc., "can be found in the intuitive judgments of sophisticated scientists," (Kahneman and Tversky 1981a, p. 46; 1981b, p. 68) then rational people might wish to avoid them, at least in some cases.

2.2 A Second Argument for the Bayesian Strategy

This discussion of whether reasonable people ever ignore subjective probabilities brings us to Harsanyi's second argument. This is that, (2) although the two different decision principles (Bayesian and maximin) often result in the same policies, whenever they differ, it is "always the maximin principle that is found to suggest unreasonable consequences." (Harsanyi 1975, p. 595)

2.21 An Example Showing Maximin Superiority

I don't think so. Consider the following example. Suppose that the night-shift foreman has discovered a leak in one of the large toxic gas canisters at the Union Carbide plant
in West Virginia. Because of past instructions, he must immediately notify both the plant's
safety engineer (who will bring a four-man crew with him to try to repair the leak within
one-half hour) and the President of the company. However, the foreman is still faced with
a problematic choice: to notify the local sheriff of the situation, so that he can begin evacu-
ation of the town surrounding the plant, or not to notify him. If he notifies the sheriff, as is
required by the *Code of Federal Regulations* and the agreement Union Carbide signed
with the town, then no townspeople will die as a result of the leak. However, he and five
other employees (the safety engineer and his crew) will lose their jobs as a result of the
adverse publicity, especially after the Bhopal accident, if they cannot fix the leak within
one half hour. Moreover, there are no other jobs available, since this is a depressed area. If
the foreman does not notify the sheriff, and if the safety crew can repair the leak during the
first half hour of their work, then he and the 5-person safety crew will each receive a slap
on the wrist for violating the law, and a $25,000 bonus from the company, not ostensibly
for violating the law, but because they solved an important problem without either fright-
ening the populace or causing the company to lose face. However, if the foreman does not
notify the sheriff, and if the safety crew cannot repair the leak during the first half hour of
their work, then the ten persons living closest to the plant (all residents of a nursing home
for the aged) will die after one half-hour's exposure to the fumes, all six of them will lose
their jobs, and he will have to notify the sheriff anyway.

The foreman uses the Bayesian rule and employs the following table. (Since he is in a
state of ignorance, he uses the principle of insufficient reason (Resnick 1987, pp. 35-37),
or what Harsanyi calls "the equiprobability assumption," (Harsanyi 1975, p. 598) to
assign equal probabilities (0.5) to both possibilities.) He decides not to notify the sheriff,
since the expected utility for this act is higher.

If crew fixes leak	If crew doesn't fix in 30 minutes	leak in 30 minutes
If I notify the sheriff now (10u)	Then 10 lives and 6 jobs are safe. (16u)	Then 6 people lose jobs but 10 townspeople safe. (4u)
If I fail to notify the sheriff now (11u)	Then 10 lives, 6 jobs safe; 6 men get bonus; people suffer no fear. (38u)	Then 10 lives and 6 jobs are lost. (-16u)

The safety engineer agrees with the foreman that the worst outcome is that in which
both the jobs and the lives are lost, but he uses the *maximin* procedure and decides that
they ought to notify the sheriff, so as to be sure to avoid this worst outcome.

In the example, the Bayesian and maximin strategies dictated different actions, and
the maximin recommendation is arguably superior, for at least three reasons: (1) The
Code of Federal Regulations establishes an obligation to notify the sheriff. (2) The lease
contract that Union Carbide made with the town establishes an obligation to notify the
sheriff. (3) The ten endangered persons have a right to know the risk facing them.

But if the maximin recommendation is arguably superior, then this case provides a counterexample to Harsanyi's (and Arrow's) argument that, whenever the recommendations of the two strategies differ, it is "always the maximin principle that is found to suggest unreasonable consequences." (Harsanyi 1975, p. 595; Arrow 1983, p. 255) A number of objections could be made to this example, but rather than answer them all, I'll simply make a general point. Although being moral might increase utility, there is no reason why a Bayesian is required to define utility in this way. Hence nothing in Bayesianism prevents something like this Union Carbide case from occurring.[6]

2.3 A Third Argument for the Bayesian Strategy: Equiprobability

Harsanyi's third argument for the Bayesian strategy is that uncertainty forces one to make the equiprobability assumption (Harsanyi 1975, p. 598), an assumption justified by "the moral principle of assigning the same *a priori* weight to every individual's interests."[7]

2.31 No Justification for the Equiprobability Assumption

Apart from the *moral* difficulties with the equiprobability assumption, e.g., confusing sameness with equality and equal interests with equal probabilities, difficulties that I don't want to address here, there are a number of problematic *epistemological* assumptions in this third argument. The most basic difficulty is that, if there is no justification for assigning a set of probabilities, because one is in a situation of uncertainty, then there is no justification for assuming that the states are equally probable. This appears to be a situation of generating knowledge out of ignorance, as Rosenkrantz put it (Rosenkrantz 1981, pp. 4.2-1; Smith 1986, pp. 325 ff.; Levi 1978, p. 382; Vail 1954, pp. 87 ff.). Moreover, to assign the states equal probabilities is to contradict the stipulation that the situation is one of uncertainty and to beg the question of whether uncertainty is really involved.[8]

2.32 The Equiprobability Assumption Could Lead to Disaster

A third difficulty with the equiprobability assumption is that it could lead to disaster. Consider the following example:

	If reactor operators make serious errors	If reactor operators make no serious errors
If we continue to use commercial nuclear plants (25)	Then more than a 100,000 could die in an accident.(-200)	Then we have a potentially good source of electricity. (250)
If we discontinue use of commercial nuclear plants (20)	Then we will have a financial loss, but no commercial nuclear disaster. (10)	Then we will have a financial loss,but no commercial/research disaster. (30)

Using the equiprobability assumption and expected utility, the Bayesian decision is to continue to use commercial nuclear reactors, since the expected utility of this option is (25), whereas the utility of its alternative is only (20). However, if the real probability associated with reactor operators making serious errors is (0.6), then the expected utility for continuing to use commercial fission is (-20). Likewise if the real probability associated with their being careful is (0.4), then the expected utility for discontinuing use of nuclear power is (18). This suggests that use of the equiprobability assumption could have devastating effects.

But if so, then why do many decisionmakers still defend use of the equiprobability assumption in situations of uncertainty? One reason is that, as Raiffa points out, situations rarely involve complete uncertainty (Luce and Raiffa 1958, p. 299). Because they do not, one often feels confident in relying on partial information (Luce and Raiffa 1958, Chapter 13, esp. p. 293). In making this defense of the equiprobability assumption, however, Luce and Raiffa explicitly noted that they were not discussing the societal case (Luce and Raiffa 1958, Chapter 13). But if not, then in *societal* situations of uncertainty, Bayesianism may fail, because of the disanalogies already noted between "event" and "process" rationalities.

3. An Alternative to Using Bayesian Rules: Calibration

This suggests that it might be more rational to calibrate particular risk assessments rather than to use either subjective estimates or the equiprobability assumption. Calibration sometimes involves the use of a scoring rule for weighting an assessor's estimates on the basis of how well later observed outcomes fit his predictions (Cooke 1988, Chapter 8). Since data on frequencies of rare events typically are not available, or we would not have to use probabilistic risk assessment in the first place, calibration also involves using experts' calibration scores for known ingredient events, e.g., ruptures of steel pipes with diameters greater than 3 inches, as a basis for weighting their probabilistic estimates of rare events, e.g.,nuclear core melts, dependent on the known ingredient events (Cooke 1988, Chapter 2, pp. 8, 18).

Empirical control is the main rationale for calibrating and weighting the predictions or the subjective probabilities of experts, as an alternative to using "raw" subjective probabilities or the equiprobability assumption. The need for such empirical control is evidenced by the frequent errors of experts, as in the case of probability estimates (for plutonium migration) used in siting the world's largest, worst, commercial nuclear dump. The subjective probabilities were proved wrong, by six orders of magnitude, when the plutonium migrated two miles offsite less than 10 years after the dump was opened.

For an example of the wide *spread* in risk assessors' probability estimates, consider the NASA engineers' subjective probability estimate, for a solid rocket booster failure causing complete destruction of the space shuttle. This figure was 1 in 100,000 as compared to 1 in 35 for Colglazier and Weatherwax for the same event (Cooke 1988, Chapter 2, p. 2). Likewise the spread for reactor-year core-melt probabilities in the reliable technical literature is four orders of magnitude (Cooke 1988, Chapter 2, p. 13). The spread in probabilities for the chance of dying in a LNG gas accident at the same location is typically three orders of magnitude. When 30 nuclear experts were asked to estimate the failure probabilities for 60 different components in a commercial boiling water reactor, average spread over the 60 components was 167,820 (the ratio of the largest to the smallest estimate) (Harsanyi 1975, p. 595).

Or, consider a recent study by hazard assessors in the Netherlands who used frequencies obtained from Oak Ridge National Laboratories. They used operating experience to determine the failure frequencies for seven reactor subsystems whose failure probabilities were subjectively calculated in WASH 1400, the Rasmussen Report, the most famous, the most extensive, and allegedly the best risk assessment ever performed. The Dutch researchers then compared these probabilities with the 90 percent confidence bounds for the same probabilities calculated in WASH-1400. The subsystem failures included things like high-pressure injection failures, and automatic depressurization-system failures for both pressurized and boiling water reactors.

Amazingly, *all* the values from operating experience fell outside the 90 percent confidence bands in the WASH-1400 study. If the authors' subjective probabilities were well calibrated, we should expect that approximately 10 percent of the true values should lie outside their respective bands. Moreover, the fact that five of the seven values fell above the upper confidence bound suggests that the WASH 1400 accident probabilities, subjective probabilities, are too low, exhibiting an overconfidence bias (Rawls 1971, pp. 75-83; see Cooke 1988, Chapter 2, pp. 14-22; Chapter 4). The Oak Ridge data, in helping to confirm charges of the expert bias of overconfidence, suggest a second reason for calibrating assessors.

A *third* reason for calibrating assessors is that their estimates appear to be dependent; the Oak Ridge data shows that an optimist about one mode of reactor failure is likely to be an optimist about another mode, and a pessimist about one failure mode is likely to be a pessimist about another such mode. The Dutch researchers rejected rank independence at the 5 % level for a majority of the WASH — 1400 experts (Cooke 1988, Chapter 2, pp. 12ff, esp. table 2-5 and 2-6). This dependence suggests both that experts' errors do not tend to cancel each other, and that calibrating assessments for ingredient events (e.g., pipe ruptures) might be a good guide to calibrating assessments of rare events (e.g., core melts).

A *fourth* reason for calibrating expert assessors is that subjective probabilities do not appear reproducible by different groups of experts using the same methodologies. For example, ten teams from different European countries recently did estimates of the failure rate of a particular auxiliary feedwater system and were unable to agree on common fault-tree analysis (Cooke 1988, Chapter 2, p. 16). Similar benchmark studies have been conducted at Sandia National Labs. In both cases, the failure of reproducibility appeared not to result merely from using different values for individual failure probabilities, but from analyzing the human reliability problems differently. (Experts identified different tasks that might be performed differently.)[9]

4. Particular Cases in Which Maximin Might Be More Rational

If calibration of expert assessors makes sense, however, then what ought one to do when calibration is not reasonable? One alternative might be to use maximin in certain situations. What specifics might characterize such a situation?

First, the most obvious situation might be one in which the risk being probabilistically assessed is so new that there is little frequency data on particular failures, as was the case in the beginning of LNG technology.

A *second* candidate situation for using maximin might be one in which probabilities are uncertain, but the consequences are potentially catastrophic. If one focused on consequences in allegedly low-probability, high-consequence cases, and if one pursued a maximin strategy, then accidents like Chernobyl might be more easily prevented. Because the Chernobyl accident was called "highly improbable" before it happened (Raloff and Silberner 1986, p. 292), and because highly improbable accidents are not a significant concern for Bayesians,[10] it might be good for rulemaking and risk assessment to create a "climate" of maximin, a climate in which decisionmakers aimed at avoiding worst cases.

A *third* type of situation in which societal risk assessors, operating under uncertainty, probably ought to use a maximin (rather than a Bayesian) rule might be cases in which human error and bureaucracy are likely to cause catastrophic societal harms. One example that illustrates the difficulty of controlling human error and bureaucracy is the fact that, even during wartime, the U.S. was unable to use available information so as to prevent the surprise attack on Pearl Harbor:

Only days prior to the attack, the Pacific Fleet was warned about possible attacks, about a change in Japanese codes (evaluated as very unusual), about Japanese ships in Camranh Bay. They had orders to be alert for Japanese action in the Pacific, messages deciphered from Japan's most secure code ordering Japanese embassies and military to destroy secret papers and all American codes in outlying islands. Despite all these warnings, (1) the U.S. fleet remained together in the harbor; (2) the island was not air patrolled; (3) the emergency warning center was not staffed; and (4) the Army was not notified under the Joint Coastal Frontier Defense Plan (Wildavsky and Douglas 1982, p. 94). This example suggests that society ought to have procedures for implementing something like the maximin rule, whenever bureaucracy and human error could cause catastrophic harm.

Fourth, it seems obvious that one ought to follow maximin and avoid a worst-case scenario, caused by a particular technological choice, to the degree that there are economically viable, technically feasible alternatives.

5. Some Objections

What are some basic objections likely to be made to my defense of calibrating assessors and using maximin whenever calibration is not possible? Luce and Raiffa claimed that few situations involve complete uncertainty, and therefore that Bayesian rules are reasonable. However, recall that even the allegedly best risk assessments vary by several orders of magnitude, and that most risk assessment takes place in conditions of uncertainty, or else experts would not need to use subjective probabilities in the first place. Moreover the 1983 statement by the U.S. National Academy of Sciences suggests that Luce and Raiffa might be overly optimistic. The Academy concluded that controversy over risk assessment was due to the "sparseness and uncertainty of the scientific knowledge of the health hazards addressed."[11]

A *second* objection is that, if expert estimates are inherently flawed, as Kahneman and Tversky suggest, then why depend on them at all, even if they are calibrated? There are several reasons. (1) Calibrated estimates are better than uncalibrated ones, even though both might be suspect. (2) The only alternative to calibration would be giving up on experts or simply using bad estimates, both of which seem unacceptable. (3) Even though experts typically fall victim to a number of heuristic biases, nevertheless some expert opinions are likely to be less biased than others; if so, then calibration could help pick out the better experts. Finally (4) the Dutch and Oak Ridge studies of calibrating expert assessors indicate that experts with the same education but more practical experience are better calibrated than those with less experience.[12] This suggests that, in at least some instances, like engineering or weather forecasting, calibration does measure something that is important about being an expert (Cooke 1988, Chapter 9, pp. 11-14).

A *third* objection to using maximin in certain situations is that it violates one of the rationality axioms. However, as my earlier remarks indicated, if Bayesians ignore ethical and democratic process and focus only on outcomes, then there are equally important grounds for doubting the adequacy of the sure-thing principle, one of the three main rationality axioms underlying Bayesianism (Harsanyi 1977c, p. 382), especially in cases of *societal* risk under *uncertainty*.

A *fourth* objection to maximin, even in certain cases, is that it appears to sanction an anti-progress, anti-science notion of risk assessment. On the contrary, one could argue that use of Bayesian methods for societal risks under uncertainty has itself resulted in highly publicized scientific failures, e.g., the Challenger disaster. This suggests that restricting use of Bayesian rules to cases in which they are successful, e.g., *individual* hazards under *certainty* and *risk*, is likely to promote respect for science and for Bayesian methods.

6. Conclusions

If the preceding arguments have been at least partially correct, then (1) a number of assessments may have erred in formulating policy conclusions based on use of the Bayesian rule under uncertainty. (2) They may also have erred in presupposing that the procedural concept of rationality underlying societal decisionmaking is not inherently different from that underlying individual decisionmaking. (3) If risk assessors are sometimes bound by a more procedural account of rationality, then *at least in cases of societal hazard assessment under uncertainty*, Rudner, Churchman, and Braithwaite may have been right in arguing that practical and ethical judgments are essentially involved in deciding which probabilistic or scientific hypothesis to accept (Eells 1982, p. 27). All this suggests that, to the degree that risk assessment is applied philosophy of science, then the ethical camel may have his nose under the philosophy of science tent. I leave it to someone else to decide how much of his body he might be able to squeeze in.

Notes

[1] Harsanyi believes that under conditions of uncertainty, we should maximize expected utility or average expected utility, where the expected utility of an act for a two-state problem is

$$u_1p + u_2(1-p)$$

where u_1 and u_2 are outcome utilities, where p is the probability of S_1 and (1-p) is the probability of S_2, and where p represents the decisionmaker's own subjective probability estimate (Harsanyi 1977a, p. 320; Otway and Peltu 1985, p. 115; Resnick 1987, p. 36; see Resnick 1987, pp. 88-91 for the Von Neumann-Morgenstern approach to expected utility). More generally, members of the dominant Bayesian school claim that expected-utility maximization is the appropriate decision rule under uncertainty (Harsanyi 1975, p. 594; see also Jeffrey 1983; Marschak 1954, pp. 187 ff.; and Ellsworth 1978, pp. 39 ff. Two of the scholars who are most vocal in the claim that utility is the way to analyze choice situations involving uncertainty are Davis 1954, p. 14 and Coombs and Beardslee 1954, pp. 255 ff. For an opposed point of view, see McClennen 1978, pp. 337 ff. McClennen argues that there is a strong reason to doubt that the expected utility principle is a basic criterion for rational behavior. For a review of a number of decisionmaking procedures, see Milnor 1954, pp. 49 ff.). They claim that we should value outcomes, or societies, in terms of the average amounts of utility (subjective determinations of welfare) realized in them. To maximize expected utility in societal choices, Harsanyi opts for using the maximum average utility level of all individuals in society (Harsanyi 1977a, p. 323; for more references to Harsanyi's work on this point, see 1977c; see 1977a, pp. 320-322 for a brief discussion of Bayesian decision theory and its associated axioms).

Proponents of the maximin school maintain that one ought to maximize the minimum, i.e., avoid the policy having the worst possible consequences (Harsanyi 1975, p. 595; for a discussion of the maximin rule, see Resnick 1987, pp. 26 ff.; for a discussion of individual decisionmaking under uncertainty, including maximin, see Luce and Raiffa 1958, Chapter 13). For Rawls, this is equivalent to the *difference principle*, according to which one society is better than another if the worst-off members of the former do better than the worst-off in the latter (Rawls 1971, pp. 75-83). In terms of social choice, the obvious problem is that often the maximin and the Bayesian/utilitarian principles recommend different actions.

[2] His first argument is that failure to use the Bayesian rule could lead to unacceptable moral consequences, such as failure to benefit those who deserve it, failure to benefit society as a whole, and failure to provide equitable treatment to persons who are most advantaged.

His second ethical argument is that using the Bayesian strategy and the equiprobability assumption is desirable because it allows one to assign equal *a priori* probability to everyone's interests. Both these arguments are problematic, it seems to me, because Harsanyi makes questionable assumptions about what it is to be less well off, because he fails to take adequate account of rights, because he provides no second-order rules of ethics, and because he confuses a number of concepts: sameness versus equality of treatment, equiprobability of states of affairs versus assigning equal *a priori* weight to everyone's interests, supererogatory actions versus justice, and preferences versus moral principles.

[3]Some of these difficulties include problems with subjective probabilities, with inter-personal comparisons of utility, with a linear and consumption-based notion of welfare, and with the notion of an idealized decisionmaker.

[4]On the *individual* level, the question is *why* the individual can define 'risk' in a certain way, if his concept of rationality is to be theoretically justifiable. On the societal level, the question is *how* a group of people can define 'risk' so as to have a concept of rationality that is *democratically* justifiable, in terms of ethical procedure.

A similar error (assuming that the individual and societal cases are relevantly similar) occurs, I might add, when philosophers attack *societal* use of benefit-cost analysis on the grounds that employing it leads to faulty individual decisions. The important point, in discussions of both Bayesianism and benefit-cost analysis, is that the societal and individual cases are relevantly dissimilar; there is no way to make societal decisions that unify individual preferences and diverse decisionmakers, via a common denominator, short of some analytic tool like benefit-cost analysis. Just as philosophers err in assuming that societal use of benefit-cost is undercut by arguments against its use in cases of individual decisionmaking, so also Bayesians err in assuming that societal use of their rule, under uncertainty, is supported by examples relevant only to the individual.

[5]This is needed to satisfy Diamond and others (Diamond 1967, pp. 765-766), in the controversy over what Sen calls "the strong independence axiom," (Sen 1977, p. 276) and over what Harsanyi calls the "sure thing principle" or the "dominance principle." This axiom essentially states that, if one strategy yields a better *outcome* than another does under some conditions, and it never yields a worse outcome under any conditions, then decisionmakers always ought to choose the first strategy over the second (Harsanyi 1977c, p. 384; 1977a, p. 321).

[6]It might be objected, at this point, that the example violates Harsanyi's criteria for moral value judgments. Someone might claim that the disutility of failing to notify the sheriff immediately, in the state where the leak is not fixed within 30 minutes, should be greater than (-16). Since a number of lives are lost in this outcome, and since one of Harsanyi's criteria for moral value judgments is that everyone's situation should be treated with equal probability, the foreman ought not to have attached so much utility to his bonus and to keeping his job, and so little disutility to the ten persons' losing their lives.

More specifically, Harsanyi claims that Bayesians should make their choices "as if they thought they would have the same probability of taking the place of any particular individual in the society." (Harsanyi 1975, p. 598) According to the objector, the foreman's decision is wrong, not because he used Bayesian principles, but because he computed the utilities in a self-interested way. Hence the example does not show that Bayesianism itself leads to wrong consequences, but only that self-interest leads to them.

There are two important responses to this objection. First, the foreman could plausibly claim that he was not acting in self interest, in assigning the utilities he did to the act of not informing the sheriff immediately, because he was trying to avoid mass panic (a frequently used defense in such situations). (This was the excuse given in the Brown's Ferry nuclear accident; for an account see Shrader-Frechette 1983, pp. 88 ff.) and needlessly troubling people. Moreover, informing the sheriff of the leak, the foreman might argue, would be like yelling "fire" in a dark and crowded auditorium, since the people most at risk were elderly and not as easily able to be evacuated.

A *second*, and more important, response to the objection is that the example is typical, in that it points up a major measurement difficulty with Harsanyi's Bayesian approach. This strategy requires agents in situations of uncertainty to rank the lives of members of various societies on a common interval scale. This requires interpersonal comparisons of utility. But interpersonal comparisons of utility are difficult to make, for reasons that we will discuss shortly (Resnick 1987, pp. 43, 205-212). This means that the foreman need not consciously have acted out of self-interest; he may merely have been unable to rank the lives of other people on a common interval scale, particularly if those being ranked were elderly and sick, and if the foreman would rather be dead than be elderly and sick. Hence, if Bayesianism requires these interpersonal comparisons, so much the worse for it, especially since Rawls' minimax strategy requires only that agents in situations of uncertainty be able to generate an ordinal ranking of the worst-off individuals in the various societies. Therefore the example is both reasonable and not necessarily in violation of any of Harsanyi's requirements.

[7]Harsanyi, 1975, p. 599. The assumption is a variant of the so-called "principle of insufficient reason," first formulated by Jacob Bernoulli. It says that, if there is no evidence leading one to believe that one event from an exhaustive set of mutually exclusive events is more likely to occur than another, then the events should be judged equally probable (Luce and Raiffa 1958, pp. 284 ff.; Arrow 1951, pp. 404-437; Nagel 1939; Savage 1954).

[8]Resnick, 1987, p. 37. Some of the other epistemological difficulties associated with the equiprobability assumption are that it is often impossible to specify a list of possible states that are mutually exclusive and exhaustive (Luce and Raiffa 1958, pp. 284-285). There could be many different ways of defining alternative "states" that determine a given outcome. Moreover, each of the different ways of defining states, and perhaps proliferating them, could conceivably result in different decision results, different accounts of how best to maximize average utility (Luce and Raiffa 1958, pp. 284-285).

[9]Cooke 1988. The failure of reproducibility, because of expert differences in the way they analyze human error and reliability, suggests that at least some of the Fischoff, Slovic, and Lichtenstein conclusions, as to causes in flaws in probability estimates, are correct. They concluded that experts systematically overlook many "pathways to disaster." These include (1) failure to consider the way human error could cause technical systems to fail, as at Three Mile Island; (2) overconfidence in current scientific knowledge, such as that causing the 1976 collapse of the Teton Dam; and (3) failure to appreciate how technical systems, as a whole, function. For example, engineers were surprised when cargo-compartment decompression destroyed control systems in some airplanes. Experts also typically overlook (4) slowness to detect chronic, cumulative effects, e.g., as in the case of acid rain; (5) failure to anticipate inadequate human responses to safety measures, e.g., failure of Chernobyl officials to evacuate immediately; and (6) failure to anticipate "common-mode" failures simultaneously afflicting systems that are designed to be independent. A simple fire at Brown's Ferry, Alabama, for example, damaged all five emergency core cooling systems for the reactor (Slovic, Fischoff, and Lichtenstein 1981, pp. 475-478). If Fischhoff *et al* are right, then they suggest some of the causes why reproducibility has failed in probability estimates of technological hazards.

[10]But on a maximin scheme, the fact that a worst-case nuclear core melt could kill 150,000 persons (Mulvihill 1965), would be a significant cause for concern. Had regulators implemented a maximin strategy, with appropriate procedures for insuring it, then the Chernobyl accident might have been less likely. (Admittedly, there is no fail-safe way to protect citizens from those who desire to violate rules, even maximin rules.) Nevertheless, regulators could attempt to encourage safety by creating a "climate" of maximin, by means of a variety of procedures, rather than a climate of complacency in which an accident is looked upon as unlikely.

[11]Otway and Peltu 1985, p. 4; Some of these difficulties include problems with subjective probabilities, with interpersonal comparisons of utility, with a linear and consumption-based notion of welfare, and with the notion of an idealized decisionmaker.

[12]Testing mechanical engineers at a Dutch training facility for operators of large technological systems, the Dutch researchers asked the engineers to estimate ten uncertain quantities, such as the maximum efficiency of a particular type of gas turbine, or the maximum admissible intake temperature for gas in an Olympus power turbine. They used the fractile calibration tests of Alpert and Raiffa (1982), and were able to illustrate that trainees (aged 20-25) at the end of their degree performed less well, on the whole, than did engineers (average age 36) who had 15 years of experience in such technological facilities (Cooke 1988, Chapter 9).

References

Arrow, K. (1951), "Alternative Approaches to the Theory of Choice in Risk-Taking Situations", *Econometrica* 19: 404-437.

_ _ _ _ _ . (1983), "Some Ordinalist-Utilitarian Notes on Rawls' Theory of Justice", *The Journal of Philosophy* 70:245-263.

Cohen, B. and Lee, I. (1979), "A Catalog of Risks", *Health Physics* 36: 707-722.

Comar, C. (1979), "Risk: A Pragmatic De Minimus Approach", *Science* 203: 319.

Cooke, R. (1982), "Risk Assessment and Rational Decision", *Dialectica* 36: 220-351.

_ _ _ _ _ . (1988), *Experts in Uncertainty*, unpublished manuscript.

Coombs, C. and Beardslee, D. (1954), "On Decision-Making Under Uncertainty", in *Decision Processes*, R. Thrall, C. Coombs, and R. Davis (eds.). London: John Wiley, pp. 255-286.

Davis, R. (1954), "Introduction", in *Decision Processes*, R. Thrall, C.Coombs, and R. Davis (eds.). London: John Wiley, pp. 1-18.

Diamond, P. (1967), "Cardinal Welfare, Individualistic Ethics, and Interpersonal Comparisons of Utility", *Journal of Political Economy* 75: 765-766.

Eells, E. (1982), *Rational Decision and Causality*. Cambridge: Cambridge University Press.

Ellsworth, L. (1978), "Decision-Theoretic Analysis of Rawls' Original Position", in *Foundations and Applications of Decision Theory*, C. Hooker, J. Leach, and E. McClennen (eds.). Dordrecht: Reidel, pp. 29-46.

Giere, R. (1979), *Understanding Scientific Reasoning*. New York: Holt, Rinehart and Winston.

Graham, J. (1982), "Some Explanations for Disparities in Lifesaving Investments", *Policy Studies Review* 1: 692-704.

Harsanyi, J.(1975), "Can the Maximin Principle Serve as a Basis for Morality? A Critique of John Rawls's Theory", *American Political Science Review* 59: 594-606.

516

_____. (1976), *Essays on Ethics, Social Behavior, and Scientific Explanation*. Dordrecht: Reidel.

_____. (1977a), "Understanding Rational Behavior", in *Foundational Problems in the Special Sciences*, R. Butts and J. Hintikka (eds.). Boston: Reidel, pp. 315-344.

_____. (1977b), *Rational Behavior and Bargaining Equilibrium in Games and Social Situations*. Cambridge: Cambridge University Press.

_____. (1977c), "On the Rationale of the Bayesian Approach", in *Foundational Problems in the Special Sciences*, R. Butts and J. Hintikka (eds.). Boston: Reidel, pp. 381-392.

_____. (1986), "Advances in Understanding Rational Behavior", in *Rational Choice*, J. Elster (ed.). New York: New York University Press, pp. 88-107.

Hushon, J. (1979), "Plenary Session Report", in *Symposium/Workshop on Nuclear and Nonnuclear Energy Systems: Risk Assessment and Governmental Decision Making*. Mclean, Virginia: Mitre Corporation, pp. 742-760.

Jeffrey, R. (1983), *The Logic of Decision*. Chicago: University of Chicago Press.

Kahneman, D. and Tversky, A. (1981a), "Subjective Probability", in *Judgment Under Uncertainty: Heuristics and Biases*, D. Kahneman, A. Tversky, and P. Slovic (eds.). Cambridge: Cambridge University Press, pp. 32-47.

_____. (1981b), "On the Psychology of Prediction", in *Judgment Under Uncertainty: Heuristics and Biases*, D. Kahneman, A. Tversky, and P. Slovic (eds.). Cambridge: Cambridge University Press, pp. 48-68.

Luce, R. and Raiffa, H. (1958), *Games and Decisions*. New York: John Wiley.

Levi, I. (1978), "Newcomb's Many Problems", in *Foundations and Applications of Decision Theory*, vol. 1, C. Hooker, J. Leach, and E. McClennen (eds.). Dordrecht: Reidel, pp. 369-384.

March, J. (1986), "Bounded Rationality", in *Rational Choice*, J. Elster (ed.). New York: New York University Press, pp. 142-170.

Marschak, J. (1954), "Towards an Economic Theory of Organization and Information", in *Decision Processes*, R. Thrall, C. Coombs, and R. Davis (eds.). London: John Wiley, pp. 187-220.

Maxey, M. (1979), "Managing Low-Level Radioactive Wastes", in *Low-Level Radioactive Waste Management*, J. Watson (ed.). Williamsburg, Virginia: Health Physics Society, pp. 400-419.

McClennen, E. (1978), "The Minimax Theory and Expected Utility Reasoning", in *Foundations and Applications of Decision Theory*, C. Hooker, J. Leach, and E. McClennen (eds.). Dordrecht: Reidel, pp. 337-368.

Milnor, J. (1954), "Games Against Nature", in *Decision Processes*, R. Thrall, C. Coombs, and R. Davis (eds.). London: John Wiley, pp. 49-59.

Mulvihill, R. (1965), *Analysis of United States Power Reactor Accident Probability*, PRC R-695. Los Angeles: Planning Research Corporation.

Nagel, E. (1939), "Principles of the Theory of Probability", in *International Encyclopedia of Unified Science*, vol. 1. Chicago: University of Chicago Press, pp. 49-60.

Nuclear Energy Policy Study Group (1977), *Nuclear Power: Issues and Choices*. Cambridge, Massachusetts: Ballinger.

Okrent, D. (1980), "Comment on Societal Risk", *Science* 208: 372-375.

Otway, H. and Peltu, M. (1985), *Regulating Industrial Risks*. London: Butterworths.

Raloff, J. and Silberner, J. (1986), "Chernobyl: Emerging Data on Accident", *Science News* 129: 292-293.

Rawls, J. (1971), *A Theory of Justice*. Cambridge: Harvard University Press.

Resnick, M. (1987), *Choices*. Minneapolis: University of Minnesota Press.

Rosenkrantz, R. (1981), *Foundations and Applications of Inductive Probability*. Atascadero, California: Ridgeview.

Savage, L. (1954), *The Foundations of Statistics*. New York: John Wiley.

Sen, A. (1977), "Welfare Inequalities and Rawlsian Axiomatics", in *Foundational Problems in the Special Sciences*, R. Butts, J. Hintikka (eds.). Boston: Reidel, pp. 271-292.

Shrader-Frechette, K. (1983), *Nuclear Power and Public Policy*. Boston: Reidel.

Slovic, P., Fischoff, B., and Lichtenstein, S. (1981), "Facts vs. Fears", in *Judgment Under Uncertainty: Heuristics and Biases*, D. Kahneman, A. Tversky, and P. Slovic (eds.). Cambridge: Cambridge University Press, pp. 463-489.

Smith, V. (1986), "Benefit Analysis for Natural Hazards", *Risk Analysis* 6: 321-329.

Starr, C. and Whipple, C. (1980), "Risks of Risk Decisions", *Science* 208: 1114-1118.

Tversky, A. and Kahneman, D. (1986), "The Framing of Decisions and the Psychology of Choice", in *Rational Choice*, J. Elster (ed.). New York: New York University Press, pp. 123-141.

Union of Concerned Scientists (1977), *The Risks of Nuclear Power Reactors, a Review of the NRC Reactor Safety Study WASH 1400*. Cambridge, Massachusetts: University of Concerned Scientists.

Vail, S. (1954), "Alternative Calculi of Subjective Probabilities", in *Decision Processes*, R. Thrall, C. Coombs, and R. Davis (eds.). London: John Wiley, pp. 87-89.

Watkins, J. (1977), "Towards a Unified Decision Theory: A Non-Bayesian Approach", in *Foundational Problems in the Special Sciences*, R. Butts, J. Hintikka (eds.). Boston: Reidel, pp. 345-380.

Wildavsky, A. and Douglas, M. (1982), *Risk and Culture*. Berkeley: University of California Press.